Coastal Altimetry

Stefano Vignudelli • Andrey G. Kostianoy
Paolo Cipollini • Jérôme Benveniste
(Editors)

Coastal Altimetry

Stefano Vignudelli
Consiglio Nazionale delle Ricerche
Istituto di Biofisica
Area Ricerca CNR
Via Moruzzi 1
56124 Pisa
Italy
vignudelli@pi.ibf.cnr.it

Andrey G. Kostianoy
P.P. Shirshov Institute of Oceanology
Russian Academy of Sciences
36, Nakhimovsky Pr.
Moscow, 117997
Russia
kostianoy@online.ru

Paolo Cipollini
National Oceanography Centre
European Way
SO14 3ZH, Southampton
United Kingdom
cipo@noc.ac.uk

Jérôme Benveniste
European Space Agency/ESRIN
Directorate of Earth Observation
Programmes, EO Science
Applications and Future
Technologies Department
Via Galileo Galilei
Frascati (Roma) 00044
Italy
jerome.benveniste@esa.int

ISBN: 978-3-642-12795-3   e-ISBN: 978-3-642-12796-0

DOI: 10.1007/978-3-642-12796-0

Springer Heidelberg Dordrecht London New York

Library of Congress Control Number: 2010930379

© Springer-Verlag Berlin Heidelberg 2011

This work is subject to copyright. All rights are reserved, whether the whole or part of the material is concerned, specifically the rights of translation, reprinting, reuse of illustrations, recitation, broadcasting, reproduction on microfilm or in any other way, and storage in data banks. Duplication of this publication or parts thereof is permitted only under the provisions of the German Copyright Law of September 9, 1965, in its current version, and permission for use must always be obtained from Springer. Violations are liable to prosecution under the German Copyright Law.

The use of general descriptive names, registered names, trademarks, etc. in this publication does not imply, even in the absence of a specific statement, that such names are exempt from the relevant protective laws and regulations and therefore free for general use.

Product liability: The publishers cannot guarantee the accuracy of any information about dosage and application contained in this book. In every individual case the user must check such information by consulting the relevant literature.

Cover design: deblik, Berlin

Printed on acid-free paper

Springer is part of Springer Science+Business Media (www.springer.com)

Preface

"I consider this the most successful ocean experiment of all times"

Walter Munk of Scripps Institution of Oceanography, University of California at San Diego speaking about the TOPEX/Poseidon altimetry mission in front of the U.S. Commission on Ocean Policy, in April 2002.

The above statement, a tribute from one of the fathers of modern oceanography and a pioneer of revolutionary techniques to sample the sea, is a convincing summary of the immense impact that satellite radar altimetry is having on our knowledge of the oceans. The TOPEX/Poseidon mission (funded by NASA and CNES), its successors Jason-1 and Jason-2, the altimeters on board ESA's ERS-1/2 and Envisat missions and the U.S. Navy's Geosat Follow-On have started and are sustaining one of the most impressive observational records in geophysics, both for its length (currently almost two decades), its synopticity, and its accuracy. The importance of altimetry in capturing the dynamics of the oceans, the intended primary application of the missions, is now paralleled by its role as a global sea level gauge – an application that has become ever more prominent in times of climate change. The by-products of the technique – wind speed and significant wave height – are important variables for climate research and for the independent validation of measurements from other sources. Altimetry's exceptional progress has been made possible by the dedication of many scientists and engineers. Their first-class research on orbit determination and instrumental, atmospheric and surface effects' corrections has improved the precision of the altimetric measurements to levels that were almost unthinkable of at the beginning (~2 cm from a 1,335-km orbit, i.e., slightly more than one part in one hundred million) and ensures accuracy and long-term stability that allow the rate of global sea level rise to be measured with a resolution accuracy better than 1 mm/y. The excellent body of research behind altimetry is comprehensively reviewed in Fu and Cazenave's book *Satellite Altimetry and Earth Sciences* (2001), whose many chapters, authored by some of the leading experts in the field, also illustrate several of the applications of altimetry. Some simple statistics will confirm the impact of altimetry on the oceanographic community: the ISI (Institute for Scientific Information) online library of refereed journals and major conference proceedings cites more than 6,000 papers on ocean altimetry; a search for "TOPEX" on Google Scholar yields more than 15,000 hits (including reports and similar grey literature). A measure of this enormous impact is also the widespread use of altimetry within the modelling community: nowadays, altimetric data are routinely assimilated into ocean models, both in near-real time for operational forecasting and in delayed time for hindcasts and reanalysis that greatly improve our understanding of many ocean processes. It is not a surprise that GCOS has included Sea Level in its list of Essential Climate Variables (ECVs) required to support the work of the UNFCCC and the IPCC. Ensuring continued support to altimetry is, now more than ever, at the centre of the general political debate on sustaining climate research.

Therefore, altimetry is mature, and is a great success story for satellite-based Earth Observation. At this point the casual reader might conclude that all the technical hurdles have been reasonably overcome, and the future is rosy. Forthcoming techniques (like the delay-Doppler technique) will just bring in more excitement, with their promise to push further the capability of altimeters, and make the picture described in the above paragraph even sharper.

While the above might be true for open-ocean altimetry, there is an important domain where altimetry remains grossly underexploited. This domain coincides, crucially, with the region where changing circulation, sea level and sea state have by far the largest impact on human society: *the coastal zone*. Altimetric data gathered over this coastal strip (up to a few tens of kilometers from the coast) remain unused in the archive, mainly as they are flagged as "bad" or "not reliable" due to contamination of the altimetric waveforms, inadequate corrections, or poorly understood surface effects. The information in the coastal altimetry records ends up being corrupted, and for a long time many satellite oceanographers simply used to throw those records away. But that information would be invaluable for studies of coastal circulation, sea level change and impact on the coastline. And we would have, already, almost two decades of data in the archives. Is that information completely unrecoverable? Many researchers believe this is not the case. They believe that not only coastal altimetry would be very *useful*, but also, and *crucially*, that it is to a large extent *feasible*, and so must be done! This book is an account of their valiant efforts.

The path to coastal altimetry crosses two crucial territories: improvement of the radar measurement retrieval algorithms (retracking) and of the geophysical corrections. First of all, the radar return signals (known as waveforms) are archived, so their conversion to geophysical measurements can be revisited by fitting models of the waveforms (which is what is called retracking) optimized for the coastal zone: this is expected to improve the estimation of height, significant wave height and wind speed, or indeed to make that estimation possible where it was previously not possible at all employing open-ocean waveform models. Then, the other aspects of the whole processing chain from sensor to product, including the various corrections and the quality control, can be largely improved integrating local knowledge; one simple example is the use of local accurate models for the correction of the high-frequency atmospheric pressure effects on the sea level.

The process of extending altimetry to the coastal zone has begun with some early papers (e.g. Crout 1997; Anzenhofer et al. 1999; Vignudelli et al. 2000) around the turn of the century, when several researchers around the world were starting to come up with new ideas and to explore some of the techniques necessary to make it work. A small but significant mark in the history of coastal altimetry was the launch of the ALBICOCCA program on 2001; this was the precursor of the European INTAS ALTICORE project (2007–2008) where this book had its genesis. Since 2006 there has been a dramatic increase in the research effort on this novel topic, and the coming together of an active international *community of researchers* who are exchanging ideas at regular Coastal Altimetry workshops (Silver Spring 2008; Pisa 2008; Frascati 2009; Porto 2010 – see http://www.coastalt.eu) and have contributed a "Community White Paper" on the role of altimetry in Coastal Observing Systems at the recent OceanObs'09 conference (Cipollini et al. 2010). The major space agencies and funding agencies have recognized the importance of the topic and are supporting it via a series of projects (e.g. COASTALT, PISTACH, RECOSETO and some OSTST projects). This book builds on many of the research efforts in the aforementioned projects and across the whole community, with the aim of presenting the current status in this new, challenging and exciting topic.

This peer-reviewed book brings together 20 chapters addressing a wide range of issues in processing and exploiting satellite altimetry from the open ocean to the coasts. The chapters presented describe the state of the art, recent developments on specific issues and practical experience. Chapter 1 gives a historic perspective and shows the current status of satellite altimetry missions. This is immediately followed in Chapter 2 by the illustration of a successful case in which coastal altimetry data have gone from research to operation. Chapter 3 discusses the user requirements for the new coastal altimetry products, i.e. what the users expect to get. The bulk of chapters from Chapter 4 to 9 deals with data processing aspects, particularly those concerning re-tracking and corrections for improving altimeter data, including editing strategy. These are the techniques that truly make coastal altimetry a reality. Chapter 10 extends the content of the book to data dissemination and services. Chapter 11 illustrates the clear link of calibration and validation to coastal altimetry. Following this introduction of how altimeter data are processed and managed is a series of Chapters (12 to 19) that show the gamut of applications in some regional seas (Mediterranean, Black, Caspian, White, Barents), around coasts of some countries (North America, China, Australia) and even lakes that have similar processing challenges. Finally, Chapter 20 explores novel technologies of forthcoming satellite altimetry missions and their expected impact on data retrieval in the coastal zone.

We intend this book for a wide audience of students, researchers, data-integrators with interest in the Coastal Zone who seek a unique, research-centred, up-to-date reference to Coastal Altimetry. Finally, we hope that this book will inspire both young and more experienced researchers, and we hope that it will be seen both as a stimulus to exploit Coastal Altimetry and an invitation to explore further the intricate details of technology, algorithms, processing, in a field that promises to bring significant and long-lasting societal benefits.

Stefano Vignudelli,
Andrey G. Kostianoy,
Paolo Cipollini,
Jérôme Benveniste

August 2010

References

Anzenhofer M, Shum CK, Rentsh M (1999) Costal altimetry and applications, Tech. Rep. n. 464, Geodetic Science and Surveying, The Ohio State University Columbus, USA

Cipollini P, Benveniste J, Bouffard J, Emery W, Gommenginger C, Griffin D, Høyer J, Madsen K, Mercier F, Miller L, Pascual A, Ravichandran M, Shillington F, Snaith H, Strub T, Vandemark D, Vignudelli S, Wilkin J, Woodworth P, Zavala-Garay J (2010) The role of altimetry in Coastal Observing Systems, in Proceedings of OceanObs'09: Sustained Ocean Observations and Information for Society (Vol. 2), Venice, Italy, 21-25 September 2009, Hall, J., Harrison, D.E. & Stammer, D., Eds., ESA Publication WPP-306. doi:10.5270/OceanObs09.cwp.16

Crout R L (1997) Coastal currents from satellite altimetry. Sea Technology 8:33–37

Fu LL, Cazenave A (eds) (2001) Satellite altimetry and earth sciences: a handbook of techniques and applications. Academic Press, San Diego, 463 pp

Vignudelli S, Cipollini P, Astraldi M, Gasparini GP, Manzella GMR (2000) Integrated use of altimeter and in situ data for understanding the water exchanges between the Tyrrhenian and Ligurian Seas. J Geophys Res 105:19,649–19,000

Abbreviations

ALBICOCCA	ALtimeter-Based Investigations in COrsica, Capraia and Contiguous Areas
ALTICORE	Value added satellite ALTImetry for COastal REgions
CNES	Centre National d'Études Spatiales
COASTALT	ESA Development of COAStal ALTimetry
ERS	European Remote Satellite
ESA	European Space Agency
GCOS	Global Climate Observing System
INTAS	International Association for the promotion of co-operation with scientists from the New Independent States of the former Soviet Union
IPCC	Intergovernmental Panel on Climate Change
ISI	Institute for Scientific Information
NASA	National Aeronautics and Space Administration
OSTST	Ocean Surface Topography Science Team
PISTACH	CNES Development of "Prototype Innovant de Système de Traitement pour les Applications Côtières et l'Hydrologie"
RECOSETO	REgional COastal SEa level change and sea surface Topography from altimetry, Oceanography, and tide gauge stations in Europe
UNFCCC	United Nations Framework Convention on Climate Change

Editorial Note and Acknowledgements

Each chapter in the book underwent rigorous peer-reviewing. The editors identified a minimum of two external reviewers for each chapter, selected on the basis of experience and knowledge of the subject. Each leading author received the formal anonymous reviews from the reviewers plus additional editorial comments, and was asked to improve or correct each one of the points raised or explain rebuttal. In some cases, where appropriate, the editorial process involved more than one review round, and this hopefully will have helped to guarantee the high quality of the final book.

The editors, also on behalf of the authors, would like to express gratitude to Springer-Verlag for the timely interest to the Coastal Altimetry topic and for supporting our efforts. The financial support of the European Space Agency and Centre National d'Etudes Spatiales (France) for publishing and disseminating the book is also gratefully acknowledged.

A sincere "thank you" from the editors goes to the chapter authors, who have played a crucial role in creating the book, for their patience during the lengthy review rounds but also for having responded in such a constructive manner to the editorial and reviewing feedback. The editors are particularly indebted to the 44 external reviewers for their invaluable help, suggestions, corrections and insights. Here is the list (including current affiliations) of those anonymous reviewers who agreed to be acknowledged here:

Arbic B., Florida State University, Tallahassee, USA
Callahan P., NASA Jet Propulsion Laboratory, Pasadena, USA
Chang X., Chinese Academy of Surveying and Mapping, Beijing, China
Cripe D., Group on Earth Observations, Geneva, Switzerland
Dinardo S., SERCO/ESRIN, Frascati, Italy
Fairhead D., University of Leeds/GETECH, Leeds, UK
Fanjul E., Puertos del Estado, Madrid, Spain
Font J., Institut de Ciències del Mar, Barcelona, Spain
Fouks V., St. Petersburg State University, St. Petersburg, Russia
Francis O., University of Luxembourg, Luxembourg
Francis R., European Space Agency, Noordwijk, The Netherlands
Gòmez-Enri J., Universidad de Cádiz, Cadiz, Spain
Heron M., James Cook University, Townsville, Australia
Jeansou E., NOVELTIS SA, Ramonville-Saint-Agne, France
Kosarev A., Moscow State University, Moscow, Russia
Le Traon P. Y., Institut Francais de Recherche pour l'Exploitation de la Mer, Brest, France

Liu Y., University of South Florida, St. Petersburg, USA
Martinez-Benjamin J. J., Technical University of Catalonia, Barcelona, Spain
Marullo S., ENEA, Frascati, Italy
Mercier F., Collecte Localisation Satellites (CLS), Toulouse, France
Maharaj A., Macquarie University, Sydney, Australia
Mertikas S., Technical University of Crete, Chania, Greece
Perosanz F., Centre National d'Etudes Spatiales, Toulouse, France
Roca M., IsardSAT, Barcelona, Spain
Sandwell D., University of California, San Diego, USA
Scozzari A., Consiglio Nazionale delle Ricerche, Pisa, Italy
Seyler F., Institut de Recherches pour le Développement, Toulouse, France
Soares A., World Meteorological Organization, Geneva, Switzerland
Stanichny S., Marine Hydrophysical Institute, Sevastopol, Ukraine
Sychev V., State Hydrometeorological University, St. Petersburg, Russia
Tuzhilkin V., Moscow State University, Moscow, Russia
Vazzana S., European Space Agency, Frascati, Italy

Contents

1 **Radar Altimetry: Past, Present and Future** 1
  J. Benveniste

2 **From Research to Operations: The USDA Global
  Reservoir and Lake Monitor** 19
  C. Birkett, C. Reynolds, B. Beckley, and B. Doorn

3 **User Requirements in the Coastal Ocean for Satellite Altimetry** 51
  C. Dufau, C. Martin-Puig, and L. Moreno

4 **Retracking Altimeter Waveforms Near the Coasts** 61
  C. Gommenginger, P. Thibaut, L. Fenoglio-Marc, G. Quartly, X. Deng,
  J. Gómez-Enri, P. Challenor, and Y. Gao

5 **Range and Geophysical Corrections in Coastal Regions:
  And Implications for Mean Sea Surface Determination** 103
  O.B. Andersen and R. Scharroo

6 **Tropospheric Corrections for Coastal Altimetry** 147
  E. Obligis, C. Desportes, L. Eymard, J. Fernandes, C. Lázaro, and A. Nunes

7 **Surge Models as Providers of Improved "Inverse Barometer
  Corrections" for Coastal Altimetry Users** 177
  P.L. Woodworth and K.J. Horsburgh

8 **Tide Predictions in Shelf and Coastal Waters: Status and Prospects** 191
  R.D. Ray, G.D. Egbert, and S.Y. Erofeeva

9 **Post-processing Altimeter Data Towards Coastal
  Applications and Integration into Coastal Models** 217
  L. Roblou, J. Lamouroux, J. Bouffard, F. Lyard, M. Le Henaff,
  A. Lombard, P. Marsaleix, P. De Mey, and F. Birol

10 **Coastal Challenges for Altimeter Data Dissemination and Services** 247
   H.M. Snaith and R. Scharroo

11	**In situ Absolute Calibration and Validation: A Link from Coastal to Open-Ocean Altimetry**	259
	P. Bonnefond, B.J. Haines, and C. Watson	
12	**Introduction and Assessment of Improved Coastal Altimetry Strategies: Case Study over the Northwestern Mediterranean Sea**	297
	J. Bouffard, L. Roblou, F. Birol, A. Pascual, L. Fenoglio-Marc, M. Cancet, R. Morrow, and Y. Ménard	
13	**Satellite Altimetry Applications in the Caspian Sea**	331
	A.V. Kouraev, J.-F. Crétaux, S.A. Lebedev, A.G. Kostianoy, A.I. Ginzburg, N.A. Sheremet, R. Mamedov, E.A. Zakharova, L. Roblou, F. Lyard, S. Calmant, and M. Bergé-Nguyen	
14	**Satellite Altimetry Applications in the Black Sea**	367
	A.I. Ginzburg, A.G. Kostianoy, N.A. Sheremet, and S.A. Lebedev	
15	**Satellite Altimetry Applications in the Barents and White Seas**	389
	S.A. Lebedev, A.G. Kostianoy, A.I. Ginzburg, D.P. Medvedev, N.A. Sheremet, and S.N. Shauro	
16	**Satellite Altimetry Applications off the Coasts of North America**	417
	W.J. Emery, T. Strub, R. Leben, M. Foreman, J.C. McWilliams, G. Han, C. Ladd, and H. Ueno	
17	**Evaluation of Retracking Algorithms over China and Adjacent Coastal Seas**	453
	L. Yang, M. Lin, Y. Zhang, L. Bao, and D. Pan	
18	**Satellite Altimetry for Geodetic, Oceanographic, and Climate Studies in the Australian Region**	473
	X. Deng, D.A. Griffin, K. Ridgway, J.A. Church, W.E. Featherstone, N.J. White, and M. Cahill	
19	**Lakes Studies from Satellite Altimetry**	509
	J.-F. Crétaux, S. Calmant, R. Abarca del Rio, A. Kouraev, M. Bergé-Nguyen, and P. Maisongrande	
20	**The Future of Coastal Altimetry**	535
	R.K. Raney and L. Phalippou	

Index 561

Radar Altimetry: Past, Present and Future

J. Benveniste

Contents

1.1 The Principle of Radar Altimetry 3
1.2 The Pioneering Era 4
1.3 The Historical Era 4
1.4 The Present Epoch 6
1.5 The Future 7
    1.5.1 CryoSat-2 Mission Data Sheet 9
    1.5.2 Sentinel-3 Mission Data Sheet 9
    1.5.3 SARAL Radar Altimeter Mission Data Sheet 9
    1.5.4 HY-2 Radar Altimeter Mission 13
1.6 Conclusions and Perspectives 14
References 15

Keywords Coastal Zone Altimetry • CryoSat-2 • Envisat • ERS-1 • ERS-2 • GEOS-3 • Geosat • GFO • GFO-2 • HY-2 • Jason-1 • Jason-2 • Jason-3 • Radar Altimetry missions • Radar Altimetry principle • SARAL • Seasat • Sentinel-3 • Skylab • SWOT • TOPEX/Poseidon

Abbreviations

AATSR	Advanced Along-Track Scanning Radiometer
ALBICOCCA	ALtimeter-Based Investigations in COrsica, Capraia and Contiguous Areas
ALTICORE	Value added satellite ALTImetry for COastal REgions
ARGOS	or "A-DCS" Advanced data collection system - data collection and location system (France) on NOAA operational satellites
CEOS	Committee for Earth Observation Satellites

J. Benveniste
Directorate of Earth Observation Programmes, Earth Observation Science,
Applications and Future Technologies Department, European Space Agency – ESRIN,
Via Galileo Galilei, Frascati (Rome), Italy
e-mail: Jerome.Benveniste@esa.int

CFRP	Carbon Fibre Reinforced Plastic
CNES	Centre National d'Études Spatiales (National Centre of Space Studies)
CNSA	China National Space Administration
COASTALT	ESA development of COASTal ALTimetry
DORIS	Détermination d'Orbite et Radiopositionement Integré par Satellite
DUACS	Data Unification and Altimeter Combination System
EC	European Commission
EP-90	Russian earth reference frame
ERS-1/2	European Remote Sensing Satellite
ESA	European Space Agency
ESOC	European Space Operation Centre, Darmstadt
ESRIN	European Space Research INstitute, Rome
EUMETSAT	European Meteorological Satellite Agency
GFO	Geosat Follow-On – geodesy satellite (US Navy)
GMES	European initiative on Global Monitoring for Environment and Security
GNSS	Global Navigation Satellite Systems
GOCE	Gravity and Ocean Circulation Explorer (ESA)
GPS	Global Positioning System
GRACE	Gravity Recovery and Climate Experiment
GSE	ESA GMES service element
HY-2	HaiYang (for Ocean in Chinese) satellite mission
INTAS	International Association for the promotion of co-operation with scientists from the New Independent States of the former Soviet Union
ISRO	Indian Space Research Organization
JGR	Journal of Geophysical Research
KaRIN	Ka-band Radar INterferometer (on SWOT)
LEO	Low Earth Orbit
LMCS	Land Monitoring Core Service LEO
LTDN	Local Time on Descending Node
MCS	EU Marine Core Service
MERIS	MEdium Resolution Imaging Spectrometer
MWR	MicroWave Radiometer
NASA	National Aeronautics and Space Administration
NOAA	National Oceanic and Atmospheric Administration
NSOAS	National Satellite Ocean Application Service
PISTACH	CNES Development of "Prototype Innovant de Système de Traitement pour les Applications Côtières et l'Hydrologie"
PSLV	Indian Polar Satellite Launch Vehicle
RA-2	Radar Altimeter 2nd generation on Envisat
SAMOSA	ESA development of SAR altimetry mode studies and applications over ocean, coastal zones and inland water
SAR	Synthetic Aperture Radar
SARAL	Satellite with ARgos and ALtika
SIRAL	SAR/Interferometric Radar ALtimeter
SPOT	Satellite Pour l'Observation de la Terre (optical imaging)
SRAL	SAR Radar ALtimeter

SRTM	Shuttle Radar Topography Mission
SSH	Sea Surface Height
SWH	Significant Wave Height
SWOT	Surface Water Ocean Topography mission
WGS84	World Geodetic System 1984

1.1 The Principle of Radar Altimetry

As the reader focuses on Coastal Zone Radar Altimetry, it is assumed that he or she has already acquired some knowledge on the principle of radar altimetry. A primer for the topic can be found in the radar altimetry Tutorial (Rosmorduc et al. 2009). More details can be found in Fu and Cazenave (2001). A brief recall at the beginning of this book may prove to be handy.

Radar altimetry measures the distance between the satellite and the surface below, transmitting radar pulses, the echoes of which are bounced back from the surface, whether ocean, ice cap, sea-ice, desert, lake, or river, see Fig. 1.1. This distance, called the range, has two ends. Above, the satellite's position is precisely known through orbit determination, referred to the ellipsoid (e.g., WGS84), with the contribution of an on-board navigation device such as DORIS or a GPS receiver, or both. Below, the absolute elevation of the sea surface, land, river, or ice sheet is derived from the difference between the orbit altitude and the range, corrected for effects of propagation through the atmosphere and reflection on the surface. The instantaneous elevation of the target is the main measurement datum. The characteristics of the echoes contain further information on the roughness of the surface, wave heights, or surface wind speed over the ocean.

Over the ocean, the measurements are averaged over 7 km (1 s of flight) along-track, and the cross-track spacing is 40–300 km (depending on the repeat period of the satellite);

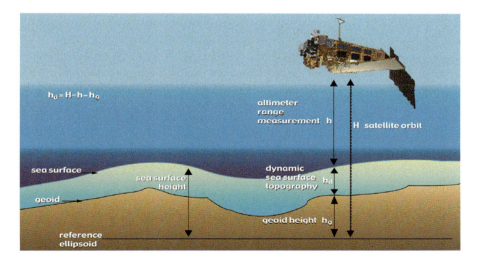

Fig. 1.1 The principle of radar altimetry

therefore, one can directly map the anomalies of the sea surface topography with a resolution resolving the eddy fields, all the more when using data from several altimeter missions. For applications in the coastal zone, it is important to recall that for most altimeters, the raw waveforms, averaged on-board by groups of 50 (ERS), 90 (Jason), or 100 (Envisat), are transmitted on-ground at a rate of 18 or 20 Hz, which means a measurement every 350 m approximately. One notable exception is that the Envisat Altimeter, RA-2, has the ability to preserve a small amount of individual echoes (unaveraged echoes, at 1795 Hz), slowly transmitted on ground, which are useful for engineering and scientific studies. See http://earth.esa.int/raies for accessing the Envisat RA-2 Individual Echoes products and detailed information; see also Gommenginger et al. (2004, 2006), Berry et al. (2006) and Gommenginger et al. (2007).

1.2
The Pioneering Era

The principle of radar altimetry measurements was envisaged in the sixties, and recognized as a high priority measurement at the Williamstown Symposium in 1969 (Kaula 1970). The first objective of the measurement was to measure the shape of the Earth, which can be considered today as a very limited ambition, but the error in the first space-borne measurement was of the order of 100 m (McGoogan et al. 1974). It was clear that a lot was to be gained with this technique, if the error could be beaten down below the level of the ocean dynamic topography. The development of altimeter technology was a constant effort, which gave birth to a series of early missions: Skylab (1973), GEOS-3 (9 April 1975–December 1978), and Seasat (June 1978–October 1978). With the advent of more precise instruments flying on a much better known trajectory, radar altimetry began to supply invaluable information in Geodesy, Oceanography, Geophysics, Glaciology, and Continental Hydrology.

GEOS-3 (Agreen 1982) emitted a long pulse with a compression technique on the received signal to reduce the noise level, albeit a still high noise level with a measurement precision of 25 cm in 1 s averages (Stanley et al. 1979), making it possible to detect the oceanic major currents and their eddy fields. Seasat further increased the measurement precision for 1 s averages to about 5 cm (Tapley et al. 1982), with a misleadingly called "pulse compression" technique (using a deramping chirp upon receive) but provided only 3 months of data, which could mean, with today's standards and expectations, a failure. But some pointed out that by having the whole radar altimetry community focusing on such a small dataset – already huge by computer capacity of the eighties – astonishing progress was rapidly made on all the components of radar altimetry, including auxiliary files providing the orbit ephemeris and the geophysical corrections, in particular the tides.

1.3
The Historical Era

The demonstration made by Seasat (Townsend 1980), later refined by Marsh et al. (1990), was an inspiration for NASA, the European Space Agency, the US Navy and the French Space Agency CNES, and led to the design of Geosat, the European Remote Sensing

Satellite Programme bearing ERS-1 and later ERS-2. Meanwhile, NASA was developing the Topography Experiment and CNES the Poseidon Programme, and they started to work together on a joint mission: TOPEX/Poseidon. Geosat, the US Navy geodetic satellite, was launched on 12 March 1985 and supplied nearly 5 years of data, but began its duty with an 18-month classified mission to measure the marine geoid, until 8 November 1986. These data were later declassified when ERS-1 had acquired a similar data set during its geodetic mission phase (from 10 April 1994 to 25 March 1995). Both datasets together produced the best ever marine geoid and sea floor topography (Smith and Sandwell 1997; Calmant et al. 2002) before the advent of space-borne gravimetry with the GRACE and GOCE missions. Geosat was the only radar altimeter flying in the eighties providing data to the science community to monitor the famous 1987 El Nino sea level anomalies and demonstrated the benefits of a long-time series focused on measuring eddy variability with a precision better than 5 cm on 1 s averages (Sailor and LeSchack 1987), but with no microwave radiometer to properly apply the wet tropospheric path delay correction.

A less well-known series of radar altimetry missions was launched by the Russians out of a programme called GEOIK. It was started in 1985 to obtain a reference system and a gravity field for the Earth, called EP-90. Initially, the data gathered within this program, run by the Ministry of Defence of the Russian Federation, were classified. They were later declassified in 1992 and given to the organizations of the Russian Academy of Science. They are claimed to be at the disposal of the international scientific community but not many labs around the world have processed these data so far. Ten GEOIK satellites, part of the Kosmos series, were launched from 1984 to 1996, all carrying a 9.5 GHz radar altimeter and space-borne geodetic instruments including Doppler system, radio range system, light signaling flash system, and laser corner reflectors. They were placed in an orbit at 1,500 km altitude with an inclination of either 74° or 83°. Some drawbacks for monitoring were that the radar altimeters were operated half of the time, providing a nonuniform coverage and the range measurement, averaged over 1-s, had a noise level of the order of 1 m. This level of performance did not permit the observation of oceanographic features. Notwithstanding, the GEOIK program delivered the geopotential models EP-90 (up to 36°) and EP-200 (up to 200°) and a marine geoid.

By that time, the radar altimetry community was solidly unified around the TOPEX/Poseidon Science Working Team, meeting regularly on both sides of the Atlantic and working together to shape the mission's attributes (Chelton 1988). In parallel, the community fully convinced of the great potential of a multi-mission radar altimetry approach, laid out the concept of a high-accuracy, low-inclination, short repeat period reference mission, focused on the large-scale ocean circulation, complemented by high-inclination, long repeat period missions, supplying increased spatial sampling essential for mesoscale eddy monitoring. This visionary strategic planning was published by Koblinsky, Gaspar, and Lagerloef, the editors of the famous and historical "purple book" (1992). ERS-1 had already been launched (17 July 1991) and was going through its commissioning and first data release (Benveniste 1993), while TOPEX/Poseidon (10 August 1992) was on its way to demonstrate the winning strategy of the "purple book".

The TOPEX/Poseidon mission was the achievement of 20 years of technological development efforts. Every effort was made to reduce the error budget, from increasing the flight altitude (to 1,336 km) to using a second altimeter frequency (C-band, 5.3 GHz) to correct for the ionospheric path delay, as well as adding a third frequency on the microwave radiometer (18 GHz) to remove the effects of wind speed on water vapor measurement. A wealth of scientific return was published in the JGR Oceans 2 years after the launch in the JGR special issue on TOPEX/Poseidon (Fu et al. 1994) and again 1 year later (TOPEX/Poseidon 1995). As the "purple book" (1992) envisioned, TOPEX/Poseidon was used to improve the ERS-1 data (Le Traon et al. 1995), and soon the data were merged together systematically (DUACS 1997) and in near-real time. The European Space Agency ERS-2 Satellite was launched on 21 April 1995 to extend the measurements of ERS-1 and it is still operational today. The Geosat Follow-On (GFO) mission was launched on 10 February 1998 on a 17-day repeat orbit, the same as Geosat, and retired in November 2008. Its mission was to provide real-time ocean topography data to the US Navy. Scientific and commercial users have access to these data through NOAA.

1.4
The Present Epoch

Jason-1 (7 December 2001) and Envisat (1 March 2002) were launched to follow-on the ERS and TOPEX/Poseidon Tandem. Envisat is the largest ever Earth Observatory and among its ten instruments carries the altimetry system: The radar altimeter of 2nd generation, a microwave radiometer, DORIS and a laser retro-reflector (Benveniste et al. 2001). Jason-1 has the same high-accuracy performance than TOPEX/Poseidon, but Jason-1 is a mini-satellite, featuring low-cost, light-weight, and low-power consumption.

A major milestone event in the history of radar altimetry was the special Symposium on "15 Years of Progress in Radar Altimetry" (Benveniste and Ménard 2006) held in Venice in March 2006. Satellite altimetry has expanded from conferences attended by a handful of scientists to this Symposium attended by around 500 scientists submitting papers with more than 1,200 authors and coauthors. The reader is encouraged to get acquainted with the proceedings, in particular with a summary of the recommendations gathered during the plenary session on the future of altimetry and the panel discussion, which brings guidance on the requirements for future radar altimetry missions. The main message was the need for continuity of a multi-mission radar altimetry system making best use of the new technology arising for a new generation of instruments in support of both Science and operational monitoring/forecasting (Fellous et al. 2006). A noteworthy subsequent event was the CEOS Ocean Surface Topography Constellation Strategic Workshop, held in Assmannshausen in January 2008 (Wilson and Parisot 2008). In 2009, this community met again within the framework of the OceanObs'09 Conference held in Venice, Italy, from 21 to 24 September and enforced these recommendations in a community White Paper (Wilson et al. 2010).

Jason-2 was launched on 8 June 2008 on the same orbit as its predecessors, TOPEX/Poseidon and Jason-1. After the Jason-2 commissioning was performed, Jason-1 has been shifted on a new orbit, mid-way between the previous one, with a 5-day time lag, to improve the space-time sampling, better satisfying the near-real time applications, and paving the way for operational oceanography.

1.5
The Future

Based on the AltiKa concept development (a single frequency radar altimeter working in Ka-band, 35 GHz, with a DORIS instrument and a microwave radiometer) by the French Space Agency CNES, a new initiative from the Indian Space Research Organization (ISRO) will launch SARAL: (Satellite with ARgos and ALtika) in late 2010. The Ka-band will enable better observation of ice, rain, wave heights, coastal zones, and land masses, with some limitation during heavy rain events. China is also joining the effort with the HY-2 (HaiYang for Ocean in Chinese) to be launched also in late 2010. The objective of HY-2 is to monitor the ocean with microwave sensors to detect sea surface wind field, sea surface height, and sea surface temperature. It will carry a dual-frequency altimeter (Ku and C-bands), a scatterometer, a microwave imager on a sun-synchronous orbit, the first 2 years with a 14-day cycle, and then 1 year with a geodetic orbit (168-day cycle with a 5-day sub-cycle). These initiatives are welcome to complement the USA/Europe efforts to develop a satellite altimetry constellation.

When the Geosat Follow-on mission ended in November 2008, the Navy's ocean forecasting capability had degraded, urging the US Navy to plan a GFO Follow-on, GFO-2. It will have essentially the same characteristics as GFO, covering the famous 17-day repeat orbit for optimal mesoscale monitoring. It should be launched by the end of 2013, and the US Navy is even taking the option to procure an additional spacecraft to foresee a swift replacement of GFO-2, which should have a 6-year lifetime.

The next-generation of radar altimeters is already flying with much higher resolution and lower noise, thanks to the delay-Doppler concept (Raney 1998; see also the chapter by Raney and Phalippou, this volume) providing for a Synthetic Aperture Radar Altimeter Mode. This improved technology is the core of ESA's Ice Mission, CryoSat-2, launched in 2010 (Wingham et al. 2006), and will fly on the ESA-built GMES operational mission, Sentinel-3, to be launched in 2013 (Aguirre et al. 2007). These missions will particularly benefit the Coastal Zone, CryoSat-2 as a demonstrator, and Sentinel-3 as a monitoring mission, as the increase of the resolution will help reduce the polluting contribution of the coast in the return echoes.

The main limitation of conventional nadir-pointing radar altimeters is the space-time coverage dilemma (either the inter-track distance is large or the time sampling is poor), being a limiting factor for monitoring the fast signals in the equatorial band, the mesoscale variability or river stage and discharge. The solution to improve this limitation is to acquire topography in two dimensions, i.e., in a swath, providing additionally the capacity to derive a high-resolution slope field, for deriving currents and river discharge. The SWOT (Surface Water Ocean Topography) mission is proposed for 2016 in a joint project including NASA and CNES.

This mission will serve both land hydrology and oceanography communities, based on a novel technology: a Ka-band Radar INterferometer (KaRIN), which works in a similar way to the Shuttle Radar Topography Mission (SRTM), which is designed to acquire elevations of ocean and terrestrial water surfaces at unprecedented spatial and temporal scales (kilometers and days) with major outcomes related to the global water cycle, water management, ocean circulation, and the impacts of climate change (see SWOT mission requirements 2009). The science goals, while providing sea surface heights and terrestrial water heights over a 120 km wide swath with a ±10 km gap at the nadir track, are to:

- over the deep oceans, provide sea surface heights on a 2 × 2 km grid with a precision not to exceed 0.5 cm when averaged over the area.
- over land, download the raw data for ground processing and produce a water mask able to resolve 100-m wide rivers and lakes, wetlands, or reservoirs of surface greater than 1-km^2. Associated with this mask will be water level elevations with an accuracy of 10 cm and a slope accuracy of 1 cm/1 km.
- cover at least 90% of the globe.

These requirements do not explicitly mention the coastal zone, but the mission design is certainly fit to improve the data retrieval for coastal zone applications. It will be very exciting to validate the swath values near the shore and apply them to coastal zone oceanographic studies.

Also it has been envisaged, under the umbrella of the Group on Earth Observation, an intergovernmental organization based in Geneva, Switzerland, that Iridium's planned next-generation constellation of 66 low-earth orbiting satellites can provide a capability for a wide range of remote sensing missions for monitoring global climate data and other environmental data, which could include radar altimeters, even though at time of writing no plans have been established or funded. The Iridium-NEXT satellites should be launched between 2014 and 2016.

In summary, the prospects for the oceanic coastal zone community to access a continuing source of radar altimetry data are quite good, with a constellation of planned missions to complement and back-up each other, as illustrated in the Fig. 1.2 reporting the missions' schedule and status.

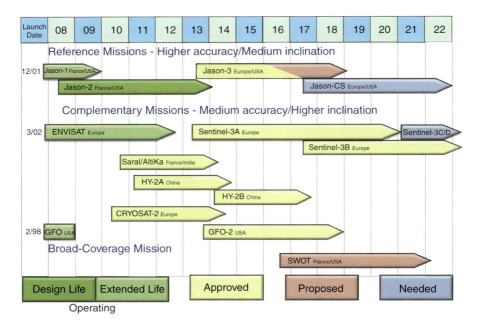

Fig. 1.2 Radar altimetry missions schedule and status from 2008 to 2022

1.5.1
CryoSat-2 Mission Data Sheet

The Scientific objectives are the following:

- To determine the fluctuations in the mass of the Earth's major land and marine ice fields (Table 1.1).

1.5.2
Sentinel-3 Mission Data Sheet

The GMES Mission Sentinel-3 is a Global Sea/Land Monitoring Mission including Altimetry. The Sentinel-3 program represents a series of operational spacecraft over the envisioned service period to guarantee access to an uninterrupted flow of robust global data products. A Full description is given in Aguirre et al. (2007) (Table 1.2).

The Scientific Objectives are as follows:

- Ocean and land colour observation data, free from sun-glint, shall have a revisit time of 4 days (2 days goal) and a quality at least equivalent to that of MERIS instrument on Envisat.
- Ocean and land surface temperature shall be acquired with at least the level of quality of AATSR on Envisat, and shall have a maximum revisit time of 4 days with dual view (high accuracy) observations and 1 day with single view.
- Surface topography observations shall primarily cover the global ocean and provide sea surface height (SSH) and significant wave height (SWH) to an accuracy and precision at least equivalent to that of RA-2 on Envisat. Additionally, Sentinel-3 shall provide surface elevation measurements – in continuity with CryoSat-2 – over ice regions covered by the selected orbit, as well as measurements of in-land water surfaces (rivers and lakes). In addition, Sentinel-3 will provide surface vegetation products derived from synergistic and colocated measurements of optical instruments, similar to those obtained from the vegetation instrument on SPOT, and with complete Earth coverage in 1–2 days.

The EU Marine Core Service (MCS) and the Land Monitoring Core Service (LMCS), together with the ESA GMES Service Element (GSE), have been consolidating those services where continuity and success depends on operational data flowing from the Sentinels.

1.5.3
SARAL Radar Altimeter Mission Data Sheet

An Indian Space Research Organization (ISRO) satellite, SARAL (Satellite with ARgos and ALtika), will embark the AltiKa altimeter (working in Ka-band, 35 GHz), built by CNES, as well as a DORIS tracking instrument (Table 1.3).

Table 1.1 CryoSat-2 data sheet

Launch	On Dnepr launcher from Baikonur (Байқоңыр) cosmodrome in Kazakhstan, 8 April 2010 (the back-up launcher was Vega)
Mission Operations	Six months commissioning followed by a 3-year operations mission
Payload	SIRAL (SAR/Interferometric Radar Altimeter) – Low-Resolution Mode provides conventional pulse-width limited altimetry over central ice caps and oceans – SAR Mode improves along-track resolution (~250 m) over sea ice by significantly increased pulse repetition frequency and complex ground processing – SAR Interferometric Mode adds a second receiving chain to measure the cross-track angle of arrival of the echo over topographic surfaces at the margins of ice caps See Fig. 1.3 for their geographical distribution Star Trackers (3) Measure the interferometric baseline orientation, as well as driving satellite attitude control DORIS Enables precise orbit determination, as well as providing on orbit position to the satellite attitude control Laser Retro Reflector Passive Optical Reflector. Enables tracking by ground-based lasers
Dimensions	4.60 × 2.34 × 2.20 m
Mass	745 kg (including 37 kg fuel and 120 kg payload)
Mission orbit	– LEO, non sun-synchronous – 369 days (30 day sub-cycle) – Mean altitude: 717 km – Inclination: 92° – Nodal regression: 0.25°/day
Mission Management	– Full mission operations with one single ground station at Kiruna – Mission control from ESOC via Kiruna ground station – Onboard measurements automatically planned according to geographical defined masks – Data processing facility at the Kiruna ground station. Local archiving of data with precision processing after 1 month following delivery of precision orbits from DORIS ground segment – Direct dissemination of data from Kiruna – Long Term Archive in CNES – Near real time ocean data products where possible – User Services coordinated via ESRIN

1 Radar Altimetry: Past, Present and Future

Table 1.2 Sentinel-3 data sheet

Launch	End 2013, on a Vega launcher (compatibility with Dnepr launcher).	
Mission Operations	Six months commissioning followed by a 7.5-year operations mission (fuel for additional 5 years), joined in space by Sentinel-3B…	
Payload	SRAL (SAR Radar Altimeter)	• SRAL is a redundant dual-frequency (Ku-band and C-band) nadir-looking SAR altimeter • Low-Resolution Mode provides conventional pulse-width limited altimetry over central ice caps and oceans • SAR Mode improves along-track resolution (~250 m) over coastal zones, sea ice, and rivers by significantly increased pulse repetition frequency and complex ground processing • These modes are associated with two tracking modes: – Closed-loop mode: refers autonomous positioning of the range window (ensures autonomous tracking of the range and gain by means of tracking loop devices implemented in the instrument). – Open-loop mode: uses a-priori knowledge of the terrain height from existing high-resolution global digital elevation models combined with real-time navigation information available the GNSS receiver
	GNSS	Enables precise orbit determination, as well as providing on orbit position to the satellite attitude control
	Laser Retro Reflector	Passive Optical Reflector. Enables tracking by ground-based lasers
	MWR	23.8 and 36.5 GHz radiometer (and optionally 18.7 GHz) Mass of 26 kg
Dimensions	Bus: 3.89 (height) x 2.2 x 2.21 m	
Mass	1,198 kg, including 100 kg fuel for 12.5 years	
Mission orbit	– Sun-synchronous orbit (14 +7/27 rev./day) – mean altitude = 814 km – inclination = 98.6° – LTDN (Local Time on Descending Node) is at 10:00 h. – The revisit time is 27 days providing a global coverage of topography data at mesoscale (inter-track distance at equator: 104 km)	
Mission Operations	– Onboard measurements automatically planned according to geographical defined masks – Data processing facility at the Kiruna ground station. Local archiving of data with precision processing after 1 month following delivery of precision orbits from DORIS ground segment – Direct dissemination of data from Kiruna – Long Term Archive in CNES – Near real time ocean data products where possible – User Services coordinated via ESRIN	

Table 1.3 SARAL data sheet

Launch	End 2010 on PSLV Launch vehicle from the Sriharikota Range	
Mission Operations	Six months commissioning followed by a 3-year operations mission – objective is 5 years	

Payload	AltiKa (Radar Altimeter + Radiometer)	• Single-frequency Ka-band altimeter (operating at 35.75 GHz, weighing less than 20 kg and consuming less than 50 Watts)
		• Two tracking modes:
		  – Closed-loop mode: refers autonomous positioning of the range window (ensures autonomous tracking of the range and gain by means of tracking loop devices implemented in the instrument).
		  – Open-loop mode: uses a-priori knowledge of the terrain height from existing high-resolution global digital elevation models combined with real-time navigation information available the DORIS receiver.
		Includes a 2-frequency radiometer function sharing the same antenna at 23.8 and 37 GHz

	DORIS	Enables precise orbit determination, as well as providing on orbit position to the satellite attitude control
	Laser Retro-Reflector	Passive Optical Reflector. Enables tracking by ground-based lasers
Spacecraft	IMS-2 Platform (Small Satellite Bus) developed by ISRO	
Dimensions	Bus: 2.7 (height) x 1 x 1 m	
Mass	415 kg, platform 250 kg , payload 165 kg (including fuel)	
Mission orbit	– Same orbit as Envisat and ERS-2 – Sun-synchronous orbit (35-day repeat) – Mean altitude = 800 km – Inclination = 98.55° – LTDN (Local Time on Descending Node) is at 18:00 h – The revisit time is 35 days providing a global coverage of topography data at mesoscale (inter-track distance at equator: 80 km) – Ground-track stability requirement: ±1 km	
Mission Operations	– Onboard measurements automatically planned according to geographical defined masks – Data processing facility at ISRO, CNES, and EUMETSAT – Direct dissemination of data from Kiruna – Long Term Archive in CNES – Near real time ocean data products where possible – User Services at ISRO and CNES	

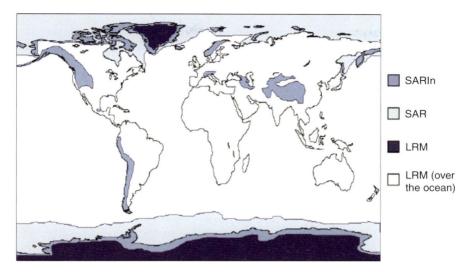

Fig. 1.3 Extent of the regions defining the reference mission. Note this is an extreme case, which illustrates maximum sea ice extent in both hemispheres. The operational mask will be more realistic (from CryoSat 2007)

The Scientific objectives are as follows:

- To study mesoscale ocean variability
- To observe coastal ocean areas
- To monitor the sea-state for analysis and forecast
- To observe inland waters
- To observe the surface of continental ice sheets
- Low rain detection and characterization

1.5.4
HY-2 Radar Altimeter Mission

HY-2 (which stands for HaiYang which means "ocean" in Chinese) is a marine remote sensing satellite series planned by China. The program foresees a series of missions: HY-2A (2010), then HY-2B (2012), HY-2C (2015), and HY-2D (2019). The objective of HY-2 is to monitor the following:

- The global ocean wind field and provide services for ocean environment forecast
- The global ocean wave field to forecast and mitigate disaster at sea
- The global ocean topography, sea level, and the gravity field with applications such as forecasting El Niño
- The sea ice fields
- The sea surface temperature

It will include a dual-frequency Radar Altimeter operating in Ku (13.58 GHz) and C-band (5.25 GHz), a scatterometer and a microwave imager. HY-2 will be launched in June 2010 by a Long March 4 launcher, with a design lifetime of 3 years, at an altitude of 963.6 km for the 14 day repeat orbit and 965 km for the 168 day repeat orbit, with an inclination of 99.3°. Its orbit radial component will have an expected 5 cm precision.

The distribution of HY-2 satellite data requires the authorization of the China National Space Administration (CNSA). Then National Satellite Ocean Application Service (NSOAS) will be responsible for the processing and distribution. For now, it is foreseen that these data will be distributed to all users only after a delay of 1 month and only in the open ocean. The reason for this is to give time to ensure a proper validation and assure the distribution of only high quality data. It seems that the Data in the Coastal Zone will therefore not be distributed as its quality is hampered by the presence of land near-by. It would be beneficial to the Coastal Altimetry community if the full rate waveform data were distributed, in near real time, so that this community could apply the novel retracking techniques specially developed to operate in the difficult terrain. The same requirement applies for the near real time monitoring of river stage and lake level. In fact there is a strong recommendation from this community to apply a free and unlimited access to HY-2 satellite data as it is foreseen for the EC and ESA Sentinel missions. In particular, HY-2 is described as an opportunity to ensure data continuity, filling gaps in existing and planned missions. Consequently, for a successful HY-2 mission, a nondiscriminatory (in delay or quality) access to data is required.

HY-1A and HY-1B (launched on May 2002 and April 2007, respectively) are designed to measure the ocean colour and temperature with the visible and infrared sensor. HY-3 will combine visible/infrared and microwave sensors.

1.6
Conclusions and Perspectives

In the history of radar altimetry, as soon as the measurement principle and the technology were demonstrated on the first two missions before the end of the 1970s, the great potential of this tool applied to oceanography became rapidly clear. As soon as mesoscale variability was being monitored over the global ocean, with Seasat, historical pioneering efforts have been spent to get the very successful open ocean altimetry into the coastal areas, where it was evident that the radar altimeter and radiometer measurements were suffering from the nearby presence of land or even still waters.

One of the first attempts to retrieve information from coastal zone altimetric data was that of Le Provost (1983) studying the tidal signal in the English Channel, where the sea surface variability is of the order of meters. Some time passed and when ERS-1 and TOPEX/Poseidon were launched, the challenge to sail nearer to the coast, with this new and improved data, was triggered. It was also clear how important were the dynamics near river estuaries, which revealed the potential of altimetry in this application (Minster et al. 1995). Manzella et al. (1997) recomputed the wet tropospheric correction for ERS-1 altimeter data over the Corsica Channel by recalibrating the model correction with the closest available radiometric estimate; they also recovered measurements that had been flagged due to the large sigma0 values over still waters. Crout (1997) reviewed the potential of TOPEX/Poseidon over the coastal ocean, focusing on 1 Hz and 10 Hz data, and emphasized that next to a coast near flat coastal topography the useful signal is retrievable closer to shore than in the case of rough terrain. Anzenhofer et al. (1999) carried out an extensive study of coastal altimetry. They described the generation of coastal altimeter data and are the first to analyze in detail various retracking algorithms and their implementation. They presented some examples based on ERS waveform data, and concluded with some recommendations on the requirement for improved local tidal modelling, a method for data screening, improvement of the wet tropospheric correction, and retracking. By that time, it was clear what needed to be improved, but the way forward was still very foggy.

A major effort in bringing radar altimetry to the coastal zone has been recently undertaken essentially through the ALBICOCCA, EU-INTAS ALTICORE, COASTALT, SAMOSA (driven by ESA), and the PISTACH (driven by CNES) Projects (Mercier et al. 2008), as well as in the USA. The results of these efforts can be found all throughout this book and in an overview of the projects in the preface. The future will see high-resolution SAR mode data coming from the Sentinel-3 mission and additional data from Jason-3, SARAL, and HY-2 missions. The community is well unified through the regular and frequent Coastal Altimetry Workshops (cf. the proceedings of these workshops accessible through http:// www.coastalt.eu). Nevertheless, there are still challenges to overcome, both at the local and global scales, as it is described in the following chapters, bringing solutions that will certainly be validated and improved in the near future.

Acknowledgments This chapter holds information communicated to me by a number of colleagues, over the years, in particular of the older generation, who were in fact the pioneers of the radar altimetry adventure, for whom I have a great admiration. They are too numerous to be all cited here, but they will recognize themselves and receive all my gratitude. In particular I thank and dedicate this chapter to my long time colleague and friend, Yves Ménard, who passed away in 2008, with whom I had so much pleasure and satisfaction to organize the "15 Years of Progress in Radar Altimetry" Symposium in Venice in 2006. I recall our precursor conversations about how to tackle the coastal zone with altimetry. Discussions with Yves were always a great inspiration. Yves would be proud of the results achieved in the past 2 years. Yves remains always present in my memory.

References

Agreen RW (1982) The 3.5-year GEOS-3 data set, NOAA technical memorandum C 55.13/2: NOS NGS 33
Aguirre M, Berruti B, Bezy JL, Drinkwater M, Heliere F, Klein U, Mavrocordatos C, Silvestrin P, Greco B, Benveniste J (2007) Sentinel-3: The ocean and medium-resolution land mission for GMES operational services, ESA Bulletin, No 131, Aug. 2007, 24–29, URL: http://www.esa.int/esapub/bulletin/bulletin131/bul131c_aguirre.pdf
Anzenhofer M, Shum CK, Rentsh M (1999) Costal altimetry and applications. Tech. Rep. n. 464, Geodetic Science and Surveying, The Ohio State University Columbus, USA, pp 1–40
Benveniste J (1993) Towards more efficient use of radar altimeter data, ESA Bulletin No. 76
Benveniste J, Ménard Y (eds) (2006) Proceedings of the "15 years of progress in radar altimetry" symposium, Venice, Italy, 13–18 March 2006, ESA Special Publication SP-614

Benveniste J, Roca M, Levrini G, Vincent P, Baker S, Zanife O, Zelli C, Bombaci O (2001) The radar altimetry mission: RA-2, MWR, DORIS and LRR, ESA Bulletin, No. 106

Berry P, Rogers C, Garlick J, Freeman JA, Benveniste J (2006) The envisat burst mode echoes – a new look from satellite radar altimetry. In: Benveniste J, Ménard Y (eds) Proceedings of the "15 years of progress in radar altimetry" symposium, Venice, Italy, 13–18 March 2006, ESA Special Publication SP-614

Calmant S, Bergé-Nguyen M, Cazenave A (2002) Global seafloor topography from a least-squares inversion of altimetry-based high-resolution mean sea surface and shipboard soundings. Geophys J Int 151(3):795–808. doi:10.1046/j.1365-246X.2002.01802.x

Chelton DB (ed) (1988) WOCE/NASA Altimeter Algorithm Workshop, Oregon State University, Corvallis, 24–26August 1987, U.S. WOCE Technical Report no. 2, U.S. Planning Office for WOCE, TX

Crout RL (1997) Coastal currents from satellite altimetry. Sea Technol 8:33–37

CryoSat Mission and Data Description (2007) Doc. No.: CS-RP-ESA-SY-0059, ESA

DUACS (1997) http://www.aviso.oceanobs.com/duacs. Accessed 1 October 2009

Fellous JL, Wilson S, Lindstrom E, Bonekamp H, Ménard Y, Benveniste J (2006) Summary of the future of altimetry session. In: Benveniste J, Ménard Y (eds) Proceedings of the "15 years of progress in radar altimetry" symposium, Venice, Italy, 13–18 March 2006, ESA Special Publication SP-614

Fu LL, Cazenave A (eds) (2001) Satellite altimetry and earth sciences: a handbook of techniques and applications. Inter Geophys Series 69:463, Academic, San Diego, ISBN: 0-12-269543-3

Fu LL, Christensen EJ, Lefebvre M, Ménard Y (1994) TOPEX/Poseidon mission overview. J Geophys Res, TOPEX/Poseidon Special Issue, 99(C12):24, 369–324, 382

Gommenginger C, Challenor P, Quartly G, Gómez-Enri J, Srokosz M, Caltabiano A, Berry P, Mathers L, Garlick J, Cotton P D, Carter D J T, LeDuc I, Rogers C, Benveniste J, Milagro M P (2004) RAIES: ENVISAT RA2 individual echoes and S-band data for new scientific applications for ocean, coastal, land and ice remote sensing. In: Proceedings of the 2004 Envisat & ERS symposium, 6–10 Sept 2004, Salzburg, Austria, ESA Special Publication SP-572

Gommenginger C, Challenor P, Gómez-Enri J, Quartly G, Srokosz M, Berry P, Garlick J, Cotton PD, Carter D, Rogers C, Haynes S, LeDuc I, Milagro MP, Benveniste J (2006) New scientific applications for ocean, land and ice remote sensing with ENVISAT altimeter individual echoes. In: Benveniste J, Ménard Y (eds) Proceedings of the "15 years of progress in radar altimetry" symposium, Venice, Italy, 13–18 March 2006, ESA Special Publication SP-614

Gommenginger C, Challenor P, Quartly G, Srokosz M, Berry P, Rogers C, Milagro MP, Benveniste J (2007) ENVISAT altimeter individual echoes: new scientific applications for ocean, land and ice remote sensing, ENVISAT Symposium, Montreux, 23–27 April 2007, ESA Special Publication SP-636

Kaula WM (1970) The terrestrial environment: solid earth and ocean physics. Williamstown Report, NASA

Koblinsky CJ, Gaspar P, Lagerloef G (eds) (1992) The future of spaceborne altimetry, oceans and climate change, a long term strategy (so-called "Purple Book"), March 1992

Le Provost C (1983) An analysis of SEASAT altimeter measurements over a coastal area: the English Channel. J Geophys Res 88(C3):1647–1654

Le Traon PY, Gaspar P, Bouyssel F, Makhmara H (1995) Using TOPEX/Poseidon data to enhance ERS-1 data. J Atmos Ocean Technol 12(1):161–170

Manzella GMR, Borzelli GL, Cipollini P, Guymer TH, Snaith HM, Vignudelli S (1997) Potential use of satellite data to infer the circulation dynamics in a marginal area of the Mediterranean Sea. In: Proceedings of 3rd ERS symposium – space at the service of our environment, Florence (Italy), 17–21 March 1997, vol 3, pp 1461–1466, European Space Agency Special Publication ESA SP-414

Marsh JG, Koblinsky CJ, Lerch F, Klosko SM, Robbins JW, Williamson RG, Patel GB (1990) Dynamic sea surface topography, gravity, and improved orbit accuracies from the direct evaluation of Seasat altimeter data. J Geophys Res 95(C8):13129–13150

McGoogan JT, Miller LS, Brown GS, Hayne GS (1974) The S-193 radar altimeter experiment, IEEE Trans 62(6):793–803

Minster JF, Genco M-L, Brossier C (1995) Variations of the sea level in the Amazon Estuary. Cont Shelf Res 15(10):1287–1302

Mercier F, Ablain M, Carrère L, Dibarboure G, Dufau C, Labroue S, Obligis E, Sicard P, Thibaut P, Commien L, Moreau T, Garcia G, Poisson JC, Rahmani A, Birol F, Bouffard J, Cazenave A, Crétaux JF, Gennero MC, Seyler F, Kosuth P, Bercher N (2008) A CNES initiative for improved altimeter products in coastal zone, PISTACH, http://www.aviso.oceanobs.com/fileadmin/documents/OSTST/2008/Mercier_PISTACH.pdf. Accessed 1 October 2009

Raney RK (1998) The delay Doppler radar altimeter. IEEE Trans Geosci Remote Sensing 36: 1578–1588

Rosmorduc V, Benveniste J, Lauret O, Maheu C, Milagro MP, Picot N (2009) Radar altimetry tutorial. In: Benveniste J, Picot N (eds) http://www.altimetry.info. Accessed 1 October 2009

Sailor RV, LeSchack AR (1987) Preliminary determination of the GEOSAT radar altimeter noise spectrum, Johns Hopkins APL Tech. Digest 8:182–183

Smith WH, Sandwell DT (1997) Global sea floor topography from satellite altimetry and ship depth soundings. Science 277(5334):1956–1962

Special Topex/Poseidon issue: Scientific results (1995) J Geophy Res 100:C12

Stanley HR (1979) The GEOS-3 project. J Geophys Res 84:3779–3783

SWOT Science requirements documents (2009) NASA/JPL. URL: http://swot.jpl.nasa.gov/files/SWOT_science_reqs_final.pdf. Accessed on 04 September 2009

Tapley BD, Born GH, Parke ME (1982) The SEASAT altimeter data and its accuracy assessment. J Geophys Res 87:3179–3188

Townsend WF (1980) An initial assessment of the performance achieved by the SEASAT altimeter. IEEE J Ocean Eng, OE5, 80–92

Wilson S, Parisot F (2008) Report of CEOS ocean surface topography constellation strategic workshop, Assmannshausen, 29–31 January 2008, EUMETSAT Report E.02, vol. 2 , 4pp

Wilson S, Parisot F, Escudier P, Fellous JL, Benveniste J, Bonekamp H, Drinkwater M, Fu L-L, Jacobs G, Lin M, Lindstrom E, Miller L, Sharma R, Thouvenot E (2010) Ocean surface topography constellation: the next 15 years in satellite altimetry. In: Hall J, Harrison DE, Stammer D (eds) Proceedings of OceanObs'09: sustained ocean observations and information for society, vol 2, Venice, Italy, 21–25 September 2009, ESA Publication WPP-306. doi:10.5270/OceanObs09.cwp.94. See also http://www.oceanobs09.net. Accessed on 1 October 2009

Wingham DJ, Francis CR, Baker S, Bouzinac C, Cullen R, de Chateau-Thierry P, Laxon SW, Mallow U, Mavrocordatos C, Phalippou L, Ratier G, Rey L, Rostan F, Viau P, Wallis D (2006) CryoSat: a mission to determine the fluctuations in earth's land and marine ice fields. Adv Space Res 37:841–871

From Research to Operations: The USDA Global Reservoir and Lake Monitor

2

C. Birkett, C. Reynolds, B. Beckley, and B. Doorn

Contents

2.1	The USDA/FAS Decision Support System	21
2.2	Satellite Radar Altimetry	22
2.3	The Creation and Implementation of the GRLM	24
2.4	Satellite Data Sets	24
2.5	Technique	25
2.6	Jason-1 Data Loss	28
2.7	Preliminary Benchmarking and Product Validation	29
2.8	T/P and Jason-1 Product Revision	32
2.9	GFO Products and Further Validation	33
2.10	Program Limitations	37
2.11	Products and Applications	38
2.12	Current Status	42
2.13	Phase IV Tasks	43
2.14	Anticipated Phase IV Results	44
2.15	Summary, Recommendations and Future Outlook	46
References		48

Keywords Lakes • Reservoirs • Satellite radar altimetry • Surface water level

---

C. Birkett (✉)
Earth System Science Interdisciplinary Center, University of Maryland, College Park, MD, USA
e-mail: cmb@essic.umd.edu

C. Reynolds and B. Doorn
International Production Assessments Branch, Office of Global Analysis, Foreign Agricultural Service, U.S. Department of Agriculture, Washington DC, USA
e-mail: curt.reynolds@fas.usda.gov
e-mail: bradley.doorn@nasa.gov

B. Beckley
SGT, NASA Goddard Space Flight Center, Greenbelt, MD, USA
e-mail: brian.o.beckley@nasa.gov

S. Vignudelli et al. (eds.), *Coastal Altimetry*,
DOI: 10.1007/978-3-642-12796-0_2, © Springer-Verlag Berlin Heidelberg 2011

Abbreviations

ALT	NASA Radar Altimeter
AVISO	Archiving, Validation and Interpretation of Satellite Oceanographic data
CNES	Centre National d'Études Spatiales
CSR	Center for Space Research (University of Texas, Austin)
DDP	Defect Detection and Prevention
DESDynI	Deformation, Ecosystem Structure and Dynamics of Ice
DORIS	Doppler Orbit Determination Radiopositioning Integrated on Satellite
DSS	Decision Support System
Envisat	Environmental Satellite
ERS	European Remote Sensing Satellite
ESA	European Space Agency
ESRI	Environmental Systems Research Institute
EUMETSAT	European Org. for the Exploitation of Meteorological Satellites
FAS	Foreign Agricultural Service
FEWS	Famine Early Warning Systems
GDR	Geophysical Data Record
GEO	United States Group on Earth Observations
GEOSS	Global Earth Observation System of Systems
GFO	Geosat Follow-On Mission
GIM	Global Ionospheric Map
GLAM	Global Agricultural Monitoring Program
GLIN	Great Lakes Information Network
GPS	Global Positioning System
GRACE	Gravity Recovery and Climate Experiment
GRLM	Global Reservoir and Lake Monitor
GSFC	Goddard Space Flight Center
GUI	Graphical User Interface
ICESat	Ice, Cloud and Land Elevation Satellite
IGDR	Intermediate Geophysical Data Record
IPCC	Intergovernmental Panel on Climate Change
IRI	International Reference Ionosphere Model
ISRO	Indian Space Research Organization
ISS	Integrated Systems Solution
ITRF	International Terrestrial Reference Frame
ITSS	Information Technology and Scientific Services
JPL	Jet Propulsion Laboratory
LAD	Least Absolute Deviation
LakeNet	World Lakes Network
LEGOS	Laboratoire d'Études en Géophysique et Océanographie Spatiales
MODIS	MODerate resolution Imaging Spectroradiometer
MOE	Medium Precision Orbit Ephemerides
NASA	National Aeronautic and Space Administration
NCEP	National Centers for Environmental Prediction
NGA	National Geospatial Intelligence Agency

NOAA	National Oceanic and Atmospheric Administration
NOGAPS	Navy Operational Global Atmospheric Prediction System
NRC	National Research Council
NRL	Naval Research Lab
OGA	Office of Global Analysis
OMB	Office of Management and Budget
OSTM	Ocean Surface Topography Mission
POE	Precise Orbit Ephemerides
RMS	Root Mean Square
SARAL	Satellite with ARgos and ALtika
SDR	Sensor Data Record
SGT	Stinger Ghaffarian Technologies company
SLR	Satellite Laser Ranging
SSALT	Solid-State ALTimeter
SWOT	Surface Water and Ocean Topography
T/P	TOPEX/Poseidon
TRMM	Tropical Rainfall Measuring Mission
UMD	University of Maryland
USAID	United States Agency for International Development
USDA	United States Department of Agriculture
USGS	United States Geological Survey
WAOB	World Agriculture Outlook Board
WAP	World Agriculture Production
WASDE	World Agriculture Supply and Demand Estimate

2.1
The USDA/FAS Decision Support System

The USDA/FAS is responsible for providing crop production estimates for all international countries and benchmark data for commodity markets and the World Agriculture Outlook Board (WAOB). These values become an Office of Management and Budget (OMB) mandated "Principle Federal Economic Indicator" published monthly in the World Agriculture Supply and Demand Estimate (WASDE) report and the World Agriculture Production (WAP) circular. Such estimates drive price discovery, trade and trade policy, farm programs, and foreign policy. In addition, the FAS is also responsible for providing an "early warning of events" to the Farm Service Agency. This service output can have an effect on the agriculture production affecting both food programs and markets.

The monthly crop estimates are produced within the FAS/Office of Global Analysis (OGA) decision support system (DSS), the only operational unit of its type in the world and the primary source of global agricultural intelligence on crop production and conditions for USDA and the US Government. The OGA uses an "all sources" methodology that varies by region and commodity but, in general, integrates the US Government and commercial satellite imagery, agro-meteorology data and crop modeling output, to provide timely, unbiased information on crop condition and production. This information is used in the monthly "lockup" process to set global production numbers. The resulting information is shared with other USDA and US Government agencies as input to their various decision-support protocols. This permits the various agencies to meet national security requirements, assess global food security needs and agriculture policy, and provide the agriculture industry/producers with commodity price discovery and an early warning of global crop production anomalies.

A number of satellite datasets are currently used within the DSS to derive information on precipitation, land cover, and soil moisture (e.g., from the NASA/TRMM, NASA/Terra, NASA/Aqua missions), which enhance the distribution of crop information. Also of prime importance is the availability of water for irrigation purposes, that is, the volume of water stored in the region's lakes and reservoirs. For many locations around the world, such knowledge is lacking because of the absence of ground-based measurements, or is difficult to obtain because of economic or political reasons. Water-deficit regions, in particular, suffer from poor reporting. Initially the DSS had to rely on vegetation response products only, since precipitation has a less direct effect on conditions, but the availability of altimetric satellite data over inland water opened up new possibilities.

2.2
Satellite Radar Altimetry

Satellite radar altimeters are primarily designed to study variations in the surface elevation of the world's oceans and ice sheets (for general information, see Fu and Cazenave 2001). Innovative use of the data, however, has additionally enabled studies of the variations in surface water level of the world's largest lakes, rivers and wetlands (see reviews in Birkett et al. 2004; Mertes et al. 2004; Crétaux and Birkett 2006). The main advantages include day/night and all-weather operation with no loss of data because of cloud coverage. With continuous operation across all surfaces, the instrument behaves like a string of pseudo-gauges, sampling the elevation at discrete intervals along a narrow satellite ground track. The presence of vegetation or canopy cover is not a hindrance to this nadir-viewing instrument, the inundated surfaces being so bright at microwave frequencies that vegetation only interferes during periods of extremely low water level. The instruments can thus observe monthly, seasonal, and inter-annual variations over the lifetime of the mission, and unlike many gauge networks that operate using a local reference frame, all altimetric height measurements are given with respect to one reference datum to form a globally consistent, uniform data set. After several decades of validated research, the altimetry data sets are mature and generally freely available via DVD or ftp.

With respect to lake/reservoir monitoring, there are several limitations to utilizing radar altimetry. The altimetric satellites are placed in a fixed repeat-cycle orbit with instruments that only have nadir-pointing ability. These result in specific ground track locations that restrict viewing to a certain percentage of the world's lakes and reservoirs. A trade-off between the temporal and spatial sampling is also in play and can impose further restrictions on target numbers. Many altimetric missions also have ocean science objectives with instrument designs not optimized for studies in rapidly changing or complex (multiple target) terrain. In such cases, surface elevation data over inland water may be degraded or non-retrievable. In addition, a number of factors place limitations on lake size (generally ≥ 100 km^2), particularly, when program accuracy requirements are taken into consideration.

Table 2.1 Selection and continuity of satellite radar altimetry missions

Satellite mission		Operation period	Temporal resolution (days)	No. of Lakes, Reservoirs
10-day repeat orbit (A)				
NASA/CNES	T/P	1992–2002	10	122, 55
NASA/CNES	Jason-1	2002–current	10	122, 55
NASA/CNES/NOAA/ EUMETSAT	Jason-2	Launch 2008	10	122, 55
NOAA/CNES/EUM	Jason-3/GFO2	Launch 2012	10, 17	
35-day repeat orbit				
ESA	ERS-1	1992–1993, 1994–1995	35	446, 165
ESA	ERS-2	1995–current[a]	35	446, 165
ESA	Envisat	2002–current	35	446, 165
ISRO/CNES	SARAL/AltiKa	Launch 2010	35	446, 165
ESA	Sentinel-3	Launch 2012	35	446, 165
17-day repeat orbit				
US NRL	Geosat	1987–1989	17	~220, ~95
US NRL	GFO	2002–current[a]	17	~220, ~95
NOAA/CNES/EUM	Jason-3/GFO2	Launch 2012	10,17	
10-day repeat orbit (B)				
NASA/CNES	TOPEX-Tandem	2002–2005	10	145, 65

1. Lakes (≥100 km^2) and in the latitude range –40 South to 52 North are potential targets. Numbers shown are approximate and reflect those targets of most interest to the USDA/FAS. Instrument tracking and current data interpretation methods have limited the 10-day repeat orbit (A) targets to ~70 at the present time. Lake number statistics are taken from Birkett and Mason (1995).
2. Except for the TOPEX-Tandem mission, satellites with the same temporal repeat cross over the same set of lakes. A lake may be crossed over by more than one satellite. Larger lakes will have multiple same-satellite crossings increasing temporal resolution.

[a]ERS-2 (from 2002) continues to operate with reduced continental coverage. GFO (from 2006) continues to operate with reduced temporal coverage over inland basins.

Current altimetric satellites cross over a selection of the world's largest lakes and reservoirs, measuring variations in the lake level with a repeat frequency ranging from 10 to 35 days (Table 2.1). Comparing measurements from the NASA/CNES TOPEX/Poseidon (T/P) mission with ground-based gauge data, for example, has shown altimetric accuracies to be variable, ranging from ~3 to ~5 cm root mean square (RMS) for the largest open lakes with wind-roughened surfaces (such as the Great Lakes, USA), to several decimeters RMS for smaller lakes or those with more sheltered waters (Morris and Gill 1994; Birkett 1995; Shum et al. 2003). With validated data sets, a number of research projects have been undertaken with applications that include fisheries and water resources, sediment transport and navigation, natural hazards (floods/droughts), basin impacts via dam impoundments, and climate change (see Crétaux and Birkett 2006, and the chapter by Crétaux et al., this volume for examples).

2.3
The Creation and Implementation of the GRLM

Observing the lake/reservoir levels via the innovative use of satellite radar altimetry combined with satellite imagery for surface area estimates does offer the potential to monitor variations in total volume storage. However, the USDA/OGA noted that water levels alone do reflect irrigation potential and could singularly help practices better understand crop production characteristics. Satellite radar altimetry was thus considered as a potential new tool that could enhance the current USDA DSS in its monthly crop assessments. The USDA program strongly emphasized its need for archival information that could reveal the general status of the lake, and the availability of near-real-time data that could quickly assess drought or high water-level conditions. With such new information, the ultimate goal was to more effectively determine the effects on downstream irrigation potential and consequences on food trade and subsistence measures. Additional system requirements also included:

(i) The monitoring of all lakes and reservoirs in regions of agricultural significance.
(ii) The production of surface water-level variations with respect to a mean reference datum and accurate to better than 10 cm RMS.
(iii) The products to be updated on a weekly basis, with a latency of no more than 2 weeks after satellite overpass (defined as "near-real time" here).
(iv) All products (graphical and ascii text) to be incorporated within the OGA Crop Explorer web site database.

In 2002, the USDA/FAS funded the implementation and operation of the near-real-time altimetric monitoring program. It became a collaborative effort between the USDA, UMD, NASA/GSFC, Raytheon ITSS and SGT. By late 2003, the Global Reservoir and Lake Monitor (GRLM) went on-line and became an additional decision support tool within the cooperative USDA/NASA GLAM program and the first program to utilize near-real-time radar altimeter data over inland water bodies in an operational manner.

The well-documented and validated data from the NASA/CNES T/P satellite and its follow-on mission Jason-1 (Table 2.1) were chosen to initiate the program. With a 10-year T/P archive (1992–2002), a 10-day time interval between lake observations, and a near-real-time Jason-1 data delivery of 2–3 days after satellite overpass, a designated set of target lakes, reservoirs, and inland seas were selected by the OGA. Originally, lakes and reservoirs on the African continent were of interest (Phase I), but the program quickly became global in outlook (Phase II).

2.4
Satellite Data Sets

Table 2.1 outlines the historical, current and future radar altimeter missions that separate into three main data sets according to satellite repeat period i.e., the temporal resolution of the lake product, 10, 17, or 35 days. The trade-off between repeat period and the number of target hits can be clearly seen. While the NASA/CNES missions have the better temporal resolution, ideal for weekly updates of lake products, the ESA missions offer a far greater quantity of potential targets.

Phases I and II of the program focused on the NASA/CNES T/P and Jason-1 data sets. The T/P mission was the first to carry two radar altimeters; the NASA radar altimeter (ALT) operating at 13.6 and 5.3 GHz (Ku and C band, respectively) and the prototype solid-state altimeter (SSALT) operating at 13.65 GHz. The SSALT was allocated ~10% of the observing time and data gathered during these periods also included some of the larger lakes and inland seas. The Jason-1 radar altimeter (POSEIDON-2) also operated at Ku and C band. Both missions performed a total of 254 ascending and descending passes over the Earth's surface with a geographical coverage extending to ±66° latitude. Each repeat-pass crossed over, to within a few kilometers, the same location on the Earth's surface. The temporal resolution of the measurements are ~10 days, and the along-track resolution of the missions have the potential for one height measurement every ~580 m (the 10 Hz T/P GDR) and every ~290m (the 20 Hz Jason-1 IGDR).

For each mission, there is a choice of data set that can be utilized to construct lake-level variations. These data sets are offered to users in several formats including Fast Delivery (generated within a few hours after satellite overpass), Intermediate Geophysical Data Records (IGDR, available a few days after satellite overpass) and Geophysical Data Records (GDR, generally available 4–6 weeks after overpass). The Sensor Data Records (SDR) that include the original radar echoes are also available but have not, to date, been utilized within the program. Notably, the lake-level accuracy depends upon the knowledge of the satellite orbit that is deduced via Global Positioning System (GPS), doppler orbit determination, and radiopositioning integrated on satellite (DORIS), and satellite laser ranging (SLR) methods. Fast delivery data may contain mean global orbit errors of ~30–40 cm; IGDR errors are ~5–10 cm and GDR errors ~2–3 cm. Striking a balance between height accuracy and operational requirements, the Jason-1 IGDR data was selected for near-real-time observations and the archive product was constructed using the T/P GDR. Both IGDR and GDR data are in binary format with data structures that contain both altimetric and geophysical parameters.

2.5
Technique

The T/P and Jason-1 data processing procedures follow methods developed by the NASA Ocean Altimeter Pathfinder Projects (Koblinsky et al. 1998), although improved algorithms are utilized and there is some adjustment to the general procedures including an allowance for more automation of the process. The methodology (outlined in McKellip et al. 2004) includes height construction, application of a repeat track technique, the derivation of a mean reference datum, and the determination of height bias between differing missions. All of these steps are performed via the creation of a time-tagged geo-referenced altimeter database.

In general, the construction of ocean surface height assumes the following two equations,

$$\text{Altimetric Height} = (\text{Altitude} - \text{Range}_{corr}) - \text{Tides} - \text{Barometric Correction} \quad (2.1)$$

$$\text{Range}_{corr} = \text{Range} + \text{Atmospheric Corrections} + \text{SSBias} + \text{CGrav Correction} \quad (2.2)$$

Here, "altitude" is the satellite orbit above a reference ellipsoid, and "range" is the distance between the altimeter and the surface (estimated from the radar echo). Both range and the resulting height must be corrected for instrument and geophysical effects. For the GRLM system, earth tide is applied, but elastic-ocean and ocean-loading tides are only applicable to the Caspian Sea (for a detailed study on the Caspian Sea, see the chapter by Kouraev et al., this volume). The inverse barometric correction is not applied because the lakes/reservoirs are closed systems. Atmospheric corrections include the dry tropospheric correction, the radiometer-based wet tropospheric correction when valid (and the model-derived correction, when not valid) and the DORIS ionospheric range correction. The sea state bias (SSBias) correction is not applied because wind effects tend to be averaged out along-track, and CGrav is a correction to offset for variations in the satellite's center of gravity (see Birkett 1995 and Fu and Cazenave 2001 for full details on the reconstruction of altimetric height).

Dedicated ocean altimeter satellites have their orbit maintained to a near-exact repeat period to facilitate geoid-independent techniques to measure changes in the surface height based on the method of collinear differences. The term "collinear" indicates that heights for a particular exact repeat orbit mission have been geo-located to a specific reference ground track. During collinear analysis, the repeat tracks are assumed to have perfect alignment to facilitate separation of sea surface height variations from the geoid. However, orbit perturbations caused by atmospheric drag and solar radiation pressure cause departures from the nominal repeat path introducing errors from the slope of the local geoid. Over most of the ocean, a departure from the nominal repeat path is typically limited to ±1 km translating into an error of 1–2 cm. In areas of steep lake bottom topography (e.g., Lake Tanganyika), these geoid-related errors can be a few centimeters. For inland water applications, data users may elect to perform both along- and across-track corrections, or just along-track corrections (as in the GRLM system) to attempt to co-align elevation measurements on various ground tracks with the reference track. In some cases, perfect co-alignment is not required and instead a type of "finding the nearest neighbor" is performed. This is achieved by calculating the distance between elevation measurements on the ground tracks.

The T/P and Jason-1 ~10 day-repeat orbit had ground track positions that varied by up to ±1 km from the nominal reference ground track. Jason-1 IGDR data are provided at the 20 Hz rate (i.e., one altimetric range measurement every 0.05s along the ground track), while the T/P GDR data are given at 10 Hz (20 Hz averaged in pairs). The construction of the T/P and Jason-1 10 Hz geo-referenced database then is as follows:

(i) Nominal 1 Hz geo-referenced locations (lat, lon) along a reference track are computed using a Hermite tenth order interpolation algorithm.
(ii) The time-tag (number of seconds along the satellite pass) for each of these 1 Hz reference locations is then calculated using the actual (I)GDR track data. Alignment is achieved by constructing perpendiculars from the reference orbit track to the actual orbital track. Locations are then linearly interpolated.
(iii) Although no across-track corrections are performed within the GRLM system, the cross-track distance from the reference orbit to the actual observation location is also stored in the reference database. In addition, a 1 Hz collinear surface height is computed from a linear fit of the 10 Hz heights with the midpoint evaluated at the 1 Hz reference location. At this point in the process, the reference track is 1 Hz and the database contains lat, long, time, and height where lat/long are fixed for all repeat passes.

(iv) The 1 Hz reference ground track is then expanded to 10 Hz by associating a 10 Hz height value to each of the ten 0.1 s intervals from the 1 Hz reference time. The closest 10 Hz height point on the neighboring ground track, rather than interpolation of adjacent 10 Hz observations is then chosen. In this way, the 1 Hz reference ground track is expanded to 10 Hz, the 10 Hz heights are indexed, and the method preserves as many lake heights as possible. For the Jason-1 IGDR data, the nearest-neighbor approach searches for the closest 20 Hz data point along the actual ground track.

The resulting reference database has a structure that is based on direct access with three-dimensional directories for each mission based upon repeat cycle number, satellite pass (or "revolution" number), and the indexed along-track 1 Hz geo-referenced locations. Each lake that the satellite flies over will have an associated revolution number and a set of along-track time indices bounding the lake traverse. Each data record is a fixed length containing the 1 and 10 Hz geo-referenced heights, along with all geophysical and environmental range corrections. This random read–write approach permits (I)GDR data to be processed upon receipt regardless of repeat cycle order, and permits immediate revisions. The organization of the geo-referenced data directories and fixed record format enables the integration of a graphical user interface (GUI) to generate near-real-time data reports and performs as a quality assurance device.

To construct the T/P and Jason-1 time series of lake height variations, the elevation measurements along one satellite overpass, from coastline to coastline, have to be compared to measurements along a reference pass for each lake. The differences in height at the 10 Hz locations are then averaged and the result represents a mean height difference (with respect to the average pass) for that particular crossing date and time. In the GRLM system, the reference pass is based on an average pass which is deduced from the 10-year TOPEX (ALT) reference archive. The along-track alignment procedures result in a maximum 10 Hz along-track alignment error of 0.05 s (0.28 km) for TOPEX (ALT) and a maximum expected error of 0.025 s (0.14 km) for the 20 Hz Jason-1 data. The estimated error of the average TOPEX (ALT) height profile at each 10 Hz location though is further reduced by virtue of the 10-year averaging (cycles 1–364).

It is at the comparison of pass with average pass stage that the rejection of erroneous height values takes place. This is done by the removal of outliers with boundary limits set according to each lake. The filtering method rejects those height measurements that are contaminated by land (coastline or island) or by additional bright objects within the altimeter footprint. Coastline data for large lakes and inland seas (e.g. Lake Ontario) are readily rejected leaving many elevation measurements over which to form an accurate average measurement. With an along-track spacing of ~580 m, many smaller lakes or narrow crossing extents (e.g., Lake Powell) will have a notable reduction in height accuracy because of a smaller number of measurements. This is often coupled with a reduction in range precision because of determining the range from a narrower radar echo profile (Birkett 1995). In these cases, filtering is relaxed with the acceptance of greater inaccuracy. A resulting time series, though with large error bars, and an inability to reveal seasonal, inter-annual or long-term trends are rejected from the system until new methods can offer improvement.

In the compilation of the T/P time series, the team did not apply any additional range (or height) bias to the prototype SSALT measurements on the GDR; however, the merger of the

T/P and Jason-1 time series requires a check on the inter-mission height bias to maintain continuity of products on the single-product output graph. During the follow-on mission validation (or tandem) phase (Jason-1 cycles 1–21 and T/P cycles 344–364), both satellites flew in formation along the same ground track separated by approximately 72 s, the satellite observations being approximately spatially and temporally coincident (Ménard et al. 2003). The instrument-independent height corrections (Eqs. 2.1 and 2.2) that do not vary significantly over the 72 s essentially cancel at the geo-referenced locations. Analysis of the global ocean collinear sea surface height differences between the Jason-1 and TOPEX (ALT) data (Chambers et al. 2003; Vincent et al. 2003; Zanife et al. 2003) revealed that a relative bias of approximately 11 cm (Jason-1 higher than TOPEX (ALT) existed between the range measurement of the two missions. A similar analysis using the Jason-1 IGDR data over a suite of large lakes generated a relative bias of ~9 cm which was applied to the GRLM combined T/P and Jason-1 graph products and these were denoted as Version 1. For some lakes, this produced a smooth transition, but for many lakes an additional offset could be observed suggesting a regional effect.

2.6
Jason-1 Data Loss

There are many factors affecting the quantity and quality of elevation measurements over inland water targets and the later section on limitations provides a summary. Here, we present details of a data loss discovered within the Jason-1 IGDR data set.

For oceanography purposes, the low-rate, 1 Hz elevation measurement is adequate for most science objectives. For inland water applications, it is the 10 Hz (T/P GDR) or 20 Hz (Jason-1 IGDR) elevations that are demanded for the smaller targets. Within the T/P GDR the user has access to one 1 Hz altimetric range value and up to ten, "range difference" values. Adding the latter to the former gives the full 10 Hz range measurements which are combined with other parameters to form lake elevation (Eqs. 2.1 and 2.2). The ground processing teams average the 20 Hz rate data in pairs to form a 10 Hz data set. The 1 Hz value is then deduced by performing a least absolute deviation fit (LAD) of the 10 Hz values with up to 20 iterations. The 1 Hz value is the fit evaluated at the mid-point (the point between the 5th and 6th range values). Range values that deviate by more than 300 mm are marked as erroneous. There are contingencies though. If the LAD fit fails to converge, or if there are more than two erroneous range values, or if the slope of the fit is too high (3,000 mm/1 Hz), then the 1 Hz value is taken as the original median value (average of the 5th and 6th range values). In this latter case, it is assumed that the logic then checks the deviations of the 10 Hz values from this new 1 Hz value. Certainly from the observation of the T/P GDR data stream in these cases (over severe terrain or narrow river regions), as few as two 10 Hz range values can be accepted and pass unhindered into the data streams for the user to examine.

The Jason-1 data sets are also based on 20 Hz measurements with assumed similar deviation and iterative methods as per T/P. However, there are subtle differences in the processing. First, the 20 Hz measurements are not averaged into pairs to form 10 Hz values. Secondly, the criteria for the formation of the 1 Hz average appear to be based simply on having more than three valid 20 Hz values. If this is not the case, then the 1 Hz and the 20 Hz values are all defaulted in the IGDR. This condition was additionally tightened during cycle 46, when the minimum number of acceptable 20 Hz values was raised to six. This change in the formulation criteria of the 1 Hz values between T/P and Jason-1 had certainly resulted in data loss over some lake targets.

The program team expressed this data-loss concern to AVISO in the summer of 2003, and suggested that the full 20 Hz range values be included in the Jason-1 data stream whether deemed valid or not by the filtering algorithms. AVISO formally acknowledged the problem at the Jason-1 Science Working Team meeting in Arles, France in November 2003, and issued a "Request for Modification" on February 24, 2004. Ultimately though the problem could not be resolved as further discussion revealed that additional onboard filtering, which rejected data according to characteristics of the radar echo shape was also operating and could not be changed. Overall, there was a considerable loss of data over smaller lakes with calm-water surfaces that lacked significant wave height formation, and for those targets that had a greater standard deviation of range values along the ground track.

2.7
Preliminary Benchmarking and Product Validation

The GRLM was initiated as a USDA-funded project, but in 2004, NASA requested that an Applied Science Program Management Group document and observe the system, noting the use of products derived from NASA satellites. An "Integrated Product for Agriculture Efficiency Team" from NASA/Stennis Space Center led the study, outlining the USDA DSS, the role of the radar altimetry and the recording the program's successes and limitations. The output of the study became the basis of the first systems engineering report (McKellip et al. 2004) and later the validation and verification report (Ross and McKellip 2006). Both reports focused on the early T/P and Jason-1 lake products that were available at the time. The team compared the original system requirements with the final T/P and Jason-1 output and assessed the latency and spatial distribution of the products. They found that during the operational phase, the latency on the near-real-time product output varied from ~20 days in 2004 to ~10 days in 2005, and although the number of acquired lakes was only 70 of the original 178 potential lakes selected by the OGA, the products revealed lake-level status on every continent with the exception of Australia. The results also highlighted the known problems of acquiring the smaller targets (<300 km^2) and those situated within narrow valleys or in rugged terrain. Other factors that could cause or affect delay on data delivery were also discussed and the unexpected demise of the Jason-1 IGDR data was highlighted, the limitation affecting half of the lakes in the GRLM program.

Historical T/P validations showed accuracies ranging from ~3 to ~5 cm RMS (e.g., Fig. 2.1a for Lake Ontario) to several tens of centimeters depending on target size, location, and surface roughness. These studies used ground-based gauge data, selecting the gauge nearest to the satellite overpass (or averaging multiple gauge measurements) and interpolating the gauge measurements to the time of the satellite overpass. Such validation methods are considered "absolute" although the altimetric process is based on averaging

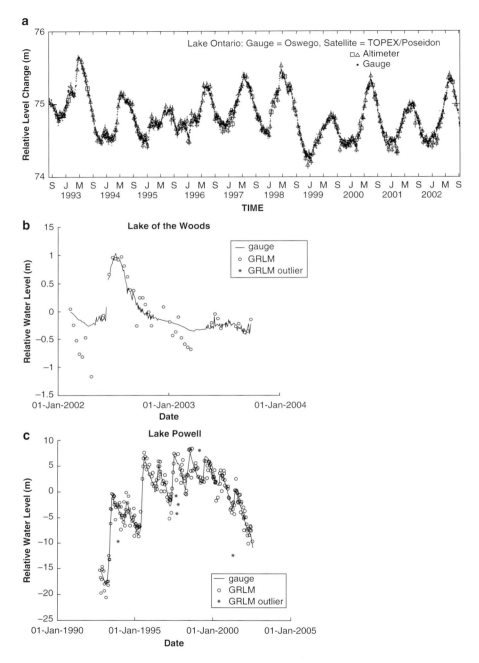

Fig. 2.1 Validation examples for (**a**) Lake Ontario (T/P), (**b**) Lake of the Woods (Jason-1), (**c**) Lake Powell (T/P), and (**d**) and (**e**) Lake Victoria (T/P and Jason-1). Note the reduction in accuracy for Jason-1 IGDR in comparison with T/P GDR for Lake Victoria (**d**) because of improved satellite orbit knowledge in the more delayed (30 days) GDR satellite data set. Lake Powell and Woods figures are courtesy of McKellip et al. (2004)

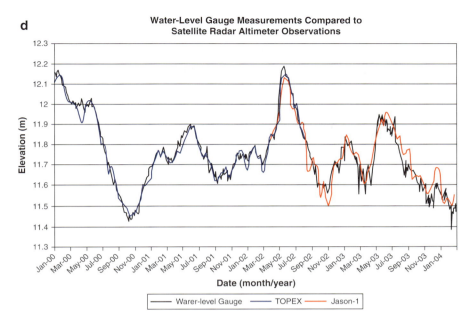

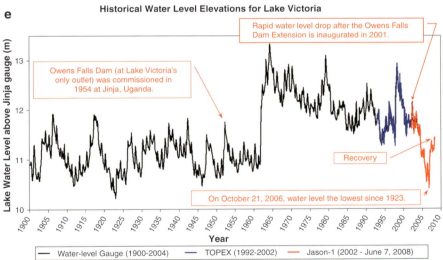

Fig. 2.1 (continued)

and there can be considerable separation distance between gauge site and ground track location. Relative validation checks can also be conducted testing one satellite product against synergistic measurements from another although errors will be introduced from comparisons of non-coincident satellite ground tracks. The GRLM benchmarking exercises revealed product errors of 5–7 cm RMS for the Great Lakes, increasing to 20–30 cm RMS for the smaller lakes (e.g., Lake of the Woods in Fig. 2.1b after removal of erroneous winter results due to the presence of lake ice). Surprisingly, the NASA/CNES instruments were able to acquire Lake Powell, and although the RMS error was ~1.6 m, seasonal and inter-annual trends could still be observed (Fig. 2.1c). The USDA/FAS also compared the T/P and Jason-1 products for Lake Victoria with gauge data from Jinja. Fig. 2.1d shows the difference in accuracy between the use of GDR (T/P) and IGDR (Jason-1) (e.g., 2.5 cm RMS compared to 5 cm RMS for the tandem phase in 2002) and Fig. 2.1e shows the merger of altimetry products onto the historical gauge data record.

Overall, the benchmarking team concluded: "So far, the program has made great strides towards meeting the immediate needs of the OGA, and the requirements of other intra-governmental and public users. Product latency typically falls with the desired limits, products span the globe touching on many crop production and crop security regions, and product accuracy is sufficient for many lakes and reservoirs in the GRLM system". The team recommended though that (a) the original accuracy requirement be relaxed for lakes with very large seasonal amplitudes, (b) the Jason-1 data drop out should be further investigated and (c) the lake coverage be increased. They additionally noted the possibility of utilizing Moderate Resolution Imaging Spectroradiometer (MODIS) derived lake area measurement to enhance the products.

2.8
T/P and Jason-1 Product Revision

As emphasized in the early benchmarking reports, the rejection of non-ocean-like Jason-1 radar echoes, totally or intermittently, affected almost half of the on-line lake products in the post-2002 time frame. While Jason-1 data recovery efforts continued some enhancement of the existing T/P and Jason-1 products took place in the form of a re-computation of the relative height bias i.e., the shift in elevation required to bring the Jason-1 mission products in-line with the T/P products.

A mean bias of 9 cm had originally been applied to the Jason-1 IGDR (Version 1) products but further investigation of the atmospheric corrections within the T/P GDR and Jason-1 IGDR data sets pointed to differences in the models used to construct these parameters. This had the potential to introduce a regional bias with respect to the dry tropospheric correction that should be similar at the same location and time period during the validation phase. In addition, results from Beckley et al. (2004) indicated regional bias variability due to orbit differences arising from inconsistencies in the use of differing terrestrial reference frames. The T/P orbits are based on the Center for Space Research CSR95 terrestrial reference frame, whereas the Jason-1 orbits are based on the more recent international terrestrial reference frame ITRF2000. The largest translation velocity differences between the two reference frame realizations occur along the Z-axis (Morel and Willis 2005) resulting in a north–south asymmetry in the orbital radial height differences. By accommodating for the differences in the terrestrial frame, and the atmospheric models, the TOPEX (ALT) and Jason-1 inter-mission bias was recalculated once again, on a lake-by-lake basis, using the mean (single iteration 3.5 sigma edit) of the collinear height differences in the mission overlap period. The enhanced products were upgraded to version 2 and placed online.

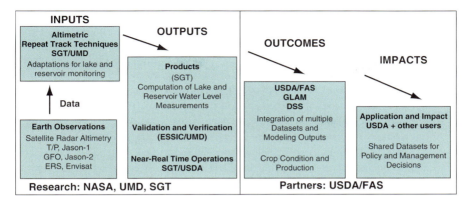

Fig. 2.2 The Integrated Systems Solutions (ISS) architecture of the GRLM

2.9
GFO Products and Further Validation

In 2006, NASA provided financial support allowing the NRL GFO mission data to be utilized and supplement the loss of data from the Jason-1 mission. Phase III of the program then came under the auspices of the NASA Applied Sciences Program where the team's collaborative effort, from data processing, product output, and utilization, were seen as an Integrated Systems Solution (ISS) (Fig. 2.2) to the USDA's DSS.

The Naval Research Lab's GFO mission was launched on February 10th, 1998 with a 17-day repeat cycle. Initial problems delayed data retrieval, and so valid operations did not commence until January 2000 (cycle 36). The mission was given a ~9 year life expectancy and by the fall of 2006 energy problems forced the instrument to be cycled off/on. The meant that the instrument was only operating during select periods of certain overpasses. As per the T/P and Jason-1 data sets, the GFO data set is freely available, but permission was sought for and granted by the National Oceanic and Atmospheric Administration (NOAA) to use the GFO data within the program. The GFO GDR data were thus ftp downloaded from NOAA (ftp://eagle.grdl.noaa.gov) for the post-2000 period and the number of on-line lake/reservoir targets crossed by this satellite was noted. Concentrating on those lakes that lacked Jason-1 data, the intersection of the GFO ground tracks with these 35 targets was estimated.

With ocean science objectives, the GFO data interpretation was assumed to be fairly straightforward following that of the T/P or Jason-1 processing chains. With minor modifications then to allow for changes in data structure between T/P, Jason-1 and GFO, the GFO data were then assembled into the time-tagged altimeter parameter database. The examination of the GDR data parameters, construction of lake water level and subsequent computation of GFO lake-level products followed the T/P, Jason-1 process with minor modifications and notes as follows:

(i) GFO Data: Two GFO products are currently available; (1) the operational data containing the Medium Precision Orbit Ephemerides (MOE) that are available 1–2 days after satellite overpass with radial orbit precision ~10–20 cm, and (2) a GDR product based on precise orbit ephemerides (POE) having a latency of ~3 weeks and a radial orbit precision ~3–5 cm. The GDR data was selected for the GRLM. It should be noted that the precise orbit for cycles 36 to 69, and for cycles 70 onward is derived by NOAA and NASA/GSFC, respectively, with more precise accuracy expected for the latter. However, there is an ongoing reprocessing of all the GFO orbits at NASA/GSFC based on (a) an improved gravity field from the NASA Gravity Recovery and Climate Experiment (GRACE) Satellite Mission, (b) an updated reference frame ITRF2005, and (c) other significant geophysical modeling improvements. In the future, this could provide high-class altimetry precision for the entire GFO mission (Lemoine et al. 2001).

(ii) GFO surface elevation: As per T/P and Jason-1, construction of the GFO surface elevation is conducted by differencing the GDR 10 Hz range parameter value from the satellite orbit parameter, and applying a number of geophysical, environmental and instrument-based corrections (Eqs. 2.1 and 2.2). Note here that the center of gravity range correction is already applied to the range parameter in the GDR (via net instrument correction). The GFO utilized a single frequency altimeter and thus the ionospheric path delay is not estimated directly as with the dual frequency altimeters onboard T/P and Jason-1. Instead, the ionosphere path delay is derived from GPS observations (from the Jet Propulsion Laboratory (JPL) Global Ionospheric Maps (GIM)), or from the international reference ionosphere model IRI95 (Bilitza et al. 1995). The dry and wet atmospheric delays are also derived differently. The dry correction stems from the Navy Operational Global Atmospheric Prediction System (NOGAPS) surface pressure data. The wet tropospheric delay is measured by a two-channel (22- and 37-GHz) microwave radiometer, or, when inoperable, the NOAA National Centers for Environmental Prediction (NCEP) model.

(iii) Geo-referencing: A 17-day reference orbit is generated from the GSFC orbit determination and geodetic parameter estimation (GEODYN) orbital software based on GFO orbital parameters and available satellite laser ranging (SLR) tracking. Geo-referenced locations along a nominal reference orbit are interpolated at 1 Hz using a Hermite 10th order interpolation algorithm. The GDR data are then aligned to these 1-s locations by constructing perpendiculars from the reference orbit to the actual orbital track location and linearly interpolating from the surrounding along-track data.

(iv) Collinear heights: The collinear surface height is computed from a linear fit of the 10 Hz GDR height values with the midpoint evaluated at the 1 Hz reference locations. The high rate 10 Hz heights are then geo-referenced with respect to time at exact 0.1 s intervals by indexing the closest 10 Hz height rather than interpolation of adjacent 10 Hz observations, to preserve as many lake heights as possible. The maximum 10 Hz along-track alignment error (at the equator) for GFO is less than 0.05 s translating to 0.28 km. The estimated error of the mean height profile at each 10 Hz location is further reduced by virtue of averaging over a period of 6 years (cycle 37–166, January 2000 to December 2005).

(v) Inter-mission bias: As previously noted, revised height bias estimates between T/P and Jason-1 were computed for each lake to minimize regional variability due to geographically correlated orbit error and path delay estimates. Since the GFO is not spatially coincident with T/P and Jason-1, a more "ad hoc" bias adjustment was performed by minimizing cycle-to-cycle mean height differences between the GFO and Jason-1 height differences with increased weight given to observed differences during the T/P,

Jason-1 verification phase. The result was an arbitrary (but constant) height-shift to the GFO results to bring them visually in line.

(vi) GFO data filtering: GFO data filtering was performed by comparison of individual 10 Hz height observations with respect to the 6-year mean reference that is constructed at each 10 Hz along-track geo-location. Individual along-track height profiles were interrogated for each cycle in an effort to identify land/island contamination to construct a representative mean profile. Note that no GDR "erroneous elevation" flag parameters were utilized.

GFO lake products, each with respect to its own 6-year reference datum, were easily calculated, but the USDA requirement to place all mission results onto one graph revealed both amplitude and phase-lag differences despite attempts to correct for inter-mission height bias. The effects were more marked for some lakes than others. One explanation centers on the fact that the satellite ground track locations differ between the instruments that are thus sampling water variations at differing locations within the lake. Without resources to explore this further, the team decided to select only the best T/P, Jason-1, GFO-merged products (15 out of the original 35) which were assigned version number 1, and uploaded to the Crop Explorer GRLM web site as a separate clickable graph and text file. Some of these targets (e.g., Lake Nasser, Fig. 2.3) benefited

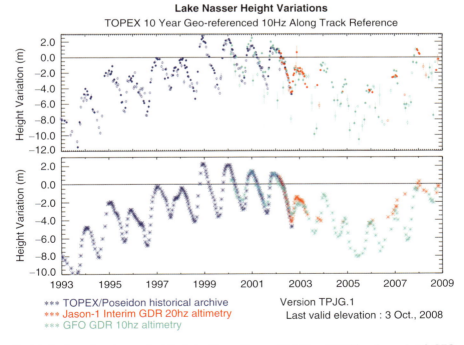

Fig. 2.3 Surface elevation product for Lake Nasser/Aswan High Dam (T/P blue, Jason-1 red, GFO green). Raw results (top), smoothed result (below)

Table 2.2 Validation of GFO Great Lakes Products

Lake	Pass	Gauge site	RMS[1]	RMS[2]	RMS[3]
Erie	069	Cleveland 9063063	14.66	14.53	14.04
Ontario	155	Rochester 9052058	23.96	23.68	13.69
Michigan	141	Calmut Harbor 9087044	12.12	12.36	12.36
Huron	227	Harbor Beach 9075014	14.01	13.88	11.40
Superior	055	Marquette 9099018	27.33	27.33	13.41

Gauge data are courtesy of NOAA and are verified 6 min or hourly products. Altimetric results are paired with one nearest gauge site. Gauge versus altimeter RMS values are for 6 min (1), hourly (2), or hourly with removal of major altimetric outliers (3). Outliers are cycles 065, 102, 131, 171 (Lake Ontario), cycle 148 (Lake Erie); cycles 104 and 105 (Lake Erie), and cycles 104 and 107 (Lake Superior)

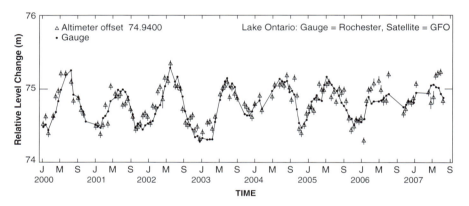

Fig. 2.4 Validating the GFO altimetric time series of height variations for the Great Lakes with ground-based hourly NOAA gauge data, example Lake Ontario

greatly from the additional GFO data, but GFO validation exercises over the Great Lakes (the results of which can be translated to other similar large bodies of water) showed accuracies ~15 cm RMS (Table 2.2, Fig. 2.4). In comparison with T/P and Jason-1, these are poorer and are themselves a cause for further investigation. Relative validation exercises between GFO and T/P, and GFO and Jason-1, were not deemed feasible because of the differences in satellite pass location, overpass time, and the computation of mean reference datum.

As per the NASA/CNES missions, other GFO target data losses are attributed to the poor acquisition of water levels for smaller targets or failure in complex terrain. Discussion with NOAA (J. Lillibridge, 2007, personal communication) also suggested that the data filtering procedures between the GFO SDR and GDR (in particular the use of the high bandwith/low variability RA Status Mode I flag) could also be placing restrictions on the amount of lake data being initially stored on the GDR. The possibility of reprocessing the

SDR/GDR and additionally updating the GFO satellite orbit to a higher precision has been suggested and noted as a future possibility.

2.10
Program Limitations

With nadir-pointing technology not dedicated to inland water monitoring, a number of data limitations were inherent at the start of the program. In addition, several other factors are important to note. At the top level, the program depends on the continuity of funding, both at the space agency level for ensuring follow-on missions, and also at the NASA/USDA program level, which maintains manpower support for product creation and delivery. The continuation of lake products is also obviously dependent on the lifetime of the satellite mission and during flight time it is not improbable to expect both satellite and instrument anomalies that result in data loss. The onboard and ground processing of the raw altimeter data prior to the formation of the IGDR/GDR may also be affected by data filtering and quality control that does not have inland water priorities. The instrument-tracking logic is also crucial to acquiring the lake surface. It determines how quickly the lake can be detected and how fast it can recover if lock is lost over the nearby terrain (Sect. 2.14). Because there can be several hundred meters separation between repeat passes, the same approach topography is not always sampled and subtleties can cause data loss if the instrument follows other smaller water bodies or surrounding topography.

As an operational program with a semi-automated system, the program team needs to ensure man-power backup to maintain continuity of the weekly updates and be readily accessible to answer queries from USDA/FAS and a wide variety of other users. Although technical information is on the GRLM web site, questions on reference datum, height accuracy, target size, and product accuracy do continue to demand further explanation. Users also make requests for additional lake targets to be included when they are not included in the USDA targets of interest list. Requests have also been made for similar products for wetlands and rivers, but as per the non-agriculture lakes, these are outside the objectives of the program. With the products in the public domain, the team has also had to consider liability and the placing of a limitation clause on the world-wide-web site. This was particularly in consideration that the products continue to be subject to further investigation and revision.

The USDA and other users are disappointed at the number of targets observable by the T/P and Jason-1 missions and at the lack of smaller targets on-line. This will improve in Phase IV (see next sections), as additional mission data sets are utilized. However, the exercise with the GFO data and consideration of the ERS/Envisat data sets does pose an interesting technical question as to how to combine results from differing missions. For those missions in the same orbit, where the follow-on mission operates synergistically with its predecessor during a "tandem-phase", the height bias between the two instruments can be deduced and the products merged. With different mission orbits the instruments are sampling lake surfaces at different locations where wind, ice, and other surface characteristics differ. In complex regions suffering from the effects of excessive abstraction of water or drought (like the Aral Sea and Lake Chad), phase and amplitude differences between observed variations found across the basin will also be common. We have no answer to this problem at the present time other than to attempt mergers based on crude "eye-balling", that is, simply shifting vertically one deduced time series to another for visualization only (as per GFO) or to deliver products on separate graphs.

Another technical issue of importance relates to the calculation of the lake mean reference datum. The T/P and Jason-1 products are based on a lake datum that is calculated from 10 years of T/P measurements. In the case of GFO, each mean reference datum is based on only 6 years of archival data. Although the GFO mean datum could be revised now that ~8 years of GDR data have been acquired, this averaging could also introduce a height bias and/or an offset that affects the match when aligning GFO with the T/P and Jason-1 results. The product accuracy can also be affected by the presence of lake ice, which we do not reject or correct for but attempt to flag on the product graph via the use of the radar backscatter coefficient. For many targets, the wet tropospheric range correction must also rely on model-derived values rather than instrument-derived. Although we assume a height error for the model-correction application, this is really a mean global value, and there is no attempt to place a magnitude on the error when the model value is absent from the IGDR or GDR data streams. Lastly, it must be stressed that there will be differences in product errors between the use of the IGDR (near real time) and GDR (archival). IGDR and GDR data sets use medium- and high-precision orbits, respectively (Fig. 2.1d), so there is loss in accuracy at the near-real-time level. Although recent science working team discussions are suggesting that with streamlined orbit-calculating processes the precision of the IGDR orbits will improve, a future phase of the program could include the revision of the IGDR-derived products with those from the GDR when available.

2.11
Products and Applications

The web–based portal, Crop Explorer, is a crucial part of the OGA operational decision support system with both an internal FAS analysis and public access functionality (http://www.pecad.fas.usda.gov/cropexplorer). On multiple occasions, it has been cited by the Office of the Secretary as a flagship example of USDA's effort to assist food-deficit countries and recently won the Environmental Systems Research Institute (ESRI) "Special Achievement in GIS" award in April 2007. Statistics show that Crop Explorer receives approximately 40,000 hits and 2000 visits per day, with 85% of the visits from the US and 15% of the visits from international visitors.

A link within Crop Explorer allows users to enter the GRLM, the front-end global map depicting target locations and the overall lake-level status (red/low water, blue/high water) with respect to a 1992–2002 mean. Each lake or reservoir target can then be selected via a series of clickable maps. The first displayed products are the combined T/P and Jason-1 in graphical form, depicting 15 years of monthly, seasonal, and inter-annual variations as well as revealing overall trends (Fig. 2.3 and 2.5). Each graph displays the raw (top) and

2 From Research to Operations: The USDA Global Reservoir and Lake Monitor

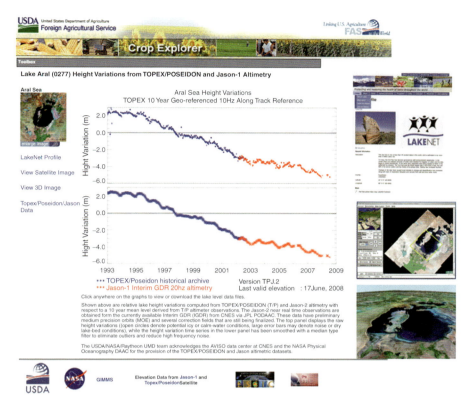

Fig. 2.5 An example of the information given in the Global Reservoir and Lake Monitor (GRLM) with focus on the Aral Sea. Through collaborative efforts, visitors to the web site can access Landsat and MODIS satellite imagery through a USGS visualization tool and connect to the LakeNet database to retrieve characteristics and biodiversity information

smoothed results (bottom), the smoothing performed with a median-type filter to eliminate outliers and reduce high-frequency noise. Error bars are given on the raw height values and estimated as per the method outlined in Birkett (1995). Clicking on a graph enables the download of the associated ASCII text file. Clicking on an additional link to the side of the page enables the display of the combined T/P, Jason-1, and GFO product. For visualization of the lake and satellite overpass location, Landsat imagery and MODIS Land Cover Classification are provided and additional imaging sources provided via a United States Geological Survey (USGS) Global Vizualization Viewer tool. An additional web link also provides access to the World Lakes Network (LakeNet) information database.

Within a short time period, the USDA DSS transitioned from a state of no direct water storage information to having access to T/P, Jason-1, and GFO water-level products for ~70 of the worlds largest lakes, reservoirs, and inland seas. In particular, the Middle Eastern and African (Lake Victoria, Nasser, Volta and Kariba) lake products have proved most useful. For example, FAS regional analysts used the GRLM products for Lakes

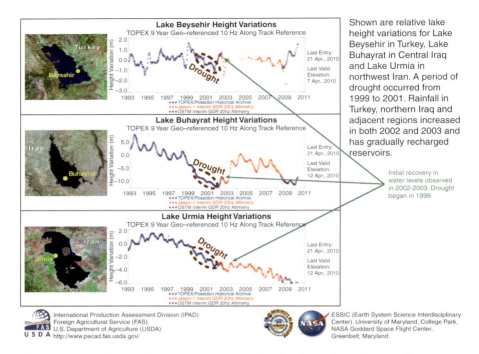

Fig. 2.6 GRLM products used to support an OGA regional analyst's discussion regarding a recovery from drought in the Middle East (Anulacion 2003). The analysts linked regional wheat production and the general state of the reservoirs to regional weather trends. (Example taken from McKellip et al. 2004)

Beysehir, Buhayrat, and Urmia (Fig. 2.6), to examine the recovery from 1999–2001 drought conditions in the regions. In these cases, it was the recharging of the reservoirs with respect to ground water reserves for winter wheat and barley production that was of concern (Anulacion 2003).

In the case of Lake Victoria, current low-level water reports led to discussions of regional drought and the Owen Falls Extension (the Kiira Dam) as potential causes. The Owen Falls dam was completed in 1954 just below the Ripon Falls, but under an agreement it was to be operated so that its outflows through both turbines and sluices were to be controlled to conform to a relation known as the "Agreed Curve". With water levels rivaling the lows of 1923, excessive withdrawals for power generation in Uganda from April 2002–October 2006 were discussed by Reynolds (2005) and Reynolds et al. (2007), with the GRLM products providing accessible and up-to-date records (Riebeck 2006). Mangeni Bennie (2006) later suggested that the Agreed Rule curve had been ignored and reservoir operation during this time had been in violation of the 1954 Nile Treaty. Sutcliffe and Petersen (2007) agreed with this result and announced that as much as 0.6 m of water level (~50%) decline was attributed to excess abstractions. The dramatic water-level drop has caused extensive environmental and economic losses and the effect on rapid population growth in the region has been highlighted (UNEP 2006).

To summarize, since implementation the GRLM has become the 9th most popular Crop Explorer page with (a relatively long) average viewing time of over two minutes. Since integration, it has received much publicity (Reynolds 2005; Riebeck 2006) with its content attracting the attention of FAS foreign resource analysts, international governments, humanitarian organizations, and conservation groups. A number of commercial, military, organization, government, and educational departments have also expressed interest along with network groups such as the Great Lakes Information Network (GLIN) and LakeNet. Other users include the World Bank, the United Nations, the USGS (both in general and through the Famine Early Warning Systems/United States Agency for International Development (FEWS/USAID)), the National Geospatial Intelligence Agency (NGA), and international organizations such as Genesys International Corporation Ltd. (India), the Lake Balaton Development Coordination Agency (Hungary), and the Research Hydraulic Institute in South Brazil. Applications cover many aspects relating to water quantity and quality. In addition to irrigation potential/agriculture impacts, users have been concerned with fish productivity, water security, vegetation ecology, and information theory metrics relating to ecological surveillance monitoring decisions.

It is interesting to note that several research groups have also utilized the lake products as a means to validate results from the GRACE mission (e.g., The Caspian Sea and Lakes Malawi, Swenson et al. 2006; Lake Victoria Malawi and Tanganyika: Swenson S.C. and Wahr J. 1997, personal communication) while other users are focused on basin or continental-scale hydrological modeling. With a continuous product series that spans more than 15 years, the data products are also attracting attention from the Intergovernmental Panel on Climate Change as a potential set of proxy climate data records.

The importance of monitoring surface water for a variety of applications is also recognized by the fact that since its inception in 2003, the GRLM has been joined by two other web-based databases containing altimeter-derived surface water levels. The ESA/De-Montfort University (UK) River and Lake web site (viewed at http://tethys.eaprs.cse.dmu.ac.uk/RiverLake/shared/main) offers near-real-time water-level products derived from the ESA Envisat mission via point and click target selection methods and product access via registration. Here, the time resolution of the product is 35 days and lake/reservoir and river channel variations are included. The Laboratoire d'Études en Géophysique et Océanographie Spatiales (LEGOS) also display lake, reservoir, river, and wetland elevations within their web-based database at http://www.legos.obs-mip.fr/soa/hydrologie/hydroweb/. Currently, there is a few months lag on product output but here the user can visualize the ERS, Envisat, T/P, Jason-1, and GFO satellite tracks across Landsat imagery and access additional location and hydrological data for each target. Point and click access to graphs and text files gives surface water variations for the 1992–2009 time period for ~100 lakes (dominated by the T/P and Jason-1 data sets) and ~250 river locations (using several altimetric data sets). There is thus similarity between the programs, although the GRLM remains primarily as a monitoring tool within a much larger decision support system. Nevertheless, multiple sites offer scope for additional checks on output, serving to verify product accuracy via cross-validation, and leads the way to future discussions on methodologies, formats, and standards.

2.12
Current Status

From mid-2008 the operational program received new funding from the NASA Applied Sciences Program and additional funding from USDA/FAS to take the program into phase IV with product expansion and enhancement objectives using both NASA/CNES and ESA satellite data sets. Because the GRLM had been on-line for several years, the USDA/FAS could examine past performance and revise the original system requirements. The objectives though remain the same with agriculture efficiency as the national (and international) priority topic of interest and maintaining relevance to the NASA Earth Science Division goals.

Past performance of the GRLM in the DSS were based on (i) findings within the McKellip et al. (2004) and Ross and McKellip (2006) reports, (ii) the compilation of verbal and written feedback from the FAS resource analysts and general users, and (iii) web statistics that were monitoring public access to Crop Explorer/GRLM. A review of these sources led to the revised 2008 DSS requirements:

(i) To observe the maximum number of lakes possible, with particular focus on acquiring the smaller (100–300 km^2) reservoirs in all terrain types.
(ii) To focus on water bodies in agriculture-sensitive regions such as Thailand, Iran, Iraq, Turkey, Argentina, India, Africa, Brazil, and Australia.
(iii) The weekly near-time product updates are to continue, with time since last satellite overpass set at no more than ~1 month.
(iv) To have products accurate to better than 20 cm RMS, or better than 10% of the expected total seasonal amplitude. The accepted accuracies must allow for monthly, inter-annual, and long-term trends to be readily discernible.
(v) To regain lake-level information that was missing from the Jason-1 data stream or Jason-1 observation period.
(vi) To assemble all the mission lake products onto one timeline graph, if permissible, within the scope of the repeat track techniques.

In general, the new FAS requirements satisfy other web user demands noting that complete global coverage of all lakes is not within the scope of the current radar altimetry capabilities (see Sect. 2.15). Overall feedback showed that a product temporal resolution of 10 or even 30 days is not being rejected, particularly for those regions where any form of current or historical gauge data cannot be acquired. Users have requested though the ability to download all lake products as one data set and have requested additional information on the construction and interpretation of the reference datum for each lake. Users, in general, also had interests in seeing additional lake basin parameters (areal extent, lake temperature and salinity, surrounding soil moisture, land cover, local precipitation etc.) being made available via a one-click map tool.

The new requirements demanded the use of additional radar altimeter mission data sets to expand lake numbers. The enhancing of existing products also focused attention on the possible merits of the SDR, the replacing of IGDR-derived products with those that were GDR-based, and on data filtering algorithms in general. Although the ISS remains the same, the GRLM system has to be modified in terms of (i) making additions to the system software to account for new data structures, (ii) expanding the on-line altimetric parameter database, and (iii) modifying the lake-level product determination software to accept multiple iterations to the altimetric range parameter based on new verification and validation exercises. There are thus several objectives within Phase IV. First, it allows the current Jason-1 and GFO products to be improved in terms of quality and quantity. It also extends the current 15-year timeline with near-real-time products derived from the follow-on Jason-2 mission (also called the Ocean Surface Topography Mission or OSTM, which was launched in June 2008 and is a joint collaboration between NASA, CNES, NOAA, and the European organization for the exploitation of meteorological satellites (EUMETSAT). With Jason-2, there are no foreseen "missing lake data" problems as per Jason-1. Also, via the inclusion of ESA ERS-1 and ERS-2 (archive 1994–2002) and Envisat (post 2002 and near-real-time) data, the number of targets in the current system will increase by at least a factor of 5. This step greatly enhances the DSS by the inclusion of a large number of smaller reservoirs (100–300 km^2), and additionally provides a means to validate the current T/P and Jason-1 products in regions where ground-based gauge data cannot be acquired.

The operational tasks of the system will of course continue with near-real-time products derived from the Jason-2 and Envisat missions and evaluation studies will run in parallel. These studies are a strong component of Phase IV under the guidance of the NASA Applied Sciences Division who request at all times that the project, the products, and the program team adhere to benchmarking and continuous evaluations that continuously assess the system and the usefulness of the products within the DSS. At the end of Phase IV, a final report will list these findings along with technical issues and validation results and the revised program system requirements will be once more be re-examined.

2.13
Phase IV Tasks

Specific GRLM system revision tasks reside in four main categories with focus on specific mission data in each,

(i) Target selection, Data ingestion, Parameter database creation: This includes the identification of all targets of opportunity in terms of exact geographical location of satellite crossing from coastline to coastline.
(ii) Parameter database revision: This pertains to the refining of the parameter database based on benchmarking exercises.
(iii) Formation of new or enhanced lake products: For coastal regions and small targets where the radiometer-derived wet tropospheric range correction is absent, improvements to model-based atmospheric corrections or combination methods will be sought to improve current lake height products (see the chapters by Andersen and Scharroo, and Obligis et al. this volume). The main focus though will be on new ERS, Envisat, and Jason-2 products which will be derived using methods based on various retracking techniques to more accurately acquire the altimetric range (see Sect. 2.14). The

Envisat products will be water-level variations based on a mean height datum calculated using ~8 years of ERS elevation measurements. Improved Jason-1 (via GDR Version C) and GFO products will be sought.

(iv) Benchmarking: This entails project and program performance measures, and includes validation and verification exercises with additions and refinements to the mission parameter database.

Overall, there are three demonstration points within Phase IV, the time at which (1) the Envisat/ERS products go on-line, (2) the Jason-2 products go on-line, and (3) the refined Jason-1/GFO products are updated. Phase IV also includes a set of project management metrics and performance measures, the latter being observed on both the product creation and delivery side, and on the product utilization side. Project management metrics, for example, include the monitoring of the number of tasks started and completed on time, the percentage of achieved deliverables within a given quarterly period, and the number of iterations a product undergoes vs. its usefulness to the DSS.

Performance measures (product creation and delivery) will follow those of the first benchmarking exercise and will include a number of tracking measures to ensure project efficiency. Such measures include the weekly noting of raw data and product latency, the spatial coverage (continent) and target type (open lake/reservoir) of products, and the monthly compilation of end-user response and feedback to assess correct prioritizing of targets compared to regional focus requirements.

The USDA/FAS performance measures (product utilization) will be defined from the 2003 NASA benchmarking process and the decision support tools evaluation report for FAS/OGA (Hutchinson et al. 2003). The expectation here is that the performance metric program for GRLM will merge with the existing program through an initial benchmark update and if needed modification to the USDA evaluation questionnaires that are utilized. Four types of questionnaires are used. Two of these cover aspects of crop analysis, one concentrates on information technology, and the fourth includes aspects pertaining to management. One of the crop analysis questionnaires uses the Defect Detection and Prevention (DDP) methodology and tool that was developed by NASA's Jet Propulsion Laboratory (JPL) (Cornford et al. 2001). This DDP tool is intended to facilitate risk management over the entire project life cycle beginning with architectural and advanced technology decisions all the way through to operations. Each questionnaire type though targets the decision makers and DSS support functions. Therefore, the metric will be defined based upon improvements to analysis and efficiency of support. Improvements in analysis can be both subjective (e.g., relevance, quality, etc) and objective (e.g., latency, frequency, accuracy, etc) assessments and the evaluation periods are based upon ad hoc events (such as disasters) and growing seasons (crop estimates).

2.14
Anticipated Phase IV Results

The NASA/CNES missions are primarily aimed at ocean applications, but the ESA missions have multiple science objectives and their methods of tracking the ever-changing and complex terrain are more sophisticated. The radar altimeters onboard the ERS-1 and

ERS-2 had two tracking modes, ocean-tracking mode (for sea surfaces) and ice-tracking mode (for ice sheets and sea ice). While ERS-1 alternated between these two modes over land, ERS-2 spent the duration of its time over land in ice-mode. The significance here is that in an attempt to capture the echoes from more highly varying terrain, the altimetric range window size (in the time domain) increases by a factor of four. Early studies on the ERS data sets revealed some loss of lake/reservoir data in both the ocean- and ice-tracking modes although the latter clearly performed better over the smaller targets (Scott et al. 1992, 1994). However, compared to T/P and Jason-1, there should be no significant loss of data with Envisat because of a number of onboard trackers that enable "guaranteed tracking continuity" (Resti 1993).

Within the data sets, there will be a selection of range values (Envisat) and range retrieval methods (ERS, Envisat) to choose from according to the radar echo, the shape of which being variable according to surface roughness and complexity. Analysis of the radar echoes and associated range extraction algorithms are therefore expected in the revised GRLM-system as are multiple iterations to the altimetric range value based on the benchmarking exercises. Although the process becomes extended for ERS and Envisat, once the algorithms are finalized the system is expected to become fairly operational. The expected gains in lake and reservoir height measurements, particularly (a) along the coastal regions, (b) within small (<300 km^2) or narrow (<1 km wide) bodies of water, and (c) for targets in complex or highly varying terrain, are significant to warrant inclusion of such post-processing methods (see the chapters on retracking by Gommenginger et al., and Yang et al. this volume).

Phase IV of the program will aim to have more automation of the weekly product updates and have a download facility to access all products at "one click". In response to public feedback, additional information on the use of individual lake reference datum will also be placed on-line. With the inclusion of the ESA and NASA/CNES Jason-2 data sets, we expect the following improvements and enhancements to the original products offered in Phases I, II and III.

(i) An increase in the baseline number of years of observations, maintaining continuity across the products.
(ii) An increase by at least a factor of 5, the total number of lakes and reservoirs in the GRLM. The ERS/Envisat missions cross over ~611 large (≥^{3}100 km^2) lakes compared to the baseline ~70 lakes via the T/P and Jason-1 missions (Fig. 2.7).
(iii) An increase (from 17 to 165) in the number of overall reservoirs.
(iv) Particularly notable is an expected gain of 70 small reservoirs (<300 km^2) and an increase (from 10 to 60) in the number of lakes that are situated within narrow valleys or are surrounded by rugged terrain.
(v) An increase of ~65% in the number of reservoirs in the specific regions of interest (a total of 39 reservoirs in India, Iraq, Iran, Turkey, Brazil, Argentina, Australia).
(vi) An increase by a factor of 10 the number of lakes being monitored in the USA (a new total of 100 of which 35 are reservoirs).
(vii) A greater spatial distribution of lakes spanning all continents with the combined synergistic efforts of NASA, NRL and ESA mission data. The products will enhance and complement each other across a 1992–2010 time span, with the potential for further extension of the baseline time frame to 2015 with future missions.

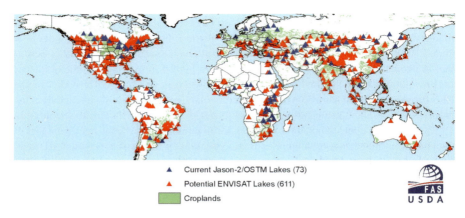

Fig. 2.7 Current lakes monitored by Jason-2 (OSTM) and potential lakes monitored by Envisat

(viii) Lake product accuracy is expected to be 10–30 cm RMS for the ERS-1 and ERS-2 products, 5–20 cm RMS for the Envisat products, and variable (5 cm to several decimeters) for Jason-2.
(ix) Enhanced Jason-1, GFO and Envisat products may improve by ~5–10 cm RMS.
(x) An indication of whether reservoirs <100 km^2 can be potentially monitored to the desired accuracy. This is most applicable to the central and northeastern regions of the USA, and the northern and eastern regions of Europe.
(xi) An increase in use of the products by USDA resource analysts and other end users. A greater efficiency and reliability within the assessments of drought or high water-level conditions across the globe, with improved downstream irrigation potential estimation. Consequences are improved knowledge of consequences on food trade and subsistence measures.

2.15
Summary, Recommendations and Future Outlook

An on-line database of water-level variations has been created for ~70 large lake, reservoir, and inland seas via a joint cooperative project between USDA, NASA and UMD. The elevation measurements have been derived using NASA/CNES and NRL satellite radar altimeter data, spanning an ~16 year time period ranging from 1992 to the present day. The focus is on the provision of near-real-time products in the form of graphical and text output and a semi-automated system, which updates the products on a weekly basis. Under new NASA sponsorship and continued USDA support, the program will also look to the incorporation of the ESA ERS/Envisat and NASA/CNES Jason-2 data sets to greatly expand the number of inland water bodies in the system and to enhance the number of smaller targets and reservoirs. The ultimate goal is the monitoring of at least 500 targets around the world, and the extension of the time line of satellite observations to ~20 years. Progress during the creation of an enhanced system will also contribute to the validation exercises of the various mission Science Working Teams and to future instrument design and data processing techniques in consideration of multi-disciplinary applications.

The GRLM has proven useful to quickly assess drought conditions in various parts of the world and in its enhanced form its use within the FAS DSS will have greater relevance to agriculture efficiency and water resources management in the future. With products in the public domain, the GRLM has also proven useful to many other users across the commercial, government, research, military, non-profit, and private sector domains. The extension of the timeline to almost 20 years, for example, is raising interest within the Intergovernmental Panel on Climate Change (IPCC), where the products are being considered as potential short-term proxy climate data records. The products have also been noted by the United States group on Earth Observations (GEO), in terms of water resources and drought records, and to the Global Earth Observation System of Systems (GEOSS) as a series of comprehensive, coordinated, and sustained observations that can serve as indicators of environmental health status.

Since inception, a number of project and data issues have arisen. To maintain product continuity, continued funding is essential at the satellite follow-on mission, product development, and routine operation levels. Manpower effort also needs to include backup for operational tasks, and the time allotted to respond to USDA and public feedback queries should not be overlooked. Many users felt the need for further detailed information on the products and maintaining a contact point between product developers and end users has been crucial. Benchmarking the program, in terms of ensuring product accuracy and delivery with respect to the original specifications, is also a task that requires regular assessment and revision, noting a level of accountability to both the USDA and other end users.

Regarding the satellite data, the quantity and quality of the products are dependent on a number of factors. With no currently operating dedicated inland water altimetry instrument, there are limitations on target size, and the lack of swath viewing does not achieve global coverage. While ground-based elevation data are reliant on gauge installation and maintenance, satellite-derived products are dependent on the lifetime of the mission and both satellite and instrument will be subject to various operating anomalies affecting data drop out and product latency. Acquiring a particular target and maximizing the number of elevation measurements across its extent (to improved range accuracy) will also depend on the tracking logic of the instrument and any onboard or ground-based data-processing steps that occur prior to delivery of the data to the project team.

Enhancement of the current technologies to allow for wide-swath viewing, and/or the reduction of effective footprint size, and an increase in along-track resolution from multiple synergistic nadir-viewing instruments would greatly assist the acquisition of additional targets. Improvements in tracking logic so that the lake surface can be more quickly acquired in highly varying terrain, or in proximity to coastlines and islands, is also recommended. Robust retracking (post-processing) methods that would uniquely identify the signal response of a small target within a complex field of view are also highly sought. Improvements to model-based wet tropospheric range corrections are also called for, particularly for small lakes lacking an instrument-based correction. As time progresses, the ability to update an average lake reference pass with additional cycles of data and the ability to replace the near-real-time (IGDR) products with archival (GDR) data becomes feasible. This is encouraged as a means of validating the near-real-time products and improving the accuracy of the on-line time series. With limitations on ground-based gauge data, both absolute (gauge) and relative (via other satellite products within the GRLM or from other teams) validation checks on product accuracy (whether near-real time or archival) are also strongly encouraged to meet the requirements of the various application programs.

Some improvements will be gained via the next generation of satellite radar altimeters that will not only ensure continuity of the GRLM program well into the 2015 time frame but also utilize enhanced technologies (see the chapter by Raney and Phalippou, this volume). Future missions include the Indian Space Research Organization's (ISRO) SARAL (Satellite with ARgos and ALtika) that will employ a Ka-band radar altimeter. Delay-Doppler (or SAR mode) altimetry will be utilized on ESA's Sentinel-3 (Table 2.1). Both of these instruments are nadir viewing but offer potential improvements via improved tracking, smaller footprints and finer range precision. Wide-swath techniques allowing for global coverage are also being considered. Current focus is on the Ka-Band Radar INterferometer (KaRIN) on the proposed NASA/CNES Surface Water and Ocean Topography (SWOT) mission (Fu 2003; Alsdorf et al. 2007). This mission was recommended within the U.S. National Research Council's (NRC) Decadal Review (NRC 2007) with a potential launch date around 2015.

While radar altimeters continue to be the focus of the program, recent attention has also turned to satellite laser altimetry (Lidar). Although the capability of lidar is limited by cloud cover, this additional tool could offer water-level information at certain resolution and accuracies. The current Ice, Cloud, and Land Elevation Satellite (ICESat) mission does offer some height retrieval capability that could be used as a Phase IV validation tool, but plans to launch an ICESat-2 follow-on mission, and a Deformation, Ecosystem Structure, and Dynamics of Ice (DESDynI) mission focused on hazards and environmental change are also being keenly noted.

With continued funding, the team hopes to achieve a multi-instrument operational lake-level observing system with the temporal and spatial resolution merits of each instrument being synergistically combined to maximize product output and consistency.

Acknowledgments The authors wish to acknowledge the USDA/FAS/OGA and NASA grants NNS06AA15G, NNX08AM72G, NNX08AT88G and NASA/JPL sub-award 4–33637(UMD) for supporting this program. Acknowledgment also goes to NASA/PODAAC, CNES/AVISO, LEGOS, NOAA, ESA UK-PAF and ESA F-PAC, for provision of the T/P, Jason-1, Jason-2, GFO, ERS and Envisat satellite altimetric data sets.

References

Alsdorf D, Rodriguez E, Lettenmaier DP (2007) Measuring surface water from space. Rev Geophys 45(2):RG2002. doi:10.1029/2006RG000197

Anulacion M (2003) Middle East and Turkey: warmer than normal and plenty of moisture. http://www.fas.usda.gov/pecad2/highlights/2003/12/me_turk_dec2003/index.htm

Beckley B, Zelensky N, Luthcke S, Callahan P (2004) Towards a seamless transition from TOPEX/Poseidon to Jason-1. Marine Geodesy Special Issue on Jason-1 Calibration/Validation Part III 27(3–4):373–389

Bilitza D, Koblinsky C, Beckley B, Zia S, Williamson R (1995) Using IRI for the computation of ionospheric corrections for altimeter data. Adv Space Res 15(2):113–120

Birkett CM (1995) The contribution of Topex/Poseidon to the global monitoring of climatically sensitive lakes. JGR-Oceans 100(C12): 25, 179–25, 204

Birkett CM, Mason IM (1995) A new global lakes database for a remote sensing programme studying climatically sensitive large lakes. J Great Lakes Res 21(3):307–318

Birkett CM, Alsdorf D, Harding D (2004) River and water body – stage, width and gradient: satellite radar altimetry, interferometric SAR, and laser altimetry. In: Anderson MG (ed) The encyclopedia of hydrological sciences, vol 2. Hydrological application of remote sensing: surface states. Wiley, Chichester, UK. http://mrw.interscience.wiley.com/emrw/9780470848944/ehs/article/hsa065/current/abstract

Chambers DP, Ries JC, Urban TJ (2003) Calibration and verification of Jason-1 using global along-track residuals with TOPEX. Mar Geod 26:305–317

Cornford SL, Feather MS, Hicks KA (2001) DDP – A tool for life-cycle risk management. In: Proc IEEE aerospace conference, 10–17 March, vol 1, pp 441–451. http://ddptool.jpl.nasa.gov/docs/f344d-slc.pdf

Crétaux J-F, Birkett CM (2006) Lake studies from satellite radar altimetry. In: Observing the earth from space, Geosciences Comptes Rendus, French Academy of Sciences. doi:10.1016/j.crte.2006.08.002

Fu L-L (ed) (2003) Wide-swath altimetric measurement of ocean surface topograph. JPL Publication 03-002. ftp://ftp-oceans.jpl.nasa.gov/pub/llf/WSOAreportFinal2.pdf

Fu L-L, Cazenave A (eds) (2001) Satellite altimetry and earth sciences: a handbook of techniques and applications. Inter Geophys Series Vol 69, Academic, San Diego, CA, ISBN 0122695453, 9780122695452

Hutchinson C, Drake S, van Leeuwen W, Kaupp V, Haithcoat T (2003) Characterization of PECAD's DSS: a zeroth-order assessment and benchmarking preparation. Report presented to NASA HQ, Washington, DC, and prepared by University of Arizona and University of Missouri.

Koblinsky CJ, Ray RD, Beckley BD, Wang YM, Tsaoussi L, Brenner LC, Williamson RG (1998) NASA ocean altimeter pathfinder project report 1: Data processing handbook. NASA Technical Memorandum NASA/TM-1998-208605. http://sealevel-lit.jpl.nasa.gov/science/search-details.cfm?ID=897

Lemoine FG, Zelensky NP, Rowlands DD, Luthcke SB, Chinn DS, Marr GC (2001) Precise orbit determination for Geosat follow-on using satellite laser ranging data and intermission altimeter crossovers. In: Proc NASA/GSFC flight mechanics symposium NASA/CP-2001-209986, pp 377–391. http://ntrs.nasa.gov/index.jsp?method=order&oaiID=20010084987

Mangeni Bennie T (2006) The dwindling Lake Victoria water level. In: Proc IASTAD environmentally sound technology in water resources management, Gaborone, Botswana, Sept 11–13, No 515–803, pp 85–90. http://www.actapress.com/PaperInfo.aspx?PaperID=28258&reason=500

McKellip R, Beckley B, Birkett C. Blonski S, Doorn B, Grant B, Estep L, Moore R, Morris K, Ross K, Terrie G, Zanoni V (2004) PECAD's global reservoir and lake monitor: a systems engineering report. Version 1.0, NASA/John C. Stennis Space Center, Mississippi

Ménard Y, Fu LL, Escudier P, Parisot F, Perbos J, Vincent P, Desai S, Haines B, Kunstmann G (2003) The Jason-1 mission. Marine Geodesy, Special Issue on Jason-1 Calibration/Validation Part I 26(3–4):131–146

Mertes LAK, Dekker AG, Brakenridge GR, Birkett CM, Létourneau G (2004) Rivers and lakes. In: Ustin SL, Rencz A (eds) Natural resources and environment manual of remote sensing, vol 5. Wiley, New York. http://www.wiley.com/WileyCDA/WileyTitle/productCd-0471317934.html

Morel L, Willis P (2005) Terrestrial reference frame effects on global sea level rise determination from TOPEX/Poseidon altimetric data. Adv Space Res 36:358–368

Morris CS, Gill SK (1994) Variation of Great Lakes water levels derived from Geosat altimetry. Water Res Res 30(4):1009–1017

NRC (2007) Earth Science and applications from space: national imperatives for the next decade and beyond, committee on earth science and applications from space: a community assessment and strategy for the future. National Research Council, Executive Summary, ISBN: 0-309-10387-8. http://www.nap.edu/catalog.php?record_id=11820

Reynolds C (2005) Low water levels observed on Lake Victoria. USDA/FAS/OGA, http://www.fas.usda.gov/pecad/highlights/2005/09/uganda_26sep2005/.

Reynolds CA, Doorn B, Birkett CM, Beckley B (2007) Monitoring reservoir and lake water heights with satellite radar altimeters. Africa GIS 2007 Conference, Ouagadougou, Burkina Faso, Sept 17–21

Resti A (1993) Envisat's radar altimeter: RA-2. ESA Bull 76:58–60. http://www.esa.int/esapub/pi/bulletinPI.htm

Riebeck H (2006) Lake Victoria's falling waters. NASA earth observatory. http://earthobservatory.nasa.gov/Study/Victoria/victoria.html

Ross K, McKellip R (2006) Verification and validation of NASA-supported enhacements to PECAD's decision support tools. NASA/John C. Stennis Space Center, Mississipi. http://ntrs.nasa.gov/archive/nasa/casi.ntrs.nasa.gov/20060025997_2006169224.pdf

Scott RF, Baker SG, Birkett CM, Cudlip W, Laxon SW, Mansley JAD, Mantripp DR, Morley JG, Munro M, Palmer D, Ridley JK, Strawbridge F, Rapley CG, Wingham DJ (1992) An Investigation of the Tracking performance of the ERS-1 radar altimeter over non-ocean surfaces. UK-PAF report to ESA No. PF-RP-MSL-AL-0100, ESA Contract No 9575/91/HGE-I

Scott RF, Baker SG, Birkett CM, Cudlip W, Laxon SW, Mantripp DR, Mansley JA, Morley JG, Rapley CG, Ridley JK, Strawbridge F, Wingham DJ (1994) A comparison of the performance of the ice and ocean tracking modes of the ERS-1 radar altimeter over non-ocean surfaces. Geophys Res Lett 21(7):553–556

Shum C, Yi Y, Cheng K, Kuo C, Braun A, Calmant S, Chambers D (2003) Calibration of Jason-1 altimeter over Lake Erie. Marine Geodesy, Special Issue on Jason-1 Calibration/Validation Part I 26(3–4):335–354

Sutcliffe JV, Petersen G (2007) Lake Victoria: derivation of a corrected natural water level series, technical note. Hydrol Sci J 52(6):1316–1321

Swenson SC, Yeh PJ, Wahr J, Famiglietti J (2006) GRACE estimates of terrestrial water storage: validation and applications. Eos Trans AGU 87(52), Fall meet. Suppl., Abstract G13C-02

UNEP (2006) Africa's lakes: atlas of our changing environment. United Nations Environment Programme, pp 28–30. http://na.unep.net/AfricaLakes/

Vincent P, Desai SD, Dorandeu J, Ablain M, Soussi B, Callahan PS, Haines BJ (2003) Jason-1 geophysical performance evaluation. Marine Geodesy Special Issue on Jason-1 Calibration/Validation Part I 26(3–4):167–186

Zanife OZ, Vincent P, Amarouche L, Dumont JP, Thibaut P, Labroue S (2003) Comparison of the Ku-Band range noise level and the relative sea-state bias of the Jason-1, TOPEX, and Poseidon-1 radar altimeters. Marine Geodesy, Special Issue on Jason-1 Calibration/Validation Part I 26(3–4):201–238

User Requirements in the Coastal Ocean for Satellite Altimetry

3

C. Dufau, C. Martin-Puig, and L. Moreno

Contents

3.1 Introduction 52
3.2 Methodology to Infer the User Requirements 53
3.3 Coastal Altimetry Users' Profile 54
3.4 Physical Requirements 55
3.5 Requirements in Accuracy and Precision 57
3.6 Requirements in Data Format and Delivery Mode 58
3.7 Conclusion 58
References 60

Keywords Coastal altimetry • COASTALT • PISTACH • User requirements

Abbreviations

ADCP	Acoustic Doppler Current Profiler
ADT	Absolute Dynamic Topography
ASCII	American Standard Code for Information Interchange
BUFR	Binary Universal Form of Representation of meteorological data
Cal/Val	CALibration/VALidation
CNES	Centre National d'Études Spatiales
COASTALT	ESA development of COASTal ALTimetry
DT	Delayed Time
ESA	European Space Agency
GDR	Geophysical Data Record
HF	High Frequency

C. Dufau (✉)
Collecte Localisation Satellites (CLS), Space Oceanography division,
8-10 rue Hermes, 31520 Ramonville, Saint Agne, France
e-mail: claire.dufau@cls.fr

C. Martin-Puig and L. Moreno
Starlab Barcelona S.L. (STARLAB), Research Department
45, Teodor Roviralta, 08022 Barcelona, Spain

MDT	Mean Dynamic Topography
netCDF	network Common Data Form (http://www.unidata.ucar.edu/software/netcdf/)
NRT	Near Real Time
OPeNDAP	Open-source Project for a Network Data Access Protocol (http://opendap.org/)
PISTACH	Prototype Innovant de Système de Traitement pour les Applications Côtières et l'Hydrologie
QC	Quality Controlled
RT	Real Time
SLA	Sea Level Anomaly
SSH	Sea Surface Height
SWH	Significant Wave Height

3.1 Introduction

Satellite altimetry over the open ocean is a mature discipline. In contrast, global altimetry data collected over the coastal ocean remain largely unexploited in the data archives, simply because intrinsic difficulties in the corrections (especially the wet tropospheric component, the high-frequency atmospheric signal, and the tides) and issues of land contamination in the footprint have so far resulted in systematic flagging and rejection of these data. In the last couple of years, significant research has been carried out to overcome these problems and extend the capabilities of current and future altimeters to the coastal zone, with the aim of integrating the altimeter-derived measurements of sea level, wind speed, and significant wave height into coastal ocean observing systems. (Cipollini et al. 2009)

The demand for improved altimetry datasets in coastal areas potentially useful in a number of applications, calls for a generation of coastal altimetry products for optimally reprocessing the radar signals in the vicinity of the coast; one particularly challenging aspect is the need for improvement in the above-mentioned corrections. Two recent European initiatives were born with the intention of meeting such demand: PISTACH and COASTALT.

As part of the Jason-2 project, the French Space Agency (CNES) has funded a project to improve altimeter products in coastal areas and inland waters, called PISTACH for "Prototype Innovant de Système de Traitement pour les Applications Côtières et l'Hydrologie". PISTACH, started in November 2007, aimed at implementing a prototype of improved Jason-2 altimetry products for coastal ocean and inland waters during the mission Cal/Val phase.

Similarly, COASTALT "ESA development of COASTal ALTimetry" is a project funded by the European Space Agency (ESA), aimed at defining, developing, and testing a prototype software processor to generate new Envisat radar altimeter products in the coastal zone.

The requirements for such products were not predetermined, but had to be decided by the PISTACH and COASTALT teams after consultation with the potential users of the coastal altimetry community; this was done by means of two questionnaires developed and distributed by the project teams. This chapter describes the collection of users' views on the basis of the two questionnaires and presents the ensuing recommendations on the definition of the new products.

The chapter is subdivided in seven sections. Sect. 3.2 presents the user requirements methodology applied to both projects. Sect. 3.3 describes the coastal users' typology and interests. Sect. 3.4–3.6 detail the user needs for physical contents, accuracy, precision requirements, data format, and delivery mode of the new products. Finally, Sect. 3.7 summarizes the main characteristics required for the coastal altimeter products that emerged both from the questionnaires analysis and from additional feedback gathered at the first and second Coastal Altimetry Workshops (Silver Spring, USA – February 2008 – and Pisa, Italy – November 2008).

3.2
Methodology to Infer the User Requirements

During the first phase of the PISTACH and COASTALT projects (winter 2007/08), dedicated teams were created to distribute, analyze, and synthesize questionnaires addressed to the coastal altimetry community with the aim of gathering user requirements. A list of key coastal altimetry users was created,[1] and each of them contacted with a questionnaire. As the COASTALT project started several weeks after the PISTACH project, it was agreed between project coordinators that COASTALT would focus on extending the users' base rather than resending the questionnaire to those experts contacted by PISTACH. Due to the nature of the sensor of each satellite (Jason-2, Envisat) two questionnaires were created by the two projects, although a few questions were common in both.

The PISTACH and COASTALT questionnaires were subdivided into five and eight main sections, respectively. Two were totally coincident (users' characterization and parameters used) with particular attention to the user's coastal zone of interest and to establish their degree of familiarization with altimetry data. The PISTACH questionnaire was focused more on the description of additional datasets in use (in situ, models), as well as on the physical processes under study by the coastal community, whereas the COASTALT questionnaire provided information about the duration of the datasets under evaluation by the coastal community.

In both questionnaires, there was an application-oriented section aimed at understanding the user's interests on: applications of interest, altimeter-derived parameters, resolutions needed, etc. In addition, the COASTALT questionnaire requested information on the product characterization in terms of accuracy and precision requirements. Furthermore, the PISTACH survey focused on studying the interest of the community for data continuity to open ocean or contiguous coastal regions. Additional sections were mostly coincident in both questionnaires (auxiliary data, data format, and distribution). Further reference to the COASTALT and PISTACH questionnaires, are available at: http://www.coastalt.eu/results/.

In summary, in a total of 28 main questions in each questionnaire, 18 were totally coincident and could be used in a joint analysis.

The questionnaires were distributed via email, and some users were directly interviewed. The COASTALT questionnaire was also distributed to the participants of the first Coastal Altimetry Workshop in February 2008. A last call for entries and comments was made during the second

---

[1]List available on request to the authors.

Coastal Altimetry Workshop in November 2008 by the PISTACH team. Overall, a total of 54 replies were gathered from the two projects; the main contribution (61%) to this study came from European experts: 33 replies came from Europe, seven from Asia, six from North America, three from Africa, two from South America, two from Russia, and one from Australia.

The PISTACH and COASTALT user requirement teams have merged their questionnaires in a consistent fashion wherever possible. The results hereafter presented include PISTACH's results when only these are available; the same for COASTALT, and the merged results wherever possible. This is clearly shown and written through the document, when considered convenient for clarification.

It is also important to notice that some questions within both questionnaires allowed for multiple answers, and thus the total number of answers (dealing with 100% in statistics) may differ from the total number of received questionnaires (54) and may be larger if multiple answers were provided.

3.3
Coastal Altimetry Users' Profile

Most questionnaires (87%) received were from oceanographers working in the public sector and respondents were mostly altimeter users (74%). In particular, 49% of the collaborators of these results belonged to public research institutions, 38% belonged to public research institutions doing operational oceanography also, and a small group – only 9% – worked only in operational oceanography.

It must be highlighted that a wide community from different sectors was contacted for contribution. However, it appears from the responses received that either operational oceanography in coastal zones is mainly conducted by the public sector, or that this sector has the highest interest in the products to be developed. The private sector is undersampled, with only five replies to the questionnaires (three from COASTALT and two from PISTACH). These few questionnaires are considered representative of the private sector in the absence of any contradictory evidence.

The broad headings "in situ data" (28%), "numerical modelling" (27%), and "satellite data" (22%) cover the majority of the data employed by the users. The category "other" (7%), where we may find statistical methodologies like statistical modelling (in the case of COASTALT questionnaire), and data assimilation (16%) cover the remaining share. It is important to notice that modelling as a whole covers a large share of the activities of the pool of users.

The in situ datasets used in coastal ocean research and applications are very diverse (Fig. 3.1). Hydrological measurements (temperature and salinity) are the most mentioned (24%), followed by sea level measurements (tide gauges – 17%) and velocity measurements using acoustic techniques (ADCP – 15%) (PISTACH result).

The user applications are of varying nature, with near real time (NRT) and delayed mode studies being more common than real-time (RT) applications among the community. The length of the datasets needed/used obviously depends on the application. In the case of operational services – some of them NRT – the data required can be as short as 1 day or even shorter, while for research studies the most requested dataset is between 1 and 10 years (COASTALT result).

3 User Requirements in the Coastal Ocean for Satellite Altimetry

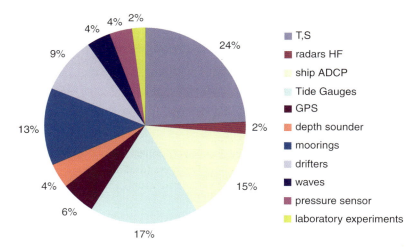

Fig. 3.1 (PISTACH) In situ datasets used by respondents in coastal ocean

Applications of altimeter products mainly concentrate on the validation of physical models (33%) and physical process analyses (27%), followed by data assimilation (23%). Scrutiny of the individual replies also reveals that the research community mainly focuses on the analysis of ocean processes, while the operational community leans more toward using altimetry for model validation, assimilation, and climate.

With regard to the definition of the coastal zone of interest, there is no clear preference between near shore (20%), shelf (26%), shelf break (28%), and open ocean (26%). As respondents are mostly altimeter users they do not work exclusively on the seashore area. In consequence, the typical distance to the shoreline varies largely: 62% of the studies are located between 0 and 100 km and 87% between 0 and 200 km.

The 0–50 km strip, where altimetry is most problematic and where dedicated altimeter processing will provide most of their improvements, received 35% of the overall multiple replies. By considering the quantity of replies over the number of total respondents instead of the total quantity of answers (multiple answers were possible), it appears that at least 67% of the respondents are interested in this strip. In addition, 24% of the respondents are only interested in this area for seashore hydrodynamics studies (waves, morphodynamic, sediment transport but also 3D circulation, upwelling, plumes and storm surge). A detail to be highlighted is that respondents who study only seashore processes do not use altimetry at all at the moment. This is an important category of potential users.

3.4 Physical Requirements

The physical processes most frequently studied within the community are: horizontal velocities (27%), surface elevation (26%), temperature (23%), and salinity (18%). Investigation of coastal oceanic processes mainly concerns shelf currents, storm surges,

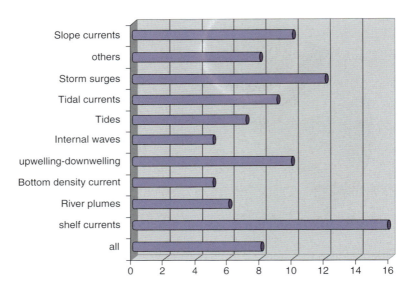

Fig. 3.2 (PISTACH) Coastal oceanic processes studied in preference

slope currents, and coastal upwelling/downwelling (Fig. 3.2). Consequently, asking for the physical parameter needed by coastal oceanographers in an altimeter product leads to a quasi-equal need of all the parameters: Sea Surface Height (SSH – 29%), Sea Level Anomaly (SLA – 31%), Absolute Dynamic Topography (ADT – 20%), and sea surface state (SWH – 20%). The preferred/required coastal altimeter product (PISTACH survey only) seems nevertheless to be an along-track SLA or ADT product with multi-mission inter-calibration (34%). A significant number of coastal oceanographers have also requested gridded products (28%). Only 18% of the replies deal with GDR datasets, and 20% with "corrSSH" (along-track SSH with corrections) products.

The current along-track resolution of the products seems to satisfy at least some users, as 38% estimate that it is adequate for their studies. The full resolution (20 Hz for Jason 2) is nevertheless required by 55% on condition that their quality is good. With or without quality flags, quality controlled (QC) data is asked for by a huge majority (44% for quality controlled (QC) data, 32% for QC data with global quality flag, and 20% for QC data with specific quality flags).

Geophysical corrections (36%), as well as atmospheric corrections (26%), and high-frequency de-aliasing corrections (21%) are the main complementary fields of interest for the community, followed by instrument corrections (17%).

An explicit need for data on both open ocean and coastal ocean has been expressed. The PISTACH survey evidences that in these cases continuity between open and coastal ocean is hugely required (90%).

The main difficulties encountered when using altimetry data combined with a numerical model and/or assimilation methods are the reference level and the steric effect (mentioned by several respondents to the PISTACH questionnaire). In fact, it seems that coastal 3D models using the Boussinesq approximation overlook a part of the mean basin-scale steric effect which is present in the altimetric dataset.

3.5
Requirements in Accuracy and Precision

We assume that the altimeter's measurements are sample values from probabilistic distributions. Then accuracy is the relationship between the mean of measurement distribution and its "true" value, whereas precision, also called reproducibility or repeatability, refers to the width of the distribution with respect to the mean.[2]

Different applications may have different requirements in terms of accuracy and/or precision. For instance, the estimation of the rate of global sea level rise from altimetry requires accuracy, but not necessarily precision given the huge numbers of measurement available to compute the mean rate. Instead, studies of El Niño require both accuracy (to discriminate the anomalous raised or lowered SSH value with respect to the mean) and precision, while the detection of fronts or bathymetric features requires only precision.

Accuracy and precision requirements were only included in the COASTALT questionnaire. Three questions were defined with regard to accuracy/precision: accuracy/precision for height measurements, accuracy/precision for significant wave height (SWH), and radiometric accuracy/precision.

Before analyzing the answers that were given to the precision and accuracy questions, it must be noted that these were left blank by a large part of the community. This clearly indicates that the users have a difficulty understanding these fundamental parameters that quantify the limitations and the range of applicability of the products, and calls for better information to the users.

Looking at the subsample of those who answered the questions, in terms of the required accuracy for height measurements and SWH, the users, perhaps not surprisingly, selected the options of highest accuracy: better than 3 cm for height measurement, and better than 5% of SWH. In practice, this would only be possible with delayed time products for sea surface height. However, the underlying message from these replies was that accuracy is an important issue and effort should be invested in its improvement.

One surprising result was that a majority of users consider the predecessor products to have an accuracy better than 3 cm, which is certainly questionable in marginal seas and when approaching the coast. This suggested that users tend to overrate the capabilities of current altimeter data in the coastal area, and, as said above, this calls for better information to the users, including a rigorous explanation of the error budget.

Seven people responded to the radiometric accuracy question. Answers were equally distributed as far as the desired accuracy was concerned. Half of the users who responded preferred a radiometric resolution better than 0.2 dB, while the other half preferred a radiometric resolution better than 0.5 dB.

With regards to precision, the majority of the users who responded are satisfied with the existing precision of non-coastal altimeter products when dealing with height measurements. For SWH, most users would like to have improved products with precision better than 5%.

Six questionnaires responded to the question on radiometric precision. Out of these six, two expressed their preferred radiometric resolution while the question about current radiometric precision was left blank.

---

[2]For further reference http://en.wikipedia.org/wiki/Accuracy_and_precision

3.6
Requirements in Data Format and Delivery Mode

Data format specifications and delivery modes are relevant in a product definition phase. The users were requested to provide information about: data format in use, delivery mode in use, and altimeter dataset upload time window needed.

According to the various responses received, the formats of higher interest in the community are: NetCDF (57%) and ASCII (26%); other formats like binary (16%) or other (1%) complete the percentage. One should perhaps also consider the BUFR format used by the Meteorological community, which was not explicitly listed as one of the choices in the questionnaires.

With regard to the delivery mode, FTP (47%) and live access server (25%) are the most desirable delivery modes; OpenDAP (22%) and DVD (6%) are also cited as of interest to a few members of the altimetry user community. The coastal dataset would need to be updated daily according to 45% of the received answers. For the remaining 55%, this updating delay can be increased to a week, month, or bigger time window. Despite this result, Delayed Time (DT-41%) and Near-Real-Time (NRT-36%) are widely desired followed by Real Time (RT-23%); this result is mainly due to the possibility of expressing multiple answers. In summary, this indicates that there is some scope for RT coastal altimetry in the future. It is also important to note that DT, NRT, and RT time delays can deal with different timescales from one application to another. Consequently, questionnaires were designed to specify this delay explicitly (<1 day for RT, some days for NRT and some months for DT).

3.7
Conclusion

After the analysis of the two surveys, the PISTACH and COASTALT user requirements team defined some product requirements that are summarized in Dufau et al. (2007) and Moreno et al. (2008), respectively. By considering the 54 answers received and also remarks that emerged from the two first workshops on coastal altimetry in Silver Spring, USA, and Pisa, Italy, the future coastal altimeter products should[3]:

- Be provided *along-track* even though a large number of respondents also asked for 2D-gridded products.
- Be provided at the *maximum posting rate* compatible *with an acceptable signal-to-noise ratio* (certainly between 1 and 10 Hz)
- Provide not only the SSH, but also *anomaly and mean value*, and *a coastal mean dynamic topography* (*MDT*).

---

[3]This list has been initially compiled with the help of Helen Snaith (NOCS), Mikis Tsimplis (NOCS), Paolo Cipollini (NOCS), and Stefano Vignudelli (CNR).

- Be provided with *individual corrections* (HF dynamics, for example) to ease its use in synergy with 2D and 3D models. Each user would then be able to apply the best combination of correction for its study.
- Include not only sea surface height, but also *significant wave height* and *wind speed*.
- *Initially* be developed as a *delayed product*, but with a processing chain *compatible with the delivery of near-real-time and real-time data*, as it is already a clear requirement for some respondents.
- Must be in *NetCDF format and distributed via FTP or OPeNDAP.*
- However *DVD distribution* should be retained for the benefit of those users with bandwidth constraints.
- Include data *as close to the coast as possible*, even when none of the main estimated parameters are considered reliable.
- Put in place all those *improvements in corrections* (including local corrections) *and retracking* so that *accuracy and precision are optimized.*
- Provide the users with *an error budget* and clear documentation on the characteristics and limitations of the products.
- Provide *quality flags*.
- Be *easy to merge across missions*, with a common correction scenario that should make possible the cross-calibration of SSH, wind, and wave information between the different altimetric missions.
- Present *continuity with* the altimeter products provided over *open ocean.*

Despite the large number of altimeter users (74%) in the 54 respondents to these questionnaires, it appears clearly that numerous questions still remain to be more accurately answered, particularly about optimal along-track sampling rate and physical contents (SSH, SLA, corrected or not, with which corrections). In fact, many respondents did not seem to be very aware of their needs, mainly due to their lack of experience in applying altimeter data to the coastal zones. For the altimeter neophyte respondents, the expression of requirements of all types of physical contents and all types of products (along-track, gridded) is symptomatic of getting the most out of the new products.

Both PISTACH and COASTALT projects have been set up to improve the GDR altimeter datasets in the vicinity of the coasts, but the two surveys have shown that 82% of the need is expressed toward higher level (i.e., more "downstream") products like along-track intercalibrated SLA or ADT or gridded products. It can be concluded that even if the new GDR products developed by PISTACH and COASTALT do not completely satisfy the user needs defined in this chapter, they constitute a huge first step toward the use of altimetry in the coastal ocean. Further user feedback on these new GDR products, as soon as they become available and are widely distributed, will permit a better understanding of the actual needs of each coastal application and steer the future development of the field.

Acknowledgments The authors of this chapter would like to acknowledge the support of the PISTACH and the COASTALT teams with special mention to Helen Snaith, Mikis Tsimplis, and Paolo Cipollini from NOCS and Stefano Vignudelli from CNR. In addition, we would like to thank R.K. Raney of the Johns Hopkins University for his contribution to the definition of the accuracy and precision terms in Sect. 3.5.

References

Cipollini P et al (2009) Progress in Coastal Altimetry: the experience of the COASTALT Project. Geophysical Research Abstracts, vol 11, EGU2009-12862. EGU General Assembly 2009

Dufau C et al. (2007) Tâche 1.1: Quantifier l'attente et la demande de la communauté utilisatrice, PISTACH CNES Contract No. 9885, CLS-DOS-NT-07-248, SALP-RP-P-EA-21480-CLS

Moreno L et al (2008) WP1 – Task 2.1 Report on user requirements for coastal altimetry products. COASTALT Technical Note, ESA/ESRIN contract No. 21201/08/I-LG, 2008. Available from http://www.coastalt.eu

Retracking Altimeter Waveforms Near the Coasts

A Review of Retracking Methods and Some Applications to Coastal Waveforms

C. Gommenginger, P. Thibaut, L. Fenoglio-Marc, G. Quartly, X. Deng, J. Gómez-Enri, P. Challenor, and Y. Gao

4

Contents

4.1	Introduction..	63
4.2	A Brief Introduction to Satellite Altimetry, On-board Trackers and Waveform Retracking	64
	4.2.1 Basic Principles of Altimetry...	64
	4.2.2 Trackers and Retrackers...	67
4.3	The Shape of Altimeter Waveforms in the Coastal Zone	68
	4.3.1 The Effect of Land on Coastal Waveforms.......................................	68
	4.3.2 Envisat RA-2 Waveforms in Coastal Areas......................................	69
	4.3.3 Jason-1 and Jason-2 Waveforms in Coastal Areas............................	71
	4.3.4 ERS-2 Waveform Shapes Over Australian Coasts	72
	4.3.5 Coastal Waveform Classification...	72
4.4	Empirical Retrackers...	75
	4.4.1 Offset Centre of Gravity Retracker (OCOG)....................................	75

C. Gommenginger (✉), G. Quartly, and P. Challenor
National Oceanography Centre, Southampton, European Way, Southampton, SO14 3ZH, United Kingdom
e-mail: cg1@noc.soton.ac.uk

P. Thibaut
Collecte Localisation Satellites, 8-10 rue Hermès, Parc technologique du canal, 31520, Ramonville Saint-Agne, France

L. Fenoglio-Marc
Institute of Physical Geodesy, Darmstadt University of Technology, Petersenstrasse 13, 64287, Darmstadt, Germany

X. Deng
Centre for Climate Impact Management and School of Engineering, The University of Newcastle, Australia

J. Gómez-Enri
Applied Physics Department, University of Cadiz, Spain

Y. Gao
College of Environment and Resources, Fuzhou University, Fuzhou, China

	4.4.2	Threshold Retracker...	76
	4.4.3	Improved Threshold Retracker	77
	4.4.4	The β-parameter Retracker	79
4.5	Physically-based Retrackers		81
	4.5.1	The Brown-Hayne Theoretical Ocean Model............................	81
	4.5.2	The NOCS Non-linear Ocean Retracker	83
	4.5.3	Practical Implementations of the Brown-Hayne Ocean Retracker............	85
4.6	Statistical Methods of Function Fitting		85
	4.6.1	Maximum Likelihood Estimator (MLE).....................................	86
	4.6.2	Weighted Least Square Estimator (WLS)..................................	86
	4.6.3	Unweighted Least Square Estimator..	87
4.7	Applications of Waveform Retracking to Coastal Waveforms............		88
	4.7.1	The CNES/PISTACH Coastal Processing System for Jason-2.................	88
	4.7.2	The COASTALT Processing System for Envisat RA-2	90
	4.7.3	Coastal Retracking Around Australia	91
	4.7.4	Application of Empirical Retrackers to Envisat RA-2 in the Mediterranean Sea	94
4.8	Innovative Retracking Techniques...		94
	4.8.1	Iterative Retracking...	96
	4.8.2	Multiple-waveform Retracking...	96
4.9	Conclusions..		97
References..			99

Keywords Altimetry • Brown ocean retracker • Empirical retracker • Innovative techniques • On-board tracker • Retracking • Waveforms

Abbreviations

BOR	Brown Ocean Retracker
CLS	Collecte Localisation Satellites
CNES	Centre National d'Études Spatiales
COASTALT	ESA development of COASTal ALTimetry
COG	Centre Of Gravity
DEM	Digital Elevation Model
DIODE	Détermination Immédiate d'Orbite par Doris Embarqué
DORIS	Doppler Orbitography and Radiopositioning Integrated by Satellite
EGM	Earth Gravitational Model
Envisat	Environmental Satellite
ERS	European Remote Sensing (satellites)
ESA	European Space Agency
FSSR	Flat Sea Surface Response
GDR	Geophysical Data Record

Geosat	GEOdetic SATellite
GFO	Geosat Follow-On
GSFC	Goddard Space Flight Center
LEP	Leading Edge Position
MBS	Mixed Brown Specular
MLE	Maximum Likelihood Estimator
MMSE	Minimum Mean Square Estimator
MSL	Mean Sea Level
MT	Median Tracker
NASA	National Aeronautics and Space Administration
NOCS	National Oceanography Centre, Southampton
OCOG	Offset Centre Of Gravity
PDF	Probability Density Function
PISTACH	CNES Development of "Prototype Innovant de Système de Traitement pour les Applications Côtières et l'Hydrologie
PR	Pulse Repetition Frequency
PTR	Point Target Response
RMSE	Root Mean Squared Error
SGT	Split Gate Tracker
SSH	Sea Surface Height
SWH	Significant Wave Height
WLS	Weighted Least Square

4.1 Introduction

Satellite altimeters provide accurate measurements of sea surface height, significant wave height and wind speed, and have done so for now over 20 years. Thanks to their high accuracy, repeatability and stability, they have become an irreplaceable tool to address a wide range of scientific questions, from global ocean circulation monitoring to long-term sea level rise or operational ocean weather forecasting. There is now increasing demand for altimeter observations in the coastal zone, where human activities are both more concentrated and more diverse. Examples of practical exploitation of altimeter data encompass a wide range of applications, from health and safety for commercial and leisure shipping, to pollutant dispersal, coastal erosion, flood risk assessment, extreme weather forecasting, etc.

Unfortunately, while satellite altimeters perform very well over the open ocean, a number of issues arise in the vicinity of land, related to poorer geophysical corrections and artifacts in the altimeter reflected signals linked to the presence of land within the instrument footprint. In this chapter, we focus on "pre-processing" issues, examining the various methods at our disposal to extract better altimetric information through ground-based waveform retracking. The reader interested in "post-processing" issues – namely, the geophysical corrections needed to correct retrieved range in the coastal zone, for example for wet tropospheric delay – is invited to read the Chapter by Roblou et al., this volume.

The chapter is arranged as follows: we begin in Sect. 4.2 with a brief introduction to satellite altimetry, aimed at a non-expert audience, to basic concepts of altimetry that are used in the rest of the chapter. We proceed in Sect. 4.3 with considering what happens when altimeters measure close to land, and we provide observational evidence of the diversity and complexity of waveforms in coastal areas as seen in data from Envisat RA-2, Jason-1, Jason-2 and ERS-2. This gives us a way to gauge the challenges we face in order to retrack altimeter waveforms contaminated by reflections from land. Sections 4.4 and 4.5 then review empirical and physically-based retracking methods, respectively, developed over many years to extract geophysical information from altimeter waveforms. In Sect. 4.6, we briefly discuss some of the statistical methods typically used to perform waveform fitting. We then proceed in Sect. 4.7 to give examples of the application of various retracking methods to altimeter waveform data from ERS-2, Envisat RA-2, Jason-1, Jason-2 at coastal sites around the world. We close the chapter with Sect. 4.8, a discussion of emerging innovative techniques and ideas put forward to help improve waveform retracking in coastal areas.

4.2
A Brief Introduction to Satellite Altimetry, On-board Trackers and Waveform Retracking

Here we begin with a brief introduction to satellite altimetry to help establish for the non-expert reader the basic concepts needed to understand the methods detailed later in this chapter. For an in-depth presentation of technical aspects of satellite altimetry, the reader is referred to Chelton et al. (1989, 2001).

4.2.1
Basic Principles of Altimetry

Satellite altimeters are nadir-pointing active microwave instruments, whose principal aim is to measure as accurately as possible the two-way travel time of short pulses reflected from the Earth's surface. For oceanographic applications, the accuracy on the derived range must be of the order of centimetres. The shape of the reflected signal, known as the "waveform," represents the time evolution of the reflected power as the radar pulse hits the surface. As well as measuring range, the shape of the waveform can provide information on the nature and properties of the reflecting surface (e.g., significant wave height – see Sect. 4.5).

Although satellite altimeters have narrow antenna beam widths (~1 degree), the altimeter footprint is actually determined by the duration of the emitted pulse. Pulse compression and deramping techniques (see e.g., Chelton et al. 2001) are used to achieve high-accuracy ranging. Hence the term "pulse-limited" altimeters – a design used for all altimeters mentioned in this chapter. Tables 4.1 and 4.2 provide a summary of the best-known recent altimeter missions, together with key mission and instrument characteristics.

For practical reasons, the instrument can only measure over a narrow range window (typically 60 m vertically), called the "analysis window." As the satellite-to-Earth distance changes as the satellite progresses along the orbit (e.g., due to Earth surface topography

4 Retracking Altimeter Waveforms Near the Coasts

Table 4.1 Satellite mission and operating characteristics of altimeters (after Quartly et al. 2001)

Altimeter	Launch - End	H (km)	Inclination	Band	Radar Frequency (GHz)	On-board range tracker
Geosat	12 Mar 1985–Jan 1990	800	108°	Ku	13.5	2nd order (α, β)
ERS-1	17 Jul 1991–10 Mar 2000	784	98°	Ku	13.8	2nd order (α, β)
ERS-2	21 Apr 1995–present	784	98°	Ku	13.8	2nd order (α, β)
TOPEX	10 Aug 1992–5 Jan 2006	1336	66°	Ku	13.6	2nd order (α, β)
				C	5.3	Slaved to Ku
Poseidon	10 Aug 1992–5 Jan 2006	1336	66°	Ku	13.65	2nd order (α, β)
GFO	10 Feb 1998–25 Sept 2008	800	108°	Ku	13.5	2nd order (α, β)
Jason-1	7 Dec 2001–present	1336	66°	Ku	13.6	Split Gate Tracker
				C	5.3	Slaved to Ku
Envisat	1 Mar 2002–present	784	98°	Ku	13.6	Model Free Tracker
				S	3.2	Slaved to Ku
Jason-2	20 Jun 2008–present	1336	66°	Ku	13.6	Median Tracker & DIODE/DEM Tracker
				C	5.3	Slaved to Ku

changes), the position of the analysis window must be dynamically adjusted to ensure that the altimeter samples at the time when the pulse hits the surface. This dynamical adjustment is done by the "on-board tracker," a predictive device, which minimizes the risk of the altimeter losing track of the surface. Further details about on-board trackers are given in the next section.

The effective footprint of a pulse-limited altimeter is controlled by the pulse duration and the width of the analysis window. Over water, after the pulse hits the water surface, the area contributing to the reflected power is an expanding disc, which, upon reaching its maximum size, continues to spread into an expanding annulus of increasing diameter but constant surface area. The diameter of the pulse-limited footprint is generally between 2 and 7 km, depending on significant wave height. This general principle of pulse-limited altimetry is well known and described in detail in Chelton et al. (1989, 2001).

Table 4.2 Operating characteristics of altimeters (after Quartly et al. 2001)

Altimeter	Band	Antenna Beamwidth (deg)	PRF (Hz)	Number of waveform gates	Nominal tracking point	Gate width (ns)	Waveform averaging frequency (Hz)	Number of waveforms averaged
Geosat	Ku	2.0°	1020	60	30.5	3.125	10	100
ERS-1	Ku	1.3°	1020	64	32.5	3.03	20	50
ERS-2	Ku	1.3°	1020	64	32.5	3.03	20	50
TOPEX	Ku	1.1°	4500	128	32.5	3.125	10	456
	C	2.7°	1200	128	35.5	3.125	5	240
Poseidon	Ku	1.1°	1700	60	29.5	3.125	20	86
GFO	Ku	1.6°	1020	128	32.5	3.125	10	100
Jason-1	Ku	1.28°	1800	104*	31	3.125	20	90
	C	3.4°	300	104*	31	3.125	20	15
Envisat	Ku	1.29°	1800	128	46.5	3.125	18	100
	S	5.5°	450	64	25.5	6.25	18	25
Jason-2	Ku	1.26°	1800	104*	31	3.125	20	90
	C	3.38°	300	104*	31	3.125	20	15

* The reception window on Jason-1 and Jason-2 contains output from 128 FFT bins; only information on the central 104 are telemetered to the ground for later reprocessing

Waveforms over water show a characteristic shape (see Fig. 4.10), often referred to as "ocean-like" or "Brown-like", featuring a sharp rise (the "leading edge") up to a maximum value, followed by a gently sloping plateau (the "trailing edge"). Waveform samples in the leading edge correspond to the time when the pulse first hits the surface, while waveform samples in the trailing edge represent the received power from annuli situated further and further away from the point of closest encounter (nadir).

Another important characteristic of pulse-limited altimeters is the pulse repetition frequency (PRF), which is chosen to ensure that successive echoes are uncorrelated. This makes it possible to reduce speckle by averaging a number of successive echoes and improve the signal to noise ratio. Averaging is done typically over 100 successive echoes (at Ku-band) corresponding to a waveform averaging frequency around 20 Hz. Table 4.2 gives the values of these parameters for recent satellite altimeter missions.

4.2.2
Trackers and Retrackers

As mentioned, waveforms are acquired thanks to a tracking system placed on-board the satellite. The purpose of the on-board tracker is to keep the reflected signal from the Earth's surface within the altimeter analysis window. The general principle of an on-board tracker is that it predicts the likely position of the next echo based on information derived from the echoes the receiver has just recorded. This computation happens autonomously on-board the satellite – hence "on-board trackers."

For most recent altimeters, the computation uses a second order filter closed loop – the (α,β) tracker – but new types of on-board trackers have started to appear on more recent altimeters (see Table 4.1). Interesting recent developments are the Envisat RA-2 Model Free Tracker, which does not rely on any assumptions about the shape of the waveform, and the new on-board tracker on Jason-2 known as "DIODE/DEM". When this new Jason-2 tracking mode is activated, the analysis window is repositioned based on information provided by the DORIS navigator system and a Digital Elevation Model introduced through the open loop. Examples of the behaviour of different on-board trackers on Jason-1 and Jason-2 are shown in Fig. 4.3.

In order to obtain the highest possible accuracy on range measurements over the ocean, today's altimeters downlink the waveforms to Earth and the final retrieval of geophysical parameters from the waveforms is performed on the ground. This is called "waveform retracking" and is the main focus of this chapter. The aim of the ground-based retracking is to fit a model or functional form to the measured waveforms, and retrieve geophysical parameters such as the range, echo power,... Functional forms can be purely empirical, or, as in the case of the Brown ocean retracker, be based on physics. Both empirical and physics-based retracking methods are presented in Sect. 4.4 and 4.5, respectively.

The final range measurement is obtained by combining the range of the analysis window (the tracker range) with the retrieved epoch obtained by retracking (the position of the leading edge with respect to the fixed nominal tracking point in the analysis window). Values of the nominal tracking point vary from 32.5 for TOPEX to 46 for Envisat RA-2 (see Table 4.2).

4.3
The Shape of Altimeter Waveforms in the Coastal Zone

P. Thibaut, J. Gómez-Enri and X. Deng

In this section, we begin by considering the many ways in which altimeter waveforms in the coastal zone are affected by the vicinity of land. We discuss the role of geometry and of the nature of the land surface to influence waveform shape, and we provide observational evidence of the diversity of waveform shapes in the coastal zone, by reporting on recent analyses of coastal waveform shape performed for Envisat RA-2, Jason-1 and Jason-2 and ERS-2 in various coastal regions of the world.

4.3.1
The Effect of Land on Coastal Waveforms

Bearing in mind the geometry of a pulse-limited altimeter footprint over open water, let us now consider what happens when an altimeter encounters a land-to-ocean or ocean-to-land transition: the altimeter footprint will now be partly over ocean and partly over land, as represented in Fig. 4.1. The power received in a given gate will thus be linked to the relative proportion of sea and land area in the corresponding footprint, and to the reflective properties (σ^0, the backscatter coefficients) of each type of surface. Taking the case of an ocean/land transition, Fig. 4.1 shows how and when the nadir of the satellite is still over ocean, it is the last samples in the waveform that are contaminated first by land returns. As the satellite nadir position approaches the coastline, more waveform samples become contaminated by non-ocean like reflections, starting from samples in the trailing edge gradually migrating towards the leading edge. In the case of a land-to-ocean transition, the closer the satellite nadir position is to the coastline, the more waveform samples will correspond to ocean cells, again beginning from the trailing edge towards the leading edge.

The lower panel in Fig. 4.1 is a top-down view of the (pulse-limited) altimeter footprint, showing the relative proportion of ocean and land area in each of the annuli away from the nadir point. The number of contaminated samples will depend on the height and surface area of the land, as well as its proximity to the nadir point. Moreover, how the waveform is affected depends on a weighted average of the surface area by the scattering coefficient of each surface i.e., "Surface$_{Land}$ × σ^0_{Land}" relative to "Surface$_{Ocean}$ × σ^0_{Ocean}" where Surface$_{Land}$ and Surface$_{Ocean}$ are the surface areas in the altimeter footprint occupied by land and ocean, respectively. If the land is much less reflective that the ocean ($\sigma^0_{Land} < \sigma^0_{Ocean}$, which is typically the case), the effect of land will be small and the waveforms will remain unaffected until close to the coast. In some environments however (e.g., coral atolls), land can be more highly reflective than the ocean ($\sigma^0_{Land} > \sigma^0_{Ocean}$), so that even a small area of highly reflective land in the footprint can have a major impact on the waveforms.

From this, it is clear that the geometry of the coastline, the relief of the land, the nature of the terrain – characteristics of the coast that are extremely diverse throughout the world – will all contribute to defining the shape of the waveform. In Fig. 4.1, the geometry was described in an ideal case, without consideration for the possible added complexity

Fig. 4.1 (*Top panel*) Schematic representation of a pulse-limited altimeter short pulse propagating from the altimeter to the sea surface in the case of an ocean to land transition. (*Lower panel*) Top-down view of the pulse-limited footprint corresponding to each waveform gate. B is the bandwidth of the altimeter, c is the speed of light

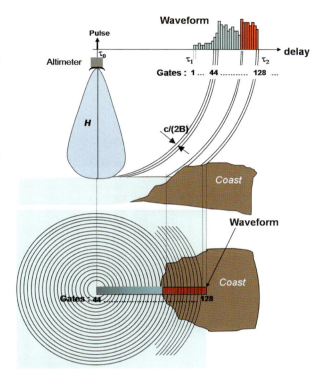

introduced by non-zero altimeter platform attitude or the behaviour of the on-board tracker. Indeed, it is assumed in Fig. 4.1 that the altimeter tracks the first return from the surface beneath. The differences of returned power between ocean and land σ^0 can make the on-board tracker shift the analysis window, thereby affecting how much of the waveform is captured. Examples of this phenomenon are presented in Sect. 4.3.3.

4.3.2
Envisat RA-2 Waveforms in Coastal Areas

Examples of land effects on coastal waveforms from the Envisat RA-2 altimeter can be seen in Fig. 4.2. The upper panel shows the amplitude of 760 successive 18 Hz Ku-band waveforms for a pass over the Tuscany coast (Italy) in the North-West Mediterranean Sea. The waveforms are plotted as latitude (x-axis), gate number (y-axis) and amplitude in FFT filters units (colour scale). The arrow denotes the flight direction. Brown-like ocean waveforms are seen in the first part of the track segment analyzed. Then, the small effect of Del Giglio Island over the shape of the waveforms is clearly observed. Particularly interesting is the effect of the Follonica Gulf as it is near an ocean to land transition. The second part of the track segment corresponds to waveforms located over land.

For any point near the sub-satellite track, the distance of that point to the spacecraft is given by $(H_0^2 + v^2 \Delta t^2)^{0.5}$, where H_0 is the distance at nearest approach (nadir), v is the spacecraft speed and Δt is the time relative to nearest approach. A small bright target may then

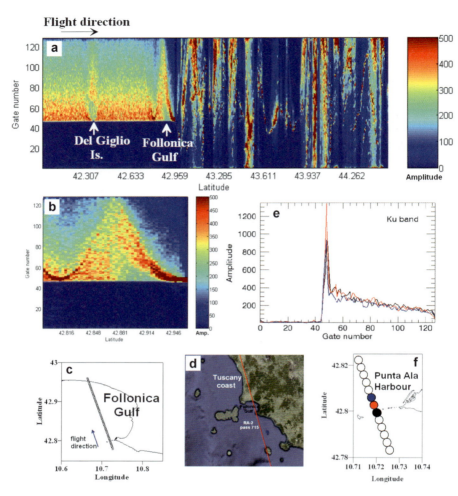

Fig. 4.2 (a) Envisat RA-2 Ku-band waveforms along ascending pass 715 over the coast of Tuscany (Italy) in the North-West Mediterranean Sea. (b) Zoom view of the waveforms in the vicinity of Follonica Gulf with (c) location map. (d) Location map of RA-2 pass 715 along the coast of Tuscany. (e) Waveform shapes in black, red and blue for 18 Hz locations shown in (f) around Punta Ala Harbour, in Follonica Gulf

trace a hyperbolic trajectory in waveform space, as it appears at different gate positions in consecutive waveforms. Fig. 4.2b shows an example from the Follonica Gulf (see location in Fig. 4.2c): a sharp feature with a complete loss of trailing edge is clearly seen migrating through the waveforms gates as the radar crosses the gulf northwards. Fig. 4.2e shows the waveform shapes as the radar crosses the Follonica Gulf, in the vicinity of Punta Ala Harbour (see location in Fig. 4.2d and 4.2f), where calmer waters are expected. The effect of the land contamination and/or calmer water in the littoral zone is most clearly seen in the significant increase of the peak amplitude. Here, the availability of accurate high-resolution coastline data play a key role in interpreting the effect of land on altimeter signal and in understanding how waveform shapes respond to land in the footprint.

4.3.3
Jason-1 and Jason-2 Waveforms in Coastal Areas

The huge diversity of waveform shapes is seen also in Jason-1 and Jason-2 data near the coasts, and this is illustrated in Fig. 4.3. The figure shows Jason-1 and Jason-2 waveforms obtained for one particular orbit segment over the Mediterranean Sea from Algeria to France via Ibiza Island (pass 187). In each subplot in Fig. 4.3, the vertical axis represents the waveform gates, latitude runs along the horizontal axis, and colour represents the waveform amplitude. The same hyperbolic trajectories - characteristic of specular reflectors migrating through waveform gate space – are seen here again here, as seen in Fig. 4.2 for Envisat RA-2.

The major differences which are seen between the waveforms in the three subplots in Fig. 4.3 are linked to the different behaviour of the three trackers used on Jason-1 and Jason-2. Fig. 4.3 top subplot shows Jason-1 waveforms acquired with a Split Gate Tracker (SGT), while the middle and bottom subplots show Jason-2 waveforms obtained with the Median Tracker (MT) and the new DIODE/DEM tracker, respectively. On land/ocean transition (plots A, B and C), the Jason-2 trackers show a much faster recovery of the signal after the transition. In this case, the Jason-2 MT tracker is better able to keep the signal in the analysis window. In all cases, whenever the altimeter has lost lock on the leading edge, the time it takes for the altimeter to regain lock is roughly three times smaller for MT than for SGT (about 0.4 s for Jason-2 compared to 1.4 s for Jason-1). We note that the

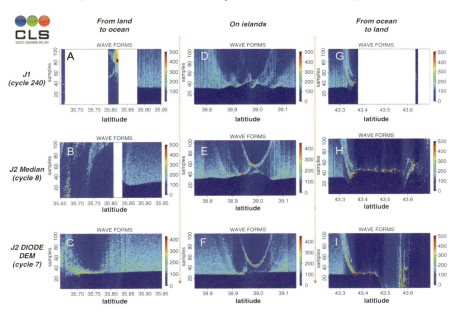

Fig. 4.3 Jason-1 and Jason-2 waveforms over various land/ocean and ocean/land transitions over the Mediterranean Sea from Algeria to France via Ibiza Island (pass 187). The top subplot shows the Jason-1 waveforms, collected with the Split Gate on-board Tracker. The middle and bottom subplots show Jason-2 waveforms, collected with the Median Tracker and the DIODE/DEM tracker, respectively. In each subplot, three transitions are considered: 1st column: a land/ocean transition over the Algerian coasts (latitude 35.73°); 2nd column: orbit segment over Ibiza Island (latitudes 38.95–38.975°); 3rd column: an ocean/land transition over the French coast (latitude 43.366°)

DIODE/DEM tracking mode on Jason-2 never loses lock on the leading edge, even though the waveforms do not conform to the Brown ocean model.

On ocean/land transitions (plots G, H and I), the performance of Jason-1 and Jason-2 MT are almost equivalent, although Jason-1 looses lock on the signal sooner than Jason-2 when the echo over land becomes too disruptive. In the case of the Jason-2 DIODE/DEM (plot I), the loss of the leading edge from the analysis window is linked to inaccuracies in the DEM used to drive the Jason-2 DIODE on-board tracker. Over Ibiza Island (plots D, E and F), all on-board trackers manage to keep good lock on the leading edge of the waveforms. We observe in this particular case that the land proximity impacts the waveform shapes in similar ways, regardless of the tracker. As the satellite progresses towards the island, the waveforms become affected first in the last few gates, then in the full trailing edge, then the echo finally become highly specular as the satellite measures over land. One also observes how the change in the position of the leading edge is compensated differently by the different trackers.

Analyses of waveform differences from cycle-to-cycle for the same track, such as shown in Fig. 4.3, can also reveal other effects, such as different cross-track positions of the altimeters (±1 km with respect to the nominal track) or seasonal variations, for example in land surface properties (moisture, vegetation) or sea state conditions.

4.3.4
ERS-2 Waveform Shapes Over Australian Coasts

Fig. 4.4 shows examples of typical waveform shapes encountered for ERS-2 over the Australian coasts (Deng 2004). Most of them (~80%) present the traditional ocean-like shape with a single rising leading edge ramp and no marked peak in the range window (Fig. 4.4a), though the waveform profiles may shift left or right with respect to the nominal tracking gate position. This type of ocean-like waveform is found over oceans up to 30 km from the coastline.

Waveform shapes shown in Fig. 4.4b–g are usually found when altimeter simultaneously illuminates both water and land surfaces. The narrow-peaked waveform, (Fig. 4.4h) corresponds to calmer waters, such as bays, gulfs, estuaries, and harbours, as well as inland lakes. These are typical highly specular radar reflections over flat surfaces, similar to those shown already in Sect. 4.3.2 and 4.3.3 for Envisat RA-2, Jason-1 and Jason-2. A distinctive feature of this type of waveform is its extremely high amplitude compared to other waveform types.

4.3.5
Coastal Waveform Classification

The previous figures have illustrated some of the many different types of waveform shapes that can be encountered in the vicinity of the coasts. It is useful to classify these waveforms by shape in order to statistically characterize the prevalence of particular waveform shapes in coastal or inland water regions. Fig. 4.5 shows the generic waveform shapes resulting from a shape classification analysis based on a Neural Network approach.

Knowing the shape class of a waveform prior to retracking can be useful to optimize the choice of retracking method. Conversely, since retrackers do not perform equally

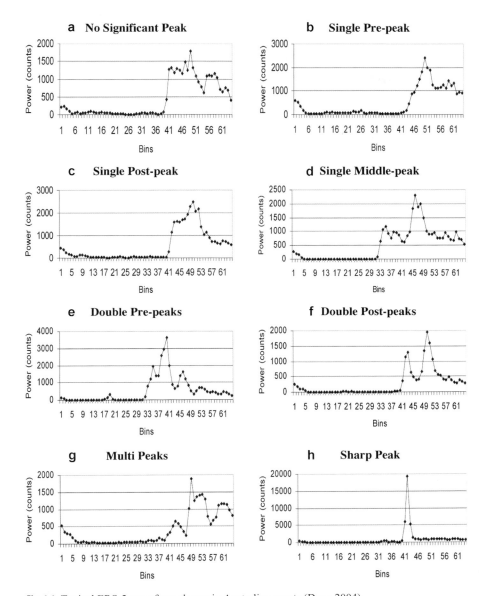

Fig. 4.4 Typical ERS-2 waveform shapes in Australian coasts (Deng 2004)

well regardless of the shape of the waveform, the shape class can serve as a post-retrieval editing criterion to discard suspicious retracker output. As an example, waveforms of class 2 or class 21 in Fig. 4.5 could not be optimally retracked with a standard Brown ocean retracker like the one used for open ocean waveforms.

Fig. 4.6 represents the distribution of Jason-2 waveforms across the waveforms classes shown in Fig. 4.5, as a function of the distance to the coast (defined as the distance to the

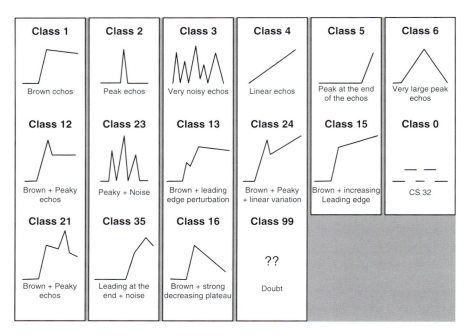

Fig. 4.5 Waveform shape classes obtained for Jason 2 using Neural Networks

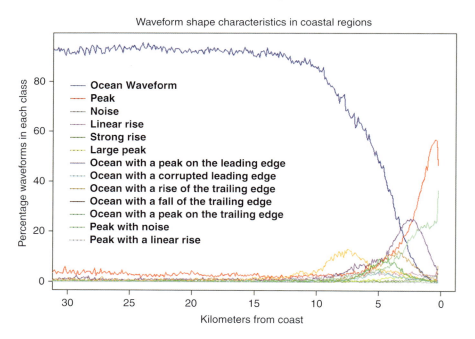

Fig. 4.6 Prevalence of various waveform shapes for Jason-2 over one full cycle as a function of distance to the nearest coastline

nearest coastal point). The data in Fig. 4.6 were computed for one complete cycle of Jason-2 data (cycle 8). We find that, on average, 94% of the waveforms are Brown-shaped waveforms up to 15 km from the coast. This percentage rapidly decreases onshore of 15 km from the coastline. On individual tracks, this distance still remains closely determined to the nature of the coast: relief, structure, vegetation. Conversely, the percentage of "peaky" waveforms rises rapidly onshore off 5–6 km from the coast.

In the 0–10 km zone, we observe two waveform shape classes corresponding to waveforms with a rise or a fall in the trailing edge. Changes in sea state in the littoral band could explain these kinds of changes in the trailing edge of the waveforms, as is seen also during σ^0 blooms events.

4.4
Empirical Retrackers

L. Fenoglio-Marc, X. Deng and Y. Gao

We now review the various retracking methods in use to retrieve geophysical information from altimeter waveforms. In this section, we consider waveform retracking methods developed over many years on the basis of empirical observations and practical experience. The empirical methods of waveform retracking may be classified into two categories: those based on the statistical properties of the waveform data and those based on fitting empirical functional forms. Both types are reported in the section that follows.

4.4.1
Offset Centre of Gravity Retracker (OCOG)

The offset centre of gravity (OCOG) retracking algorithm was developed by Wingham et al. (1986) to achieve robust retracking. The goal is to find the centre of gravity of each waveform based on the power levels within the gates (Fig. 4.7). It is a purely statistical approach which does not depend on a functional form.

Based on the definition of a rectangle about the effective centre of gravity of the waveform, the amplitude (A) and width (W) of the waveforms and the gate position of the waveform centre of gravity (COG) are estimated from the waveform data using:

$$A = \sqrt{\sum_{i=1+n_1}^{N-n_2} P_i^4(t) \bigg/ \sum_{i=1+n_1}^{N-n_2} P_i^2(t)} \qquad (4.1)$$

$$W = \left(\sum_{i=1+n_1}^{N-n_2} P_i^2(t) \right)^2 \bigg/ \sum_{i=1+n_1}^{N-n_2} P_i^4(t) \qquad (4.2)$$

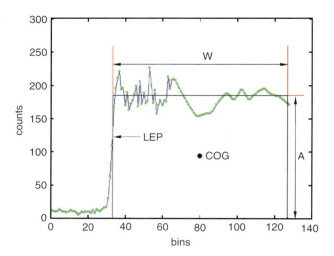

Fig. 4.7 Schematic diagram of the OCOG method

$$COG = \sum_{i=1+n_1}^{N-n_2} iP_i^2(t) \bigg/ \sum_{i=1+n_1}^{N-n_2} P_i^2(t) \qquad (4.3)$$

where, P_i is the waveform power, N is the total number of samples in the waveform, and n_1 and n_2 are the numbers of bins affected by aliasing at the beginning and end of the waveform, respectively (e.g., $n_1 = n_2 = 4$ in Hwang et al. 2006).

Finally the leading edge position (LEP) is given by:

$$LEP = COG - \frac{W}{2} \qquad (4.4)$$

OCOG is an easy-to-implement, robust waveform retracker, which depends solely on the statistics of the waveform samples. It is the algorithm behind the Ice-1 retracker for the Envisat RA-2 altimeter. It is well adapted to surfaces where range varies rapidly (such as continental ice) but its use for retrieving range accurately is limited because its formulation is not related to any physical properties of the reflecting surfaces. OCOG is sometimes used to calculate the initial values for the Threshold retracker (Sect. 4.4.2) and Improved Threshold retracker (Sect. 4.4.3) and the β-parameter function fitting method (Sect. 4.4.4).

4.4.2
Threshold Retracker

To improve the estimation of range, the threshold retracking method was developed by Davis (1995, 1997). This method is based on the dimensions of the rectangle computed using the OCOG method. The threshold value is referenced with respect to the OCOG amplitude or the maximum waveform amplitude as 25%, 50% and 75% of the amplitude. The retrieved range gate is determined by linearly interpolating between adjacent samples of the threshold crossing the steep part of the leading edge slope of waveform. Davis

(1997) suggests using a 50% threshold level for waveforms dominated by surface scattering and a 10–20% threshold for volume-scattering signals.

The steps for the method are:

- Calculate the thermal noise:

$$P_N = \frac{1}{5}\sum_i^5 P_i \qquad (4.5)$$

- Compute the threshold level:

$$T_h = P_N + q \cdot (A - P_N) \qquad (4.6)$$

- The retracked range on the leading edge of the waveform is computed by linearly interpolating between the bins adjacent to T_h using

$$G_r = G_{k-1} + \frac{T_h - P_{k-1}}{P_k - P_{k-1}} \qquad (4.7)$$

Where A is determined by (4.1), P_N is the averaged value of the power in the first five gates, q is the threshold value (e.g., 50%), G_k is the power at the kth gate, where k is the location of the first gate exceeding T_h.

4.4.3
Improved Threshold Retracker

In the case of complex waveforms, these simple retrackers cannot properly determine the surface ranging gate. Improved threshold retrackers have been developed and we distinguish here between methods using external data to select the best ranging gate (Hwang et al. 2006) and methods that find the leading sub-waveform directly without additional data (Lee et al. 2008; Bao et al. 2009). Improved threshold retrackers were applied by Hwang et al. (2006) to Geosat/Geodetic Mission data along the coasts of Taiwan, by Fenoglio-Marc et al. (2009) to Envisat RA-2 data along the Mediterranean coasts (see also the chapter by Bouffard et al., this volume), and by Lee et al. (2008) to TOPEX data over land surfaces.

The method used in Hwang et al. (2006) consists in first identifying sub-waveforms within the measured waveform. The leading edge and the corresponding retracked range and sea surface height (SSH) are computed for each subwaveforms. The "best" retracked range is then selected based on comparison with external data. Presently, the "best" retracked range is that which yields the smallest difference from the previous retracked SSH. The method is applied to data starting over the open ocean and proceeding towards the land, as the SSH can be determined more accurately in the open ocean than in coastal waters.

Fig. 4.8 shows a flowchart of the improved threshold retracker. First, one computes the mean difference between the powers at every other gate, which is computed as:

$$d_2^i = \frac{1}{2}(P_{i+2} - P_i) \qquad (4.8)$$

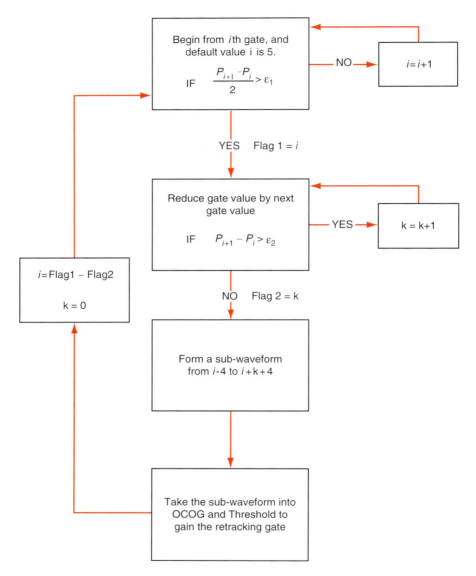

Fig. 4.8 Flow chart of the improved threshold retracker

where P_i is the returned power for the ith gate. If half the difference is greater than a given value ε_1, it means the power shows a peak and a sub-waveform is detected. The difference between two successive powers is then computed, and if this difference is greater than a given value ε_2, the corresponding gate is included in the actual sub-waveform. A given sub-waveform terminates when the difference is smaller than ε_2, where the surface tracking gates of the sub-waveforms are determined using Eq. 4.1 to determine the amplitude A and by applying the threshold retracker Eqs. 4.5 to 4.7.

4 Retracking Altimeter Waveforms Near the Coasts

Hwang et al. (2006) use $\varepsilon_1 = 8$ and $\varepsilon_2 = 2$ and the default value for the ith gate is 5. Fenoglio-Marc et al. (2009) use $\varepsilon_1 = 0.2S$ and $\varepsilon_2 = 0.2S_1$, where S is the standard deviation of all power differences for a gate separation of 1, given by:

$$S = \sqrt{\frac{(N-2)\sum_{i=1}^{N-2}(d_2^i)^2 - \left(\sum_{i=1}^{N-2}d_2^i\right)^2}{(N-2)(N-3)}} \qquad (4.9)$$

and S_1 is the standard deviation of all power differences at neighbouring gates, given by:

$$S_1 = \sqrt{\frac{(N-1)\sum_{i=1}^{N-1}(d_1^i)^2 - \left(\sum_{i=1}^{N-1}d_1^i\right)^2}{(N-1)(N-2)}} \qquad (4.10)$$

where N is the number of gates in the waveform. The power difference between neighbouring gates is defined as:

$$d_1^k = P_{k+1} - P_k; \qquad (4.11)$$

With this method, several leading edges can be detected in one waveform. However, only one of those will be the correct retracked range, and one must therefore compare with a reference height in order to determine which the correct retracked range is. Lee et al. (2008) and Bao et al. (2009) use slightly different methods to find the leading sub-waveform. Both methods use two types of differences between waveforms samples, the so-called Differences I and Differences II. The first are power differences between two successive gates, p(i+1) – p(i), the second are power differences between a gate and the gate after next, p(i+2) – p(i).

Bao et al. (2009) find the start and end gate of the leading edge, G_{start} and G_{end}, and use the OCOG method between G_{start} and G_{end+4} to compute the corresponding amplitude (Eq. 4.1). Lee et al. (2009) estimate first the noise level and the maximum value of the leading edge to avoid the bump before the leading edge or spike after the leading edge, then estimate the retracking gate as in the original threshold retracker method using the chosen threshold level.

4.4.4
The β-parameter Retracker

The β-parameter retracker was the first algorithm developed to retrieve ranges from the Seasat radar altimeter over continental ice sheets (Martin et al. 1983). This algorithm comes as a 5- and a 9-parameters functional form to fit single or double-ramped waveforms, respectively. The Ice Altimetry Group of the NASA's Goddard Space Flight Center (GSFC) has developed algorithms based on these functions, which are referred to in the literature as the NASA GSFC V4 retracker (NASA GSFC 2006).

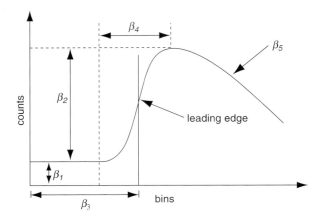

Fig. 4.9 Schematic diagram of 5-parameter β-retracker waveform

The 5-parameter retracker is used to fit single-ramp return waveforms as shown in Fig. 4.9. The general expression for the 5-parameter functional form of the returned power y(t) is (Zwally and Brenner 2001):

$$y(t) = \beta_1 + \beta_2(1+\beta_5 Q)P\left(\frac{t-\beta_3}{\beta_4}\right) \quad (4.12)$$

$$Q = \begin{cases} 0 & \text{for } t < \beta_3 + 0.5\beta_4 \\ t - (\beta_3 + 0.5\beta_4) & \text{for } t \geq \beta_3 + 0.5\beta_4 \end{cases} \quad (4.13)$$

$$P(x) = \int_{-\infty}^{x} \frac{1}{\sqrt{2\pi}} \exp\left(\frac{-q^2}{2}\right) dq \quad (4.14)$$

The unknown parameters β relate to the shape of the waveform as shown in Fig. 4.9 and as follows: β_1 is the thermal noise level, β_2 the return signal amplitude, β_3 the mid-point on the leading edge, β_4 the waveform rise-time, and β_5 the slope of the trailing edge.

The linear trailing edge can be replaced by an exponential decay term, as in the NASA GSFC V4 retracker. A slightly different expression of the 5-parameter β-retracker with an exponential trailing edge is given in Deng and Featherstone (2006) and reads:

$$y(t) = \beta_1 + \beta_2 \exp(-\beta_5 Q)P\left(\frac{t-\beta_3}{\beta_4}\right) \quad (4.15)$$

with

$$Q = \begin{cases} 0 & \text{for } t < \beta_3 - 2\beta_4 \\ t - (\beta_3 + 0.5\beta_4) & \text{for } t \geq \beta_3 - 2\beta_4 \end{cases} \quad (4.16)$$

The 9-parameter form of the β-retracker with exponential trailing edge reads:

$$y(t) = \beta_1 + \sum_{i=1}^{2} \beta_{2i} \exp(-\beta_{5i} Q_i) P\left(\frac{t - \beta_{3i}}{\beta_{4i}}\right) \quad (4.17)$$

where

$$Q_i = \begin{cases} 0 & \text{for } t < \beta_{3i} - 2\beta_{4i} \\ t - (\beta_{3i} + 0.5\beta_{4i}) & \text{for } t \geq \beta_{3i} - 2\beta_{4i} \end{cases} \quad (4.18)$$

The waveforms created with the empirical β-parameter function show marked similarities with the waveforms produced with the theoretical Brown ocean waveform model (see Sect. 4.5.1). However, the empirical β parameters remain simple fitted parameters, not related to physical properties. Deng and Featherstone (2006) shows that while β_5 controls the slope of the trailing edge in the β-parameter model, in the Brown (1977) model, the slope of the trailing edge can be exploited to extract information on platform mispointing (see Sect. 4.5.1). The slope of the trailing edge of the β-parameter model can be larger than that of the Brown (1977) model, making the β-parameter model able to fit more complex waveforms over coastal oceans.

4.5
Physically-based Retrackers

P. Thibaut, X. Deng, J. Gómez-Enri

In this section, we present waveform retrackers derived from theoretical knowledge of microwave scattering at nadir. This includes the well-known Brown ocean waveform model, widely used for the operational processing of altimeter waveforms over the open ocean. This section also includes, for completeness, some other variants of the Brown-type waveform retracker that are sometimes seen in the literature.

4.5.1
The Brown-Hayne Theoretical Ocean Model

Starting from microwave scattering theory, Moore and Williams (1957), Barrick (1972) and Barrick and Lipa (1985) demonstrated that for a rough scattering surface, the average return power as a function of time delay (t) could be expressed as a convolution of three terms:

$$W(t) = FSSR(t) * PTR(t) * PDF(t) \quad (4.19)$$

where *FSSR* is the flat sea surface response, *PTR* is the radar point target response, and *PDF* is the ocean surface elevation probability density function of specular points. The

PTR function is a (sinx/x)² function which is usually approximated by a Gaussian function in order to perform the convolution of the three terms. Hence:

$$PTR(t) \approx exp\left(\frac{-t^2}{2\sigma_p^2}\right) \quad (4.20)$$

where σ_p is the width of the radar point target response function. Barrick (1972) and Brown (1977) used:

$$\sigma_p = \frac{1}{2\sqrt{2ln2}} r_t \approx 0.425 . r_t \quad (4.21)$$

with r_t the time resolution. MacArthur (1978) suggested that the width of the radar point target response function is better approximated by $\sigma_p = 0.513 r_t$. This assumption is the one used by Hayne (1980) in its refined analytical ocean return model derived from the Brown model. This approximation is also the one used in the Jason ground processing, while the Envisat RA-2 retracking uses $\sigma_p = 0.53\ r_t$.

The formulation of the theoretical shape of an echo over an ocean surface was given by Brown (1977) and refined by Hayne (1980) to include a fixed skewness parameter, λ_s. This is the model implemented for example as the Ocean-1 retracker on Envisat RA-2. The analytical formula reads:

$$V_m(t) = T_n + a_\xi \frac{P_u}{2} \exp(-v)$$

$$\left\{[1 + erf(u)] + \frac{\lambda_s}{6}\left(\frac{\sigma_s}{\sigma_c}\right)^2 \left\{[1 + erf(u)]c_\xi^3 \sigma_c^3 - \frac{\sqrt{2}}{\sqrt{\pi}}[2u^2 + 3\sqrt{2}c_\xi \sigma_c u + 3c_\xi^2 \sigma_c^2 - 1]\exp(-u^2)\right\}\right\}$$

$$(4.22)$$

where

$$erf(x) = \left(\frac{2}{\sqrt{\pi}}\right) . \int_0^x e^{-t^2} dt \qquad \gamma = \frac{\sin^2(\theta_0)}{2\ln(2)}$$

$$a_\xi = \exp\left(\frac{-4\sin^2 \xi}{\gamma}\right) \qquad b_\xi = \cos(2\xi) - \frac{\sin^2(2\xi)}{\gamma}$$

$$c_\xi = b_\xi a \qquad\qquad a = \frac{4c}{\gamma h\left(1 + \frac{h}{R_e}\right)} \quad (4.23)$$

$$u = \frac{t - \tau - c_\xi \sigma_c^2}{\sqrt{2}\sigma_c} \qquad v = c_\xi\left(t - \tau - \frac{c_\xi \sigma_c^2}{2}\right)$$

$$\sigma_c^2 = \sigma_p^2 + \sigma_s^2$$

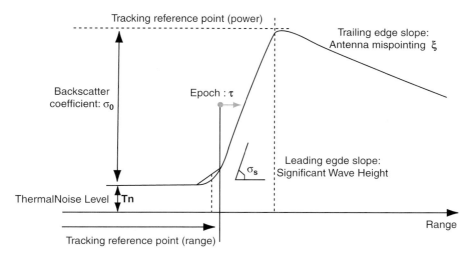

Fig. 4.10 Theoretical Brown ocean waveform shape and corresponding retrieved ocean parameters

with c the velocity of light, h the satellite altitude, R_e the Earth radius, SWH the significant wave height ($=2c\sigma_s$), ξ the off-nadir mispointing angle, and θ_0 the antenna beam width.

Fig. 4.10 shows a representation of a typical ocean waveform with the corresponding parameters that can be retrieved. The waveform is characterised by:

- τ the epoch or time delay i.e. the position of the waveform in the analysis window, with respect to the nominal tracking reference point.
- σ_s the slope of the leading edge which relates to the significant wave height (SWH).
- P_u the amplitude of the signal, which relates to the backscatter coefficient (σ_0).
- T_n the thermal noise level, usually estimated from an arithmetic average of waveform samples from the noise floor plateau before the leading edge.
- Other parameters, such as the altimeter mispointing (ξ) and skewness (λ_s).

When the *PDF* of the ocean surface elevation probability density function of specular points is assumed Gaussian (see Eq. 4.19), the skewness term λ_s equals zero and (Eq. 4.22) reduces accordingly to the Brown model.

4.5.2
The NOCS Non-linear Ocean Retracker

A slightly different theoretical formulation for the shape of ocean waveforms was proposed by Challenor and Srokosz (1989) to account explicitly for the non-linearity of ocean waves (the fact that real waves have flatter troughs and peakier crests). In this case, the probability density function of the height of specular points on the sea surface PDF(t) is given by Srokosz (1986) for weakly non-linear ocean waves as:

$$PDF(t) = \frac{1}{\sqrt{2\pi}\sigma_s} \exp\left\{\frac{-c(t-\tau)^2}{2\sigma_s^2}\right\} \left\{1 + \frac{1}{6}\lambda_s H_3\left(\frac{c(t-\tau)}{\sigma_s}\right) - \frac{1}{2}\delta H_1\left(\frac{c(t-\tau)}{\sigma_s}\right)\right\} \quad (4.24)$$

where σ_s is the standard deviation of sea surface elevation; λ_s the ocean skewness (defined as λ_{300} in Srokosz 1986), H_3 is the Hermite polynomial of order 3, δ is the cross-skewness (defined as γ in Srokosz 1986) and H_1 is the Hermite polynomial of order 1.

The skewness is a measure of the non-linearity of the ocean waves and is related to the third moment of the ocean wave spectrum. The cross-skewness is defined as the normalised expectation of the elevation and slope squared and has no clear physical meaning. This term is usually set to zero as Tokmakian et al. (1994) previously showed that it is not possible to estimate both cross-skewness and epoch (range) accurately due to their high correlation. Setting the cross-skewness term δ to zero, the final expression of the returned power is (Gómez-Enri et al. 2006):

$$P_r(t) = \frac{const1 * \sigma_0}{h^3}\exp\left\{-\frac{4}{\gamma}\sin^2\xi\right\} \left\{erfc\left(\frac{-(t-\tau)}{\sqrt{2}\sigma_c}\right) - \frac{1}{\sqrt{2\pi}}\exp\left(\frac{-(t-\tau)^2}{2\sigma_c^2}\right)\frac{\lambda_s SWH^3}{24c^3\sigma_c^3}H_2\left(\frac{(t-\tau)}{\sigma_c}\right)\right\}$$

$$\times \begin{cases} 1 & \text{for } t < \tau \\ \exp\left\{-\frac{4c}{\gamma h}(t-\tau)\cos 2\xi\right\} I_0\left(\frac{4}{\gamma}\sqrt{\frac{c(t-\tau)}{h}}\sin 2\xi\right) & \text{for } t \geq \tau \end{cases} \quad (4.25)$$

where

$$const1 = \frac{G_0^2 \lambda_R^2 c \eta P_T \sigma_P}{4(4\pi)^2 L_p}\sqrt{\frac{\pi}{2}} \qquad \gamma = \frac{\sin^2\Psi_B}{2\ln 2} \qquad \sigma_c = \sqrt{\sigma_p^2 \frac{4\sigma_s^2}{c^2}} \quad (4.26)$$

with Ψ_B the half power (3dB) antenna beam-width, G_0 the antenna gain parameter, λ_R the radar wavelength, c the velocity of light in vacuo, η the pulse compression ratio, P_T the transmitted power, σ_P is related to the pulse width and L_p the two-way propagation loss over and above the free-space loss. erfc(x) represents the complementary error function and $I_0(t)$ the modified Bessel function of the first kind (Abramowitz and Stegun 1968), and h the satellite height (not including here the Earth radius term suggested by Rodriguez 1988). The epoch, τ, equal to $2h/c$, represents the position in time of the ocean surface, σ_0 is the radar backscatter cross-section, ξ the mispointing angle (measured from nadir). If the probability density function of reflected facets is considered Gaussian, the skewness coefficient λ_s vanishes and we find again the model radar returned power described by Brown (1977).

4.5.3
Practical Implementations of the Brown-Hayne Ocean Retracker

Despite the small number of theoretical models available to describe ocean waveforms, there are many flavours of ground-based retracking, which differ in very subtle (and usually poorly documented) ways. The most obvious difference is the number of parameters retrieved, with some schemes retrieving thermal noise directly from the fitting, while others estimate the noise power independently by averaging the power in the thermal noise plateau.

Another example is how one deals with altimeter mispointing, a parameter linked to the slope of trailing edge with strong impact on retrieved amplitude. Studies have shown that, in the Brown model, mispointing is strongly correlated with σ^0 and that it is not possible to retrieve both mispointing and σ^0 simultaneously (Tokmakian et al. 1994). One solution, as adopted on Envisat, is to retrieve the mispointing separately first (using the RA-2 Ice-2 retracker) then use the retrieved value as input to the (3-parameters) Hayne model (retrieving epoch, SWH and σ^0). For Jason-1 however, where large platform mispointing values were occasionally experienced, the basic geometrical simplifications behind the Brown/Hayne model are no longer tenable and a second order four-parameter model had to be developed to fit Jason-1 waveforms and allow all four parameters (epoch, SWH, $\sigma 0$ and mispointing) to be retrieved simultaneously (Amarouche et al. 2004). However, application of this processing to Jason-2 yields highly correlated errors between σ^0 and mispointing (Quartly 2009a), prompting suggestions that the 3-parameter fit is more appropriate. The second-order four-parameter model must be retained for Jason-1, but empirical corrections are then required to improve the accuracy of σ^0 estimates and their consistency with Jason-2 (Quartly 2009b).

In the current Jason-1 and Jason-2 ground-based retracking algorithm, the Gaussian expression of the point target response (PTR, see Eq. 4.20) is retained and the value of σ_p is then kept constant (0.513 r_t). The biases resulting from such an approximation are corrected with look-up correction tables (Thibaut et al. 2004). These look-up correction tables are built using a Jason-1 altimeter simulator that: (i) accounts for the true (measured) PTR to simulate the altimeter waveforms (via convolution), (ii) uses the Gaussian approximation for the PTR, to derive the average return power given by the convolution formulation in the retracking algorithm. Thus, tables integrate all the differences between the Gaussian approximation and the real shape of the PTR.

4.6
Statistical Methods of Function Fitting

The aim of this section is to provide a short overview of some of the most common statistical methods used to fit a model to measured waveforms. Since these methods are seldom documented in full, it seems worthwhile to pull this information together in one place. It has been shown that the choice of fitting method can have a major impact on the retrieval of altimetric measurements, in particular range.

4.6.1
Maximum Likelihood Estimator (MLE)

The maximum Likelihood estimator is one of a number of statistical methods used to fit the theoretical model return power (u_i) to the measured return power ($\hat{u}_i$). Assuming gate-to-gate and pulse-to-pulse independence, one can make appropriate assumptions about the (multiplicative, negative exponential) noise on the measurements and it becomes possible to express the measure of misfit, χ^2, as (Challenor and Srokosz 1989; Tokmakian et al. 1994):

$$\chi^2 = \sum_{i=1}^{n} \left(-N \frac{\hat{u}_i}{u_i} - N \ln u_i \right)^2 \qquad (4.27)$$

where N is the number of individual echoes averaged to form the measured waveform (in the case of Envisat RA-2, N = 100 for 18 Hz waveforms) and n is the number of gates (n is typically around 100 out of 128 available gates, to avoid parts of the waveforms affected by aliasing). The objective is to optimise the value of the estimated parameters, θ_j, in order to minimise χ^2. The partial derivatives of χ^2 against θ_j read:

$$\frac{\partial \chi^2}{\partial \theta_j} = N \sum_{i=1}^{n} \frac{\partial u_i(\theta_j)}{\partial \theta_j} \left(\frac{\hat{u}_i}{u_i^2} - \frac{1}{u_i} \right) \qquad (4.28)$$

where θ_j represent the estimated parameters e.g., epoch, significant wave height,... The MLE estimation gives, asymptotically, the minimum variance unbiased estimators of the parameters. The main advantage of MLE, compared to ordinary or weighted least-squares (see below), is that it is possible to derive the variance-covariance matrix to give a measure of the error on the estimates of the retrieved parameters (Tokmakian et al. 1994). The variance-covariance matrix, V, is given as the inverse of the Fisher information matrix, F, (Challenor and Srokosz 1989):

$$V = F^{-1} \qquad (4.29)$$

with

$$F = \left\{ N \sum_{i=1}^{n} \frac{1}{u_i^2} \frac{\partial u_i}{\partial \theta_j} \frac{\partial u_i}{\partial \theta_k} \right\} \qquad (4.30)$$

where $j, k = 1,...npar$, with *npar* the number of parameters to be retrieved (typically *npar* = 3, 4 or 5).

4.6.2
Weighted Least Square Estimator (WLS)

In the weighted least square approach, the χ^2 measure of misfit is simply defined as:

$$\chi^2 = \sum_{i=1}^{n} \left(\frac{\hat{u}_i - u_i}{W_i} \right)^2 \qquad (4.31)$$

where W_i represents the weight given to each gate in the waveform. Typically, the weights are equal to the measured power, $\hat{u}_i$. When the weights are all set to 1, we get the ordinary (unweighted) least square solution. It can be shown that the choice of weights gives more or less importance to gates with larger noise and that this has a significant impact on the retrieval of the altimetric range (Maus et al. 1998; Sandwell and Smith 2005).

4.6.3
Unweighted Least Square Estimator

The estimation of the optimal parameters for Envisat RA-2, Jason-1 and Jason-2 is performed by making the measured waveform coincide with a return power model according to an unweighted Least Square Estimator derived from a Maximum Likelihood Estimator (Dumont 1985; Rodriguez 1988; Callahan and Rodriguez 2004; Deng and Featherstone 2005, 2006).

Assuming decorrelation of the sample noise in each waveform and decorrelation from one individual echo to the other, an iterative solution is obtained by developing the total cost function in a Taylor series at the first order from an initial set of estimates. This estimator is also called the MMSE (Minimum Mean Square Estimator) and the estimates are given by the following equation (Zanifé et al. 2003):

$$\theta_{m,n} = \theta_{m,n-1} - g(BB^T)^{-1}_{\theta_{m,n-1}} (BD)_{\theta_{m,n-1}} \qquad (4.32)$$

where $\theta_{m,n}$ is the estimated parameter at iteration n, m is the index of estimated parameters and g is the loop gain (varying between 0 and 1). The matrix of weighted partial derivatives (B) and the vector of weighted differences (D) are given by:

$$B_{m,k} = \frac{1}{P_u} \frac{\partial u_k}{\partial \theta_m} \qquad (4.33)$$

$$D_{k,1} = \frac{u_k - \hat{u}_k}{P_u} \qquad (4.34)$$

with k the gate index, u_k the model waveform sample at gate k, $\hat{u}_k$ the measured waveform sample at gate k, and P_u the estimated power. Here, the original weight $1/u_k$ of the Weighted Least Square Estimator method has been replaced by $1/P_u$ for the three following reasons: (i) to avoid giving too much weight to the noise region, which contains no signal; (ii) to ensure that all samples have the same weight; (iii) to normalize the residual, thus providing more robust estimates.

4.7
Applications of Waveform Retracking to Coastal Waveforms

In this section, we now present some recent examples where waveform retracking methods were applied to coastal waveforms from various altimeters. Within the past few years, much effort has been devoted to understanding how to improve the acquisition and retrieval of altimeter measurements closer to the coasts and over inland waters. Even though different physical processes affect altimeter signals in coastal areas and over inland waters, in both cases, signals from land severely contaminate the Brown-like waveforms and impact the ability to retrieve accurate altimeter data over water close to land. The exploitation of altimetry measurements over these regions is now a well-identified goal for present and future altimetry missions (pulse-limited and others).

The common objective of all these investigations is to provide valid altimetric measurement as close as possible to the coast, aiming for altimetric retrieval accuracy comparable to what is achieved in the open-ocean.

4.7.1
The CNES/PISTACH Coastal Processing System for Jason-2

The PISTACH coastal processing system was developed by Collecte Localisation Satellites (CLS) under funding from CNES, and is dedicated mainly to reprocessing Jason-2 data over coastal areas and inland water, and to the production of new coastal and inland Jason-2 Level-2 products.

In the frame of the CNES/PISTACH project (Mercier 2008) and accounting for the special shapes of the waveforms near the coasts, a modified solution of the conventional Jason-2 retracking scheme was developed and successfully tested. The main idea of the scheme was to restrict the analysis window to retrack only those gates in the waveform not corrupted by land effects. Many retracking window widths and positions were tested (adaptative or not). The retained solution consisted in fixing the width of the window to a constant value, in order that the results could be compared on the same basis, and to enable the relative biases and trends between the various retracking algorithms to be computed. Indeed, establishing the consistency between different algorithms is essential. It is crucial to characterise biases between different fitting algorithms (particularly as a function of the significant wave height) and to apply corrections if needed. The overall objective is to minimise discontinuities between "open-ocean" and "coastal-ocean" solutions.

While the Jason-2 deep-ocean retracker is based on a 4-parameters Brown/Hayne model (range, amplitude, significant wave height and mispointing angle), the Jason-2 coastal retracking only solves for the first three parameters. The reason for this is that the reduced window analysis does not allow the retrieval of mispointing angle, which is retrieved from the slope of the trailing edge of the waveform. Fig. 4.11 shows the output from the open-ocean and the coastal retracking schemes for Jason-2 waveforms over Ibiza Island in the Mediterranean Sea (see also Fig. 4.3). For the epoch and the backscatter coefficient σ^0, the PISTACH coastal scheme is able to return results nearer to the coast, where the Geophysical

4 Retracking Altimeter Waveforms Near the Coasts

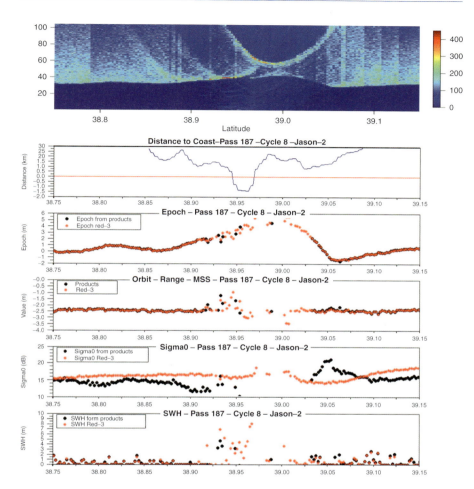

Fig. 4.11 Output from the CNES/PISTACH coastal retracker system for Jason-2 data in cycle 10, pass 187 (ascending) when the satellite flies over Ibiza Island. Showing from top to bottom: Ku band waveforms, distance to the nearest coast, retrieved epoch, uncorrected sea surface height (no geophysical corrections applied), backscatter coefficient and significant wave height. Red diamonds: 20Hz output of the PISTACH coastal retracking scheme. Black dots: 20 Hz Jason-2 Geophysical Data Record data

Data Record (GDR) products contain no data (Thibaut and Poisson 2008). It is also interesting to note that the new σ^0 coefficient is much more stable than the one in the GDR (not only in the coastal zone but in the entire data segment). The reason for this is that data bins in the trailing edge of the waveforms, often contaminated by land returns (as seen on the top subplot of Fig. 4.11), are no longer taken into account in the retracking and no longer impact the estimation of other parameters. The new retracking scheme also returns more values of the significant wave height, especially between latitude 39 and 39.04 where the leading edge of the waveforms is not impacted by land effects, as opposed to waveforms between 38.92° and 38.95°. In this latitude band, which corresponds to the end of the Ibiza overflight, all the altimetric estimates from the PISTACH coastal scheme are improved compared to the GDR product. In total, around 15 additional 20 Hz measurements are retrieved with this new retracking scheme, corresponding to over 5 km of ground track for which data in the GDR products are missing.

4.7.2
The COASTALT Processing System for Envisat RA-2

The COASTALT processing system was developed as a pan-European effort led by the National Oceanography Centre, Southampton, with funding from the European Space Agency. Its aim was to develop a prototype Level 2 software processor for the computation and production of new coastal altimetry products for Envisat RA-2.

The retracking scheme under COASTALT relies on physically-based waveform models and consists of three waveform retrackers run in parallel (Gommenginger et al. 2008). These are:

- The conventional Brown Ocean Retracker (BOR), implemented in the form of the NOCS Ocean retracking scheme in its linear form presented in Sect. 4.5.2, to retrack ocean-like waveforms.
- The Beta-parameter retracker with Exponential trailing edge (SBE), as outlined in Sect. 4.4.4 and Deng and Featherstone (2006). This functional form is well suited to fit high intensity peaky waveforms with a rapidly decaying trailing edge characteristic of specular echoes in the coastal zone. Its 5- or 9-parameter implementation permits echoes from one or two scattering surfaces to be detected and tracked.
- An experimental Mixed Brown-Specular retracker (MBS), consisting of the superposition of a Brown ocean waveform and one or more Beta-retracker specular echo (es).

The performance in the coastal zone of different retrackers can be estimated by a variety of means, including the statistics of the residuals, crossover analyses or comparisons with in-situ measurements. In COASTALT, the goodness of fit of the three retrackers is quantified using the root mean squared error (RMSE) of the residuals between the measured waveforms and the best-fit model waveforms. This permits an objective classification and identification of those waveforms that depart from typical ocean waveform shape. High values of RMSE will thus objectively identify regions where waveforms deviate strongly from Brown-type ocean waveforms.

Fig. 4.12 presents the RMSE for Envisat RA-2 tracks (ascending 00357 and 00128, and descending 00022) in the North-Western Mediterranean retracked with the Brown Ocean Retracker. Ascending track 00357 shows low RMSE over open waters but the effect of Del Giglio Island is quite evident through an increase in the value of RMSE. Large RMSE values are observed also when encountering high-amplitude specular echoes around Follonica Gulf associated with calm inshore waters. Obviously, very large RMSE values are obtained over land although it is encouraging that the NOCS ocean retracker should be sufficiently robust to track even those very non-Brown waveform over land.

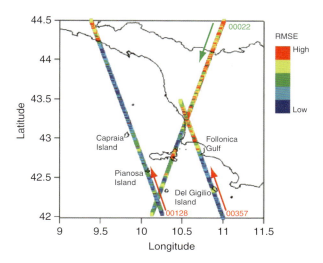

Fig. 4.12 Root-mean-square error of the residuals between the best-fit Brown ocean waveform model and the 20Hz Envisat RA-2 waveforms over the coast of Tuscany in the North-West Mediterranean Sea. Low values of RMSE indicate regions where the fit is good

4.7.3
Coastal Retracking Around Australia

ERS-2 waveforms were retracked by Deng (2004) over the Australian coasts using some of the empirical and theoretical retrackers described in Sect. 4.4 and 4.5. Fig. 4.13 shows six typical coastal waveforms and the functional forms fitted to them. Most of the waveforms are single-peaked ocean-like shapes (Fig. 4.13b and c), which can be fitted using the conventional Brown ocean model without skewness (see Sect. 4.5.1).

Waveforms nearer to shore with higher peak amplitude and double ramps were fitted successfully with the five-parameter β-retracker with an exponential trailing edge (Fig. 4.13a, see Eq. 4.15), and the nine-parameter β-retracker (see Eq. 4.17) with either a linear trailing edge Fig. 4.13d and e) or an exponential trailing edge (Fig. 4.13e).

4.7.3.1
Iteratively Reweighted Least-squares Procedure

Waveforms are intrinsically noisy, due to the "fading" noise that results from the superposition of signals from individual reflecting facets in the altimeter footprint (Quartly et al. 2001). The noise is clearly seen in the trailing edge (e.g., Fig. 4.13b), but is more difficult to identify in the leading edge. The estimation of parameters affected by the noise in the trailing edge (e.g., mispointing) will impact the estimation of parameter in the leading edge (e.g., range) if a simple least squares fitting procedure is applied to an individual waveform. In addition, altimeters will display non-ocean features in the trailing edge as they approach the coast, (see Sect. 4.3.1) which can significantly affect the estimation and make it difficult to determine the accuracy of the results.

To improve the accuracy of parameter estimation, Deng and Featherstone (2006) proposed an iteratively reweighted least-squares procedure based on the robust statistical algorithm of

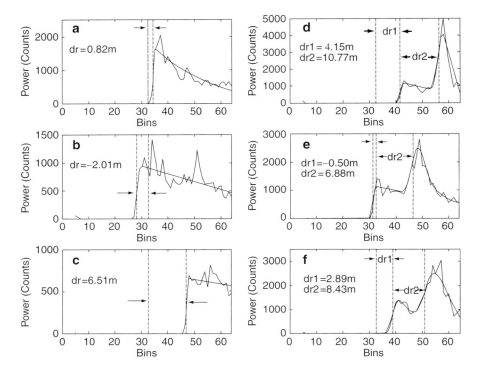

Fig. 4.13 Coastal waveforms for ERS-2 retracked using (**a**) the five-parameter β-retracker with exponential trailing edge, (**b**) and (**c**) the Brown ocean retracker (without skewness), (**d**) and (**f**) the nine-parameter β-retracker with linear trailing edge, and (**e**) the nine-parameter β-retracker with exponential trailing edge. Here 'dr' is the range correction inferred by the offset of the tracking gate (or bin) estimate with respect to the nominal tracking point at gate 32.5. (Deng, 2004)

outlier detection, in order to reduce the effect of both noise and unexpected features in the trailing edge of coastal waveforms. The iteratively reweighted least-squares procedure minimizes the impact of possible errors in the waveform on estimated parameters by down-weighting waveform samples that do not agree with the majority of the data. An initial least squares fitting is conducted using approximate values of the unknown parameters and *a priori* uniform weights, p_i^0, where *i* is the number of waveform samples considered. After iteration, a weight function w_i Comment: re-align "wi" with text is computed as (Li 1988; Deng and Featherstone 2006):

$$w_i = \begin{cases} \dfrac{1}{|v_i|} & for\ |v_i| > k\sigma \\ 1 & for\ |v_i| \le k\sigma \end{cases} \qquad (4.35)$$

where v_i is the residual of waveform samples, σ^2 is the posterior estimation of the unit weight variance, and *k* is the down-weighting factor with $k > 0$. The weights, p_i^{j+1}, for the next iteration are then determined using w_i as follows:

$$p_i^{j+1} = p_i^j w_i \qquad (4.36)$$

where j means the j-th iteration in the procedure. The weights, together with the estimated parameters, are then used in the next iteration of the least-squares fitting to produce new estimates. This way, waveform samples considered to be "outliers" will be down-weighted, while "non-outlier" samples will maintain their full weight during the iterative fitting.

4.7.3.2
Application of Iteratively Reweighted Least-squares Fitting to ERS-2 and Jason-1 Coastal Waveforms

The iteratively reweighted least-squares fitting procedure was applied to ERS-2 and Jason-1 waveforms over the Australian coast (Deng and Featherstone 2005, 2006; Deng et al. 2008). Fig. 4.14 shows profiles of the retracked 20 Hz ERS-2 along-track SSH (track 21085 and 21364 from cycle 42) together with the GDR SSH and the AUSGeoid98 geoid height versus along-track distance from the coastline. As the waveforms are obtained over Australian coastal regions, the regional gravimetric geoid model AUSGeoid98 (Featherstone et al. 2001) was used in preference to the global EGM96 provided in the altimeter GDR products. It is acknowledged that the geoid is the equipotential surface of the Earth's gravity field, and loosely the mean sea level, which is different from the instantaneous SSH. The verification of any improvement in the retrieved SSH will therefore rely on whether or not the method leads to a reduction in the standard deviations.

Fig. 4.14 shows that the typical near-land effects on the un-retracked SSH profiles from the GDR begin around 20 km from the coast for track 21085 (left subplot) and around 17.5 km from the coast for track 21364 (right subplot). As, the satellite approaches the coast, the un-retracked GDR SSH decreases and gradually increases, showing a "V" pattern for both tracks. In both cases, the SSH obtained with the iteratively re-weighted least square fitting differ from the GDR SSH by up to –5 m within 10 km of the coast, and do not display the "V" pattern seen in the GDR SSH. The geoid height also shows a near-linear behaviour over the same distance, and since altimeter SSH should follow the same struc-

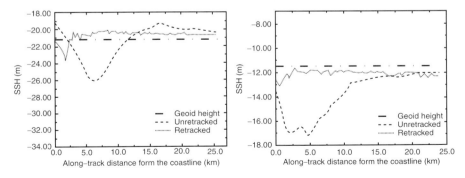

Fig. 4.14 AUSGeoid98 height and SSH for ERS-2 along ground track (*left*) 21085 and (*right*) 21364 (cycle 42) from GDR and retracked waveforms. (Deng and Featherstone 2006)

ture as the geoid, this confirms the impact of waveform land contamination on the unretracked SSH data and the merit of the new fitting procedure over the coastal region.

The reweighted least-squares approach was also applied to Jason-1 Ku-band waveforms (cycle 142) offshore of the Western Australian coast, as shown in Fig. 4.15. The results are compared in the middle subplot in Fig. 4.15 to the Jason-1 MLE-retracked SSH from the GDR products. Using again the geoid as a reference, it showns that the iteratively reweighted least-squares algorithm reduces the standard deviation from 17.8 cm for GDR MLE SSH to 16.2 cm for the reweighted fitting SSH. This is a 9% improvement in the error on SSHs compared to the GDR products.

The main improvement with this method appears to occur near the coast and in high sea state areas. The retrieval accuracy is also assessed by computing the along-track sea surface slope for track a253 (bottom subplot in Fig. 4.15), which crosses a region of generally high SWH, in the Leeuwin Current system. The sea surface slope is computed at 20 Hz without filtering, for both MLE GDR and re-retracked SSH. Slopes based on the MLE and iteratively reweighted SSH have rms deviations of 18.4 and 17.0 mm/km, respectively, corresponding an 8% improvement in range error with the iteratively reweighted least-squares approach compared to the GDR MLE retracker.

Further details about the methods and performance of coastal retracking in the Australian coastal regions can be found in the chapter by Deng et al., this volume.

4.7.4
Application of Empirical Retrackers to Envisat RA-2 in the Mediterranean Sea

Another example of applying waveform retracking in the coastal zone is provided by Fenoglio-Marc et al. (2008), who retracked Envisat RA-2 waveforms over the coasts of the Mediterranean Sea using four empirical retrackers: OCOG, threshold, improved threshold and the 5-parameter β-retracker (see Sect. 4.4). The quality of the retracking was quantified using the standard deviation of the differences between the sea surface heights (SSH) calculated from both raw and retracked ranges and the geoid height, as well as the improvement percentage of the SSH standard deviation compared to standard products (Hwang et al. 2006). The 5-parameter β-retracker and the improved threshold methods turned out to be the most suitable methods, yielding a small improvement with respect to the standard Envisat RA-2 products.

Further details about this work in the Mediterranean Sea, and about similar investigations in the China Seas and its adjacent regions, are presented in the chapters by Bouffard et al. and by Yang et al. in this volume.

4.8
Innovative Retracking Techniques

G. Quartly, P. Challenor, J. Gómez-Enri

The previous sections showed some examples of methods and applications that have recently been employed to exploit waveforms in the coastal zone. The examples provided in the previous section form by no means an exhaustive list, but simply highlight the

4 Retracking Altimeter Waveforms Near the Coasts

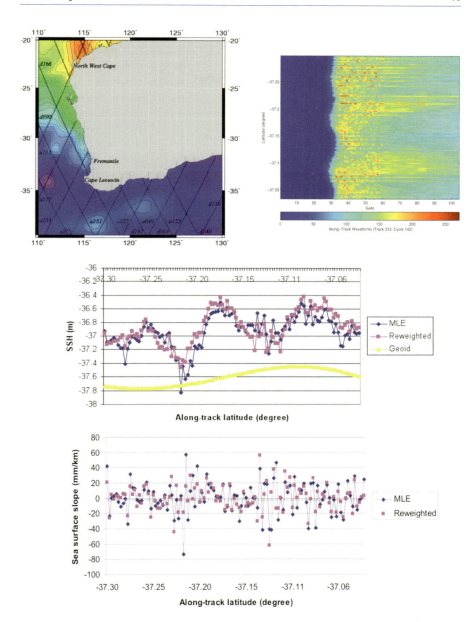

Fig. 4.15 (*top left*) Jason-1 ground tracks and track numbers over the Western Australian coast, where the affix 'a' denotes ascending tracks. The colour map and contours lines show the mean dynamic topography. (*top right*) Waveforms along track a253 in a region of generally high significant wave height where the waveforms LEP depart from the nominal tracking point (gate 32). (*middle*) MLE and retracked along-track SSH with geoid height. (*bottom*) SSH-derived sea surface slope, for the waveforms in top right subplot

intense level of interest in this topic at this time. Following on from this surge in activity, new ideas are continuously emerging on how to improve the retracking of coastal waveforms. This section introduces some of these innovative ideas, commenting on some of the underlying assumptions and the benefit they can bring to coastal retracking.

4.8.1
Iterative Retracking

Most current retracking systems treat each waveform independently. There may be a different degree of averaging but in essence, even though the fitted form may vary, the methods are the same. Yet we know that the geophysical parameters of interest generally vary smoothly across the ocean surface, but this information is usually ignored. In the coastal zone we have the additional complexity introduced by bright targets, such as sandbanks, entering and propagating through the waveforms. Such targets move through successive waveforms as a quadratic and this predictable pattern has been used to model the returns from bright phenomena from transponders to icebergs and rain (Powell et al. 1993; Quartly 1998; Tournadre et al. 2008).

One example where the smooth variation along-track of geophysical parameters is exploited is that of Sandwell and Smith (2005), who propose a two-pass iterative retracking method. In the first pass, each waveform is retracked independently in the conventional manner. A smooth curve is fitted to the derived significant wave heights and, in a second retracking iteration, sea surface height is retrieved from these waveforms assuming that the significant wave height is that taken from the smooth curve. This reportedly greatly reduces the uncertainty on the value of sea surface height, although part of this reduction in uncertainty may be deceptive. While assuming significant wave height varies smoothly (a reasonable assumption) it is also assumed that in its smoothed incarnation, significant wave height is known without error (an unreasonable one). This latter problem could be resolved by using as input to the retracking, significant wave height values sampled from a probabilistic distribution centred on the smoothed wave height value. However this would be computationally expensive. In what follows, we explore novel ideas that go beyond retracking individual waveforms separately, to exploit instead the inter-waveform properties to help improve geophysical retrievals.

4.8.2
Multiple-waveform Retracking

What might a non-independent retracker look like? We can group possible methods into two classes. The first approach would still process waveforms individually and sequentially but would use information from the previously tracked waveforms to help in the process. This is what the α–β tracker on-board the satellite does, but such trackers are known (Walter Smith, 2008, personal communication) to introduce false correlation into their estimates of the parameters. Our aim is to include the "true" geophysical correlation while avoiding spurious induced correlation. One possible method would be to replace the current maximum likelihood estimation (MLE) with a Bayesian method. The likelihood would be the same as in MLE but we would add an informative prior from our previous estimates.

The alternative approach to these sequential retrackers would be simultaneous retrackers. Here, instead of processing independent waveforms sequentially, the waveforms are now processed in batches. Ideally a waveform batch would consist of waveforms from when the altimeter track leaves one landmass to when it hits another. However such large batches would be computationally expensive and it would be difficult, but not impossible, to fit all possible variations, from rain to small islands, to the waveforms automatically. More realistically, smaller batches could consist of segments where the altimeter was affected by a rainstorm or a coastal transition. The advantage of such simultaneous processing is that we could then model the underlying variation of the geophysical phenomena as a smooth process (for example an auto-regressive process). Maus et al. (1998) provide an exemplar of this technique, fitting a smoothly varying sea surface height profile to batches of 408 ERS-1 waveforms. Their model, which assumes other geophysical variables (SWH, wind speed) remain constant and that SSH undulations can be mapped by a polynomial with 41 unknowns, improves the spatial resolution of the ERS-1-derived geoid from 41 to 31 km.

Similarly, bright targets, once identified, could be fitted in those batches at the same time. Fig. 4.16 shows an example of simulated Envisat RA-2 waveforms in the vicinity of Pianosa Island in the Mediterranean Sea (where intermittent bright targets are known to occur – see Gómez-Enri et al. 2010) for two different idealised surface scenarios. In the two scenarios, the Earth's surface is represented through a Digital Elevation Map of Pianosa Island and either a constant or a smoothly varying roughness field (represented in top right subplot by changes in the backscatter coefficient, σ^0, in the lee of the island). The real waveforms observed at this same location with Envisat RA-2 are shown in Fig. 4.17. The resemblance between the simulated waveforms in Fig. 4.16 and the measured ones in Fig. 4.17 is striking and gives some confidence that unravelling land effects from smooth roughness changes in coastal areas should become possible in future.

4.9
Conclusions

In this chapter, we presented observational evidence of the huge diversity and complexity of altimeter waveform shapes observed by today's satellite altimeters in coastal regions. Altimeter waveforms comply with open-ocean type echo shape typically up to 10–15 km from land, beyond which peakier echoes and waveforms with multiple specular reflectors become predominant. These complex composite waveforms pose a real challenge to today's conventional approach to waveform retracking. We reviewed the various retracking methods currently available, covering both empirical and theoretical approaches, developed and refined over many years. We highlighted some of the important but all-too-often ignored subtleties of ground-based retracking, such as the role in range retrieval accuracy of the choice of statistical fitting, or the impact of artefacts in key parts of the waveforms. Changes

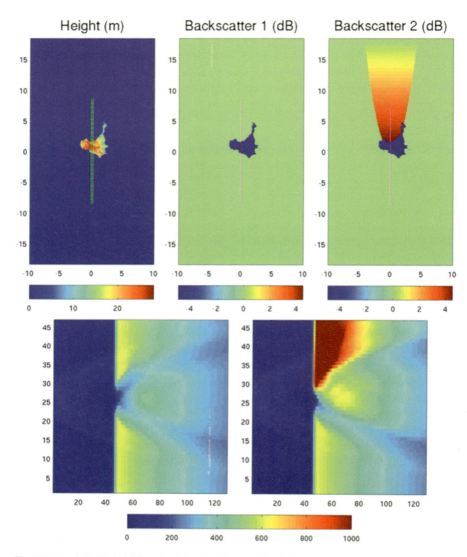

Fig. 4.16 (*top left*) Digital Elevation Map for Pianosa Island, in the Mediterranean Sea, with location of 46 simulated waveforms; (*top middle*) and (*top right*) Ocean roughness scenarios represented by two different backscatter (σ^0) fields; (*bottom left*) and (*bottom right*) Simulated waveforms for Envisat RA-2 near Pianosa Island for two ocean roughness scenarios shown above. The significant wave height is set to 1.4 m throughout the scene

in the operation and behaviour of on-board trackers, such as those present on Jason-1 and Jason-2, are also likely to affect retrieval accuracy and will need to be examined with care. All these issues are now re-emerging and coming to the fore when dealing with complex, rapidly varying waveforms in coastal areas. Examples were presented of applications of

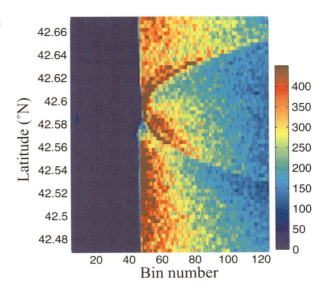

Fig. 4.17 Envisat RA-2 18 Hz waveforms for ascending track 000128 in the vicinity of Pianosa Island in the Mediterranean Sea, for the waveform locations shown in previous figure

waveform retracking methods to coastal waveforms using a number of coastal retracking schemes across the world. However, in order to bridge the final few kilometres to the coast – the area most relevant to human activities – it is likely that altogether different approaches will need to be conceived. A wealth of innovative ideas is continuously emerging, including iterative sequential retracking or simultaneous batch processing, to offer new ways forward towards achieving high accuracy coastal altimetry.

References

Abramowitz M, Stegun IA (1968) Handbook of mathematical functions. Dover Publications, Chap 9, 374pp.

Amarouche L, Thibaut P, Zanife OZ, Dumont J-P, Vincent P, Steunou N (2004) Improving the Jason-1 ground retracking to better account for attitude effects. Marine Geodesy 27:171–197

Bao L, Lu Y, Wang Y (2009) Improved retracking algorithm for oceanic altimeter waveforms. Prog Nat Sci. doi:10.1016/j.pnsc.2008.06.017

Barrick DE (1972): Remote sensing of the sea state by radar. In: Derr VE (ed) Remote sensing of the troposphere, Chap 12. US Govt. Printing Office, Washington, DC

Barrick DE, Lipa BJ (1985) Analysis and interpretation of altimeter sea echo. Adv Geophys 27:61–100

Brown GS (1977) The average impulse response of a rough surface and its applications. IEEE Trans Antennas Propag AP-25(1):67–74

Callahan PS, Rodriguez E (2004) Retracking Jason-1 Data. Mar Geod 27:391–407

Challenor PG, Srokosz MA (1989) The extraction of geophysical parameters from radar altimeter return from a non-linear sea surface. In: mathematics in remote sensing. Clarendon Press, Oxford, pp 257–268

Chelton DB, Walsh E, MacArthur J (1989) Pulse compression and sea level tracking in satellite altimetry. J Atmos Ocean Technol 6:407–438

Chelton DB, Ries JC, Haines BJ, Fu L-L, Callahan PS (2001) Satellite altimetry. In: Fu L-L, Cazenave A (eds) Satellite altimetry and Earth sciences: a handbook of techniques and applications. Academic, San Diego, USA, pp 1–132

Davis CH (1995) Growth of the Greenland ice sheet: a performance assessment of altimeter retracking algorithms. IEEE Trans Geosci Remote Sens 33(5):1108–1116

Davis CH (1997) A robust threshold retracking algorithm for measuring ice-sheet surface elevation change from satellite radar altimeter. IEEE Trans Geosci Remote Sensing 35(4): 974–979

Deng X (2004) Improvement of geodetic parameter estimation in coastal regions from satellite radar altimetry. PhD thesis, Curtin University of Technology, Perth, Australia

Deng X, Featherstone WE (2005) Improved determination of sea surface heights close to the Australian Coast from ERS-2 satellite radar altimetry. In: Sansò F (ed) A window on the future of geodesy. Springer, Berlin, pp 314–319

Deng X, Featherstone WE (2006) A coastal retracking system for satellite radar altimeter waveforms: application to ERS-2 around Australia. J Geophys Res 111:C06012. doi:10.1029/2005JC003039

Deng, X, Lee HK, Shum CK, Roesler C, Emery W (2008) Retracking of radar altimetry for coastal applications. http://www.coastalt.eu/pisaworkshop08/pres/. Accessed 14 January 2009

Dumont JP (1985) Estimation Optimale des Paramètres Altimétriques des Signaux Radar Poseidon, PhD thesis from Institut National Polytechnique de Toulouse, Toulouse, France

Featherstone, WE, Kirby JF, Kearsley AHW, Gilliland JR, Johnston GM, Steed J, Forsberg R, Sideris MG (2001) The AUSGeoid98 geoid model of Australia: data treatment, computations and comparisons with GPS-levelling data. J Geod 75:13–330

Fenoglio-Marc L (2008) Retracking in the North-Western Mediterranean Sea, Oral presentation during 2nd coastal altimetry workshop, Pisa, Italy, Nov 2008. Available from http://www.coastalt.eu/pisaworkshop08/pres/.

Fenoglio-Marc L, Fehlau M, Ferri L, Becker M, Gao Y, Vignudelli S (2009) Coastal sea surface heights from improved altimeter data in the Mediterranean Sea, Proceedings GGEO2008. Springer Verlag, IAG Symposia

Gómez-Enri J, Gommenginger C, Srokosz M, Challenor PG, Drinkwater M (2006) Envisat radar altimeter tracker bias. Mar Geod 29:19–38

Gómez-Enri J, Vignudelli S, Quartly GD, Gommenginger CP, Cipollini PG, Challenor PG and Benveniste J (2010) Modeling Envisat RA-2 waveforms in the coastal zone: case-study of calm water contamination, IEEE Geosc. Remote Sens Lett 7(3): 474–478. doi:10.1109/LGRS.2009.2039193

Gommenginger CP, Gleason S, Snaith H, Challenor P, Quartly G, Gomez-Enri J, COASTALT processor and products, Oral presentation during 2nd coastal altimetry workshop, Pisa, Italy, Nov 2008. Available from http://www.coastalt.eu/pisaworkshop08/pres/

Hayne GS (1980) Radar altimeter mean return waveform from near-normal-incidence ocean surface scattering, IEEE Trans Antennas Propag AP-28 (5):687–692

Hwang C, Guo JY, Deng XL, Hsu HY, Liu YT (2006) Coastal gravity anomalies from retracked Geosat/GM altimetry: improvement, limitation and the role of airborne gravity data. J Geod 80:204–216

Lee Hyongki CK, Shum Y Yi, Brau A, Kuo C-Y (2008) Laurentia crustal motion observed using TOPEX/Poseidon radar altimetry over land. J Geodyn 46:182–193

Li, D. R. (1988), The error and reliability theory, 331pp Surv. and Mapp., Beijing, China.

MacArthur JL (1978) Seasat, a radar altimeter design description, Rep. SDO-5232. Applied Physics Lab, Johns Hopkins University, Baltimore, MD

Martin TV, Zwally HJ, Brenner AC et al (1983) Analysis and retracking of continental ice sheet radar altimeter waveforms. J Geophys Res 88:1608–1616

Maus S, Green CM, Fairhead JD (1998) Improved ocean-geoid resolution from retracked ERS-1 satellite altimeter waveforms. Geophys J Int 134(N1):243–253

Mercier F (2008) Improved Jason-2 altimetry products for coastal zones and continental waters (Pistach project), OSTST Nice 2008. Poster available on http://www.aviso.oceanobs.com/fileadmin/documents/OSTST/2008/Mercier_PISTACH.pdf

Moore RK, Williams CS Jr (1957) Radar terrain return at near-vertical incidence. Proc IRE 45(2):228–238

NASA GSFC (2006) The GSFC retracking algorithms. http://icesat4.gsfc.nasa.gov/data_processing/gsfcretrackdoc.960725.html. Last updated 12/11/2006

Powell RJ, Birks AR, Bradford WJ, Wrench CL, Biddiscombe J (1993) Using transponders with the ERS-1 altimeter to measure orbit altitude to ± 3 cm. In: Proceedings of the first ERS-1 symposium, 4–6 November 1992, Cannes, France, ESA SP-359, pp 511–516

Quartly GD (1998) Determination of oceanic rain rate and rain cell structure from altimeter waveform data. Part I: Theory. J Atmos Ocean Technol 15(6):1361–1378

Quartly GD (2009a) Optimizing σ^0 information from the Jason-2 altimeter. IEEE Geosci Remote Sens Lett 6:398–402. doi:10.1109/LGRS.2009.2013973

Quartly GD (2009b) Improving the intercalibration of σ^0 values for the Jason-1 and Jason-2 altimeters. IEEE Geosci Remote Sens Lett 6:538–542. doi:10.1109/LGRS.2009.2020921

Quartly GD, Srokosz MA, McMillan AC (2001) Analyzing altimeter artifacts: statistical properties of ocean waveforms. J Atmos Ocean Technol 18:2074–2091

Rodriguez E (1988) Altimetry for non-Gaussian oceans: height biases and estimation of parameters. J Geophys Res 93(C11):14107–14120

Sandwell D, Smith W (2005) Retracking ERS-1 altimeter waveforms for optimal gravity field recovery. Geophys J Int. doi:10.1111/j.1365-246X.2005.02724.x

Srokosz, MA, (1986) On the joint distribution of surface elevation and slopes for a nonlinear random sea, with an application to radar altimetry. J Geophys. Res 91:995–1006.

Thibaut P, Poisson JC (2008) Waveforms processing in Pistach project, Oral presentation during 2nd coastal altimetry workshop, Pisa, Italy, Nov 2008. Available from http://www.coastalt.eu/pisaworkshop08/pres/

Thibaut P, Amarouche L, Zanife OZ, Steunou N, Vincent P, Raizonville P (2004) Jason-1 altimeter ground processing look-up correction tables. Mar Geod 27:409–431

Tokmakian RT, Challenor PG, Guymer H, Srokosz MA (1994) The U.K. EODC ERS-1 altimeter oceans processing scheme. Int J Remote Sensing 15:939–962

Tournadre J, Whitmer K, Girard-Ardhuin F (2008) Iceberg detection in open water by altimeter waveform analysis. J Geophys Res Oceans 113(C8):C08040

Wingham DJ, Rapley CG, Griffiths H (1986) New techniques in satellite tracking system. In: Proceedings of IGARSS' 86 symposium, Zurich, pp 1339–1344

Zanifé O-Z, Vincent P, Amarouche L, Dumont J-P, Thibaut P, Labroue S (2003) Comparison of the Ku range noise level and the relative sea state bias of the Jason-1, TOPEX and Poseidon-1 radar altimeters. Marine Geodesy, Special Issue on Jason-1 Calibration/Validation 26(1):3–4

Zwally HJ, Brenner AC (2001) Ice sheet dynamics and mass balance. In: Fu LL, Cazenave A (eds) Satellite altimetry and earth sciences: a handbook of techniques and applications. Elsevier, New York, pp 351–369

Range and Geophysical Corrections in Coastal Regions: And Implications for Mean Sea Surface Determination

O.B. Andersen and R. Scharroo

Contents

5.1	Introduction	104
5.2	Sea Level Anomalies	108
5.3	Altimetric Sea Level Corrections	109
5.4	Dry Troposphere Refraction	111
5.5	Wet Troposphere Refraction	113
5.6	Ionosphere Refraction	116
5.7	Sea-State Bias Correction	121
5.8	Tide Corrections	124
	5.8.1 Ocean Tide	125
	5.8.2 Solid Earth Tide Correction	128
	5.8.3 Pole Tide Correction	128
5.9	Dynamic Atmosphere Correction	129
5.10	Geoid and Mean Sea Surface Models	132
5.11	Reference Frame Offsets	134
5.12	Range Corrections and Mean Sea Surface Determination	134
5.13	Summary	141
References		142

Keywords Altimeter data corrections • Altimetry • Geoid • Tides

Abbreviations

AVISO	Archiving, Validation and Interpretation of Satellite Oceanographic data
BM4	4-Parameter SSB model
CLS	Collecte Localisation Satellites
DEOS	Department for Earth Observation and space Systems

O.B. Andersen (✉)
DTU-Space, National Space Institute, Copenhagen, Denmark
e-mail: oa@space.dtu.dk

R. Scharroo
Altimetrics LLC, Cornish, New Hampshire, USA
e-mail: remko@altimetrics.com

DLM	Dynamically Linked model
DNSC	Danish National Space Centre
DORIS	Doppler Orbitography and Radiopositioning Integrated by Satellite
ECMWF	European Centre for Medium-Range Weather Forecasts
EGM	Earth Gravity Model
EM	ElectroMagnetic
ERS	European Remote-sensing Satellite
EUMETSAT	European Organization for the Exploitation of Meteorological Satellites
FES	Finite Element Solution
GFO	Geosat follow-on
GIM	Global Ionosphere Map
GOT	Goddard Ocean Tide model
GPS/GNSS	Global Positioning System/Global Navigation Satellite System
GRACE	Gravity Recovery And Climate Experiment
IB	Inverse Barometer
IGS	International GPS Service for geodynamics
IRI	International Reference Ionosphere
ITRF	International Terrestrial Reference Frame
JMR	Jason-1 Microwave Radiometer
JPL	Jet Propulsion Laboratory
MDT	Mean Dynamic Topography
MOG 2D	Two-dimensional Gravity Waves model
MSS	Mean Sea Surface
NCEP	National Centers for Environmental Prediction
NIC	NOAA Ionosphere Climatology
NOAA	National Oceanic and Atmospheric Administration
OSTST	Ocean Surface Topography Science Team
PO.DAAC	Physical Oceanography Distributed Active Archiving Center
QWG	Quality Working Group
RADS	Radar Altimeter Database System (http://rads.tudelft.nl/)
RMS	Root Mean Square
SLP	Sea Level Pressure
SSB	Sea-State Bias
SWH	Significant Wave Height
SWT	TOPEX science Working Team
TEC	Total Electron Content
TECU	Total Electron Content Unit
TMR	TOPEX Microwave Radiometer
TOPEX	TOPography EXperiment

5.1
Introduction

Satellite altimetry works conceptually by the satellite transmitting a short pulse of microwave radiation with known power towards the sea surface, where it interacts with the sea surface and part of the signal is returned to the altimeter where the travel time is measured accurately. The determination of sea surface height from the altimeter range measurement involves a number of corrections: those expressing the behavior of the radar pulse through the atmosphere, and those correcting for sea state and other geophysical signals. A number of these corrections need special attention close to the coast and in shallow water regions.

Global description of sea level and its variation with time to high accuracy is the ultimate derivative from a satellite altimeter's range measurement. However, the accuracy with which this can be made is directly linked with the accuracy of the corrections applied to derive the sea surface height.

The corrections fall into two groups. Range corrections deal with the modifications of the radar speed and actual scattering surface of the radar pulse. Geophysical corrections adjust observed sea surface height for the largest time-variable contribution like ocean tide and atmospheric pressure in order to isolate the ocean dynamic height contributors to sea surface height variations.

If the atmosphere were a perfect vacuum, and the wave distribution of the ocean had a well-known distribution, then the range between the altimeter and the sea surface would have been easily determined from the two-way travel time and the speed of the emitted radar pulse. However, the presence of dry gasses, water vapor, and free electrons in the atmosphere reduces the speed of the radar pulse causing the observed range to become longer and the sea surface height to be too low, if this is not accounted for.

The corrections established to model and adjust for the refraction and delay of the radar signal in the atmosphere are correspondingly split into three components. The dry tropospheric correction accounts for dry gasses (mainly oxygen and nitrogen), the wet tropospheric correction accounts for the water vapor, and the ionospheric correction accounts for the presence of free electrons in the upper atmosphere.

Both the wave distribution and the scattering of the radar signal by the sea surface are non-Gaussian: not only are wave troughs more prevalent, but they also reflect more of the radar signal back to the satellite than the crests, causing the sea level detection to be biased low. This bias is related to the local sea state (wind and wave conditions) and is therefore called the sea-state bias. The sea-state bias correction attempts to account for the difference between the actual scattering surface and the actual mean sea surface within the altimeter footprint which is the parameter sought for sea surface height studies. The correction is a combined effect of electromagnetic, skewness, and tracker biases.

Following the schematic illustration in Fig. 5.1 and using similar notation to Fu and Cazenave (2001) the corrected range $R_{corrected}$ is related to the observed range R_{obs} as

$$R_{corrected} = R_{obs} - \Delta R_{dry} - \Delta R_{wet} - \Delta R_{iono} - \Delta R_{ssb}$$

where $R_{obs} = c\,t/2$ is the computed range from the travel time t observed by the onboard ultra-stable oscillator (USO), and c is the speed of the radar pulse neglecting refraction.

The height, h, of the sea surface above the reference ellipsoid is given as

$$h = H - R_{corrected} = H - (R_{obs} - \Delta R_{dry} - \Delta R_{wet} - \Delta R_{iono} - \Delta R_{ssb})$$

where H is the height of the spacecraft determined through orbit determination. Consequently, the sea surface height accuracy is directly related to the accuracy of the orbit determination. Fortunately, the orbit accuracy has improved from tens of meters on the first altimeters to about 1 cm on the most recent satellites (e.g., Bertiger et al. 2008).

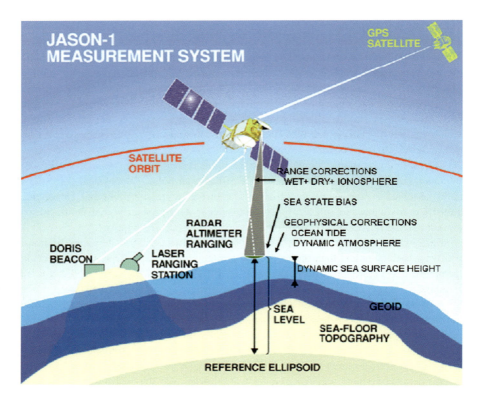

Fig. 5.1 A schematic illustration of the principle of satellite altimetry and the corrections applied to the altimeter observations of sea surface height. The range corrections affect the range through the speed of the radar pulse and sea-state bias. The geophysical corrections removed the largest known contributor to sea level in order to enhance the oceanographic contributor. (Figure modified from AVISO)

All values must be given in a fixed coordinate system based on a mathematical determinable ellipsoid model of the Earth. Most altimetric products are consistently processed in the International Terrestrial Reference Frame (ITRF) (McCarthy 1996). Furthermore, one should be aware of which tide system (mean tide, zero tide, or tide free) applies to the reference ellipsoid, in order to use this consistently in comparison with other sea surface height data like buoy or gauge data referenced using GPS data, as GPS data are normally processed in a tide-free system.

After the (satellite-specific) range corrections have been applied, the resulting sea surface heights vary spatially between ±100 m with temporal variations up to ±10 m.

The primary focus of satellite altimetry is the study of dynamic sea surface height signals related to oceanographic processes, which are normally on the submeter scale. In order to study/isolate these, it is necessary to remove the dominant geophysical contributors to sea surface height variations, which are:

- h_{geoid} the geoid correction. This gives – by far – the largest contribution to the measured sea surface height. The geoid correction arises from the fact that, in absence of all forces other than gravity and centrifugal forces, the sea surface would coincide with the equipotential surface called the geoid. This correction has negligible temporal variation and acts like a change in the reference system on the observations and the heights are subsequently given relative to the geoid rather than relative to the reference ellipsoid. By removing the permanent geoid signal and referencing the sea surface height to a given geoid model the sea surface height is reduced to meter scale. Geostrophic ocean currents are subsequently computed from this quantity.

- h_{tides} the tide correction, which is the dominant contributor to temporal sea surface height variations. Ocean tides are the largest tidal component, but the correction also accounts for solid earth tides, loading tides, and pole tides. Tidal signals have the advantage that they can be modeled and described relatively accurately in the deep ocean from a combination of the astronomical forces of the Sun and Moon and hydrodynamic time-stepping models. In coastal regions the situation is considerably more complicated.
- h_{atm} the dynamic atmosphere correction, which corrects the sea surface height variations due to the time varying atmospheric pressure loading. The atmosphere exerts a downward force on the sea surface and lowers the sea surface when the pressure is high and vice versa. This correction normally involves a static response (inverse barometer) of the ocean to atmospheric forcing for low-frequency signals (longer than 20 days) combined with a correction for dynamic high-frequency variations (shorter than 20 days) in sea surface height that are aliased into the altimetric measurements because of the subsampling by the altimeter.

The actual sea surface height is a superposition of these geophysical signals and the dynamic sea surface height, h_D, such that

$$h = h_D + h_{geoid} + h_{tides} + h_{atm}$$

or

$$h_D = h - h_{geoid} - h_{tides} - h_{atm}$$

which means that these geophysical corrections should be subtracted from the sea surface height observations. Note that these are corrections for genuine geophysical signals, but they act like corrections to the range.

The next step is to combine the range and geophysical corrections into a combined set of corrections. To avoid confusion, which generally arises as the range corrections have to be applied to the range and that the geophysical corrections have to be applied to the sea surface height, the space agencies have decided to define that *all corrections* are to be *subtracted* from the observed sea surface height. The range and sea-state bias corrections have the effect of making the range estimation shorter, which corresponds to raising the sea surface height. By convention, this is implemented as a negative "height" correction that is removed from the sea surface height.

Combining the range and geophysical corrections using this convention, the dynamic sea surface height, h_D, is derived from the height, H, of the spacecraft, and the range, R_{obs} as:

$$h_D = H - R_{obs} - \Delta h_{dry} - \Delta h_{wet} - \Delta h_{iono} - \Delta h_{ssb} - h_{tides} - h_{atm} - h_{geoid}$$

where all corrections are sea surface height corrections and, i.e., $\Delta R_{dry} = -\Delta h_{dry}$, etc.

A summary of all range corrections and corrections for geophysical signals is shown in Table 5.1 with typical values for mean and standard deviation of the various corrections.

Table 5.1 Typical values for the mean and standard deviation of all the time-variable corrections applied to sea surface height observed by the satellite altimeter. All values have been computed from 6 years of Jason-1 altimeter data. The values have been split into deep ocean and coastal regions by the 500 m depth contour. The geoid correction is assumed to have insignificant temporal variation and is not shown in the Table (e.g., Wahr et al. 2004; Andersen et al. 2005)

	Mean[a] (cm)	Time-variable deep ocean (std dev) (cm)	Time variable coastal (std dev) (cm)
Dry troposphere	−231	0–2	0–2
Wet troposphere	−16	5–6	5–8
Ionosphere	−8	2–5	2–5
Sea-state bias	−5	1–4	2–5
Tides	∼ 0–2	0–80	0–500
Dynamic atmosphere	∼ 0–2	5–15	5–15

[a]The mean of the corrections are by convention shown with negative numbers corresponding to the fact that the corrections should be subtracted from the sea surface height

For each of the range and geophysical corrections several models are available, and standards to be applied by the space agencies are regularly decided and updated by the international scientific community. These standards are generally decided by the OSTST (Ocean Surface Topography Science Team) for the TOPEX and Jason missions and by the QWG (Quality Working Group) for the ERS and Envisat missions.

One of the keys to the great success of satellite altimetry and the constant improvement of the quality of sea surface height observations from satellite altimetry lies in the constant improvement of orbit determination and analysis and improvement of the corrections applied to the sea surface height observations by the science community. With current state of the art altimeters the accuracy of each 1-s sea surface height observations is currently at the 2–3 cm level in the open ocean, whereas this value is somewhat higher towards the coast (e.g., Strub 2008).

Altimeter data are distributed through agencies like, EUMETSAT, AVISO, PO.DAAC, and NOAA. In addition to these two operational data centers, RADS (Radar Altimeter Database System) is a joint effort by the Delft University of Technology Department for Earth Observation and Space Systems (DEOS) and the NOAA Laboratory for Satellite Altimetry in establishing a harmonized, validated, and cross-calibrated sea level database from all altimetric missions. RADS users have access to the most recent range and geophysical corrections and create their own altimetric products for their particular purpose.

5.2
Sea Level Anomalies

By applying the geophysical corrections, and particularly the geoid correction, the sea surface height, h_D, is referring to the chosen geoid model and the range of sea surface height values are typically reduced from ranging up to 100 m to range up to a few meters.

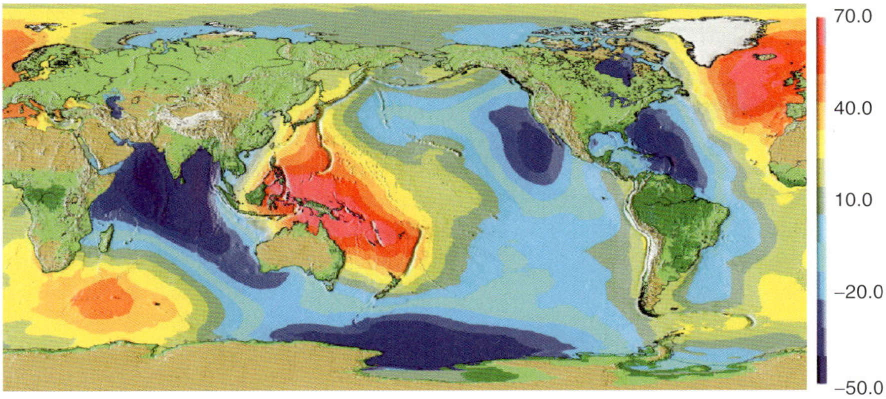

Fig. 5.2 The DNSC08MSS mean sea surface model. The figure has been significantly smoothed in order to illustrate the main global mean sea surface height features. Units are meters

The local sea level height observations with the geoid removed will seldom have zero temporal mean because the dynamic sea surface height, h_D, has both a permanent and a time-variable component. The permanent component reflects, among others, the steric expansion of seawater as well as the ocean circulation that is very nearly in geostrophic balance. The temporal average of the dynamic topography is called the mean dynamic topography.

For studies of sea surface height variations it is often more convenient to refer the sea surface height to the mean sea surface height rather than to the geoid, thus creating the *sea level anomalies* h_{sla}, determined as

$$h_{sla} = H - R_{obs} - \Delta h_{dry} - \Delta h_{wet} - \Delta h_{iono} - \Delta h_{ssb} - h_{tides} - h_{atm} - h_{MSS}$$

Subtraction of the mean sea surface conveniently removes the temporal mean of the dynamic sea surface height and creates sea level anomalies that, in principle, have zero mean. This is so, because mean sea surfaces are normally computed from averaging altimetric observations over a long time period and preferably combining data from several exact repeat missions. Examples are the DNSC08 MSS computed from an average of 12 years of satellite altimetry and the CLS01 MSS computed from an average of 7 years of satellite altimetry. The DNSC08MSS is shown in Fig. 5.2. When removing the temporal mean, also the temporal mean of the corrections are removed (noticing the large mean of the dry troposphere correction) and only the time-variable part of the correction is then a concern.

5.3
Altimetric Sea Level Corrections

In the following sections each correction is introduced with focus on coastal regions. Readers who are interested into a thoroughly physical description of each correction should consult Fu and Cazenave (2001). For each correction, an investigation of the accuracy will

Table 5.2 The two state of the art range and geophysical corrections selected by the OSTST and available in the RADS database applied in this study for investigating the accuracy of the various range and geophysical corrections

Correction	Observation or model
Dry troposphere	ECMWF (model)
	NCEP (model)
Wet troposphere	Radiometer
	ECMWF (model)
Ionosphere	Radiometer – Smoothed Dual Frequency
	JPL GIM (model)
Dynamic atmosphere	IB (model, local pressure)
	MOG2D_IB model
Tides	FES2004 (model)
	GOT4.7 (model)
Sea-state bias	BM4 (model)
	CLS NPARAM-GDRC (model)
Geoid/mean sea surface	EGM2008 (geoid model)
	DNSC08MSS (mean sea surface)

be given, by evaluating two state of the art corrections available for Jason-1 satellite altimetry with focus on coastal regions. Table 5.2 shows the two most recent state of the art corrections selected by the OSTST and these are the corrections that have been used for this investigation.

Each satellite will have a number of features that make it unique, and which require special considerations with respect to the range and geophysical corrections that can be applied for that particular satellite. Various models might not be available for the correct time and location and instrument failure might call for altimeter specific corrections (e.g., Naeije et al. 2002). Satellite-specific investigations of various range and geophysical corrections are a huge subject and will not be discussed in this chapter. Interested readers can consult the web site of the T/P-OSTST community (http://sealevel.jpl.nasa.gov/).

The chapter will be completed with a section into available mean sea surfaces for the derivation of sea level anomalies. With the increasing use of satellite altimetry for climate purposes and the constant improvement in the accuracy of sea surface height and sea level anomalies relative to a mean sea surface, this subject is becoming an area of increasing importance. Small differences between recent mean sea surface models of the order of ±10 cm are directly related to the range corrections applied at the time of derivation. Consequently, users requiring highly accurate satellite altimetry observations should be sure to use consistently derived altimetric products. Users should also be aware of the fact that some large-scale signals detected in sea level anomalies can be related to the use of *different* range and geophysical corrections available on state of the art altimetric products and state of the art available mean sea surfaces.

5.4
Dry Troposphere Refraction

The correction for refraction from dry atmospheric gases is by far the largest adjustment that must be applied to the range as shown by its mean value in Table 5.1. The vertical integration of the air density is closely related to the pressure, and the dry troposphere correction is for most practical purposes approximated using information about the atmospheric pressure at sea level (or sea level pressure, SLP) and a refractivity constant of 0.2277 cm^3/g. Formulations for the dry troposphere refraction is described and derived in length by (Smith and Weintraub 1953; Saastamoinen 1972) and is conveniently given in units of cm as

$$\Delta h_{\text{dry}} \approx -0.2277\, P_0\, (1 + 0.0026 \cos 2\varphi)$$

where P_0 is the SLP in hecto-Pascal (hPa) and the coefficient in the parenthesis is the first order Taylor expansion of the latitude (φ) dependence of standard gravity evaluated at the location of observation to account for the oblateness of the Earth.

Since direct observations of SLP are infrequent and sparsely distributed, data from operational weather models are generally used. The two most widely used models are the European Centre for Medium-Range Weather Forecasts (ECMWF) and the U.S. National Centers for Environmental Prediction (NCEP). Both of these are delivered on regular grids at regular intervals and the surface pressure is then interpolated from these grids (Fig. 5.3).

The spatial distribution of the dry troposphere correction computed from the ECMWF model features a latitudinal dependency, with highest values (−2.33 m and a standard deviation of a few centimeters) around the subtropical gyres and smallest values at high latitudes (−2.27 m and standard deviation around 4–6 cm).

The dry troposphere correction has long spatial scales and it varies only slowly compared with the wet troposphere. This means that the dry correction is relatively unaffected by the presence of land and is not expected to degrade significantly close to the coast. Fu et al. (1994) reported an accuracy of the dry troposphere correction of 0.7 cm, degrading slightly towards the coast.

Fig. 5.4 shows the standard deviation of residual sea surface height variations applying the dry troposphere correction based on ECMWF and NCEP as a function of the distance to the coast. Six years of Jason-1 data have been used and averaged in bins of 2 km as a function of the distance to the coast in order to investigate the quality and difference between the corrections in coastal regions. Comparing the standard deviation of the residual signal while leaving all other corrections unchanged evaluates the two models. A lower standard deviation will indicate that the correction removes more "signal," indicating that the correction is more efficient (better). Data for this and the following analysis of other corrections have been extracted from RADS and editing of data close to the coast has been disabled, as this would otherwise have removed too many data closer than around 30 km to the coast for Jason-1.

Interpreting the curves in Fig. 5.4 for the ECMWF and NCEP dry troposphere correction shows that it is basically impossible to tell the two curves apart indicating that the two models have removed the same amount of signal and are presumably of same accuracy. The investigation also confirms that there is only very minor difference (<5 mm) between

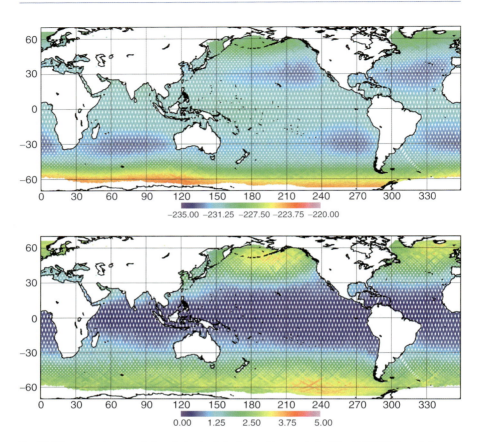

Fig. 5.3 The mean (*upper plot*) and standard deviation (*lower plot*) of the ECMWF based dry troposphere correction. The values have been computed from 6 years of Jason-1 data. Range corrections are given in centimeters

Fig. 5.4 The residual sea surface height variation (in cm) from 6 years of Jason-1 observations corrected using the ECMWF and NCEP based dry troposphere correction and then shown as a function of the distance to the coast (in km)

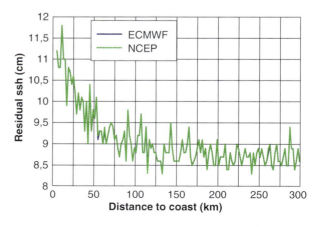

these two corrections, which is naturally also a consequence of the low temporal variability of the correction. The increase in residual sea surface height closer to the coast is due to increased sea level variability in shallow water depth and not related to the corrections.

A commonly made error in the production of altimeter products (e.g., Jason-2) is to replace the sea level pressure grids by surface pressure grids. The difference between the two is the level at which the pressure is calculated: the former is the pressure at sea level, the latter at the surface. Over oceans the difference is, in principle, irrelevant, since both refer to sea level. However, most atmospheric models are currently based on the linear combination of Gaussian distributions of several spatial scales. As a result, coastal grid points are determined in part by Gaussians centered on points in-land. When the coast is of significant elevation, the surface pressure in-land will be much lower, and thus artificially lower the surface pressure over coastal seas. For example, near Greece errors of a few kPa (several cm of path delay) arise from this error. When using the dry tropospheric correction near the coastline, it should be verified that it was based on sea level pressure grids and not surface pressure grids.

5.5
Wet Troposphere Refraction

The wet troposphere refraction is related to water vapor in the troposphere, and cloud liquid water droplets. The water vapor dominates the wet tropospheric correction by several factors, and the liquid water droplet from small to moderate clouds is generally smaller than one centimeter (Goldhirsh and Rowland 1992).

Although smaller than the dry tropospheric range correction in magnitude, the wet troposphere correction is more complex with higher temporal variations, with rapid variations in both time and space and therefore also needs careful attention in the coastal region. The correction can vary from just a few millimeters in dry, cold air to more than 30 cm in hot, wet air.

The water vapor range correction is in its most simple form is given from the following equation in units of cm as

$$\Delta h_{wet} \approx 636 (cm^3/g) \int \sigma_{vap}(z) dz$$

which involves a vertical integration of the water vapor density σ_{vap} in grams per cubic centimeters (g/cm^3). The equation has been simplified by replacing the altitude dependent temperature T of the troposphere with a constant temperature. More advanced formulas are often considered (e.g., Eymard and Obligis 2006).

The columnar water vapor can be estimated using downward looking passive microwave radiometers onboard the satellite or from a collection of ground-based observations integrated into a time-stepping model like ECMWF. As shown in Fig. 5.5, the correction has significant temporal and spatial variability, which stresses the importance of having simultaneous observations of the columnar water vapor and radar altimetric observations of the sea surface height.

Accurate water vapor estimates can be obtained from three-frequency microwave radiometers available on satellites like TOPEX/Poseidon, Jason-1, and Jason-2. The onboard

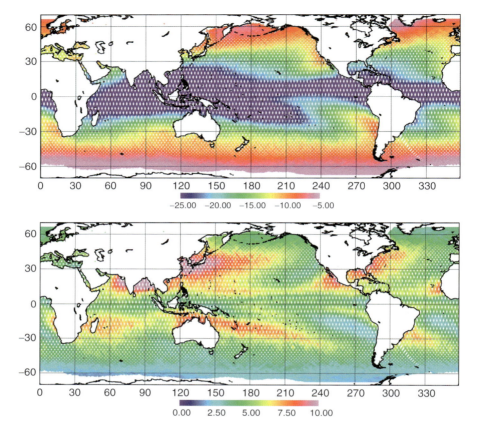

Fig. 5.5 The mean (*upper plot*) and standard deviation (*lower plot*) of the wet troposphere correction from radiometer observations of columnar water vapor from 6 years of Jason-1 data. The values are given in centimeters

nadir looking microwave radiometer on TOPEX (TMR) or Jason (JMR) uses the microwave radiation received by the satellite to determine the brightness temperature of the ocean surface around 18.7, 23.8, and 34 GHz (for Jason-1) and the instrument provides nearly direct measurements of the wet troposphere correction by monitoring the strong water vapor absorption line centered at 22.235 GHz and using the two other frequencies to separate this from the liquid water and surface winds. High accuracy relies both on the retrieval method but also on instrument calibration and drift calibration (e.g., Ruf 2002)

Keihm et al. (2000) demonstrated from T/P data, that the accuracy of the wet troposphere correction is around 1.1 cm in the open ocean, even when heavy clouds and wind are present in the microwave foot print. For two-frequency microwave radiometers onboard the ERS, GFO, and Envisat satellites, the accuracy is somewhat less than for three-frequency radiometers (e.g., Schrama et al. 1997). Microwave radiometers are prone to drift and calibration errors are an important issue, so a huge effort is devoted to limit this effect on the range corrections, in order to obtain the best long-term sea level estimate (e.g., Keihm et al. 2000).

The mean of the wet troposphere signal taken from the onboard radiometer on Jason-1 and averaged over 6 years is shown in Fig. 5.5. It has a strong dependency on latitude, with highest values in the equatorial band (−30 cm) and smallest values in the Antarctic circumpolar current (−5 cm). The lower part of the figure shows that the average standard deviation of the signal is around 5 cm. The lowest variation is found close to Antarctica and the highest temporal variability is found at mid-latitude, where it reaches up to 10 cm in the western part of the Pacific Ocean.

The footprint of the radiometer is dependent on the height of the spacecraft and the scanning frequency of the radiometer, but typical values of the footprint of the main beam range between 20 and 30 km. This is considerably larger than the 4–10 km footprint of the altimeter as illustrated in Fig. 5.6 for a pass across the Western Mediterranean Sea. Consequently, the radiometer is contaminated by the presence of land much earlier than the altimeter, as the spacecraft approaches the coast and generally the main beam is affected up to 30 km from the coast. The wet troposphere correction derived from the onboard radiometer is similarly affected, and a lot of research is put into improving the wet troposphere correction in coastal regions (i.e., Obligis et al. 2009).

The effect of land on the accuracy of the radiometer derived wet troposphere correction, can be as large as 2–4 cm (Brown 2008) close to the coast within the main beam. However, effects can be seen out to 60 km when the side lobes of the main beam get affected (effects up to 2–4 mm). Land has larger emissivity than ocean, and the presence of normally, much warmer land degrade the humidity retrieval methods thus degrading the accuracy of the correction. A number of studies are currently ongoing in order to improve the correction close to the coast.

Fig. 5.7 illustrates how the various available wet troposphere corrections are affected in the coastal zone, by showing the residual sea surface height variation applying two

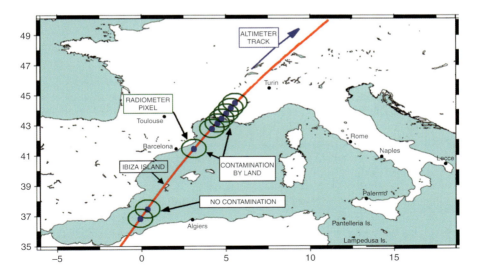

Fig. 5.6 An example of a Jason-1 track crossing the western Mediterranean Sea. Blue dots indicate the footprint of the altimeter and green circles show the extent of the main beam of the radiometer. The figure illustrates where the radiometer observations are contaminated by land. (Modified from Eymard and Obligis 2006)

Fig. 5.7 The residual sea surface height variation (in cm) applying the wet troposphere correction from the onboard radiometer and interpolated from ECMWF columnar water vapor estimates averaged in 2 km bins as a function of the distance to the coast (in km)

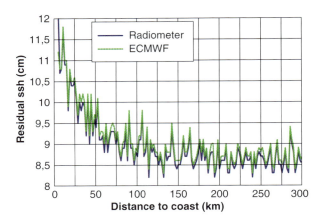

different wet troposphere corrections. As before 6 years of Jason-1 data have been averaged with respect to the nearest distance to the coast.

The onboard radiometer reduces the sea surface height variability slightly more than the interpolated ECMWF correction everywhere in the deep ocean to roughly 50 km from the coast, with the radiometer reducing the sea surface height variability between 10% and 20% more than the ECMWF corresponding to between 5 and 10 mm.

Closer than 50 km to the coast, where the radiometer observations are expected to degrade, the radiometer correction still performs equally accurate to the ECMWF model. This is most likely due to the fact that the ECMWF model also degrades close to the coast where the correction changes rapidly and observations are lacking. All the way until around 10 km from the coast, where the radiometer corrections becomes seriously contaminated by land, it still performs equally accurate to the ECMWF model values. Overall the analysis shows that the accuracy of the correction degrades from around 1.1 cm in the open ocean to roughly half the accuracy at around 30 km from the coast. This number is hard to confirm exactly, as the result showed that both the radiometer and the ECMWF model based correction degrades equally all the way to 10 km from the coast.

The chapter by Obligis et al., this volume, investigates two methods to improve the retrieval of the wet troposphere correction in the coastal zone. One uses the Dynamically Linked model (DLM) based on the dynamic combination of radiometer measurements, model data, and GPS/GNSS-derived path delay estimates. The second approach investigates the measured brightness temperatures in order to remove the contamination coming from the surrounding land and hereby improving the accuracy of the radiometer observations in the coastal zone.

5.6
Ionosphere Refraction

The refraction of electromagnetic waves in the earth's ionosphere is directly linked to the presence of free electrons and ions at altitudes above about 100 km. At this altitude high-energy photons emanating from the sun are able to strip atomic and molecular gasses of an electron. At lower altitudes, up to about 300 km, NO^+ and O_2^+ are the most prevalent species, while at higher altitudes, stretching even far above the altitude of the radar altimeters, we find H^+, O^+, N^+, and He^+ (e.g., Bilitza 2001).

The interaction between the electromagnetic waves and the ions causes the waves to slow down by an amount proportional to the electron density in the ionosphere. Hence, the total delay incurred on the altimeter radar pulse along its path through the ionosphere is proportional to the number of electrons per unit area in a column extending from the earth surface to the altimeter. This columnar electron density count is referred to as the Total Electron Content (TEC). The TEC unit, or TECU, equals 10^{16} electrons/m^2.

The ionospheric correction (the negative of the delay) is also inversely proportional to the square of the radar frequency:

$$\Delta h_{iono} = -k\text{TEC}/f^2$$

where k is a constant of 0.40250 m GHz^2/TECU. At the Ku-band frequency of approximately 13.6 GHz, this means that the ionospheric delay amounts to about 2 mm for each TEC unit (Schreiner et al. 1997).

Since the delay is dispersive, the delay at the two frequencies will be different. If the frequencies differ significantly, then the difference in their path delay (and hence the difference in their total travel time) can be used as a measure for the TEC value. This difference in path delay is used most prominently by geodetic GPS systems, measuring at two frequencies (L1 and L2) to eliminate the effect of the ionospheric delay on the GPS positions. The more recent high-precision altimeters (TOPEX, Jason-1, Envisat, and Jason-2) also carry dual-frequency altimeters with the same purpose: estimating TEC to be able to correct for the ionospheric path delay in the primary Ku-band range measurement.

The secondary frequencies used in altimetry are in the C-band (TOPEX and Jason: 5.3 GHz) and S-band (Envisat: 3.2 GHz). The range measurements at these frequencies are inherently less precise than the Ku-band ranges. Differencing the range measurements at two frequencies creates a rather noisy estimate of the ionospheric delay, which needs to be smoothed over about 200 km along the altimeter tracks in order to not adversely impact the altimeter measurement precision, while still capturing most of the important features of the global distribution of the total electron content (Imel 1994).

As Fig. 5.8 shows, the ionospheric correction is largest, both in mean magnitude and in variation, along two bands paralleling the geomagnetic equator (approximately −10 and 6 cm, for the mean and standard deviation, respectively). Towards the poles, the magnitude becomes smaller and also much more predictable, as its variation drops. One should not be deceived by the relative homogeneity of the mean: the variations are in places almost as large as the mean. The local variation of the ionosphere depends on a number of factors: season of year, time of day, and solar activity.

The annual variation is generally relatively small and is mainly due to the changing orientation of the sun with respect to the geomagnetic field. Seasonal changes in TEC levels are most prominent in Polar Regions where summer levels go up to about five times the winter levels. The variation during the day is quite marked: TEC levels can increase about ten-fold between their lowest level around 2:00 at night local time and their highest level 12 h later at 14:00 local time. Finally, the dependency on solar activity closely follows the 11-year cycle in solar radio flux ($F_{10.7}$, as depicted in Fig. 5.9). This gives a variation by a factor of 5 between low and high solar activity.

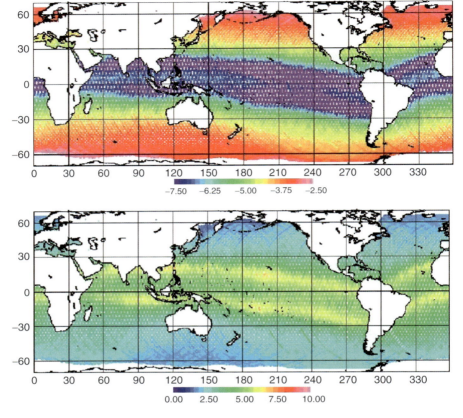

Fig. 5.8 Mean (*upper plot*) and standard deviation (*lower plot*) of the ionosphere range correction from smoothed dual-frequency altimeter observations based on 6 years of Jason-1 data. Values are given in centimeters

Fig. 5.9 Variation of the solar 10.7-cm radio flux during the last two solar cycles. The duration of the altimeter missions is shown by rectangles. The outlined rectangles of TOPEX, Jason-1, Jason-2, and Envisat indicate the availability of dual-frequency measurements. TEC models are shown at the bottom

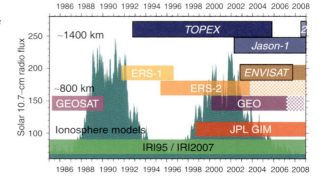

Since the ionosphere is insensitive to coastlines, unlike the way the wet troposphere is strongly influenced by coastal dry or wet atmospheric currents, one would not expect that modelling ionospheric path delay near coasts is any different from doing so anywhere on the open ocean. But be not deceived: the majority of the Indian Ocean and Western Pacific coastline is located beneath the most volatile region of the ionosphere. Secondly, when based on dual-frequency altimeter measurements, the ionospheric correction will eventually be affected by land in the altimeter footprint. This, however, will largely coincide with the general deterioration of the altimeter range measurement.

As an alternative to dual-frequency altimeter measurements, a number of climatological models have been in use, like the Bent model (Llewellyn and Bent 1973) or the International Reference Ionosphere (IRI) (e.g., Bilitza 2001; Bilitza and Reinisch 2008). These climatological models attempt to find systematic behavior in the temporal and spatial variation of historical TEC data and correlate those with one or two global ionospheric forcing parameters, like solar radio flux or sun spot number, to estimate TEC at any location on the globe.

More recently, observational models are being produced based on alternative satellite dual-frequency measurements. Since 1998, both the Jet Propulsion Laboratory and the University of Berne produce 6-hourly maps of the TEC based on observations by over a hundred GPS stations in the IGS network (e.g., Komjathy and Born 1999; Komjathy et al. 2005). These GPS-derived global ionosphere maps (GIM) can be interpolated in space and time to the altimeter ground track and come close to the accuracy of the dual-frequency altimeters. A complication is that the GPS measurements integrate TEC to about 20,000 km altitude, while altimeters fly only at 800 or 1,350 km altitude, and thus see only a portion (even though the major portion) of the ionosphere (Iijima et al. 1999).

The NIC09 ionosphere climatological model (Scharroo and Smith 2010) is based on the GIMs for 1998–2008 and can also be applied to all single frequency altimeter data prior to 1998. Just as other TEC models, its accuracy is more or less proportional to the TEC itself. At 18% error, NIC09 comes close to the JPL GIM model accuracy of 14% and is much better than the 35% error of the most recent IRI model (IRI2007).

Another way of estimating TEC is using the Doppler Orbitography and Radiopositioning Integrated by Satellite (DORIS) system, which is used primarily for orbit determination for T/P and Jason. This ground-based set of beacons can also be used to determine the integrated electron content along the line of sight between the satellite and the DORIS transmitter which can then be combined with a model to derive TEC. However, given the lack of accuracy compared to the GPS-based ionosphere maps, the production of DORIS ionosphere maps has ceased and their values are no longer reported on the altimeter data products.

Since all of these ionosphere models are global, they are not affected by the presence of coastlines. This gives us an opportunity to look at the impact of land in the altimeter footprint on the determination of the ionospheric correction.

Fig. 5.10 shows the RMS sea level anomaly as a function of the distance to the coast while applying two different ionospheric corrections: based on the dual-frequency altimeter measurements or the JPL GPS ionosphere maps (Web1). At this level the differences are minute. A more detailed portrayal of the accuracy of the ionospheric corrections is given in Fig. 5.11, where the various models are compared to Jason-1. Since this comparison was made during low solar activity, the model errors are rather low (less than 1 cm), but still far exceeding the comparison between the dual-frequency altimeter measurements

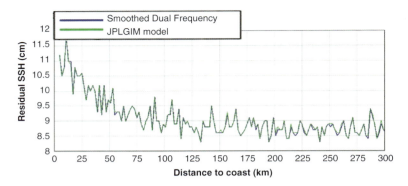

Fig. 5.10 The residual sea surface height variation (in cm) applying the ionospheric correction from the dual-frequency altimeter or the interpolated JPL GIM estimates averaged in 2 km bins as a function of the distance to the coast (in km)

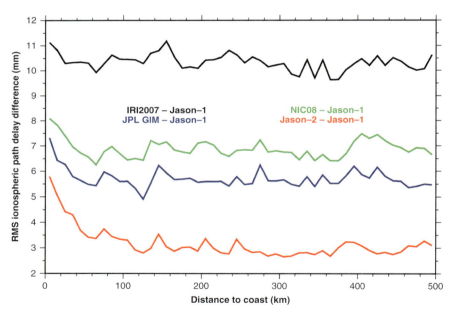

Fig. 5.11 The root-mean-square difference of several ionosphere correction models compared to the Jason-1 dual-frequency correction during Jason-1 Cycle 246 (9–19 September 2008). The RMS difference between the dual-frequency corrections as measured by Jason-1 and Jason-2 along their collinear tracks is depicted as well

of Jason-1 and Jason-2. The semblance of an increased error at distances shorter than 100 km from the coast is certainly not due to contamination of the altimeter footprint. Rather, the increased error of the dual-frequency correction should be sought in prevailing conditions near the coast that could make the measurements less accurate: such as unusual sea states (lower than common wave heights) and short-wavelength variations in the wind fields. In addition, the smoothing process of the ionosphere correction will also become one-sided, producing a noisier smoothed value.

In general, the dual-frequency ionosphere correction applies well to coastal data. Where there is no direct measurement of the electron content available, the JPL GIM or NIC09 models will suffice.

5.7
Sea-State Bias Correction

The sea-state bias (SSB) correction compensates for the bias of the altimeter range measurement toward the troughs of ocean waves. This bias is thought to arise from three interrelated effects: an electromagnetic (EM) bias, a skewness bias, and an instrument tracker bias. Radar scattering (EM) bias is physically related to the distribution of the specular facets. The elevation skewness bias is related to the fact that the altimeter uses a median tracker, while the mean is the desired parameter. The instrument tracker bias is related to the specific tracker used to estimate the significant wave height (SWH) derived from the waveform (e.g., Brown 1977), where the SWH represents the average height of the 1/3 highest waves considered.

The SSB was originally modeled as a simple percentage of the SWH (e.g., $\Delta h_{ssb} = -0.04$ SWH), explaining that, with increasing SWH, altimeter ranges longer or more below the mean sea surface within the altimeter footprint.

The SSB also depends on the wind field and the different wave types (e.g., fully developed swell waves versus wind waves). Therefore, a more advanced parametric model consisting of four parameters has been introduced to describe the SSB (Chelton 1994; Gaspar et al. 1994). The model is generally referred to as the BM4 model:

$$\Delta h_{ssb} = \text{SWH}(a_1 + a_2 U + a_3 U^2 + a_4 \text{SWH})$$

where U is the wind speed derived from the backscatter coefficient. This formula gives the total SSB correction including all the three contributions to SSB mentioned above, as all of these are dependent on the SWH.

Each altimeter will have a different set of coefficients for the parametric estimation of the four parameters (a_1, a_2, a_3, and a_4). Table 5.3 illustrates how different the empirically estimated parameters are for six different altimeters (Scharroo and Lillibridge 2005). The big spread seen in the a_1 parameter represents the linear scaling of the SWH parameter on the SSB correction.

In principle, the dependence on wind speed, U, and significant wave height of the SSB should be estimated nonparametrically rather than by a preassumed functional form like the one given above. Several authors have experimented with nonparametric estimations of the SSB (e.g., Chelton 1994; Rodrigues and Martin 1994). The nonparametric method for estimating the SSB developed by Gaspar and Florens (1998) represents a significant improvement compared to the classical parametric methods and better retrieves wind and

Table 5.3 Estimation of the four BM4 SSB parameters for six different altimeters. An additional bias coefficient a_0 is introduced in order to ensure that $\Delta h_{ssb} = 0$ when SWH = 0 (The table is taken from Scharroo and Lillibridge 2005)

	Jason-1	TOPEX (side A)	TOPEX (side B)	ERS-1	ERS-2	GFO
a_0	0.110106	0.012450	0.028889	0.054265	0.107618	0.092034
a_1	−0.034376	−0.030578	−0.032113	−0.075043	−0.068219	−0.055742
a_2	0.001145	0.002776	0.002992	0.001413	0.001465	0.002743
a_3	−0.001969	−0.002962	−0.002780	−0.001790	−0.001701	−0.003756
a_4	0.000083	0.000127	0.000101	0.000098	0.000082	0.000153

wave related variations. It improved the estimation of rare sea-state events, improving the SSB accuracy up to a few centimeters for some of the sea states.

To avoid specification of a parametric formulation of the SSB model, Gaspar and Florens reformulated and solved the SSB estimation problem in a totally nonparametric way based on the statistical technique of kernel smoothing. This method has since been refined and enhanced significantly (Labroue et al. 2006); and the uncertainty of the new TOPEX SSB estimate is now close to 1.1 cm (Labroue et al. 2008).

The spatial pattern from 6 years of Jason-1 observations is shown in Fig. 5.12. The non-parametric SSB correction has smallest magnitude at low to mid latitudes (from −5 to −10 cm mean value with a standard deviation around 1 cm) and highest values at high latitudes (up to −20 cm in the mean value with standard deviation variability greater than 5 cm).

In order to evaluate the BM4 parametric and the nonparametric sea-state model in the coastal zone, Fig. 5.13 shows the standard deviation of residual sea surface height variation from 6 years of TOPEX observations applying the BM4 and the CLS nonparametric SSB model (Jason-1 data unavailable with both corrections).

The temporal variation of the SSB correction is only a few cm in most places. Therefore it is hard to determine which correction model reduces the variability of the sea surface height the most and both seem to be performing equally accurate in coastal regions.

In the coastal zone there are several complications to SSB computation. One is the fact that the changing shape of the waves and the wind propagation, as well as interaction with bathymetry and the coastal shape will create more noisy waveforms and hence more noisy SWH estimates.

Furthermore, the retracking bias is related to the specific waveform shape (mostly Brown) and the retracker used to derive the height from the waveform, will need to be changed or made more tolerant close to the coast as the waveform shape changes (see the chapter by Gommenginger et al., this volume).

To examine the effects of changing waveforms, Fig. 5.14 shows how Envisat waveform changes when the altimeter approaches the coast. The plots shows a track going from land to sea in the left panel and a track going from sea to land in the right panel. The along-track distance to the coast is shown on the vertical axis and the figure should be read horizontally with one line for each distance to the coast. Each horizontal line represents the power in each of the 128-Envisat recording bins for each 18-Hz observation, with red colours representing high power and blue colours representing low power. Normal Brown waveforms

5 Range and Geophysical Corrections in Coastal Regions: And Implications for Mean Sea Surface Determination

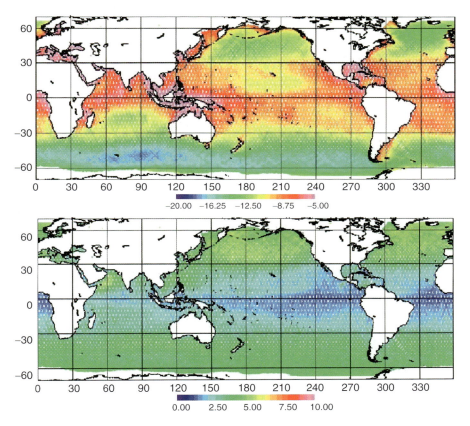

Fig. 5.12 The mean (*upper plot*) and standard deviation (*lower plot*) of the nonparametric sea-state bias correction for Jason-1. Values are given in centimeters

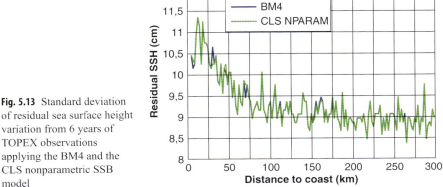

Fig. 5.13 Standard deviation of residual sea surface height variation from 6 years of TOPEX observations applying the BM4 and the CLS nonparametric SSB model

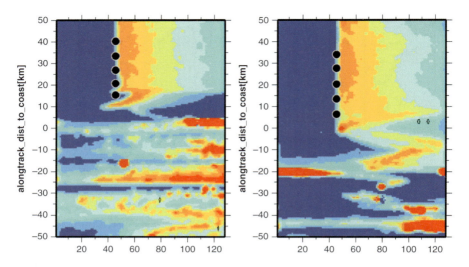

Fig. 5.14 The waveform power (*red high* and *blue low*) as a function of the distance to the coast. Two Envisat tracks close to Italy in the northeast part of the Western Mediterranean Sea are shown. The along-track distance to the coast is shown on the vertical axis and the waveform power in each 128 bins along the horizontal axis. Normal Brown waveforms are seen away from the coast in the top of the figure, but close to the coast and on land the waveforms are distorted (Fenoglio-Marc 2008). Black dots indicate track point and location of available 1-Hz SSH observations

are seen away from the coast in the top of the figure. In the left panel, the waveforms are more or less Brown (Brown 1977) all the way until 15 km from the coast. However, the waveforms on the track approaching the coast (right panel) are less affected by the coast and more or less Brown all the way to 8 km from the coast. Closer than this the waveform shapes change and retracking using more tolerant retrackers is needed if sea surface height shall be retrieved (Berry et al. 2005).

SWH, and thus SSB parameters, can currently only be estimated for Brown waveforms which limits SSB estimation to roughly 10–15 km from the coast. However in the future more research is needed into retracking and how to estimate SSH as well as SWH and SSB parameters closer than roughly 10–15 km from the coast.

Fig. 5.14 also illustrates how dramatically the waveform shape changes within 1 s. This also illustrates the importance of applying the corrections on the individual 18-Hz data and not on the 1-s averaged values in future altimetric products close to the coast.

5.8
Tide Corrections

The ocean tide correction is by far the correction that reduces the temporal sea surface height variance the most. In an analysis of collinear differences of sea surface heights, Ray et al. (1991) found that the ocean tides were responsible for more than 80% of the total signal variance in most regions.

Besides the dominating ocean tide signal, the tidal correction includes correction for several smaller tidal signals: the loading tide, the solid earth tide, and the pole tide. The sum of the tidal corrections can be written as

$$\Delta h_{tides} = \Delta h_{ocean\ tide} + \Delta h_{load\ tide} + \Delta h_{solid\ earth\ tide} + \Delta h_{pole\ tide}$$

Below each of the tidal contributions are described. However, only the ocean has been analyzed in coastal regions. The solid earth and pole tide correction are independent of coastal regions and are normally derived using closed mathematical formulas.

5.8.1
Ocean Tide

The altimeter senses the geocentric or elastic ocean tide, which is the sum of the ocean and load tide:

$$\Delta h_{elastic\ ocean\ tide} = \Delta h_{ocean\ tide} + \Delta h_{load\ tide}$$

This is in contrast to tide gauges fixed to the sea bottom which only measure the ocean tide. As the altimeter measures from space, it observes the sum of ocean tide and small loading displacement of the ocean's bottom due to the loading by the water column. The load tide has a magnitude of 4–6% of the ocean tide and can be determined by a convolution of the ocean tide and the response of the upper lithosphere to the ocean loading by that model (i.e., Scherneck 1990; Agnew 1997).

The determination of the ocean tides have dramatically improved since the launch of TOPEX/Poseidon and most recent investigations indicate that global models are now accurate to around 1–2 cm in the global ocean. The tidal signal has much higher frequency than the sampling of the satellite (0.5 days versus 9.9 days sampling for TOPEX/Poseidon). Therefore, the tidal signal is aliased into periods longer than 20 days because of the sub-sampling of the satellite (i.e., Andersen 1995). One example is the largest semidiurnal constituent M_2, which is aliased into a signal with a period of 62 days by the 9.9156 days sampling of TOPEX/Poseidon and Jason.

The amplitude of the M_2 constituent is shown in Fig. 5.15 from the global ocean tide model called GOT4.7. This model is from 2006 and is the latest in a sequence of empirical ocean tide models derived from satellite altimetry (Ray 2008). The model corrects for the major eight diurnal and semidiurnal constituents (K_1, O_1, P_1, Q_1, M_2, S_2, K_2, N_2) along with a number of smaller constituents. Furthermore, a number of local tide models have been patched into GOT4.7 for several coastal regions like the Bay of Fundy on the Canadian and US east coast.

The 2D Finite Element Solution FES2004 (Lyard et al. 2006) is another widely used global ocean tide model based on assimilation of satellite altimetry into a time-stepping finite element hydrodynamic model. The FES models were pioneered by Christian Le Provost in the early 1990s (Le Provost et al. 1994) and have been developed since to include 15 tidal constituents distributed on 1/8° grids. Both the GOT4.7 and FES2004 include corrections for long period tides and also the largest quarter diurnal shallow water constituent M_4.

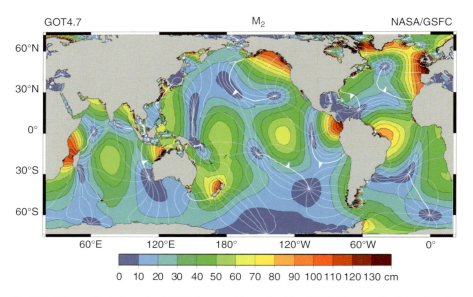

Fig. 5.15 M2 amplitude from the GOT4.7 ocean tide model (Ray 2008). The amplitude is given in cm and the largest amplitudes are found in the English Channel and Bay of Fundy on the Canadian and US east coast

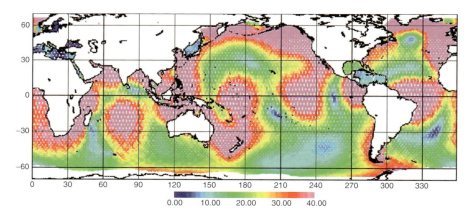

Fig. 5.16 The standard deviation of the ocean tide correction given in centimeters

Fig. 5.16 shows the standard deviation of the ocean tides correction, which closely resembles the amplitude of the largest ocean tide constituent M_2 in Fig. 5.15, indicating that in large parts of the ocean, the standard deviation of the ocean tide signal is larger than 50 cm.

The tidal range is larger in coastal regions than in the open ocean, and coastal tidal waves are much more complex. The pattern of the tidal waves is scaled down as the speed of the tidal wave reduces, because bottom friction modifies the progressions of the

Fig. 5.17 Standard deviation of residual sea surface height variation from 6 years of Jason-1 observations applying the FES2004 and GOT4.7 ocean tide models

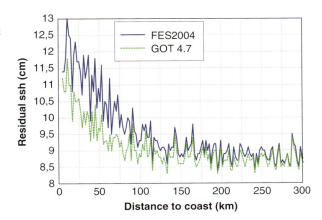

tidal waves. Similarly, resonance and near-resonance responses add to the complexity of the tidal pattern and produce some of the world's largest tidal amplitudes like in the Irish Sea (Britain) or in the Bay of Fundy, where amplitudes reach 10 m. This also means that mis-modelling or omission of tidal constituents in particularly coastal regions, will potentially create large differences between the ocean tide models like FES2004 and GOT4.7.

In order to investigate the correction in coastal zones, Fig. 5.17 shows the standard deviation of residual sea surface height variation applying the FES2004 and GOT4.7 ocean tide models. It clearly illustrates, that in the deep ocean there is virtually no difference between the two models. However in coastal region there is significant difference in the models and here GOT4.7 is clearly reducing more sea surface height variability than FES2004.

In order to explore the difference between the FES2004 and GOT4.7 ocean tide model a bit further, the spatial pattern of the standard deviation of residual sea surface height variations of FES04 minus GOT4.7 is given in Fig. 5.18. Yellow and red regions indicate that GOT4.7 reduces the residual sea surface variability more than FES2004. The improvement ranges up to 2–4 cm in some regions and the largest improvement is clearly seen in coastal regions which are known for very large coastal tides like the North Sea, the Patagonian shelf, the Yellow Sea, and around Indonesia.

This investigation stresses the importance of further improving ocean tide modelling in coastal regions. Improving the ocean tide model in particular shallow water region is still the single correction where the largest accuracy gain can still be made.

Recent investigations by, e.g., Ray (2008) and Roblou and Lyard (2008), also demonstrate the potential improvement of future ocean tide models by improving bathymetry, improved modelling (3D), improved inversion strategies, modelling of wetting and drying, an inclusion of more shallow water constituents (e.g., Andersen et al. 2006b) and inclusion of data from the TOPEX and Jason-1 interleaved missions.

One approach is using data assimilation with nested high-resolution models tied to lower resolution global models. This approach appears feasible, but it places severe demands on the accuracy and resolution of bathymetry data. However, Ray et al. (2009)

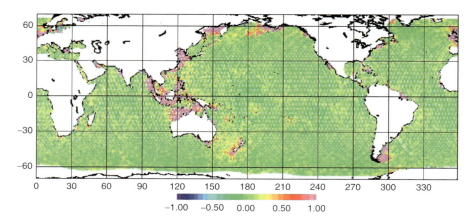

Fig. 5.18 The standard deviation of residual sea surface height variations computed from the difference between FES2004 and GOT4.7 (in cm). Where the signal is yellow and red GOT4.7 reduces the residual sea surface variability more than FES2004

demonstrate the importance of using this approach for future development of future global ocean tide models.

In the deep ocean, recent investigations showed that ocean tide has an accuracy of around 1.4 cm (e.g., Bosch 2008). However, the global ocean tide model still have errors exceeding 10–20 cm close to the coast as also demonstrated by Ray (2008) in a comparison with coastal tide gauges.

5.8.2
Solid Earth Tide Correction

The solid Earth tide is the response of the solid Earth to gravitational forces of the Sun and the Moon. As the response is fast enough to be considered in equilibrium with the tide-generating forces, the tidal elevation is proportional to the tidal potential. The proportionality is determined by the Love numbers, and the solid Earth tide is computed using closed formulas as described by Cartwright and Taylor (1971) and Cartwright and Edden (1973). The magnitude of the solid earth tide ranges up to ±20 cm and the correction is assumed to be highly accurate. The permanent part of the Earth tide, due to the simple presence of Sun and Moon, is explicitly *excluded* since the solid earth tide correction only deals with temporal height variations.

5.8.3
Pole Tide Correction

The variation of the Earth's axis of rotation with respect to its mean geographic pole has a period around 428–435 days and amplitude varying between 0.002″ and 0.30″. This variation is called the *Chandler Wobble* after Seth Carlo Chandler, who confirmed its existence in 1891 through observations of latitude. The changes in the centrifugal forces due to the variation of the Earth's axis produce a signal in sea surface height at the same frequency, called the pole tide. Unlike other luni-solar tides, the pole tide has nothing to do with the gravitational attraction of the Sun and Moon. Furthermore, the tide has the ability to modify what is presumably its own generation mechanism (i.e., the Chandler Wobble). The pole tide is computed as described by, e.g., Wahr (1985) to high accuracy.

5.9
Dynamic Atmosphere Correction

The ocean reacts roughly as a huge inverted barometer, coming up when atmospheric pressure is low, and down when pressure rises. The sea surface height correction due to variations in the atmosphere, is divided into low-frequency contribution (periods longer than 20 days), and a high-frequency contribution (periods shorter than 20 days).

For the low-frequency contribution the classical inverse barometer correction is used to account for the *presumed* hydrostatic response of the sea surface to changes in atmospheric pressure (Wunsch and Stammer 1997). One hecto-Pascal increase in atmospheric pressure depresses the sea surface by about 1 cm. Consequently, the instantaneous correction to sea level can be computed directly from the surface pressure and in units of cm it is given like

$$\Delta h_{ib} \approx -0.99484(P_0 - P_{ref})$$

where P_0 can be determined from the dry troposphere correction. P_{ref} is the global "mean" pressure (reference pressure). Traditionally a constant global value of 1,013.3 hPa has been used as this value is the average pressure over the globe.

However, the mean pressure over the globe is not identical to the mean pressure over the ocean. Fig. 5.19 shows that the mean pressure over the ocean is closer to 1,011 hPa. Furthermore, the figure shows that the global mean pressure is not constant, but has annual amplitude of roughly 0.6 hPa. When applying an inverse barometer correction with a

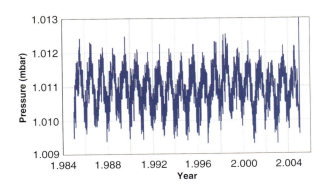

Fig. 5.19 Global mean pressure integrated over the oceans from the ECMWF model for the period 1985–2005. The global mean pressure is around 1,011 mbar (hPa) and it has a clear annual signal with amplitude of roughly 0.6 mbar

non-constant reference pressure, the standard deviation of the residual sea surface height signal is lower than when applying an inverse barometer correction with a constant reference pressure. This shows that using a non-constant reference pressure for the low-frequency part of the correction better corrects atmospheric pressure effects.

The ocean has a dynamic response to pressure forcing at high frequencies (periods below 5 days) and at high latitudes. Furthermore, investigations have shown that wind effects prevail around the 10-day period (Fukumori et al. 1998; Ponte and Gaspar 1999; Fu et al. 2003; Pascual et al. 2008) and that signals up to this frequency should be removed as they might be aliased by the sub-Nyquist sampling of the satellite (9.9156 days for Jason-1, 35 days for Envisat) into lower frequency signals of longer periods greater than twice the sampling interval for the satellite (i.e., longer than 20 days for Jason-1), and this will pollute the sea surface height signal if not corrected.

Global adjustment for high-frequency sea level variability has been implemented using a MOG2D model (Carrère and Lyard 2003). The MOG2D (two-dimensional Gravity Waves model) is a barotropic, nonlinear, and time-stepping model with the model's governing equations being classical shallow water continuity and momentum equations. The finite elements space discretization, allows for increasing the resolution in shallow water areas and other areas of interest like strong topographic gradients-areas. The grid size ranges from 100 km in the deep ocean to 10 km in coastal, shallow seas. MOG2D_IB is a combined dynamic atmosphere correction that combines an inverse barometer correction for time scales longer than 20 days with a MOG2D model for high-frequency variability with timescales shorter than 20 days.

The spatial distribution of the mean and variance of the MOG2D_IB correction is given in Fig. 5.20 and is quite similar to the spatial pattern of the dry troposphere correction in Fig. 5.3, but with quite different amplitudes: mean range is from −10 cm around the 30° parallels to 15 cm at high latitude. Similarly, the standard deviation of the correction changes dramatically from a few cm around the Equator to 10–15 cm at high latitudes.

Investigation of the inverse barometer correction and the MOG2D_IB dynamic atmosphere correction in coastal regions is shown in Fig. 5.21. Again, the standard deviation of residual sea surface height from 6 years of Jason-1 data applying the two models is shown. The figure reveals that MOG2D_IB removes considerably more sea surface variability in coastal regions compared with the traditional inverse barometer corrections. This is a direct effect of the inclusion of correction for temporal frequencies shorter than 20 days in the MOG2D_IB correction. In many regions, the improvement is as large as 10–15 mm, corresponding to 10–30% reduction of the residual sea surface height signal.

A closer investigation into which coastal regions benefit from the use of the MOG2D_IB model can be seen in Fig. 5.22. This shows the difference between residual sea surface height variation from MOG2D_IB and from a local inverse barometer correction. The largest improvement is found in the North Sea and in the Yellow Sea where the improvement is more than 30% compared with the inverse barometer correction, but also large regions in the Antarctic Circumpolar Current region benefit from using the MOG2D_IB model.

The accuracy of the dynamic atmosphere correction is naturally limited by the uncertainty in the sea surface atmosphere pressure with global accuracy estimate of roughly 1 cm (Fu and Cazenave 2001). The error is somewhat higher closer to the coast where the

5 Range and Geophysical Corrections in Coastal Regions: And Implications for Mean Sea Surface Determination   131

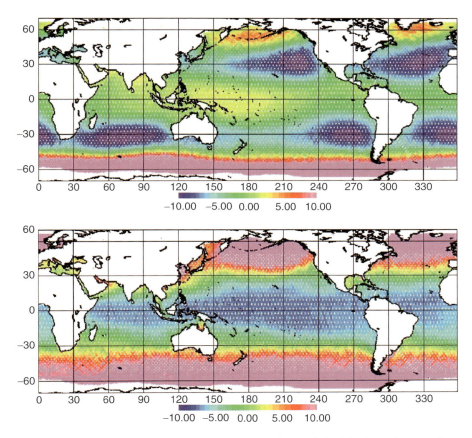

Fig. 5.20 The mean (*upper plot*) and standard deviation (*lower plot*) of the dynamic atmosphere correction based on MOG2D_IB correction. Values are given in centimeters

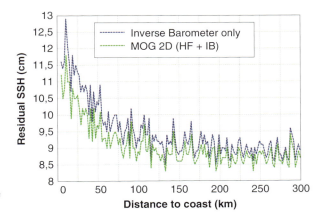

Fig. 5.21 Standard deviation of residual sea surface height variation (in cm) applying the inverse barometer correction and the MOG2D_IB

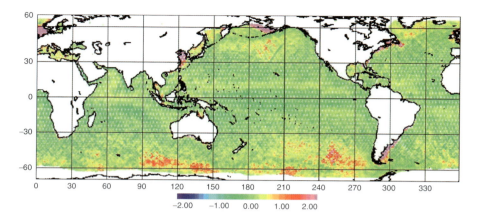

Fig. 5.22 Difference between standard deviation of residual sea surface height variation from MOG2D_IB and a local inverse barometer correction. Region of red and yellow colours indicate, that MOG2D_IB removes more sea surface height variability by including a correction for periods shorter than 20 days. Units are in centimeters

inclusion of the high-frequency component will increase the error, even though it still removes more dynamic atmospheric related signal.

The presented comparison clearly demonstrates the importance of including the high-frequency variability extension to the local inverse barometer correction in shallow water regions for further improvement of the dynamic atmosphere correction in the future. Numerous local surge models are available for the high-frequency part of the dynamic atmosphere correction. The use of high-frequency sea level modelling by applying high-resolution storm surge models is further pursued in the chapter by Woodworth and Horsburgh, this volume, who demonstrate the successful application of storm surge modelling results on the Northwest European shelf. The application of local models has both pros and cons but the patching of these models into a global model is a huge task and the largest obstacle for using this approach in the future. Woodworth and Horsburgh also make clear that the continued development of global barotropic models is important to provide corrections for all coastal areas in future coastal altimetric product.

5.10
Geoid and Mean Sea Surface Models

The geoid height dominates the set of geophysical corrections with variation ranging from −105 m south of India to +85 m north east of New Guinea and spatial variations in the geoid are due to spatial inhomogeneities in the density of the Earth interior and crust as well as variations in the depth of the ocean. The mean sea surface mimics the geoid to within a few meters and consequently the two look alike when plotted (see Fig. 5.2).

For different applications either the geoid or the mean sea surface can be removed from the sea surface height observation. If the study concerns temporal variations of the sea surface from the mean as, e.g., storm surge studies, then the mean sea surface should be removed and not the geoid variations. If the issue is studies of large-scale ocean circulation, then the geoid should be removed. Both of these surfaces can be determined with an accuracy of 10–20 cm (e.g., Pavlis et al. 2008; Andersen and Knudsen 2009).

The global geoid is mainly determined from a compilation of gravity derived from GRACE, satellite altimetric, terrestrial, airborne, and marine observations along with numerous other data sources. The accuracy of the geoid is also fundamental to the precision of orbit determination and therefore in two ways fundamental to the accuracy of sea surface height observations. With the release of the EGM2008 the RMS uncertainty is around 10 cm for the ocean (Pavlis et al. 2008).

The mean sea surface height is a geometrical description of the averaged sea surface height. The DNSC08MSS model is derived from a combination of 12 years of satellite altimetry from a total of nine different satellite missions covering the period 1993–2004 and the RMS uncertainty is between 5 and 10 cm. The DNSC08MSS height was shown previously in Fig. 5.2.

The mean sea surface and the geoid do not coincide because the dynamic sea surface height h_D has both a permanent and a time-variable component. The permanent component reflects the steric expansion of seawater as well as the ocean circulation that is very nearly in geostrophic balance. The temporal average of the dynamic topography is called the mean dynamic topography h_{MDT}. The mean dynamic topography is the quantity that bridges the mean sea surface with the geoid height, since

$$h_{MSS} = h_{geoid} + h_{MDT}$$

Each of the three quantities can, in principle, be derived from the other two quantities. Long-term mean dynamic topography MDT (Fig. 5.23) can be derived from compilation of 100 years of ship data (Loziers et al. 1994, 1995) but it can also be computed from

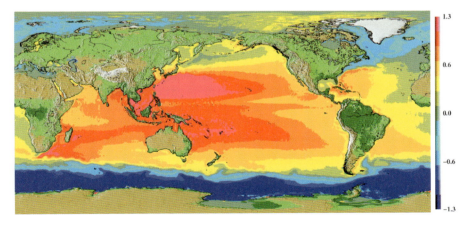

Fig. 5.23 An example of the DNSC08MDT created from filtered differences between the DNSC08MSS and the EGM2008 geoid model. The colour scale is in meters

altimetry and geoid information (e.g., Andersen and Knudsen 2009) or from a merger of both (e.g., Rio et al. 2005). In principle these mean dynamic topography models should agree, but in reality they show very interesting differences (Andersen and Knudsen 2009). With the increased accuracy of new geoid and new mean sea surface models the agreement has been improved significantly, but there are still inconsistencies to be explained in the equation due to inaccuracies in each quantity (Andersen et al. 2006a).

5.11
Reference Frame Offsets

For joint use of multi-mission studies of sea level variability using satellite altimetry it is important to correct for reference frame differences between the various altimeters, which includes differences between the orbits as well as some other geographical differences in the altimeter-dependent models (like sea-state bias).

One example of these reference frame offsets are models expressed by five spherical harmonic coefficients: C_{00} (a constant bias), C_{10}, C_{11}, S_{11} (shifts in Z, X, and Y direction), and C_{20} (a difference in flattening). Though these are not standard range or geophysical corrections, they can be treated as a "standard" correction along with the others and this way multi-mission altimetry will be in the same reference frame. Reference frame offsets are implemented in the RADS data archive like a standard sea surface height correction and can easily be applied by the user.

Because it is eventually arbitrary which offset to trust as reference, we chose TOPEX. So all these spherical harmonic coefficients are 0 for TOPEX and the coefficients for the other satellites are relative to the TOPEX coefficients (see Fig. 5.24).

5.12
Range Corrections and Mean Sea Surface Determination

For oceanographic applications of satellite altimetry, a MSS is used along with the sea surface height observations to create sea level anomalies (h_{sla}) as the MSS can be given with higher accuracy than the geoid along track. For ocean circulation studies, the Mean Dynamic Topography (MDT) is the fundamental parameter. The MSS is determined by averaging satellite-derived sea surface height observations over time. In many cases, the MDT model is determined from the mean sea surface and the geoid (e.g., Rio and Andersen 2009). In some Level-3 products from, e.g., AVISO, the MDT computed from an MSS minus a geoid model is even added back to the altimetric anomalies (Level-3 products) to compute absolute altimetric heights.

It is therefore important that the suite of standard corrections applied to determine the MSS is the same as those used to compute the altimetric anomalies. There is otherwise a possibility that the differences in the corrections will show up the altimetric signal. The issue is equally important for the use of satellite altimetry in both open ocean and in shelf and shallow water regions. This means that the user should be aware of checking that the sea level anomalies (h_{sla}) has been calculated using the same set of corrections that was

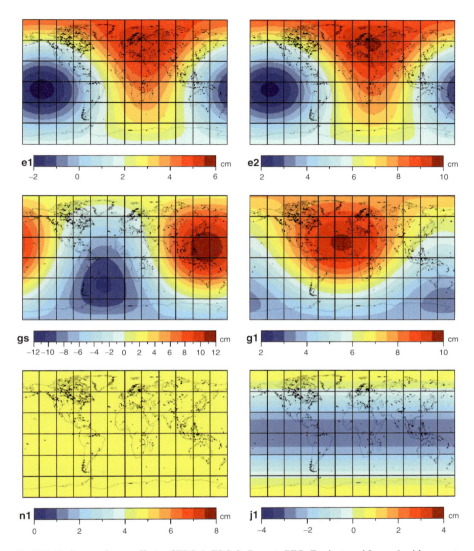

Fig. 5.24 Reference frame offsets of ERS-1, ERS-2, Geosat, GFO, Envisat, and Jason-1 with respect to TOPEX

used for the MSS. If not, the user might get erroneous results, caused by the difference in corrections rather than real ocean dynamic topography.

In order to investigate the scale of the problem, three recent global MSS models were studied for their differences. The MSS height above the reference ellipsoid generally ranges between ±100 m and the difference between various MSS models seldom exceeds 10 cm. This is less than 0.1% of the total range. However, the magnitude of the error can be critical for studies of ocean dynamics.

The differences can be related to three factors: different range corrections applied; different orbital solutions; or a different averaging period for the mean sea surface determination.

The presently two most widely used global MSS models are the DNSC08MSS (Andersen and Knudsen 2009) and the CLS01 MSS model (Hernandez and Schaeffer 2000). The CLS01 MSS model is based on 7 years of satellite altimetry data covering the period 1993–2000 whereas the DNSC08MSS is based on 12 years of data based on the 1993–2004 period. Both of these mean sea surface models are actually already "outdated" with respect to state of the art range corrections, as these have been updated since the models were developed. Therefore an updated version of DNSC08MSS using state of the art range and geophysical corrections was introduced in order to further explore the impact of using an updated set of corrections. Table 5.4 shows the range and geophysical corrections applied for the derivation of the various MSS models.

The difference between the DNSC08MSS and CLS01 mean sea surfaces is shown in Fig. 5.25, The difference between the two mean sea surfaces has a standard deviation of 2.3 cm.

The first important contributor to the difference between the two MSS models is the fact that CLS01 and DNSC08MSS have adopted a different mean reference pressure for the inverse barometer correction (it is otherwise identical). The CLS01 model uses the mean average pressure over the ocean of 1,011 mbar, whereas the DNSC08MSS uses the global mean of 1,013 mbar. This generates a 2 cm global offset or bias between the two mean sea surfaces. This bias has been removed from Fig. 5.25 in order to center the colour scale, so this difference will not show up.

The second feature, which shows up in the difference in Fig. 5.25, is the north–south oriented striations. This striation originates from a combination of different range corrections and is most prominently seen in the central Pacific Ocean. This effect is subscribed to

Table 5.4 The set of range and geophysical corrections applied for the generation of CLS01, DNSC08, and the updated DNSC08 MSS. Only corrections applied to the TOPEX/Poseidon and Jason-1 have been shown. The corrections applied to different satellites missions (ERS-1, ERS-2, Geosat, GFO, etc.) included in the determination of the MSS will be different

Standards	CLS01	DNSC08MSS	Updated DNSC08MSS
Reference period	1993–1999	1993–2004	1993–2004
Orbit	ORB_POE_N	GGM02/ITRF2000	EIGEN-GL04C
Dry troposphere	ECMWF	ECMWF	ECMWF
Wet troposphere	Radiometer	ECMWF	Radiometer
Ionosphere	Altimeter	Altimeter	Altimeter
Dynamic atmosphere	IB (1,011 mbar)	IB (1,013 mbar)	MOG-2D_IB
Ocean tides	GOT 99	GOT 00.2	GOT4.7
Sea-state bias	BM4	BM4	Non-PARAM

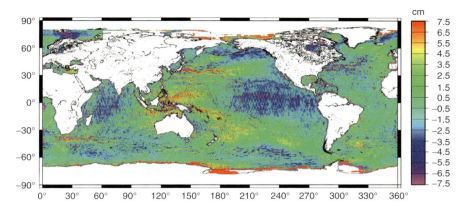

Fig. 5.25 The difference between DNSC08MSS and CLS01 Mean Sea surfaces. An offset of 2 cm has been removed from the difference due to different reference pressure used for the Dynamic Atmosphere correction (Figure courtesy of S. Holmes and N. Pavlis [NGA])

the way that different corrections are averaged in the computation of TOPEX and particularly ERS-2 mean profiles for the MSS models. It is related to averaging of sun-synchronous altimetric observations by particularly the ERS-2 satellite, as ERS-2 data is an important data source in both CLS01 and the DNSC08MSS models.*

The third contributor is the oceanwide difference of ±5 cm from east to west in the Pacific Ocean. This reflects interannual ocean variability that has been averaged out differently for the two MSS. DNSC08MSS is averaged over 12 years (1993–2004) whereas the CLS01 MSS is referenced to the 7-year (1993–1999) period.

This can be seen by computing the differences between identical processed datasets averaged effect of inter-annual ocean variability to 7 years (1993–1999) or 12 years (1993–2004) as shown in Fig. 5.26. The east-west dipole in the difference between CLS01 and DNSC08MSS in the Pacific Ocean is highlighted in the figure. The feature is largely caused by the 1997–1998 El Niño being averaged differently into the 7-year CLS01 period than in the 12-year period for DNSC08MSS. Besides the Pacific dipole structure, interannual ocean variability related to the location of the large current systems like the Kuroshio Current is seen in the figure.

An examination of the difference between the older set of range corrections used for derivation of the DNSC08MSS and the new standard set of corrections (shown in Table 5.4) and used for the updated DNSC08MSS is shown in Fig. 5.27. The figure is smoothed with a Gaussian smoother to 1,000 km and the same averaging period of 12 years was used.

The difference has a mean of 2.2 cm and a standard deviation of 1.9 cm. This reflects the combined effect of different gravity models applied for the orbit computation and different range and geophysical corrections. Most of the mean difference of 2.2 cm is readily explained by the difference in mean pressure for the dynamic atmosphere correction as seen before.

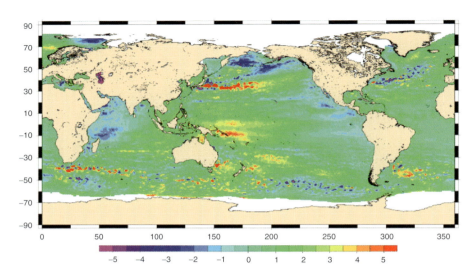

Fig. 5.26 The difference between the temporal mean of 7 and 12 years of TOPEX/Poseidon altimetric data. Interannual ocean variability related to the location of the Kuroshio Current and the El Niño signal is very clearly seen in the figure. The differences range up to ±5 cm with standard deviation of 1.9 cm

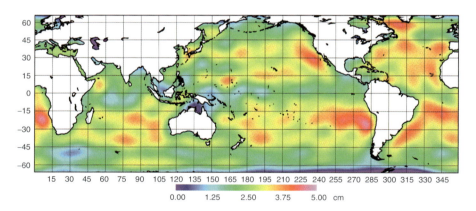

Fig. 5.27 Difference between DNSC08MSS and the updated DNDSC08MSS based on state of the art range corrections. Both MSS have been computed over 12 years for the 1993–2004 period

Only the ocean tide correction, the dynamic atmosphere correction, and the sea-state correction have been updated between the two MSS versions, and it was shown previously, that there is no influence on the mean from the ocean tide correction. Therefore the mean differences must be subscribed to differences in the orbits applied and the dynamic atmosphere and sea-state bias corrections applied.

The difference, due to the use of different sets of orbits creates a difference ranging between 0 and 1 cm with a mean value of 0.15 cm, so the major contributor to the difference must be subscribed to the use of the different range and geophysical corrections.

5 Range and Geophysical Corrections in Coastal Regions: And Implications for Mean Sea Surface Determination 139

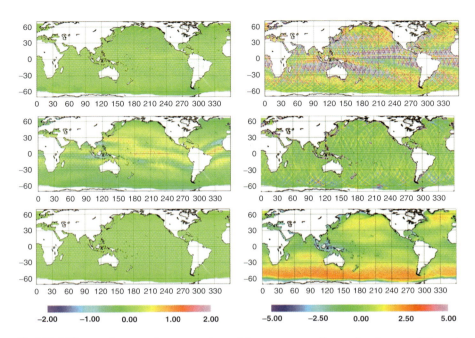

Fig. 5.28 Difference between the mean values of the range corrections using 6 years of Jason-1 altimetry. Upper row is dry and wet troposphere correction. Middle row is ionosphere and dynamic atmosphere correction. Lower row is ocean tide and sea-state bias correction (corresponding to the models listed in Table 5.2). All figures except the sea-state bias correction are shown using a colour scale ranting between ±2 cm. The sea-state bias correction uses ±5 cm

In order to investigate the striation seen in Fig. 5.25 and its relationship to the application of different range and geophysical correction, the difference between 6 years mean profiles using Jason-1 sampled data are shown in Fig. 5.28.

Only the wet troposphere correction in the upper right panel of Fig. 5.28 seems to have striation patterns in the difference between the averaging of the ECMWF model and the radiometer correction. This also corresponds very well with the DNSC08MSS model being produced using the ECMWF wet troposphere correction and the CLS01 MSS being developed using the radiometer wet troposphere correction.

For the dry troposphere, the ionosphere, and the tide ocean tide correction the difference on a six year mean is insignificant. Large differences emerge between the BM4 parametric SSB model and the CLS nonparametric SSB model in the lower right panel of Fig. 5.28. The difference ranges between ±5 cm with some latitudinal pattern. Relatively large differences are found in coastal regions, which indicate that this is an important issue for future research and the users should be aware of consistent use of SSB model for studies of sea level anomalies in coastal regions.

If the user wants to use data from sun-synchronous satellites like ERS-1, ERS-2, and Envisat, the situation is more complicated and more care should be taken.

Fig. 5.3, 5.5, 5.8, 5.10, and 5.12 illustrate the mean value over time of a particular correction computed from six Jason-1 data. If the value was computed using 6 years of Envisat or ERS-2 data a more complicated and different result would emerge. This is because the ERS and Envisat satellites are sun-synchronous. Consequently, sun-synchronous contributions to the range and geophysical corrections will be observed at the same phase and averaging the satellite altimeter observations over time will not average them out. As such these sun-synchronous contributions might end up in the sea level anomalies.

For mean sea surface computation it is important to include the sun-synchronous satellites ERS and Envisat as these satellites have denser ground-tracks and furthermore are the only satellites covering the high latitude regions between the 65 and 82 parallels. However, here the ERS mean profiles are normally fitted to the TOPEX profiles in order to minimize the striation (e.g., Andersen and Knudsen 2009). However striations might still be seen due to the use of ERS and Envisat data.

Fig. 5.29 illustrates a 6-year mean of the difference in state of the art ocean tide and inverse barometer corrections. These plots are "identical" to the plots shown in Fig. 5.28, but computed from ERS-2 rather than Jason-1.

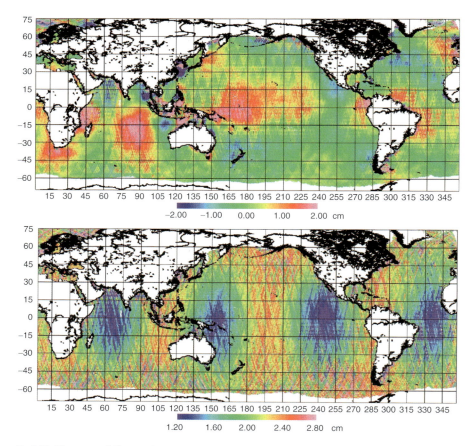

Fig. 5.29 The mean difference between the FES2004 and GOT4.7 ocean tide correction computed from 6 years of ERS-2 data (*upper plot*) and the mean difference between the inverse barometer correction and MOG2D_IB correction (*lower plot*). Values are given in centimeters

Interesting global patterns of highs and lows are seen in both the difference between the applied tide models and difference in the applied dynamic atmosphere correction. Both correspond to aliasing of the S_2 constituent. For the ocean tide the effect is most likely due to the difference in the amplitude of the S_2 tide between the FES2004 and GOT4.7 ocean tide models.

For the dynamic atmosphere correction in the lower panel of Fig. 5.29, the cause is the fact that the S_2 atmospheric tide is included in the MOG2D_IB correction, but not in the inverse barometer correction. Interested readers are referred to Rio and Andersen (2009) for a more detailed investigation of the subject.

It is important that users of satellite altimetry using particularly sea level anomalies (h_{sla}) make sure that the suite of standard corrections applied to the MSS determination is identical to (or at least compatible with) those used to compute the altimetric anomalies. This is equally important in the open ocean and the coastal zone. Otherwise the differences between the applied correction models might contaminate the altimetric sea surface height signal. This is naturally more important for users of sun-synchronous data from ERS-1, ERS-2, and Envisat than for users of the TOPEX/Poseidon, Jason-1, and Jason-2 data.

5.13
Summary

Most range and geophysical corrections need special attention in proximity to the coast: either because the signal is much larger or the correction is less accurate in coastal regions.

The wet troposphere correction is complex and shows high temporal variations, with rapid variations in both time and space and therefore also needs careful attention in the coastal region. However, closer than 50 km to the coast, where the radiometer observations are expected to degrade, the radiometer correction still performs equally accurate to the ECMWF model. This is most likely due to the fact that the ECMWF model also degrades close to the coast where the correction changes rapidly and observations are lacking. Several approaches aim at improving the correction. One uses the Dynamically Linked Model (DLM) based on the dynamic combination of radiometer measurements, model data, and GPS/GNSS-derived path delay estimates. The second approach investigates the measured brightness temperatures in order to remove the contamination coming from the surrounding land, thereby improving the accuracy of the radiometer observations in the coastal zone.

The sea-state bias correction is related to the waveform shape and cannot be estimated close to the coast (<10 km) where changing waveforms prevent the estimation of significant wave height essential to derive the correction. Further away from the coast, the temporal variation of the SSB correction is only a few cm in most places and it is hard to determine if the parametric or non-parametric corrections are performing better in coastal regions.

The high-frequency extension to the inverse barometer correction using a time-stepping finite element model with high resolution in coastal regions showed large improvement in coastal regions like the North Sea and in the Yellow Sea where the improvement is more than 30% compared with the classical inverse barometer correction. However, also large regions in the Antarctic Circumpolar Current region benefited from using the MOG2D_IB model.

Tides are by far the largest contributor to sea level variations and the investigation stresses the importance of further improving ocean tide modelling in coastal regions. In the deep ocean, recent investigations showed that ocean tide models have an accuracy of around 1.4 cm (e.g., Bosch 2008). However, the global ocean tide models still have errors exceeding 10–20 cm close to the coast as also demonstrated by Ray (2008) in a comparison with coastal tide gauges. Particularly in shallow water regions, the largest accuracy gain in altimetric sea level determination is still expected to come from improvement of tidal models.

Altimetric sea level anomalies are determined by removing the mean sea surface from the sea surface height observations. When computing those anomalies, the same suite of range and geophysical corrections should be applied, as those used during the process of determining the mean sea surface. With the accuracy of current sea level observations, differences between the applied correction model used for the sea surface height and mean sea surface models might contaminate the altimetric sea level anomalies. This was shown to be more important for users of sun-synchronous data from ERS-1, ERS-2, and Envisat than for users of the TOPEX/Poseidon, Jason-1, and Jason-2 data.

If determined properly, the dry troposphere and ionosphere corrections are barely contaminated or degraded in coastal regions. When determined by dual-frequency altimeters, land in the altimeter footprint could contaminate the ionosphere correction, but when smoothed properly, this effect is negligible. The dry troposphere correction could suffer land effects only when it is erroneously based on surface pressure instead of sea level pressure grids.

Acknowledgment The authors would like to acknowledge the space agencies for providing altimetric data. M.-H. Rio provided valuable input to the section on range corrections and mean sea surface determination as member of the GOCE User Toolbox development team. Several authors (i.e., Fenoglio-Marc, Eymard, and Pavlis) who provided figures are also acknowledged.

References

Agnew D (1997) NLOADF: a program for computing ocean tide loading. J Geophys Res 102(B3):5109–5110

Andersen OB (1995) Global Ocean tides from ERS-1 and TOPEX/Poseidon altimetry. J Geophys Res 100(C12):25,249-25,260

Andersen OB, Knudsen P (2009) The DNSC08 mean sea surface and mean dynamic topography. J Geophys Res 114:C11001. doi:10.1029/2008JC005179

Andersen OB, Knudsen P, Trimmer R (2005) Improved high resolution gravity field mapping (the KMS02 global marine gravity field). In: Sanso F (ed) A window on the future of geodesy. IAG symposium, vol 128, pp 326–331

Andersen OB, Vest AL, Knudsen P (2006a) The KMS04 multi-mission mean sea surface. In: Knudsen P (ed) GOCINA: improving modelling of ocean transport and climate prediction in the North Atlantic region using GOCE gravimetry. Cahiers du Centre Européen de Géodynamique et de Séismologie, vol 25, pp 103–106, Luxembourg

Andersen OB, Egbert GD, Erofeeva SY, Ray RD (2006b) Mapping nonlinear shallow-water tides: a look at the past and future (Honorary symposium for C. Le Provost). Ocean Dyn 56:416–429

Berry PAM, Garlick JD, Freeman JA, Mathers EL (2005) Global inland water monitoring from multi-mission altimetry. Geophys Res Lett 32(16):L16401. doi:10.1029/2005GL022814

Bertiger W, Desai S, Dorsey A, Haines B, Harvey N, Kuang D, Lane C, Weiss J, Sibores A (2008) Jason-2, Precision orbit determination status, OSTST workshop, Nice, Nov 2008, http://www.aviso.oceanobs.com/fileadmin/documents/OSTST/2008/oral/bertiger.pdf

Bilitza D (2001) International reference ionosphere 2000. Radio Sci 36(2):261–275. doi:10.1029/2000RS002432

Bilitza D, Reinisch BW (2008) International reference ionosphere 2007: Improvements and new parameters. Adv Space Res 42(4):599–609. doi:10.1016/j.asr.2007.07.048

Bosch W (2008) EOT08a model performances near coasts. Presented at Second coastal altimetry Workshop, Pisa, Italy, 6–7 November 2008

Brown GS (1977) The average impulse response of a rough surface and its applications. IEEE Trans Antennas Propag AP-25(1):67–74

Brown S (2008) Novel near-land radiometer wet path delay retrieval algorithm: Application to the advanced microwave radiometer. Presented at Second coastal altimetry workshop, Pisa, Italy, 6–7 November 2008

Carrère L, Lyard F (2003) Modelling the barotropic response of the global ocean to atmospheric wind and pressure forcing – comparisons with observations. Geophys Res Lett 30(6):1275

Cartwright DE, Edden AC (1973) Corrected tables of tidal harmonics. Geophys J Royal Astron Soc 33:253–264

Cartwright DE, Taylor RJ (1971) New computations of the tide-generating potential. Geophys J Royal Astron Soc 23:45–74

Chelton DB (1994) The sea state bias in altimeter estimates of sea level from collinear analysis of TOPEX data. J Geophys Res 99(C12):24,995–25,008

Eymard L, Obligis E (2006) The altimetric wet troposphere correction: progress since the ERS-1 mission. In: Symposium 15 years of progress in satellite altimetry, Venice, Italy, 2006

Fenoglio-Marc L (2008) Retracking in the (NW) Mediterranean Sea. Presented at Second coastal altimetry workshop, Pisa, Italy, 6–7 November 2008

Fu LL, Cazenave A (2001) Satellite altimetry and earth sciences: a handbook of techniques and applications. Academic, San Diego, CA, 624pp

Fu LL, Christensen EJ, Yamaron CA, Lefebvre M, Ménard Y, Dorrer M, Escudier P (1994) TOPEX/Poseidon mission overview. J Geophys Res 99(C12):24,369–24,382

Fu LL, Stammer D, Leben D, Chelton R (2003) Improved spatial resolution of ocean surface topography from the T/P-Jason-1 altimeter mission. EOS Trans AGU 84(26):241–242

Fukumori J, Raghunathand R, Fu LL (1998) Nature of global large scale sea level variability in relation to atmospheric forcing: a modelling study. J Geophys Res 99:24369–24381

Gaspar P, Florens JP (1998) Estimation of the sea state bias in radar altimeter measurements of sea level: results from a new nonparametric method. J Geophys Res 103(C8):15803–15814

Gaspar P, Ogor F, Le Traon PY, Zanifé OZ (1994) Estimating the sea state bias of the TOPEX and Poseidon altimeters from crossover differences. J Geophys Res 99(C12):24981–24994

Goldhirsh J, Rowland JR (1992) A tutorial assessment of atmospheric height uncertainties for high-precision satellite altimeter missions to monitor ocean currents. IEEE Trans Geosci Remote Sensing 20:418–434

Hernandez F, Schaeffer P (2000) Altimetric mean sea surfaces and gravity anomaly maps inter-comparisons AVI-NT-011-5242-CLS. CLS Ramonville St Agne, France, 48pp

Iijima BA, Harris IL, Ho CM, Lindqwister UJ, Mannucci AJ, Pi X, Reyes MJ, Sparks LC, Wilson BD (1999) Automated daily process for global ionospheric total electron content maps and satellite ocean altimeter ionospheric calibration based on global positioning system data. J Atmos Solar-Terr Phys 61:1205–1218. doi:10.1016/S1364-6826(99)00067-X

Imel DA (1994) Evaluation of the dual-frequency ionosphere correction. J Geophys Res 99(C12):24895–24906

Keihm SJ, Zlotnicki V, Ruf CS (2000) TOPEX microwave radiometer performance evaluation, 1992–1998. IEEE Trans Geosci Remote Sensing 38(3):1379–1386

Komjathy A, Born GH (1999) GPS-based ionospheric corrections for single frequency radar altimetry. J Atmos Solar-Terr Phys 61(16):1197–1203. doi:10.1016/S1364-6826(99)00051-6

Komjathy A, Sparks L, Wilson BD, Mannucci AJ (2005) Automated daily processing of more than 1000 ground-based GPS receivers for studying intense ionospheric storms. Radio Sci 40:RS6006. doi:10.1029/2005RS003279

Labroue S, Gaspar P, Dorandeu J, Mertz F, Zanifé OZ (2006) Overview of the Improvements made on the empirical determination of the sea state bias correction. In: Symposium 15 years of progress in radar altimetry, Venice, Italy, 2006

Labroue S, Tran N, Vandermark D, Scharroo R, Feng H (2008) Sea state bias correction in coastal waters. In: Second coastal altimetry workshop, Pisa, Italy, 6–7 November 2008

Le Provost C, Genco ML, Lyard F, Vincent P, Canceil P (1994) Spectroscopy of the world ocean tides from a finite element hydrodynamic model. J Geophys Res 99(C12):24777–24,797

Llewellyn SK, Bent RB (1973) Documentation and description of the Bent ionospheric model. Rep. AFCRL-TR-73-0657. Air Force Cambridge Research Laboratory, Hanscom Air Force Base, Massachusetts

Loziers MS, McCarthy MS, Owens WB (1994) Anomalous anomalies in averaged hydrographic data. J Phys Oceanog 24:2624–2638

Loziers MS, Owens WB, Curry RG (1995) The climatology of the North Atlantic. Proc Oceanogr 36:1–44

Lyard F, Lefèvre F, Letellier T, Francis O (2006) Modelling the global ocean tides: modern insights from FES2004. Ocean Dyn 56(5–6):394–415. doi:10.1007/s10236-006-0086-x

McCarthy DD (1996) IERS Standards (1996) IERS Technical Note 21, Central Bureau of IERS, Observatoire de Paris, France

Naeije MC, Doornbos E, Mathers L, Scharroo R, Schrama EJO, Visser P (2002) RADSxx: exploitation and extension, NUSP-2 report 02-06, SRON/NIVR, Delft, the Netherlands

Obligis E, Desportes C, Eymard L, Fernandes J, Lazaro G, Nunes A (2009) Tropospheric corrections for coastal altimetry, ibid

Pascual A, Marcos M, Gomis D (2008) Comparing the sea level response to pressure and wind forcing of two barotropic models: validation with tide gauge and altimetry data. J Geophys Res 113:C07,011. doi:10.1029/2007JC004459

Pavlis NK, Holmes SA, Kenyon SC, Factor JK (2008) An Earth gravitational model to degree 2160: EGM2008. Presented at the 2008 General Assembly of the European Geosciences Union, Vienna, Austria, 13–18 April

Ponte RM, Gaspar P (1999) Regional analysis of the inverted barometer effect over the global ocean forced by barometric pressure. J Geophys Res 104(C7):15587–15602

Ray RD (2008) Tide corrections for shallow-water altimetry: a quick overview. Presented at Second coastal altimetry workshop, Pisa, Italy, 6–7 November 2008

Ray RD, Koblinsky CJ, Beckley BD (1991) On the effectiveness of Geosat altimeter corrections. Int J Remote Sensing 12:1979–1984

Ray RD, Egbert G, Erofeeva SY (2009) Tide prediction in shelf and shallow water; Status and prospects, ibid

Rio MH, Andersen O (2009) GOCE user toolbox – standards and recommended models. ESA publication. http://earth.esa.int/gut/, 70pp

Rio MH, Schaeffer P, Hernandez F, Lemoine JM (2005) The estimation of the ocean mean dynamic topography through the combination of altimetric data, in-situ measurements and Grace geoid: from global to regional studies. In: Proceedings of the GOCINA workshop in Luxemburg, 13–15 April 2005

Roblou L, Lyard F (2008) Tidal modelling in coastal and shelf seas: methodology and application to the Persian Gulf. Presented at Second coastal altimetry workshop, Pisa, Italy, 6–7 November 2008

Rodrigues E, Martin JM (1994) Estimation of the electromagnetic bias from retracked TOPEX data. J Geophys Res 99:24957–24969

Ruf CS (2002) Characterization and correction of a drift in calibration of the TOPEX microwave radiometer. IEEE Trans Geosci Remote Sens 40(2):509–511

Saastamoinen J (1972) Atmospheric corrections for the troposphere and stratosphere in radio ranging of satellites. In: Hendriksen SW, Mancini A, Chovitz BH (eds) The use of artificial satellites for geodesy. Geophysical Monograph Series 15:247–251, AGU, Washington, DC

Scharroo R, Lillibridge JL (2005) Non-parametric sea-state bias models and their relevance to sea level change studies. In: Lacoste H, Ouwehand L (eds) Proceedings of the 2004 Envisat & ERS symposium, European Space Agency Special Publication, ESA SP-572

Scharroo R, Smith WHF (2010) A GPS-based climatology for the total electron content in the ionosphere. J Geophys Res (in press)

Scherneck HG (1990) Loading Greens function for a continental shelf with a Q structure for the mantle and density constraints from the geoid. Bulletin d'Information des Marées Terrestres 108:7775–7792

Schrama EJO, Scharroo R, Naeije M (1997) Radar altimeter database system (RADS): towards a generic multi-satellite altimeter database system, Final Report. 88p. SRON/BCRS publication, USP-2 report 00-11, ISBN 90-54-11-319-7

Schreiner WS, Markin RE, Born GH (1997) Correction of single frequency altimeter measurements for ionosphere delay. IEEE Trans Geosci Remote Sensing 35(2):271–277. doi:10.1109/36.563266

Smith EK, Weintraub S (1953) The constants in the equation for atmospheric refractive index at radio frequencies. Proc IRE 41:1035–1037

Strub T (2008) Coastal altimeter evaluation of Jason-1 and Jason-2. Presented at OSTST meeting, Nice 2008

Wahr J (1985) Deformation of the Earth induced by polar motion. J Geophys Res 90(B11): 9363–9368

Wahr J, Swenson S, Zlotnicki V, Velicogna I (2004) Time-variable gravity from GRACE: first results. Geophys Res Lett 31(L11):501. doi:10.1029/2004GL019779

Web1: JPL Global Ionosphere Model, http://iono.jpl.nasa.gov/gim.html

Wunsch C, Stammer D (1997) Atmospheric loading and the oceanic "inverted barometer" effect. Rev Geophys 35:79–107

Tropospheric Corrections for Coastal Altimetry   6

E. Obligis, C. Desportes, L. Eymard, J. Fernandes, C. Lázaro, and A. Nunes

Contents

6.1 Introduction 148
6.2 Combination and Processing of Available Information 149
    6.2.1 Dynamically Linked Model Approach 150
    6.2.2 GNSS-derived Path Delay Method 152
6.3 Brightness Temperature Contamination and Correction... 162
    6.3.1 Instrumental Configuration and Reference Field 163
    6.3.2 An Analytical Correction of TBs: The "erf Method" 164
    6.3.3 Using the Proportion of Land in the Footprint.. 167
    6.3.4 Performance Analysis over Real Measurements 168
    6.3.5 Discussion 170
6.4 Dry Tropospheric Correction 170
6.5 Conclusions and Perspectives 172
References 174

Keywords Altimetry • GPS • Microwave radiometry • Water vapor • Wet tropospheric correction

---

E. Obligis (✉)
Collecte Localisation Satellites, 8-10 rue Hermes, 31520, Ramonville St-Agne, France

C. Desportes
Mercator Ocean, 8-10 rue Hermès, 31520 Ramonville, St Agne, France

L. Eymard
IPSL/LOCEAN, Université Pierre et Marie Curie, 4 Place Jussieu, 75252, Paris, France

M. J. Fernandes and C. Lázaro
Faculdade de Ciências da Universidade do Porto, Departamento de Matemática Aplicada, Rua do Campo Alegre, 687, 4169-007, Porto, Portugal

A. L. Nunes
Instituto Superior de Engenharia do Porto, Rua Dr. António Bernardino de Almeida, 431, 4200-072, Porto, Portugal

Abbreviations

2T	2-meter Temperature
COASTALT	ESA development of COASTal ALTimetry
CorSSH	corrected Sea Surface Height
dB	Decibel
DLM	Dynamically Linked Model
ECMWF	European Centre for Medium-Range Weather Forecasts
EPN	EUREF Permanent Network
ESA	European Space Agency
GDR	Geophysical Data Record
GFO	Geosat Follow-On
GMF	Global Mapping Functions
GNSS	Global Navigation Satellite System
GPD	GNSS-derived Path Delay
GPS	Global Positioning System
IGS	International GNSS Service
MWR	Microwave Radiometer
NCEP	U.S. National Centers for Environmental Prediction
NWM	Numerical Weather Model
PD	Path Delay
SSM/I	Special Sensor Microwave Imager
STD	Slant Total Delay
T/P	TOPEX/Poseidon
TB	Brightness Temperature
TCWV	Total Column Water Vapour
TMR	TOPEX Microwave Radiometer
VMF1	Vienna Mapping Functions 1
ZHD	Zenith Hydrostatic Delays
ZWD	Zenith Wet Delays

6.1
Introduction

The exploitation of altimetric measurements over ocean relies on the capability to correct the altimeter range from all external perturbations. Two of them are related to the troposphere characteristics and should be estimated to properly correct the altimeter range: the wet and the dry path delays.

The wet tropospheric path delay is almost proportional to the integrated water vapor content of the atmosphere, and strongly affects the range measured by the altimeter. It varies between 0 and 50 cm but with a high variability in space and time. Up to now, meteorological models do not satisfactorily describe water vapour variations (pattern location and extension, magnitude, time evolution, etc…), so a dedicated instrument is added to the mission, a microwave radiometer. However, the radiometer footprint size is much larger than the altimeter one (about

25 km and 5 km for the radiometer and the altimeter, respectively). In open ocean, the combination altimeter/radiometer is satisfactory both in terms of accuracy and spatial resolution, because of the mean scale of major meteorological events (from a few tens of kilometers to several hundreds of kilometers). This is not the case for transition areas (sea/land), because of the radiometer measurement specificities: the given footprint is the 3 dB diameter, meaning that roughly half the signal is coming from outside this footprint. Consequently, the signal coming from the surrounding land surfaces (with a strong and highly time-varying emissivity compared to ocean) contaminates the measurement and makes the humidity retrieval method unsuitable. Furthermore, the spatial and temporal resolutions of meteorological models are not sufficient enough to handle the specific problem related to coastal atmospheric variability: surface temperature gradients between land and sea or relief, for example, may generate specific small-scale atmospheric features that could not be represented in the current model products. This means that, today, for example, Envisat and Jason-1 altimeter measurements located at less than about 50 km from land cannot benefit from a suitable correction.

Nevertheless, the exploitation of altimeter measurements in coastal areas becomes a challenge for oceanography. Different dedicated studies have been conducted recently and promising methods are identified today to propose a wet tropospheric correction in these transition areas, which could fulfil the constraint related to altimeter measurement processing, in terms of both accuracy and spatial resolution. Two different strategies have been considered to handle the problem specific to the wet tropospheric correction for coastal altimetry.

The first one, explained in Sect. 6.2, consists of combining information in the coastal area to update and improve the radiometer wet tropospheric correction in the coastal band. We present two different methodologies based on the combination of available information: the dynamically linked model (DLM) approach, based on the dynamic combination of radiometer and model data, and the GNSS-derived Path Delay (GPD) method, based on GNSS (Global Navigation Satellite System)-derived path delays combined with radiometer valid measurements and model-based information.

The second one, presented in Sect. 6.3, consists of the correction of the measured brightness temperatures in order to remove the contamination coming from the surrounding land. An accurate correction allows for the subsequent application of a standard ocean algorithm.

The dry tropospheric path delay is proportional to the sea level pressure and is by far the largest correction to be applied to the altimeter range (higher than 2 m). However, its main characteristic is that its time-variable part is low (lower than 5 mm), so it is quite well estimated using meteorological models. In Sect. 6.4, the main issues related to the computation of the dry correction for open ocean are reviewed. For coastal areas, a first assessment of the accuracy of ECMWF (European Centre for Medium-Range Weather Forecasts)-derived dry tropospheric path delay is presented.

6.2
Combination and Processing of Available Information

In order to improve the computation of the wet tropospheric correction in coastal areas, two methods making use of data other than the microwave radiometer measurements have been developed and are described here. The first one only requires numerical weather model

(NWM) data and radiometer wet tropospheric corrections available on the Geophysical Data Record (GDR) product and has the advantage of being mission-independent. The second approach combines information from tropospheric path delays determined at a network of GNSS stations, valid radiometer measurements and NWM-based information.

6.2.1
Dynamically Linked Model Approach

This section describes the dynamically linked model (DLM) approach, which consists of replacing, in coastal regions, the microwave radiometer (MWR) derived correction with a large-scale atmospheric reanalysis model-derived correction, such as ECMWF, dynamically linked to the closest points with valid MWR correction.

It should be emphasized that this approach is significantly different from the use of the ECMWF model correction everywhere. It is well-known that the wet tropospheric models are not suitable for processing long time series of satellite altimetry since they possess long-term errors and discontinuities (Scharroo et al. 2004). In addition, they possess small-scale errors, which may be due to mislocation and absence of some atmospheric features. Although this method does not compensate for the small-scale errors, it significantly reduces the large-scale ones. This happens because the large-scale errors for small regions, such as land-contaminated tracks in coastal zones, are quasi-linear and will be removed by locally adjusting the NWM to the MWR correction.

The method only requires information available on the GDR – optionally, information from a distance-to-land global grid – and can be implemented globally for any satellite. The question is then which particular strategy should be adopted to ensure the continuity in the transition between the MWR and NWM wet tropospheric correction fields.

The definition of the coastal track segments to be corrected can be based on the microwave radiometer land flag, provided this is reliable. Comparisons with GNSS-derived tropospheric delays show that in some close-to-land regions, at distances less than 30 km from the coastline, the MWR correction can be quite noisy, while the corresponding land MWR flag is still off. This problem can be overcome by using distance-to-land information and ensuring that all points with a distance-to-land smaller than a specified value, e.g., 20 km, are also considered contaminated points and must be corrected. The first approach has the advantage that it only requires information present on the GDR (MWR and NWM corrections and the land MWR flag). The second approach depends on the quality of the coastline used to compute the distance-to-land grid and reduces the possibility that land-contaminated points are left uncorrected, but might lead to the unnecessary correction of valid radiometer points.

Considering a simple but efficient implementation, two types of algorithms can be adopted: an island type and a continental coastline type. The first case is suggested by the land contaminated segments formed around relatively small islands or peninsulas, where a land contaminated segment is formed and there are valid microwave radiometer points on each side of the segment. The second case occurs around the continental coastlines, typically when a satellite is approaching or receding from a large land mass. In this case, there is only a valid MWR correction on one of the sides of the contaminated segment.

In the island type of algorithm, the NWM field is adjusted to the MWR field at the beginning and end of the land contaminated segment by using a linear adjustment. The NWM correction is then dynamically linked to the MWR field at the beginning and end of the segment, for example, at the first (A) and last (B) points with valid MWR correction. For all contaminated points between A and B, the final field will be equal to the NWM field plus a linear correction, ensuring that at both A and B the microwave radiometer and adjusted numerical weather model fields are the same, achieving the required continuity.

In the continental coastline type of algorithm, there is only valid information on one side of the contaminated segment and therefore it is not possible to perform an adjustment of the NWM and MWR fields by using a linear adjustment. In this case, the adjustment is performed only at the beginning or end of the segment by using a bias correction computed as the difference between the microwave radiometer and numerical weather model fields at the first (A) or last (B) point with valid MWR field. For all contaminated points between the beginning of the segment and A (or between B and the end of the segment), the final field will be equal to the NWM field plus a bias correction, ensuring that at A or B both fields will be the same and therefore continuous.

Prior to this implementation, GDR data shall be separated into segments where a new segment is considered whenever a data gap greater than a specified value, e.g., 20 s, exists. The algorithm shall then be applied to each data segment. In this way, the only points that will not be recovered are segments for which there are no valid microwave radiometer points close enough to perform the adjustment.

Studies conducted by Fernandes et al. (2003) show that this simple approach leads to a data recovery of 80–90% of the invalid measurements in the coastal regions, does not introduce discontinuities in the correction and can be used to generate coastal products in an operational processing scheme. Mercier (2004) has also applied a similar type of algorithm to improve data recovery near the coast.

Fig. 6.1 illustrates two examples of the application of this algorithm to two regions: a continental coastline and a mixed region. In the map on the right-hand side of the figure, the blue points represent land-contaminated points in the microwave radiometer wet tropospheric correction that must be adjusted by the algorithm. In the graphs on the left, the cyan points represent the ECMWF model correction, the red points the original MWR correction and the blue points the final correction after applying the DLM algorithm. In the coastal regions, the original noisy MWR correction (in red) is replaced by the continuous smoother correction (in blue). The green dots represent the algorithm flag: 2 for points which were contaminated and ended corrected, 1 for points which were contaminated but were not corrected and 0 for points that were not contaminated originally.

The performance of the method is directly related to the accuracy of the adopted NWM. Usually, the GDRs provide the wet tropospheric correction from two models: ECMWF and NCEP (U.S. National Centers for Environmental Prediction).

To inspect the differences between the wet tropospheric correction from the onboard microwave radiometer and from the ECMWF and NCEP models, these have been compared with Envisat cycles 48 to 75, corresponding to the period June 2006 to January 2009. Considering only points with valid MWR measurement, covering the whole world, the difference between the MWR and ECMWF corrections is 2 ± 18 mm, while the corresponding difference between the MWR and NCEP is 7 ± 33 mm.

To evaluate the sensitivity of the DLM output to the adopted NWM, the algorithm has been applied to the above-mentioned 28 Envisat cycles, to the whole range of latitudes and longitudes, using each of the above mentioned NWM. A total number of 1,062,058 coastal "land contaminated points" have been corrected (representing 94% of the total number of land contaminated points). For these points, the difference, which before adjustment was

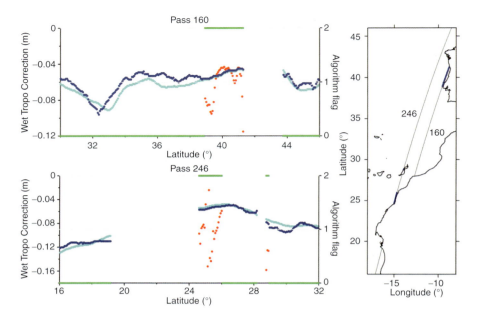

Fig. 6.1 Examples of the implementation of the DLM algorithm for two Envisat passes (Cycle 54). The top left figure shows path 160 crossing the Iberian Peninsula; the bottom left shows pass 246 crossing the Canary Islands and African coastline. The right plot illustrates the location of the passes

−8 ± 27 mm (difference between ECMWF and NCEP), reduces to −1 ± 7 mm after the DLM adjustment to the microwave radiometer field.

Being impossible to perform an accuracy assessment of the method using microwave radiometer measurements, this analysis gives an indication of the variability of the results when two different numerical weather models are used. This shows that ECMWF is more precise than NCEP and that the DLM adjustment to ECMWF will provide wet tropospheric corrections in the coastal region within an accuracy close to 1 cm.

6.2.2
GNSS-derived Path Delay Method

In recent years, an increasing number of inland GNSS (Global Navigation Satellite System) stations became available. In the scope of GPS (Global Positioning System) positioning, the tropospheric delay is often seen more as a nuisance factor, thus requiring effective estimation. Consequently, a vast number of developments have been pursued, aiming at the modelling of this effect (e.g., Hopfield 1971; Saastamoinen 1972; Niell 1996). GNSS data can be used to determine zenith tropospheric delays at the station location with an accuracy of a few millimetres (e.g., Niell et al. 2001; Snajdrova et al. 2006). The potential for the remote determination of atmospheric water vapour and integrated precipitable water also led to the development of several models mainly for meteorological purposes (e.g., Bevis et al. 1992). Within the scope of altimetry range correction, such data have been used with the purpose of calibration of the wet tropospheric correction derived from the onboard microwave radiometers over offshore oil platforms (e.g., Bar-Sever et al. 1998; Edwards et al. 2004).

This section describes a method for deriving the wet tropospheric correction for coastal altimetry from GNSS-derived tropospheric delays, the so-called GPD (GNSS-derived Path Delay) approach. The method is based on tropospheric path delays (zenith hydrostatic delays, ZHD, and zenith wet delays, ZWD) precisely determined at a network of land-based or offshore GNSS stations, further combined with additional MWR measurements and data from a NWM, such as those produced by ECMWF. The following subsections describe the major issues and developments related to this approach. This is mainly based on work performed at the University of Porto (UPorto), Portugal, in the framework of the ESA-funded project COASTALT – Development of Radar Altimetry Data processing in the Oceanic Coastal Zone.

6.2.2.1
Determination of Tropospheric Path Delays at GNSS Stations

Methodologies for estimating the wet tropospheric delay from GNSS data are, at present, well-established and have been used by a vast number of authors; details can be found in Haines and Bar-Sever (1998), Desai and Haines (2004), Edwards et al. (2004) and Moore et al. (2005). A number of suitable software packages have been developed for the processing of GPS networks with the purpose of determining the tropospheric parameters, such as GAMIT (Herring et al. 2006), GIPSY/OASIS (Webb and Zumberge 1995) and Bernese (Dach et al. 2007).

Here, reference is given to the freeware GAMIT package (Herring et al. 2006), which is able to estimate a zenith path delay and its atmospheric gradient for each station, modelled in both cases by a piecewise-linear function over the span of the observations.

The tropospheric propagation delay is determined by GAMIT according to the following equation (Herring et al. 2006):

$$STD(E) = ZHD \times mf_h(E) + ZWD \times mf_w(E) \tag{6.1}$$

where STD is the slant total delay measured by GNSS, E is the elevation angle of the GNSS satellite and mf_h and mf_w are the mapping functions for hydrostatic and wet components, respectively.

A priori ZHD is evaluated from meteorological data using the modified Saastamoinen zenith hydrostatic delay model (Saastamoinen 1972; Davis et al. 1985) described by Eq. 6.8 (details in Sect. 6.4). In this way, for each slant total delay (STD) observation a combined zenith total delay (ZTD) is determined as the sum of ZHD and ZWD. At a given step interval (e.g., 1 h) and for each station of the defined network, a combined ZTD is estimated from the observations to all visible satellites.

For the determination of tropospheric parameters, a wide span of satellite elevation angles is advisable (Niell et al. 2001). For this reason, a relatively low (e.g., 7°) elevation cutoff angle can be adopted.

Since all stations belonging to the same regional network shall observe a given satellite with similar viewing angles, the corresponding zenith delays will be highly correlated. For this reason, networks must include stations with a good global distribution in order to provide stability to the solution.

In the estimation procedure, mapping functions play a major role since they are responsible for the conversion between zenith and slant delays. The most commonly used mapping functions are all based on the same equation first proposed by Marini (1972): a continued fractions form in terms of sin(E) with coefficients a, b and c. What distinguishes the several mapping functions that have been used over the last decade are the values of the adopted coefficients, their derivation, and the knowledge of atmospheric composition and structure they express. Radiosonde data (Niell 1996), ray tracing through NWM (Niell 2001; Boehm and Schuh 2006) or climatologies (e.g., Boehm et al. 2006) have been used to derive the referred sets of coefficients.

The Vienna Mapping Functions 1 (VMF1, Boehm and Schuh 2004) are based on direct ray tracing through NWM. The VMF1 coefficients are derived from the pressure level data calculated by ECMWF and are given on a global $2.0° \times 2.5°$ latitude–longitude grid, four times a day, at 0, 6, 12 and 18 h UTC.

VMF1 are at present the mapping functions that allow a description of the atmosphere with the finest detail, leading to the highest precision in the derived tropospheric parameters. The climatology-based global mapping functions (GMF, Boehm et al. 2006) are used in GAMIT whenever VMF1 are not available.

In summary, the determined parameter is the combined ZTD. The separation of this quantity into a sum of ZHD and ZWD depends on the accuracy of the surface pressure data used to compute the ZHD component. The source of the hydrostatic corrections can be, for example, either the pressure values input to the model as *a priori* (every 6 h using VMF1 mapping functions) or measurements of station pressure recorded in a meteorological file, when available.

Comparisons made at four European stations (GAIA, CASCais and LAGOs, in the coast of Portugal; MATEra in Italy), where ZHD values were computed from local surface pressure data and from VMF1 grids, show an agreement within 2–3 mm accuracy (1-σ). Therefore, the corresponding wet correction (ZWD) can be separated from the dry correction (ZHD) with the same accuracy.

The GNSS-derived tropospheric delays determined by GAMIT are quantities which refer to station location and therefore to station height. To be able to use these fields to correct altimeter measurements, the computed ZHD and ZWD have to be reduced to sea level by applying separate corrections for the two components. A procedure for computing the correction for station height is described in (Kouba 2008).

6.2.2.2
Comparison of GNSS-derived Tropospheric Fields with GDR Corrections

With the purpose of inspecting the suitability of GNSS-derived tropospheric parameters for correcting coastal altimetry, a comparison study between two types of tropospheric fields has been performed and is presented here. These fields are the GNSS-derived tropospheric corrections (dry or ZHD and wet or ZWD) at a network of European stations near the coast and the corresponding altimeter tropospheric corrections, usually present on GDR products, at the station's nearby points with valid radiometer correction.

The 3-year period of the analysis (July 2002 to June 2005) has been selected as the unique period for which there were four altimeter missions with different ground tracks:

TOPEX/Poseidon (T/P), Jason-1, Envisat and Geosat Follow-On (GFO). Altimeter data from these four missions have been used.

GNSS data were available for the same period, for a global network of about 55 IGS (International GNSS Service) and EPN (EUREF Permanent Network) stations. 30-s phase measurements were processed with the GAMIT software, using double differences and an elevation cutoff angle of 7°. IGS precise satellite orbits and clock parameters have been used (Dow et al. 2005).

Tropospheric parameters (ZHD + ZWD) were derived at 1-hour interval using VMF1 mapping functions and corrected for station height using the procedure described in (Kouba 2008). Although a shorter interval, e.g., 30 min, would be preferable, the 1-hour interval was adopted as a compromise between precision and computation time. The ZHD has been computed from *in situ* pressure data at each station (where available) or from the VMF1 grids, using the modified Saastamoinen model (Eq. 6.8 – Saastamoinen 1972; Davis et al. 1985). Tropospheric delays from a subset of 13 stations in the West European region ($30° \leq \varphi \leq 55°$, $-20° \leq \lambda \leq 5°$) have been selected for this study.

Altimeter tropospheric fields have been obtained from AVISO-corrected sea surface height (CorSSH, AVISO 2005). ZHD is the dry correction from ECMWF and ZWD is the wet microwave radiometer correction. Previous to any analysis, altimeter data have been stacked, i.e., interpolated into reference points along altimeter reference tracks.

For each point along each altimeter ground track, GNSS data of the surrounding stations have been interpolated for the altimeter measurement time. Therefore, for each altimeter point on the reference ground tracks, a set of time series of tropospheric fields have been generated: for the dry (ZHD) and wet (ZWD) altimeter tropospheric corrections and for the corresponding field determined at each GNSS station.

Various statistics have been computed for each pair of altimeter and station fields: correlation, mean and standard deviation (sigma) of the differences between the altimeter and the corresponding GNSS-derived field, for the station with maximum correlation. Results for the NW Iberia are shown on Fig. 6.2 and 6.3, for ZHD and ZWD, respectively.

Tables 6.1 and 6.2 show the statistics of the aforementioned fields represented in these figures, but for the whole study region ($30° \leq \varphi \leq 55°$, $-20° \leq \lambda \leq 5°$). Although points at larger distances have also been considered, in this statistical analysis only points up to 100 km distance of the GNSS station and at distances larger than 20 km from the coast have been included.

One of the first results that came out of this study was the importance of correctly performing the height corrections to the GNSS tropospheric fields, since the parameters determined at each station refer to station height and must be correctly reduced to sea level for use in satellite altimetry.

For ZHD, the height correction is almost a bias, a function of station height, and can reach several decimetres, as is the case of BELL station (h = 803 m) for which the mean correction reaches 0.212 m. For ZWD, the correction is smaller, a function of ZWD itself, and can reach several centimetres (mean correction is 0.043 m for BELL station).

The statistics presented show that the height correction is performed precisely by using the procedure in (Kouba 2008). For the comparison between GNSS-derived and Envisat ZHD, the statistics for the mean difference after correction are, in metres, −0.004, 0.004, −0.001 and 0.002 for minimum, maximum, mean and standard deviation, respectively. The corresponding statistics for ZWD are −0.013, 0.019, 0.002 and 0.007 (in metres). These values are a clear indication of the agreement between the altimeter and GNSS-derived fields.

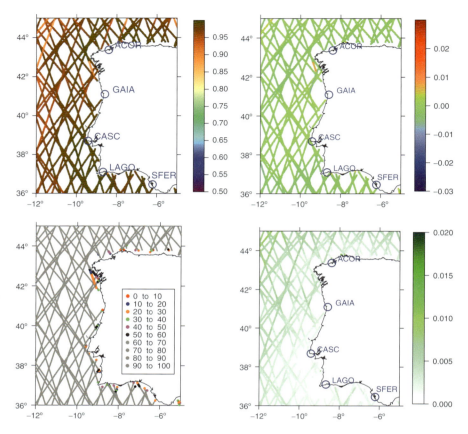

Fig. 6.2 Comparison between GDR ZHD at altimeter ground-track points and GNSS-derived ZHD at land-based stations. From *top* to *bottom* and *left* to *right* – for each ground-track point: maximum correlation, percentage of valid cycle points, mean and standard deviation of the differences between GDR ZHD and GNSS-derived ZHD (in metres) for the station with maximum correlation (x- and y-axis refer to longitude and latitude, respectively). Results refer to UPorto GNSS-derived solutions for period July 2002 to June 2005 and the NW Iberian region. GNSS-derived ZHD has been reduced to sea level

As expected, results show that the distance to the station decreases the correlations increase (top left maps on Fig. 6.2 and 6.3) and the mean and sigma of the differences decrease in a consistent way (right maps on Fig. 6.2 and 6.3; top and bottom, respectively).

The station showing maximum correlation with each altimeter track point is, in general, one of the closest stations, but not necessarily the closest one, because of local variations of the atmospheric fields, as shown in the bottom left map on Fig. 6.3.

The bottom left map on Fig. 6.2, showing the percentage of valid cycle points for the study period, explains why for some close-to-land tracks the maximum correlation occurs for a station which is not in the vicinity of the point. In these cases, the time series have a small number of points and, therefore, the corresponding correlations are not statistically

6 Tropospheric Corrections for Coastal Altimetry 157

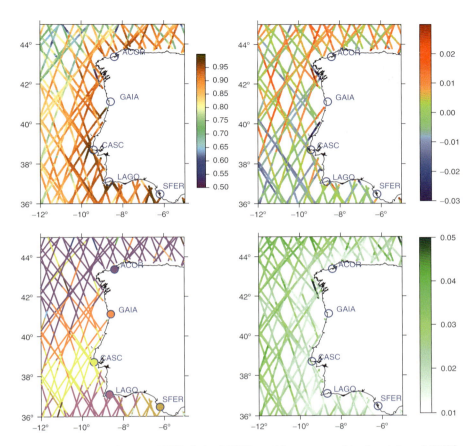

Fig. 6.3 Comparison between MWR-derived ZWD at altimeter ground-track points and GNSS-derived ZWD at land-based stations. From *top* to *bottom* and *left* to *right* – for each ground track-point: maximum correlation, GNSS station with which the correlation is maximum, mean and standard deviation of the differences between MWR- and GNSS-derived ZWD (in metres) for the station with maximum correlation (x- and y-axis refer to longitude and latitude, respectively). Results refer to UPorto GNSS-derived solutions for the period from July 2002 to June 2005 and the NW Iberian region. GNSS-derived ZWD has been reduced to sea level

significant. This is clear for the ascending 917 Envisat track almost parallel to the NW Spanish coast (percentage of valid cycle points below 30%).

When comparing the two fields (ZHD and ZWD), the highest correlations and the smallest differences are shown for ZHD since this field is easier to model and suffers the smallest variations; on the contrary, ZWD reveals the lowest correlations and the highest differences as it undergoes the largest spatial and temporal variations.

Overall, the results show that for distances up to 100 km the correlations are high (typically around 0.98 for ZHD and 0.93 for ZWD). For ZHD (Table 6.1), the mean difference is below 1 mm and the standard deviation of the differences has values around 2 mm for all missions. For ZWD (Table 6.2), although for all satellites the absolute mean difference is below 1 cm, actual values depend on the mission: around 2 mm for Envisat, 5 mm for

Table 6.1 Statistics for the comparison (altimetry-GPD) of sea level ZHD time series for altimetry track points (number in brackets) distant 20–100 km from maximum correlation GNSS station (difference and sigma in meters)

		Min	Max	Mean	Sigma
Envisat (461)	Difference	−0.004	0.004	−0.001	0.002
	Correlation	0.903	0.999	0.985	0.015
	Sigma	0.001	0.005	0.002	0.001
Jason-1 (157)	Difference	−0.002	0.004	0.000	0.002
	Correlation	0.944	0.997	0.986	0.012
	Sigma	0.001	0.004	0.003	0.001
T/P (120)	Difference	−0.003	0.005	0.000	0.002
	Correlation	0.967	0.997	0.985	0.008
	Sigma	0.001	0.005	0.003	0.001
GFO (298)	Difference	−0.003	0.004	0.000	0.002
	Correlation	0.846	0.996	0.977	0.036
	Sigma	0.001	0.008	0.003	0.001

Jason-1, −9 mm for T/P and −5 mm for GFO. The standard deviation of the differences has values around 16 mm for all satellites, which is within the expected variability for this field.

The different values for the mean differences in the ZWD field, shown on Table 6.2, suggest that there may be small biases between the microwave radiometers of the various satellites, a fact which is not unexpected. It appears that AVISO uses different editing criteria, resulting in quite different data amounts near the coast for the various missions. While for Envisat, Jason-1 and GFO, points are found down to the vicinity of the coast, for T/P the minimum distance is around 30 km. To mitigate this effect, only points at distances from the coast larger than 20 km have been considered. It should be highlighted that a precise comparison between the various missions solely based on the comparison with the GNSS data is difficult to achieve.

This study shows the suitability of the GNSS-derived tropospheric fields for use in the correction of the coastal altimeter measurements. However, owing to the relatively scarce number of GNSS stations, these fields have to be combined with additional available data to obtain the required spatial information.

As far as the tropospheric correction for coastal altimetry is concerned, providing GNSS-derived total correction (ZTD) seems to be as appropriate as providing the wet correction (ZWD) alone, since the dry or hydrostatic component (ZHD) is independently derived with very high precision.

6.2.2.3
Data Combination Methodology

An innovative approach to generate wet delay (ZWD) reliable estimates at satellite ground-track positions with invalid radiometer corrections (land-flagged) is by using a linear

Table 6.2 The same as Table 6.1 for ZWD

		Min	Max	Mean	Sigma
Envisat (595)	Difference	−0.013	0.019	0.002	0.007
	Correlation	0.766	0.999	0.933	0.039
	Sigma	0.005	0.029	0.016	0.004
Jason-1 (181)	Difference	−0.012	0.027	0.005	0.007
	Correlation	0.819	0.976	0.933	0.038
	Sigma	0.010	0.029	0.016	0.005
T/P (146)	Difference	−0.024	0.002	−0.009	0.006
	Correlation	0.835	0.982	0.946	0.027
	Sigma	0.008	0.028	0.015	0.004
GFO (326)	Difference	−0.017	0.008	−0.005	0.006
	Correlation	0.792	0.988	0.926	0.040
	Sigma	0.008	0.034	0.018	0.005

space-time objective analysis technique (Bretherton et al. 1976). The technique combines the available independent ZWD values derived from ECMWF model atmospheric fields with those from GNSS stations and with the microwave radiometer measurements (only those with valid MWR flag) to update a first guess value known *a priori* at each altimeter ground-track location with invalid MWR measurements. The statistical technique interpolates the wet correction measurements at the latter locations and epochs from the nearby (in space and time) MWR-, ECMWF- and GNSS-derived independent measurements, which are selected by imposing a data selection criterion, takes into account the respective accuracy of each data set, and updates the first guess value providing simultaneously a quantification of the interpolation errors. Therefore, the selected independent observations are weighted according to statistical information regarding their accuracy.

The combination of ZWD, rather than ZTD, independent quantities is preferred, since, as described before, the dry (hydrostatic component, ZHD) can be accurately derived from independent sources.

Besides the white noise associated with the measurements of each data set, the objective analysis technique requires *a priori* information on the ZWD signal variability and knowledge of the space-time analytical function that approximates the empirical covariance estimate.

Given these parameters, the generated estimates are optimal and no other accurate linear combinations of the observations, based on a least squares criterion, exist (Bretherton et al. 1976).

Spatial covariances between each pair of observations (valid MWR-, GNSS- or ECMWF-derived ZWD) and each observation and the location at which an estimate is required can be derived from a Gauss–Markov function (Schüler 2001), provided that the spatial correlation scale is known. Bosser et al. (2007) state that the ZWD varies spatially and temporally with typical scales of 1 to 100 km and 1 to 100 min, respectively, and these ranges allow a preliminary establishment of the spatial and temporal correlation scales. In the results shown here, both the temporal and spatial correlation scales are assumed to remain constant over the analysed region (30°N≤ φ≤ 55°N, 20W°≤λ≤ 20°E). More accurate values are expected by fitting the empirical auto-covariance function.

In the absence of the knowledge of an empirical covariance model of the background field, covariance functions that decrease exponentially with the square of the distance (or time) between acquisitions can be adopted, although other analytical functions to model the observed covariance should be exploited.

The space variability of the ZWD field may be expressed by

$$F(r) = e^{-\frac{r^2}{C^2}} \tag{6.2}$$

where r is the distance between each pair of points and C is the spatial correlation scale. The temporal variability of the field is also taken into account by a stationary Gaussian decay (Leeuwenburgh 2000):

$$D(\Delta t) = e^{-\frac{\Delta t^2}{T^2}} \tag{6.3}$$

where Δt is the time interval between the acquisition of the measurements associated with each pair of locations and T is the temporal correlation scale. The covariance function is represented by the space-time analytical function $G(r,\Delta t)$ that is obtained by multiplying the space correlation function F(r) by the time correlation function $D(\Delta t)$.

The covariance matrix between all the pairs of observations and the covariance vector between observations and the location and epoch at which an estimation is required can be computed from $G(r,\Delta t)$ to within the signal variance. The function G is normalized so that the correlation equals unity when r = 0 and Δt =0.

In the absence of sufficient data to perform the objective analysis, the algorithm gives the average value of the field calculated using the selected observations.

An example of implementation of this methodology is presented for Envisat cycle 58 (7 May 2007 to 11 June 2007) for the aforementioned geographical region. A MWR measurement has been considered invalid whenever its GDR land-ocean flag is set to 1 (land-flagged) or its quality interpolation flag is larger than 0 (the most comprehensive case). For these measurements, an estimation of the wet delay with an associated formal error is given by the implemented mapping routine. The observations include: valid MWR measurements acquired along the altimeter ground track, ZWD from GNSS land-based coastal stations (derived hourly and interpolated to a 30-min interval, further reduced to sea level) and ECMWF model-derived wet delay estimates (provided every 6 h at a regular 0.25° grid spacing). The latter were determined from two single-level parameter fields of the ECMWF deterministic atmospheric model: integrated water vapour (total column water vapour, TCWV) and surface temperature (2-m temperature, 2T) (ECMWF 2009). The formulation presented by Askne and Nordius (1987) was followed for the estimation of ZWD from TCWV, in which the mean temperature of the troposphere was modelled from 2T according to Mendes et al. (2000).

For the implementation of the objective analysis, the time and space correlation scales were set to 100 min and 100 km, respectively, and assumed constant over the whole period and study region. MWR-and GNSS-derived observations were assumed to have the same white measurement noise of 5 mm, while a value of 1 cm was assigned to white noise of the ECMWF-derived estimates. Since the observations have quite different spatial and temporal samplings, some issues must be addressed in the data selection criteria for the establishment of the influence domains to be considered in the merging procedures. The space domain radius was set equal to the space correlation scale, but given the temporal sampling of the ECMWF model, the temporal influence interval has been set to 3 h to guarantee that, at least, one model sample grid contributes to the estimation of the background wet delay field.

For each invalid MWR measurement, a first guess value has been computed as the mean value of the selected observations. The objective analysis procedure updates this value with the information added by the measurements themselves. The white noise of each data set and the data selection criteria were set as described above.

In order to compute the noise-to-signal ratio to set up the covariance matrix, the *a priori* knowledge of the signal variability is required. The *a priori* ZWD variability was computed using the data combination methodology and data from the three independent sources for the year 2007.

Using the information on the variability of the ZWD field, the data combination methodology generates a ZWD estimate along with a realistic relative mapping error value for the locations where the MWR measurements have been considered as invalid.

A more detailed analysis, not presented here, was performed for some Envisat passes very close to the Portuguese coast – passes 1, 74, 160, 917 – and allows a more elaborated discussion on the performance of the method. In general, the results show that when GNSS-derived wet delay values and valid MWR measurements are included in the selected observations, the wet delay estimates that result from the application of the methodology are clearly influenced by them. Output values show clear departures from the ECMWF values present in the GDR product. However, in most of the analysed cases, the variability of the field depicted by the ECMWF-derived values remains unchanged, i.e., when the ECMWF-derived values increase or decrease with time, the output field shows the same tendency. Moreover, results show that there are no significant biases between the computed wet delay values and the immediately adjacent valid MWR measurements. The continuity of the wet delay values is considered to be an added value of the implemented methodology.

Fig. 6.4 shows that the accuracy of the final GPD product is highly dependent on the availability and distribution of the three data sets used, in both time and space. In the worst case, the estimation is solely based on NWM measurements. Concerning the accessibility of present NWM, the ECMWF temporal sampling of 6 h is far from ideal, the required sampling being at least 3 h. Concerning the availability of valid MWR measurements, the worst cases take place when: (i) an isolated segment with all points with invalid radiometer measurements occurs; (ii) the track is parallel to the coastline, where a contaminated segment of several hundreds of kilometers length may occur. When the track is almost perpendicular to the coast there will be valid MWR measurements usually within a distance of 30–100 km. Considering the GNSS-derived path delays, various regions can be identified, particularly around European and USA coastlines, where relatively dense networks of coastal stations can be found. However, there are many regions without known available

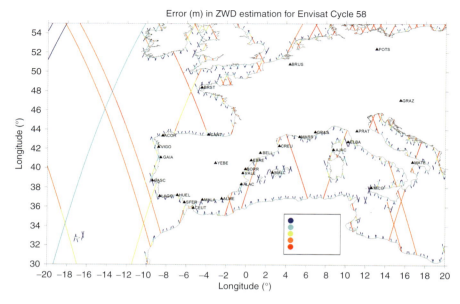

Fig. 6.4 Formal error (in metres) associated with the ZWD estimates calculated, using the datacombination methodology, for all invalid MWR measurements present in Envisat Cycle 58

GNSS stations for distances of several hundreds of kilometers. For this purpose, a densification of the network of coastal GNSS stations is advisable, with a station approximately every 100 km, preferably with meteorological sensors. Emphasis should be given to the merging of data derived from offshore GNSS stations (e.g., buoys and oil platforms) and local/regional NWM, whenever available.

6.3
Brightness Temperature Contamination and Correction

The algorithms used to retrieve the wet path delay from the measured brightness temperatures (TBs) implicitly assume that the satellite is flying over an oceanic surface. This is why the nominal retrieval method is unsuitable as soon as the satellite approaches the coasts, where the footprint contains land. The signal coming from the land surface is thus considered a "contamination". The study of this section is based on the following idea: if we know how to simulate this contamination, then we can remove it from the TBs to obtain a "clean" signal, i.e., without land-relative signal. From these corrected brightness temperatures, "decontaminated" measurements, we could then use the same oceanic retrieval everywhere, up to the coast.

To solve this problem, different methods have been proposed so far: Ruf proposed (Ruf 1999) an analytical and theoretical correction of TBs before retrieval. He assumed a track perpendicular to a straight coastline. Bennartz (1999) tackled the problem of mixed land/water measurements ("footprints") in SSM/I data by taking the fraction of water surface within each footprint.

In this section, the TB contamination by land is analyzed in details and two correction methods are presented and evaluated.

6.3.1
Instrumental Configuration and Reference Field

Sensitivity studies are performed in TOPEX microwave radiometer (TMR) configuration, but the proposed methodologies are not dedicated to TOPEX; they are applicable to any other similar instruments onboard altimetry missions. The TMR operates at 18, 21 and 37 GHz with a footprint diameter of 44.6, 37.4 and 23.5 km, respectively (Ruf et al. 1995). The first TB is mainly sensitive to the surface roughness, the second to atmospheric humidity and the third to cloud liquid water (but all receive a significant contribution from the surface and atmosphere). TBs are available every second (which means about 7 km along track between two measurements).

It is almost impossible to provide a reliable evaluation of the proposed correction methods with real measurements, because of the spare availability of reference information available globally. To quantitatively evaluate the new methods, we therefore decided to use simulated brightness temperatures over reference meteorological fields with related instrumental characteristics. This way, it becomes possible to (i) characterize TBs' contamination by land when the satellite approaches the coasts, and (ii) evaluate quantitatively the performances of the different algorithms.

Details about the simulator development and its evaluation can be found in (Desportes et al. 2007). To sum up briefly, TBs are simulated with a radiative transfer model; over sea, the surface emissivity is derived from sea surface wind speed and sea surface temperature over land, monthly mean emissivity atlases are used, provided by Karbou et al. (2005). After computing local TBs (on each grid point), the map is then smoothed by a Gaussian function to simulate the antenna lobe, using the main lobe width that is function of the frequency. With this tool, we simulate the TBs that would have been measured by the TMR on each point of the grid (0.1° per 0.1°). Meteorological analyses are provided by the ALADIN model (see Hauser et al. 2003), the operational mesoscale forecast model of Météo-France. Surface fields of pressure, temperature, wind, as well as temperature and humidity profiles on 15 pressure levels are available at a 0.1° spatial resolution. The ALADIN meteorological analyses are used as geophysical reference fields. Two different cases were selected for this study: 16 March and 15 April 1998 because the ALADIN area was overpassed by TOPEX on the same track but in different atmospheric conditions. The first day corresponds to an off-shore dry wind, whereas for the second day of study, the atmosphere moisture over Mediterranean Sea is higher. The chosen track, number 187 (see Fig. 6.5), presents various interesting cases of contamination: clear land/sea and sea/land transitions (Algerian and French coasts), overflight of an island (Ibiza in the Balearic Islands), and track tangent to the coastline (Creus Cape in Spain).

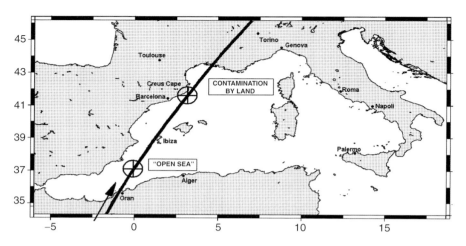

Fig. 6.5 TOPEX track number 187, 16 March around 12:00 (cycle 202) or 15 April around 06:00 (cycle 205)

6.3.2
An Analytical Correction of TBs: The "erf Method"

Ruf (1999) proposed an analytical correction of the measured TBs, function of the TB difference between sea and land. The track is assumed to be perpendicular to a straight coastline. TBs over sea and land are assumed to be constant. The major limitation of this method is the assumption that the track is perpendicular to the coastline. We propose, in the following, an improvement of the method by taking into account the angle between the track and the coast. Then, we test the sensitivity of this method.

6.3.2.1
Improved Method

Ruf used TBland and TBsea (taken 200 km after and before the transition) and a table of coefficients calculated for a track perpendicular to the coast, to calculate the following correction for a frequency ν and the distance d to the coast:

$$\mathrm{corr}(d,\nu) = [\mathrm{TBland}\,(\nu) - \mathrm{TBsea}\,(\nu)] \times \mathrm{table}\,(d,\nu) \tag{6.4}$$

First, we replaced the coefficients table by an erf function, primitive of a Gaussian function, which was found to fit well the actual sea-land TB evolution. The new correction is the following function:

$$\mathrm{corr}(d,\nu) = \frac{\mathrm{TBland}\,(\nu) - \mathrm{TBsea}\,(\nu)}{2} \times \left(1 + \mathrm{erf}\left(\frac{d}{\alpha(\nu,\theta)}\right)\right) \tag{6.5}$$

6 Tropospheric Corrections for Coastal Altimetry 165

where d is the distance to the coast and α is a parameter conditioning the curve's slope at d = 0 (α only depends on the frequency ν and on the angle θ between the track and the coast). Fig. 6.6b shows the simulated TBs along an arbitrary track (Fig. 6.6a). The more the track is perpendicular to the coast, the shorter is the contamination by land. Simulated TBs are corrected using this function before applying the path delay (PD) retrieval algorithm. Whereas the PD error for uncorrected TBs on sea reaches values far greater than 1 cm (our reference), PDs from corrected TBs are obviously almost perfect, as we assumed perfectly known TBsea, TBland and θ.

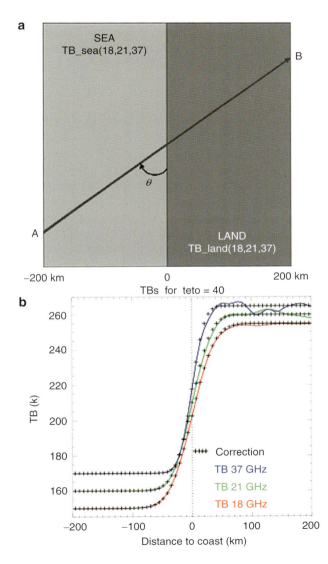

Fig. 6.6 (**a**) *Top*. Created track. The satellite goes from sea to land, crossing the coastline with a given angle; (**b**) *Bottom*. Simulated TBs along the created track presented above, from sea to land. A random noise is added to the simulated TBs over land, while the correction is unchanged

6.3.2.2
Sensitivity to Errors in θ or in TBland and TBsea

To evaluate the method in realistic conditions, we introduced different error types in TBland for simulated TBs, without altering the correction. We could have introduced this error on TBsea but TBs over land are more variable in space than over sea, and anyway the problem, as the correction function, is fully symmetrical.

By introducing a strictly equal bias on each frequency, we can accept a bias in TBland up to ±20 K to reach a 1-cm error on PD after correction. This can be explained by a compensation effect on the three channels. On the other hand, if we introduce an error only in one channel, assuming that TBland at the two other frequencies are perfectly known, the acceptable error bias is reduced to ±3 K to reach the limit of 1 cm. In other words, a relative error of 3 K is as damaging as a systematic bias of 20 K.

To generalize the error analysis, we introduced a Gaussian random error on TBland, independently on each channel, on one thousand samples. The maximum acceptable standard deviation on TBland is 2.7 K, to reach a mean error lower or equal to 1 cm on PD after correction (Table 6.3).

Finally, we introduced a random error on θ. The same error on θ will have a greater impact on PD if θ is small. For a θ value of 60°, a 40° error in θ estimation is allowed (to reach the limit of 1-cm error on PD). For a θ value of 20°, it decreases to 7°.

The land simulation case of the last line in Table 6.3 is the closest to reality: we estimated on the northwestern Mediterranean coast that the standard deviation of TBland is about 2.5 K for a 50-km segment coast. Fifty kilometers is the length of contamination for an angle between track and coast of about 25° at 18 GHz (or less in higher frequencies). This corresponds, therefore, to a mean error on PD of about 1 cm, which is already our limit. To this error, we have to add the one due to θ estimation and the one due to TBsea estimation. A θ angle of about 25° and a standard deviation of error on θ of about 10° lead to the limit of a 1-cm error on PD. If we estimate the error due to TBsea estimation to be 1 cm, the total error is about 1.7 cm. This is too much considering that the coast will never be actually rectilinear.

This method thus appears too sensitive to the geometry. Furthermore, it does not allow us to process complex cases like tangent tracks (what is θ in this case?) or islands (it is difficult to estimate θ and impossible to estimate TBland) or even small angles between track and coast. The complex pattern of the contaminating coast is not taken into account. As a consequence, this algorithm seems not adapted to a global operational processing.

Table 6.3 Error on PD after correction, when a Gaussian error is introduced on TBland

TBland standard deviation (K)	Mean error on PD after correction, on sea (cm)
20	6.5
10	3.5
5	1.6
2.7	1
2.5	0.8

6.3.3
Using the Proportion of Land in the Footprint

6.3.3.1
Description of the Proportion Method

This second approach is similar to (Bennartz 1999) in which Bennartz tackled the problem of mixed land/water measurements in SSM/I data by deriving the fraction of land surface within each measurement from a high-resolution land-sea mask. In this section, we describe the method we used to correct the TBs.

The correction function uses p, the actual proportion of land in the footprint. Far from coasts, p is zero on sea and 1 on land. The proportion p is calculated by means of a 0.01° resolution land-sea mask, taking into account the radiometer field of view characteristics. Therefore, p depends on frequency: at high frequencies the footprint is smaller. That is why, in the case of the island on track 187 (see Fig. 6.7a, first spike), contamination is greater at high frequencies. On the contrary, in the case of the tangent track, the smallest footprint hardly reaches the coast: it contains less land and the contamination is lower.

TBland and TBsea are estimated along the satellite track. For a complete sea-land transition, TBsea is the last uncontaminated TB (the last encountered TB with p = 0), and TBland is the first uncontaminated TB (p = 1). For incomplete transitions, we take the closest encountered TB (always along track) with p = 0 or 1.

This leads to the following correction function:

$$\text{corr}(p,\nu) = [\text{TBland}(\nu) - \text{TBsea}(\nu)] \times p(\nu). \quad (6.6)$$

A comparison between simulated and measured TBs is shown in Fig. 6.7b. TBs are satisfactorily simulated near coasts: the simulated TBs obtained by adding the land contribution to the TBsea values fit well the real measurements performed by the TMR, even in the most complex configurations.

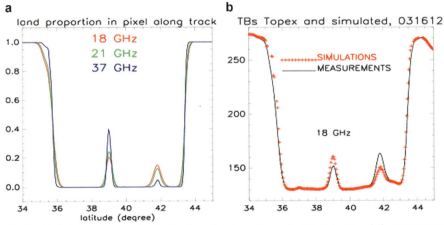

Fig. 6.7 (**a**) *Left*. Land proportion in the footprint along the track; (**b**) *right*. Simulations compared to actual measurements on the same track

6.3.3.2
Performance of the Method over Simulated Data

To test this correction on a large number of data, we simulate TBs along real TOPEX tracks. As the horizontal spacing of TOPEX tracks is very wide (about one track every 230 km), we add translated tracks to increase the horizontal resolution and number of cases.

We then apply the TB correction algorithm on each track, calculate the corrected PD and compare it, over sea with PDref. The PD obtained from contaminated TBs and the PD obtained propagating the last uncontaminated PD are also compared to PDref. In order to characterize the error in the coastal strip between the different obtained PDs and the ALADIN PD, we calculate the rms error for each transition case (262 cases among sea-land, land-sea, flying over an island or near a coast). Then, we calculate the quadratic mean of these rms errors. In this way, no case is favored: every transition has the same influence, whatever its length (if the angle between the track and the coast is small, the transition is longer).

Results are given in Table 6.4. The error near coasts is significantly reduced. We assume a linear dependency between the land proportion and the observed TBs. Nevertheless, this assumption is not always valid especially at 37 GHz, because of nonlinearity of the atmospheric radiative transfer for atmosphere-sensitive channels. Discrepancies with respect to the mean linear dependency come from atmospheric humidity variations above the surface (which is the signal we want to catch) but also from emissivity variations over sea and over land along the track (that are neglected in the proportion method).

This method is quite sensitive to the choice of TBsea and TBland, especially when the satellite overpasses an island. If the nearest point over land where $p = 1$ can be found at less than 200 km (in view of the characteristic atmospheric structures dimensions) the corresponding TBland value is assumed to be similar.

6.3.4
Performance Analysis over Real Measurements

By applying the retrieval algorithm to measured TBs without any correction, we calculate a "contaminated path delay". Then, using either the erf method (Sect. 6.3.2) or the proportion method (Sect. 6.3.3), we calculate two different "corrected path delays". The three

Table 6.4 RMS errors on the coastal strip between contaminated PD (initial PDsim)/propagated PDsim/corrected PDsim and PDref

	Contaminated PD (cm)	Propagated PD (cm)	Corrected PD (cm)
16 March	12.4	5.2	2.3
15 April	10.9	4.6	2.6

6 Tropospheric Corrections for Coastal Altimetry

obtained PDs are compared with our reference, the ALADIN PD. We can benefit from the wide set of coastal transitions encountered within the ALADIN area to statistically evaluate the performances of the two methods.

Results are shown for the two different cases in Fig. 6.8. The obtained variations around coasts are consistent with the previous comparison. The signal near coasts is better corrected than over the island and tangent track cases. Again, the reason is the use of a distant TB. This appears as the major limitation of the method, since there is no way to estimate the adequate TB to use when there is no pure land footprint available.

Note that ALADIN PD cannot be used as a reference here, because of its negative bias with respect to TMR, and to its too smooth variations, compared with the actual ones.

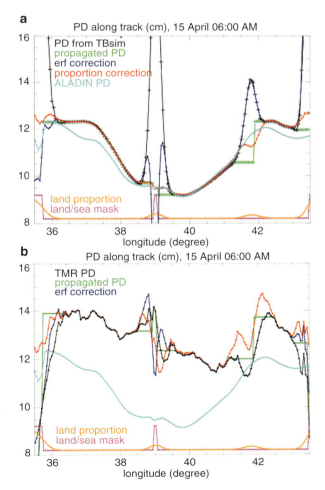

Fig. 6.8 Comparison between the PDs obtained from TBs after the different correction methods for track 187. (**a**) *Top*. Simulated TBs, (**b**) *Bottom*. Measured TBs

6.3.5
Discussion

The objective of this study was to analyze in detail the contamination of the brightness temperatures by land, and to propose a new approach. The validation of the different methods we have tested in this study is almost impossible using real brightness temperatures. We do not have enough in situ measurements (radiosondes, GPS) in coastal areas and no accurate enough model to fully evaluate the proposed corrections. We, therefore, chose to develop and use a simulator to perform sensitivity tests and to quantitatively evaluate the different methods for a large number of geometric and meteorological situations.

We first suggested refinements to the approach proposed by Ruf (analytical correction) (Ruf 1999). The results are satisfactory in a very optimal case (reduction by more than 60% of the error with respect to no correction). But sensitivity tests showed that the brightness temperature over land and the angle between the satellite track and the coast should be known with very good accuracy. However, these parameters are difficult to estimate, especially in complex geometries. Therefore, this method seems difficult to use globally in an operational processing.

The approach proposed by Bennartz (1999), developed for SSM/I mixed land-water footprints, has been adapted to the TOPEX/TMR case. It mainly uses the proportion of land in the footprint. This method is robust and seems more apt, because it allows the processing of any configuration. Results obtained on simulations are satisfactory, the error is 50% lower than with the previous method. But the hypothesis of a linear dependency between the land proportion and the observed TBs, not completely valid, leads to ignore atmospheric variations in the transition area. An additional limitation is the lack of information on TBland in the case of islands and tangent tracks (as in the previous method).

In this study, we used measurements from the TMR, collocated with the TOPEX/Poseidon altimeter measurements. Nevertheless, the proposed methodology is applicable to other radiometers, just taking into account the corresponding instrumental characteristics (frequency, footprint size, incidence angle). This method of decontamination of the brightness temperature also improves the retrieval of other radiometer products like the atmospheric attenuation of the altimeter backscattering coefficients and the cloud liquid water content.

6.4
Dry Tropospheric Correction

The dry tropospheric correction (or Zenith Hydrostatic Delay, ZHD) is responsible for about 90% of the total path delay caused by the troposphere on the altimeter measurement of the two-way travel time from the satellite to the nadir sea level (e.g., Chelton et al. 2001). It varies slowly in space and time – typical scales of 100–1,000 km and 3–30 h, respectively (Bosser et al. 2007). In fact, according to Bevis et al. (1992), the use of the term "dry" to refer to the hydrostatic component of the troposphere is actually misleading since it omits the fact that the water vapour indeed slightly contributes to the hydrostatic path delay as well.

Although it is by far the largest range correction to be performed, with values ranging from around 2.25 to 2.35 m, the estimation of the dry tropospheric correction is accurately performed solely based on sea level pressure (980–1035 hPa) (Chelton et al. 2001). Smith and Weintraub (1953) developed a semiempirical expression for the evaluation of the refractivity of the non-dispersive dry troposphere, later improved by Thayer (1974) and Liebe (1985), by accounting for the small contribution (~1%) of water vapour to the total pressure (and the compressibility factor of air). On the basis of the above approach, the dry tropospheric range correction (in cm) has been routinely estimated as

$$\Delta R_{dry} \approx 222.74 \, P_0 \, / \, g_0(\varphi) \tag{6.7}$$

(e.g., Chelton et al. 2001), where P_0 is the sea level pressure (in hPa) and $g_0(\varphi)$ is the acceleration due to gravity (in cm/s^2) at latitude φ. Several authors (e.g., Hopfield 1971; Saastamoinen 1972) worked upon the model expressed by Eq. 6.7 to develop the most commonly used expressions in geodesy and radio astronomy. The model in Saastamoinen (1972) was later revisited by Davis et al. (1985), yielding the expression

$$\Delta R_{dry} = \frac{0.0022768 \, P_0}{1 - 0.00266 \cos(2\varphi) - 0.00028 \, h} \tag{6.8}$$

where h is the ellipsoidal altitude (in km) and all the other variables have been defined above.

Although the dry tropospheric correction is only moderately sensitive to errors in the surface pressure, a 5 hPa accuracy would be needed to secure a 1-cm accuracy (Chelton et al. 2001). As direct measurements of surface pressure are seldom available over open ocean, the dry tropospheric correction estimations have to rely on pressure values derived from numerical weather models (NWM) as those from ECMWF or NCEP.

At present, the use of the modified Saastamoinen model together with NWM sea level pressure (e.g., ECMWF global 0.25° × 0.25° grids generated every 6 h) results in an uncertainty of less than 1 cm in the dry tropospheric range correction (Chelton et al. 2001). Recent studies (e.g., Bosser et al. 2007) further refined the Saastamoinen model (Eq. 6.8) by using an updated global Earth gravity model and a global climatology for air density (instead of the standard atmosphere of the original formulation). From the latter, it is claimed that an accuracy of 0.1 mm can be achieved for the dry correction providing surface pressure measurements can be guaranteed within an uncertainty of 0.1 hPa (requiring the use of high-accuracy barometers during quiet meteorological conditions, which does not usually happen).

For coastal altimetry applications, no degradation of the accuracy for the dry tropospheric correction is expected, since land-ocean transition has not been referred to as specifically affecting the time or space scales of surface pressure variation. Nevertheless, some work could be addressed to inspect possible shear at the land/sea interface and effects of night/day mass transport. As the accuracy of the dry correction basically relies on that of the surface pressure, coastal altimetry would surely benefit from the vicinity of the overland meteorological stations network (usually denser on the more populated coastal areas) eventually providing surface pressure data with higher temporal and spatial resolution than that currently obtained from global NWM. Although these data are, in principle, already assimilated into global NWM, the use of local models with higher spatial and temporal resolution (e.g., ALADIN from Météo France/ECMWF) should be tested for the computation of improved dry tropospheric correction.

Any improvement on the estimation of the dry tropospheric correction will also impact the quality of the wet correction as derived from GNSS measurements, as the latter is obtained by subtracting the hydrostatic component to the GNSS-derived total correction, as previously described in this chapter. A number of studies (e.g., Bai and Feng 2003; Hagemann et al. 2003; Wang et al. 2007) have been conducted to assess the accuracy of the NWM-derived surface pressure for overland GPS stations locations. A comparison of the accuracy of the NWM-derived surface pressure with that obtained by collocated synoptic measurements for a set of land-based GPS stations can be found in (Chelton et al. 2001). Differences up to 3 hPa were reported by the authors to frequently occur, varying with station location and time of the year (significantly larger deviations were also found for some locations). Those differences did not present a specific and easy to model space or time pattern. As, unfortunately, only a very limited number of GPS stations are equipped with meteorological sensors, (Bai and Feng 2003) also stressed the need for seeking alternative sources for surface meteorological data or the use of solutions, which bypass the need for such data, if one wants to take advantage of the existing GPS networks. Hagemann et al. (2003) refer to a less than 1 hPa mean bias between collocated surface pressure measurements and those horizontally and vertically interpolated from nearby (up to a 100 km distance) World Meteorological Organization (WMO) stations. Wang et al. (2007) also found good results when using spatially and temporally interpolated synoptic surface pressure values from the 3-hourly WMO stations measurements.

Comparisons between dry tropospheric correction values present in Envisat Geophysical Data Records (GDR) files (cycles 30 to 64 – September 2004 to December 2007) and those derived from *in situ* surface pressure data at three GPS stations (GAIA, CASC and LAGO, in the coast of Portugal) show differences with a mean value of 0.002 ± 0.003 m that range from -0.005 to 0.020 m. In this analysis, only points within 100 km from each station and 50 km from the coast have been considered. Although the extreme values occur at ground-track points closer than 10 km from the coast, there is no clear degradation pattern associated with their distance to land.

6.5
Conclusions and Perspectives

To get workable altimetry products in coastal areas, specific corrections are needed. In the case of the wet tropospheric correction, different studies have been conducted recently and the first results are encouraging. The use of external information to describe more accurately the atmospheric humidity in the coastal band should allow a significant improvement in the quality of the altimeter products. Radiometer product deficiencies in these transition areas can be overcome by the use of valid MWR measurements in the vicinity of the points, of GNSS measurements from nearby stations, of meteorological models assuming a dedicated processing, or of land information (surface emissivity and temperature) to correct the brightness temperatures.

The DLM is a simple method that only requires GDR data, with optional distance-to-land information, is mission-independent and can be used as a backup method whenever a more precise one is not available. A global implementation to 28 Envisat cycles, using two different NWM (ECMWF and NCEP), shows that the method reduces the rms of the differences between the NWM corrections at the coastal land-contaminated points from 28 to 7 mm, suggesting that the DLM implementation with ECMWF will provide wet tropospheric corrections in the coastal band with an rms accuracy close to 1 cm. The method fails when there are no valid radiometer measurements close enough to perform the adjustment (at about 6% of the points). In this case, the output can be the original model correction, adequately flagged.

GPD is an approach based on the combination of GNSS-derived path delays, valid MWR measurements and NWM-based information. The GPD estimated corrections are a combined value of all available measurements within the specified spatial (100 km) and temporal (100 min) scales. Therefore, the result is highly dependent on the spatial and temporal distribution of the three data types used. The most critical cases occur for isolated segments containing only invalid MWR measurements (the closest valid MWR measurements are at distances longer than 100 km) or tracks almost parallel to the coastline, for which there are no GNSS stations within a distance of 100 km. In this case, the estimated values are solely based on NWM measurements, assumed less accurate. A considerable number of configurations can be found, which allow the estimation of the wet delays within 1cm error: points at distances < 50 km from a GNSS station, points for which there are valid MWR measurements within a distance < 50 km or passes with an associated measurement time very close to the time of the closest NWM grid. To achieve this accuracy everywhere, an augmentation of the GNSS networks is advisable, ensuring a coastal station approximately every 100 km and, more importantly, in the locations where isolated segments occur containing all MWR measurements invalid. Considering a global implementation of the method, the accessibility of NWM grids at a higher temporal sampling, ideally 1 h, is of crucial importance.

The correction of brightness temperatures using the land proportion has been developed and tested in TMR configuration. Results are satisfactory and an operational version of the algorithm has been proposed and implemented in the Pistach prototype (Mercier et al. 2008) dedicated to a specific processing of the Jason-2 mission for coastal areas and inland waters. The implementation is quite simple, if auxiliary tables containing the proportion of land along the satellite track are computed first. The decontamination method is powerful but requires instrumental characteristics and very good *a priori* knowledge of the overflown surfaces. Performances are difficult to assess quantitatively because no reference values are available so far. Nevertheless, this method provides a radiometer wet tropospheric correction everywhere in the coastal band, and the improvement with respect to other estimations based on NWP models is evident in cases of high atmospheric variability (which is often the case near the coast) and/or inaccuracies in NWM estimations (smooth or wrong trends).

In conclusion, depending on the GNSS network density, on the atmospheric variability, which may condition the meteorological model accuracy, and on the geography knowledge (accuracy of the proportion of land), one method can be better than another. In a near future, it will be necessary to provide for each altimeter ground-track point the estimated value by the different methods and a quality flag or the associated accuracy, so that the user can decide which is the most suitable method for his application.

In parallel to the proposition of these new processing strategies, it is a necessity to think about new instruments with a much better spatial resolution and, therefore, a much smaller contamination area. In this context, the potential of high-resolution radiometers (higher than 150 GHz) with a much better spatial resolution and a much smaller land impact is obvious and should be considered for future altimetry missions. Moreover, the estimation of GNSS-derived path delays will benefit from the augmentation of the GNSS coastal stations preferably equipped with meteorological sensors.

The dry correction can be accurately computed from pressure measurements. The present accuracy of the ECMWF-model-derived correction fulfils the requirements for open ocean. Results presented in this chapter indicate that in the coastal region the precision of this correction is still well within the 1-cm accuracy.

Acknowledgements Studies presented in this chapter have been funded by the CNES research and technology program, by ESA-funded project COASTALT (ESA/ESRIN Contract No. 21201/08/I-LG) and by FCT project POCUS (PDCTE/CTA/50388/2003). Authors would like to acknowledge the European Centre for Medium-Range Weather Forecasts (ECMWF) for providing the reanalysis data on the single-level atmospheric fields of the Deterministic Atmospheric Model.

References

Askne J, Nordius H (1987) Estimation of tropospheric delay for microwaves from surface weather data. Radio Sci 22:379–386

AVISO (2005) Archiving, validation and interpretation of satellite oceanographic data. AVISO DT-CorSSH data are http://www.aviso.oceanobs.com/index.php?id=1267. Accessed June 2008

Bai Z, Feng Y (2003) GPS water vapor estimation using interpolated surface meteorological data from Australian automatic weather stations. J Global Position Syst 2(2):83–89

Bar-Sever YE, Kroger PM, Borjesson JA (1998) Estimating horizontal gradients of tropospheric path delay with a single GPS receiver. J Geophys Res 103(B3):5019–5035

Bennartz R (1999) On the use of SSM/I measurements in coastal regions. J Atmos Ocean Technol 16:417–431

Bevis M, Businger S, Herring TA, Rocken C, Anthes RA, Ware RH (1992) GPS meteorology: remote sensing of atmospheric water vapor using the global positioning system. J Geophys Res 97(D14):15787–15801

Boehm J, Schuh H (2004) Vienna mapping functions in VLBI analyses. Geophys Res Lett 31(L01603). doi:10.1029/2003GL018984

Boehm J, Niell A, Tregoning P, Schuh H (2006) Global mapping functions (GMF): a new empirical mapping function based on numerical weather model data. Geophys Res Lett 33(L07304). doi:10.1029/2005GL025546

Bosser P, Bock O, Pelon J, Thom C (2007) An improved mean-gravity model for GPS hydrostatic delay calibration. IEEE Geosci Remote Sens Lett 4(1):3–7

Bretherton FP, Davis RE, Fandry CB (1976) A technique for objective analysis and design of oceanographic experiment applied to MODE-73. Deep-Sea Res 23:559–582

Chelton DB, Ries JC, Haines BJ, Fu LL, Callahan PS (2001) Satellite altimetry. In: Fu LL, Cazenave A (eds) Satellite altimetry and earth sciences. International Geophysics Series, vol 69. Academic, pp 1–131, ISBN: 0-12-269543-3

Dach R, Hugentobler U, Fridez P, Meindl M (eds) (2007) Bernese GPS software - version 5.0. Astronomical Institute, University of Bern

Davis JL, Herring TA, Shapiro II, Rogers AEE, Elgered G (1985) Geodesy by radio interferometry: effects of atmospheric modelling errors on estimates of baseline length. Radio Sci 20(6):1593–1607

Desai SD, Haines BJ (2004) Monitoring measurements from the Jason-1 microwave radiometer and independent validation with GPS. Mar Geod 27(1):221–240. doi:10.1080/01490410490465337

Desportes C, Obligis E, Eymard L (2007) On the wet tropospheric correction for altimetry in coastal regions. IEEE Trans Geosci Remote Sensing 45(7):2139–2149

Dow JM, Neilan RE, Gendt G (2005) The international GPS service (IGS): celebrating the 10th anniversary and looking to the next decade. Adv Space Res 36 (3):320–326, 2005. doi:10.1016/j.asr.2005.05.125

ECMWF (2009) http://www.ecmwf.int/products/catalogue/pseta.html

Edwards S, Moore P, King M (2004) Assessment of the Jason-1 and TOPEX/Poseidon microwave radiometer performance using GPS from offshore sites in the North Sea. Mar Geod 27(3):717–727. doi:10.1080/01490410490883388

Fernandes MJ, Bastos L, Antunes M (2003) Coastal satellite altimetry – methods for data recovery and validation. In: Tziavo IN (ed) Proceedings of the 3rd meeting of the international gravity & geoid commission (GG2002), Editions ZITI, pp 302–307

Hagemann S, Bengtsson L, Gendt G (2003) On the determination of atmospheric water vapor from GPS measurements. J Geophys Res 108(D21):4678. doi:10.1029/2002JD003235

Haines BJ, Bar-Sever YE (1998) Monitoring the TOPEX microwave radiometer with GPS: stability of columnar water vapour measurements. Geophysic Res Lett 25(19):3563–3566

Hauser D et al. (2003) The FETCH experiment: an overview. J Geophys Res 108(C3):8053. doi:10.1029/2001JC001202

Herring T, King R, McClusky S (2006) GAMIT reference manual – GPS analysis at MIT – Release 10.3. Department of Earth, Atmospheric and Planetary Sciences, Massachusetts Institute of Technology

Hopfield HS (1971) Tropospheric effect on electromagnetic measured range: prediction from surface weather data. Radio Sci 6:357–367

Karbou F, Prigent C, Eymard L, Pardo JR (2005) Microwave land emissivity calculations using AMSU measurements. IEEE Trans Geosci Remote Sensing 43(5):948–959

Kouba J (2008) Implementation and testing of the gridded Vienna mapping function 1 (VMF1). J Geod 82:193–205. doi:10.1007/s00190-007-0170-0

Leeuwenburgh O (2000), Covariance modelling for merging of multi-sensor ocean surface data, methods and applications of inversion. In: Hansen PC, Jacobsen BH, Mosegaard H (eds) Lecture notes in earth sciences, vol 92. Springer, pp 203–216. doi:10.1007/BFb0010278 is Berlin/Heidelberg

Liebe HJ (1985) An updated model for milimeter wave propagation in moist air. Radio Sci 20(5):1069–1089

Marini JW (1972) Correction of satellite tracking data for an arbitrary tropospheric profile. Radio Sci 7(2):223–231

Mendes VB, Prates G, Santos L, Langley RB (2000) An Evaluation of the accuracy of models of the determination of the weighted mean temperature of the atmosphere. In: Proceedings of ION, 2000 national technical meeting, January 26–28, 2000, Pacific Hotel Disneyland, Anaheim, CA

Mercier F (2004) Amélioration de la correction de troposphère humide en zone côtière. Rapport Gocina, CLS-DOS-NT-04-086

Mercier F, Ablain M, Carrère L, Dibarboure G., Dufau C, Labroue S, Obligis E, Sicard P, Thibaut P, Commien L, Moreau T, Garcia G, Poisson JC, Rahmani A, Birol F, Bouffard J, Cazenave A, Crétaux JF, Gennero MC, Seyler F, Kosuth P, Bercher N (2008) A CNES initiative for improved altimeter products in coastal zone: PISTACH. http://www.aviso.oceanobs.com/fileadmin/documents/OSTST/2008/Mercier_PISTACH.pdf

Moore P, Edwards E, King M (2005) Radiometric path delay calibration of ERS-2 with application to altimetric range. In: Proceedings of the 2004 Envisat & ERS symposium, Salzburg, Austria 6–10 September 2004

Niell AE (1996) Global mapping functions for the atmosphere delay at radio wavelengths. J Geophys Res 101(B2):3277–3246

Niell AE (2001) Preliminary evaluation of atmospheric mapping functions based on numerical weather models. Phys Chem Earth 26:475–480

Niell AE, Coster AJ, Solheim FS, Mendes VB, Toor PC, Langley RB, Upham CA (2001) Comparison of measurements of atmospheric wet delay by radiosonde, water vapor radiometer, GPS, and VLBI. J Atmos Ocean Technol 18:830–850

Ruf C (1999) Jason microwave radiometer – land contamination of path delay retrieval. Technical note for the Jason project

Ruf C, Keihm S, Janssen MA (1995) TOPEX/Poseidon microwave radiometer (TMR): I. Instrument description and antenna temperature calibration. IEEE Trans Geosci Remote Sensing 33(1):125–137

Saastamoinen J (1972) Atmospheric correction for troposphere and stratosphere in radio ranging of satellites. In: Henriksen S, Mancini A, Chovitz B (eds) The use of artificial satellites for geodesy, vol 15. Geophysics Monograph Series, AGU, Washington, DC, pp 247–251

Scharroo R, Lillibridge JL, Smith WHF, Schrama EJO (2004) Cross-calibration and long-term monitoring of the microwave radiometers of ERS, TOPEX, GFO, Jason and Envisat. Mar Geod 27:279–297

Schüler T (2001) On ground-based gps tropospheric delay estimation. PhD thesis, Universität der Bundeswehr München, Studiengang Geodäsie und Geoinformation available at http://ub.unibw-muenchen.de/dissertationen/ediss/schueler-torben/inhalt.pdf. Accessed 13 January 2009

Smith EK, Weintraub S (1953) The constants in the equation for atmospheric refractive index at radio frequencies. Proc Inst Radio Eng 41(8):1035–1037

Snajdrova K, Boehm J, Willis P, Haas R, Schuh H (2006) Multi-technique comparison of tropospheric zenith delays derived during the CONT02 campaign. J Geod 79:613–623. doi:10.1007/s00190-005-0010-z

Thayer GD (1974) An improved equation for the radio refractive index of air. Radio Sci 9(10): 803–807

Wang J, Zhang L, Dai A, Van Hove T, Van Baelen J (2007) A near-global, 2-hourly data set of atmospheric precipitable water from ground-based GPS measurements. J Geophys Res 112(D11107). doi:10.1029/2006JD007529

Webb FH, Zumberge JF (1995) An introduction to GIPSY/OASIS-II. JPL Publication D-11088, Jet Propulsion Laboratory, Pasadena

Surge Models as Providers of Improved "Inverse Barometer Corrections" for Coastal Altimetry Users

7

P.L. Woodworth and K.J. Horsburgh

Contents

7.1	Introduction	178
7.2	Sea Level Variability due to Meteorological Forcing in the Deep Ocean and Coastal Seas	179
7.3	Regional Storm Surge Models	180
7.4	Global Barotropic Models	185
7.5	COASTALT Validation Tests	186
7.6	Conclusions	187
References		188

Keywords Altimetric corrections • Inverse barometer • Storm surges • Tide-surge modelling

Abbreviations

CNES	Centre National d'Études Spatiales, France
COASTALT	ESA development of COASTal ALTimetry
COOP	Coastal Ocean Observations Panel of the Intergovernmental Oceanographic Commission
CS3X	Continental Shelf model (at 3 times former resolution)
ECMWF	European Centre for Medium-Range Weather Forecasts
FEMA	Federal Emergency Management Agency, USA
GDR	Geophysical Data Record
GOOS	Global Ocean Observing System
HANSOM	Hamburg Shelf Ocean Model
JCOMM	Joint Technical Commission on Oceanography and Marine Meteorology
LIB	Local Inverse Barometer
MOG2D	2-D Gravity Waves Model

P.L. Woodworth (✉) and K.J. Horsburgh
National Oceanography Centre, 6 Brownlow Street, Liverpool, L3 5DA, UK
e-mail: plw@pol.ac.uk

S. Vignudelli et al. (eds.), *Coastal Altimetry*,
DOI: 10.1007/978-3-642-12796-0_7, © Springer-Verlag Berlin Heidelberg 2011

MOHID	MOdelo HIDrodinãmico (Hydrodynamic Model)
MSS	Mean Sea Surface
NAE	North Atlantic and European atmospheric model
NCEP-NCAR	National Centers for Environmental Prediction – National Center for Atmospheric Research, USA
NOC	National Oceanography Centre
OCCAM	Ocean Circulation and Climate Advanced Modelling Project
PISTACH	Prototype Innovant de Système de Traitment pour l'Altimétrie Côtière et l'Hydrologie
SEPRISE	Sustained, Efficient Production of Required Information and Services within Europe
T/P	TOPEX/Poseidon
T-UGO	Toulouse – Unstructured Grid Ocean model
WMO	World Meteorological Organisation

7.1 Introduction

The dawn of the "Age of Precise Altimetry" can be dated from the launch of TOPEX/Poseidon (T/P) in 1992 followed by Jason-1 in 2001 and Jason-2 in 2008, and in this one and a half decades the altimeter community has acquired data sets which have revolutionised the science of oceanography (Fu and Cazenave 2001). However, as others have explained (e.g., see the chapter by Dufau et al., this volume and Cipollini et al. 2009), altimetry in coastal areas has been relatively under-exploited. This has been primarily due to the coarse spatial-temporal sampling provided by one, or a small number of, nadir-pointing altimeters, but also to technical limitations near the coast (e.g., altimeter and radiometer footprints), and to larger uncertainties in the various correction terms to be applied to the altimeter data.

One of the most important requirements for the effective use of altimetric sea surface heights in coastal areas is an improved "inverse barometer correction." This correction term represents the "storm surge" signal that is due to the action of both winds and air pressures, and can be several decimetres in magnitude, or even be measured in metres at certain times and in some shallow-water coastal areas. It will be argued below that the only feasible way to provide a suitable estimate of this correction is through numerical storm surge modelling, and in particular by exploiting the modelling already undertaken at operational flood forecasting centres.

In the next section, we review the role of winds and air pressures in producing changes in sea level, and the contrasting situation of meteorological forcing in the deep ocean and coastal seas. In Sect. 7.3, we focus on regional storm surge models used in many parts of the world, using examples from the UK and Denmark, and refer to findings from a survey of national and regional models undertaken by the World Meteorological Organisation (WMO). Sect. 7.4 notes the small number of suitable global barotropic (i.e. two-dimensional [2-D], depth-averaged) models which might be employed as an alternative to the collection of regional ones. Sect. 7.5 refers to the immediate objectives of the COASTALT (ESA development of COASTal ALTimetry) project insofar as the coastal improved "inverse barometer correction" is concerned. Finally, brief conclusions are given in Sect. 7.6.

7.2
Sea Level Variability due to Meteorological Forcing in the Deep Ocean and Coastal Seas

An understanding of the role of winds and air pressure changes is essential towards any explanation of the variability of the ocean circulation and of sea level. At long timescales (weeks to seasons and longer), the winds determine the strength of the major ocean currents and, through dynamical adjustments, modify the ocean circulation at depth (and its vertical density structure), as well as altering the shape of the ocean surface (the ocean dynamic topography). At seasonal and much longer timescales, such wind-driven processes play a major part in climate change. A proper appreciation of these processes requires the use of a three-dimensional (3-D, depth dependent) ocean general circulation model (e.g., that of the Ocean Circulation and Climate Advanced Modelling Project, OCCAM, Webb et al. 1997), or coupled atmosphere-ocean general circulation models, and a detailed analysis of their outputs, which can present as complex a task as study of the real ocean itself.

At shorter timescales (hours to weeks), a determination of the effect of winds and air pressure changes on the ocean circulation and sea level can be much simpler, in that the perturbations in the ocean due to the meteorological changes can be represented to good approximation by a 2-D model scheme, such as those discussed below. The relevant physics in such numerical models is represented in a set of hydrodynamic equations (Pugh 1987, Chapter by Dufau et al., this volume) in exactly the same way as for 2-D modelling of the ocean tide, with accelerations due to the tidal potentials of the Moon and Sun replaced by those due to wind stress and air pressure gradient (Pugh 1987; Flather 1988, Chapter by Obligis et al., this volume). In the deep ocean (depth > 200 m), the acceleration related to wind stress can be neglected to first order because of it being multiplied by (1/water depth), and resulting changes in sea level will thereby be largely proportional to those in air pressure. This is the "local inverse barometer (LIB)" model wherein a rise/fall of 1 mbar in air pressure results in a fall/rise of 1 cm in sea level (to within approximately half a percent).

The LIB model is used extensively in satellite altimeter studies as an "LIB correction" of sea surface height (sea level) into sub-surface pressure, which is the main parameter of interest to oceanographers in ocean circulation studies. Such a correction, insofar as it goes, is a rigorous one. However, it is also sometimes used imprecisely in the sense of a sea surface height correction term (i.e. removal of "meteorological noise") and in this context it is imperfect as it omits consideration of the dynamical adjustments which have to occur when the ocean is perturbed. These adjustments can be large at short timescales. Detailed discussions of the timescales when 2-D models can be used instead of 3-D ones, and of the importance of the dynamical terms at short timescales, can be found in many scientific papers (e.g., Ponte et al. 1991; Ponte 1994; Wunsch and Stammer 1997; Mathers and Woodworth 2001; Carrère and Lyard 2003).

In the shallow waters of coastal seas, the acceleration term related to wind stress becomes more important because of its water depth denominator (although not to the complete exclusion of the air pressure gradient), and can be the dominant forcing in shallow waters such as the south and eastern North Sea, Bay of Bengal and Gulf of Mexico. At its most extreme, this "wind setup" forcing can result in considerable damage and loss of life in a storm surge of many metres (e.g., Bay of Bengal surges, Murty et al. 1986, Dube et al. 1997; the 1953 North Sea flood, Wolf and Flather 2005; Hurricane Katrina surge, FEMA 2006: for a review of such extreme events see Lowe et al. 2010). However, even under much less extreme weather conditions, surges can provide a signal of several decimetres in a tide gauge or altimeter time series, which can be considerably larger than other signals of interest to coastal oceanographers.

Therefore, it is quite clear that the LIB model as applied (correctly or imprecisely) in the deep ocean cannot be employed as a "correction" for altimetry in coastal areas. Nevertheless, a number of schemes can be found in the literature, which have attempted to preserve something of the simplicity of the LIB for storm surge applications, in employing air pressure data alone and not wind fields. These schemes made use of distant and lagged air pressure measurements in order to parameterise the wind field, and a regression parameterisation between such meteorological data and local sea surface elevations (e.g., Amin 1982). Indeed, such a scheme formed the basis for flood warning in the UK for many years (Flather 2000). However, since the computer age, a more rigorous approach has been possible through the use of numerical surge models forced by air and wind fields provided by meteorological agencies (Heaps 1983).

Consequently, the only feasible approach in studies such as COASTALT (a collaborative project of European laboratories funded by the European Space Agency) or PISTACH (Prototype Innovant de Système de Traitment pour l'Altimétrie Côtière et l'Hydrologie, a project similar to COASTALT undertaken by the Centre National d'Études Spatiales (CNES) in France) is to employ data sets of sea surface height provided by numerical models forced by accurate fields of winds and air pressures. The main question now becomes, which of the alternative schemes available are most suitable for application to the altimeter projects.

7.3
Regional Storm Surge Models

Storm surge models (sometimes called tide-surge models) now exist for many parts of the world. These models are used for operational flood forecasting and employ forecast wind and air pressure information in order to predict likely surge elevations and currents up to several days ahead. A good review of storm surge modelling is given by Gonnert et al. (2001), and operational coastal flood warning systems for European seas are described in detail by Flather (2000). To achieve high resolution around coastlines, it is possible to nest finer grids (e.g., Greenberg 1979) or use finite-element techniques (e.g., Jones and Davies 2006). Such two-dimensional depth-averaged models have been proved to be successful as predictive tools of wind-driven currents as well as surge elevations in shelf seas. Models typically solve the following equations in discretised form, where Cartesian formulations are used here for simplicity (the equations are more usually expressed in spherical coordinates for regular latitude-longitude grids):

$$\frac{\partial U}{\partial t} + U\frac{\partial U}{\partial x} + V\frac{\partial U}{\partial y} - fV = -g\frac{\partial \eta}{\partial x} - \frac{1}{\rho}\frac{\partial P_a}{\partial x} + \frac{1}{\rho H}(\tau_{sx} - \tau_{bx}) \quad (7.1)$$

$$\frac{\partial V}{\partial t} + U\frac{\partial V}{\partial x} + V\frac{\partial V}{\partial y} + fU = -g\frac{\partial \eta}{\partial y} - \frac{1}{\rho}\frac{\partial P_a}{\partial y} + \frac{1}{\rho H}(\tau_{sy} - \tau_{by}) \quad (7.2)$$

$$\frac{1}{H}\frac{\partial \eta}{\partial t} + \frac{\partial U}{\partial x} + \frac{\partial V}{\partial y} = 0 \quad (7.3)$$

where U and V are vertically averaged velocity fields in the x and y directions, η is the free surface elevation, H is the total depth, f is the Coriolis parameter, g is Earth's gravity, ρ is the uniform density of water, P_a is the atmospheric pressure, and τ_{sx} and τ_{bx} are respectively surface and bed stresses in the x direction, with similar terms for the y direction in Eq. 7.2.

As an example, the current UK operational surge model (CS3X) covers the entire northwest European continental shelf at 12 km horizontal resolution. Its surface boundary conditions are the sea level pressure and 10 m wind fields from the Met Office North Atlantic and European (NAE) atmospheric model, at a similar spatial resolution (0.11° longitude by 0.11° latitude). Tidal input at the model open boundaries consists of the largest 26 constituents. Fig. 7.1a indicates the meteorological and surge model grids employed at present in the

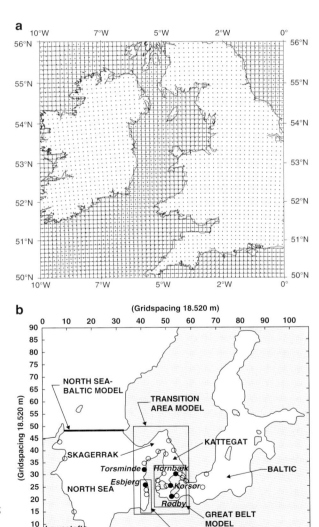

Fig. 7.1 (a) The grid used for the numerical tide-surge model (CS3X) employed in the current UK operational surge forecasting system. Only the Irish Sea section of the grid is shown to give an impression of model resolution and matching of a finite-difference grid to a coastline. The complete grid covers the entire NW European continental shelf from 40–63°N and eastwards of 20°W. The model is forced by winds and air pressures covering the entire North Atlantic and Europe on a 0.11° grid indicated by dots. (b) An example of a nested modelling scheme used for operational surge forecasting by the Danish Meteorological Institute. (From Flather 2000)

UK forecast scheme. Finer resolution models of critical areas can be nested within any shelf-wide model. Examples include the Bristol Channel and the south coast for the UK system, or the Wadden Sea and Great Belt models within the Danish Meteorological Institute shelf scheme (Fig. 7.1b). A number of examples of other surge modelling activities can be found in recent issues of *Marine Geodesy* (volume 32, number 2, 2009) and *Natural Hazards* (volume 51, number 1, 2009).

In shallow areas with large tides, the models are usually run once in tide-only mode and then a second time with both tidal and meteorological forcings, so as to account for tide−surge interaction, with the difference between the two model outputs defined as the "surge." Such interactions are particularly important in the southern North Sea (Prandle and Wolf 1978). (It is probably good practice to only use the word "surge" to imply a genuine meteorological contribution to sea level, in contrast to the overall difference between observed sea levels and tidal predictions, which could contain additional contributions from water density changes and measurement errors, and which is more properly referred to as the "residual.") In the UK example, the model suite runs four times each day and simulations consist of a 6-h hindcast portion (where the model is forced with meteorological analysis) followed by a 48-h forecast.

The ongoing performance of such modelling can be validated by means of comparison to real-time tide gauge data, and, if performance is satisfactory, then model predictions of water levels for several days ahead (i.e. the timescale for which weather forecasts have skill) can be provided to local and national governmental agencies in order to issue warnings to coastal industry and populations (e.g., in the UK coastal flood warning system described in Fig. 7.2). Validation of the UK model is performed monthly by comparison

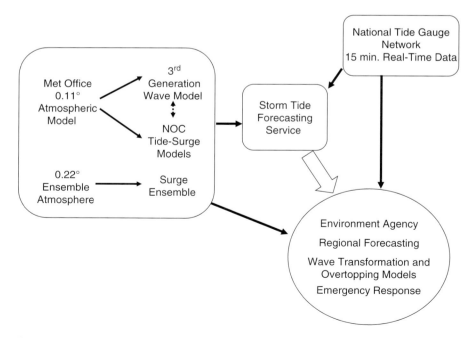

Fig. 7.2 Components of the UK coastal flood warning system

with observed sea level data from the national tide gauge network (see comparison plots in http://www.pol.ac.uk/ntslf/surgemonthlyplots/). Typical monthly mean root-mean-square errors in the model accuracy, derived from comparison to 15 min tide gauge data, are of the order of 10 cm (Wortley et al. 2007).

The same type of models have also been used successfully in hindcast-mode for research purposes, using extended sets (order 50 years) of meteorological reanalysis to compute variations in surge levels over a long period. Examples include Flather et al. (1998) and Bernier and Thompson (2006). The accuracy of such computed surge fields is probably of the order of 10 cm, although accuracy would be expected to be much worse where model spatial and temporal resolution (meteorological model and/or surge model) is insufficient. Examples would include surges initiated by tropical cyclones or by polar lows.

A review of numerical modelling activities some years ago (WMO 1996) found that 75% of models used by the various agencies were 2-D and depth-averaged, and that their forecast skill was comparable to 3-D models that resolved the vertical structure of current. Table 7.1 provides a list of the areas for which regional surge models are available at the

Table 7.1 A list of operational storm surge models based on the survey of JCOMM (2009) supplemented by other models known to us. Note that there can be more than one model for a given area

North Atlantic:
• North West European Continental Shelf (including North and Irish Seas, Baltic Sea, Iberian Peninsula, and North East Atlantic), Mediterranean Sea, Black Sea
• Grand Banks, Newfoundland, Labrador, USA (parts of)
South Atlantic:
• Argentina coast including Rio de la Plata
Pacific:
• NE Pacific, USA (parts of)
Asia:
• Japan, Korea, Hong Kong, Arabian Sea (India, Pakistan), Bay of Bengal (India, Bangladesh, Burma, Sri Lanka), Russia, Caspian Sea (Kazakhstan)
Australasia:
• Western, central and eastern Australia regional models (tropical cyclone and surge models)
• New Zealand (for tides, surges and waves)
Africa:
• South Africa has an operational wave forecast system to which tides and surges may be added in future
Arctic:
• Arctic models from Russian groups
French Activities:
• Meteo-France uses a depth-averaged, numerical model to provide stand-alone systems for forecasting tropical cyclone storm-surges in the French Antilles, New Caledonia, French Polynesia and in La Réunion

present time. This list has been extracted from a more recent survey by JCOMM (2009), which presented approximately 18 models. As found in the WMO (1996) survey, approximately three-quarters are 2-D models with resolution ranging from 10 to 20 km for regional models and 3-D applications, to 1 km for finer nested grids. Surge forecasts range up to 120 h, although most limit themselves to 36–72 h, with only 12 h or less useful for tropical cyclones. The survey contained notable omissions (e.g., from USA and Australia) which we have supplemented from our recent work (Horsburgh 2009) and personal knowledge of the subject; these additional models have also been included in Table 7.1. Altogether, one sees that a large part of the coastal ocean is covered by surge models of various designs. One suspects that the accuracy of operational computed surge fields in most cases will again be of the order of a decimetre, although probably not significantly better than that (that is the typical accuracy of hindcast modelling on the NW European continental shelf which is probably more accurate than most in spite of surges being large in this area).

One can see that with regard to provision of an "improved inverse barometer correction" for coastal altimetry, the most practical option is as follows:

(i) To take advantage of the archived regional surge fields from these operational agencies based on the most recent forecast meteorological information (typically a few hours old) or, if available, from hindcast data. In principle, it would be feasible to combine together in one place the outputs from operational surge models from different regions. Indeed, such an activity was proposed long ago as part of Coastal Component of the Global Ocean Observing System (GOOS), now subsumed within the Intergovernmental Oceanographic Commission Coastal Ocean Observations Panel (COOP), but this proposal has never been taken forward. Surge forecasts from several groups are available for Europe through the EuroGOOS SEPRISE (Sustained, Efficient Production of Required Information and Services within Europe) demonstration project http://www.eurogoos.org/sepdemo/), but only for tide gauge locations and not for the entire model grids. Information for the entire grids remains with the individual agencies. A reservation about this collective approach is that, while it might be feasible in some areas, it is clear that large regions would not be represented (e.g., the African coastline).

Africa serves to mention two particular issues with the modelling: one is that for adequate surge accuracy (say 10 cm) then ~10 km meteorological regional forcing is needed in most places and that resolution of meteorological fields may not be available; another concerns problems of resolving narrow shelves with models of sufficiently high spatial resolution.

A second option might also be possible in some regions:

(ii) To recompute surge fields at some later time when higher-quality reanalysis meteorological information has become available, as undertaken for the research activities referred to above. This possibility will be restricted to a few regions for which meteorological reanalyses are performed within a period adequate for use with the coastal altimetric product. One notes that some operational agencies do not provide analysis or reanalysis meteorological products at all, and, of those that do, not all archive them. An alternative to regional reanalyses in some areas might be those provided by the global reanalysis projects including those of the National Centers for Environmental Prediction – National Center for Atmospheric Research (NCEP-NCAR) (Kistler et al. 2001) and the European Centre for Medium-Range Weather Forecasts (ECMWF) (e.g., Uppala et al. 2004).

7.4
Global Barotropic Models

An alternative to the use of outputs from many regional surge models is to use a single global barotropic (2-D) model forced with global meteorological information. An advantage of a global model is that it leads to a global coastal altimetric product, even if that product is inevitably better in some regions than others. Such global models presently available include the following:

- MOG2D (2-D Gravity Waves Model) (T-UGO) (Toulouse – Unstructured Grid Ocean model) – this is a finite-element 2-D model with a high spatial resolution at the coastline (Carrère and Lyard 2003) (e.g., 15 km elements for the global model, 4 km for regional models). This model is now used for routine processing of Jason and Envisat altimetry to provide estimates of variability with timescales shorter than 20 days, lower frequencies being assumed to be describable by the inverse barometer. The T-UGO incarnation can also be used in 3-D mode. More details of barotropic modelling by French groups are provided in the chapter by Roblou et al., this volume.
- OCCAM barotropic version from NOC, the highest resolution version at present being 0.25°. This is a high resolution development of the 1° barotropic model which was used by Hughes and Stepanov (2004) and Stepanov and Hughes (2004) and which in turn was developed from the code of Webb et al. (1997).
- The barotropic model of Hirose et al. (2001; and see papers referenced therein) implemented to provide high-frequency (periods shorter than 20 days) corrections to earlier T/P and Jason Geophysical Data Records (GDRs) (AVISO 2003) with a primary emphasis on the deep ocean. This model has since been replaced in the T/P and Jason GDRs by MOG2D.

However, there are disadvantages to the use of a global model over regional ones in that the latter are often based on the basis of considerable local knowledge and experience, especially with regard to the accuracy of the essential bathymetric information, and in the quality of regionally-derived meteorological fields over globally-derived fields, and tend to be of higher resolution. They will also tend to have been subject to more extensive validation.

Such considerations concerning regional versus global approaches to future modelling are similar to those with regard to the coastal ocean tide correction (see the chapter by Ray et al., this volume). In both cases, the provision of the best possible bathymetric information is an essential requirement for accurate modelling.

7.5
COASTALT Validation Tests

The COASTALT project is currently assessing the utility of altimetry in coastal areas with the use of data from three test areas in Europe. In this phase of the project, the regional model approach will be followed:

- NW Mediterranean study area – in this area the regional finite-element MOG2D-MedSea surge model will be used with meteorological forcing data sets based on short-delay hindcast information. The model outputs are available in finite-element form as described above or in a square 0.25° grid. This model was used previously for Mediterranean oceanography by Cipollini et al. (2008). It has been shown by Pascual et al. (2008) to provide broadly similar information for surges to that of the HAMburg Shelf Ocean Model (HANSOM) which employed the 1/4° by 1/6° Mediterranean grid of Álvarez Fanjul et al. (1997).
- West of Britain – in this area the NOC/Met Office regional finite-difference surge model (CS3X) described above will be used with meteorological data sets based on operational (forecast) information.
- West of Portugal – in the area east of 12° W the MOHID (MOdelo HIDrodinâmico) finite-difference model developed at the Instituto Superior Técnico of the Technical University of Lisbon will be used. This model has been applied previously to a number of coastal areas (see http://www.mohid.com for many examples). It is available in 2-D (for tides) and 3-D (for full hydrodynamic model) versions and can be applied to a range of coastal spatial scales by means of nested models. Model resolution will be of the order of 4 km off-shore and 1 km near the coast. Reanalysis meteorological data will be used for COASTALT applications (i.e. the model will be run in delayed mode and not operational mode).

These three models can be considered as representative of those given in Table 7.1, at least as far as Europe is concerned, and while the project is focused on the test areas. However, it will be important in a further stage of the project, as considerations move towards provision of a global coastal product, that both the regional-combination and global model approaches be followed.

Given the provision of an improved "inverse barometer correction" and other corrections for coastal areas, attention will turn towards exploitation of the coastal altimetry data. The correction will be needed wherever it is necessary to reduce variance in the altimetric sea surface heights prior to study of other signals. Other applications include where it is required to adjust altimetric sea surface gradients closer to "level" (geoid) ones, either on an individual satellite pass basis such as in altimeter calibration using coastal tide gauges (e.g., Woodworth et al. 2004) or averaged over many passes as in the computation of an altimetric quasi-geoid.

The following may seem a trivial comment but we believe it is important to keep in mind. It is clear that the coastal surge correction (and also the ocean tide correction, for example) can be large, with a typical uncertainty at any one location of the order of a decimetre. (As the most significant source of uncertainty for storm surge magnitude is the accuracy of the meteorological forcing, the errors associated with the surge correction and their spatial scales can be estimated most effectively using a suite of surge models forced by a meteorological ensemble, see Horsburgh et al. 2008). This sizeable uncertainty may dishearten some potential users of altimetry in coastal areas. However, it is important to realise that it is best not to consider the use of altimetry in coastal areas in a general sense (although some chapters in this book will have that aim), but take into account always the particular applications in which the altimeter data are to be employed, and the magnitudes and spatial scales of the physical processes likely to be involved in such studies, together with the magnitudes and scales of the correction terms.

In fact, one notes that a large amount of coastal work is probably already possible with existing ("global") corrections. For example, the study of slope currents requires a precision of the order of 1 cm/10 km over perhaps 100 km. At first sight this is impossible to address given that some correction terms (e.g., the surge correction) have uncertainties much larger in magnitude. However, the uncertainty in the correction in this case will have a much larger spatial scale than the signal of interest. Another example is the computation of a mean sea surface (MSS) quasi-geoid, which is an important reference surface for many applications, in which one applies tide and surge corrections to the altimeter data. In this case the inaccuracy of the tide and surge modelling can be considerably reduced by time-averaging. Consequently, it is clear that a coastal user must always judge if it is possible to tolerate the uncertainties of correction terms in any analysis.

7.6 Conclusions

This paper has shown that a large number of coastal areas are covered by operational storm surge models. Therefore, it should be technically feasible to collect the operational sea surface height computations together for application to quasi-global coastal altimetry such as in the COASTALT and PISTACH projects. Indeed, such an exercise has been suggested before in different contexts, without the idea being pursued. Models for other areas, ideally using the same core modelling scheme, could be readily developed, if effort could be devoted to that task and if the essential high-quality bathymetric information was accessible. Where hindcast surge information was available, then that could be used instead of the operational products. An alternative to this collective regional approach is the further development of global barotropic models. However, to be useful for present purposes, the global models would require similar high-quality local bathymetry, and they would probably not be able to accommodate some of the ad hoc "tuning" of regional surge models which some groups have been able to perform. In our opinion, the coastal altimetry community should pursue both approaches.

Acknowledgements We thank Jane Williams for help with many aspects of research into storm surges.

References

Álvarez Fanjul E, Pérez B, Rodriguez I (1997) A description of the tides in the eastern North Atlantic. Prog Oceanogr 40:217–244. doi:10.1016/S0079-6611(98)00003-2

Amin M (1982) On analysis and forecasting of surges on the west coast of Great Britain. Geophys J R Astron Soc 68:79–94

AVISO (2003) AVISO and PoDaac user handbook – IGDR and GDR Jason-1 products, AVISO report SMM-MU-M5-OP-13184-CN, Edition 2.0, April 2003. Archivage, Validation et Interprétation de données des Satellites Océanographiques, Toulouse, France

Bernier NB, Thompson KR (2006) Predicting the frequency of storm surges and extreme sea levels in the northwest Atlantic. J Geophys Res 111:C10009. doi:10.1029/2005JC003168

Carrère L, Lyard F (2003) Modeling the barotropic response of the global ocean to atmospheric wind and pressure forcing – comparisons with observations. Geophys Res Lett 30(6):1275

Cipollini P, Vignudelli S, Lyard F, Roblou L (2008) 15 Years of altimetry at various scales over the Mediterranean. In: Vittorio Barale (JRC Ispra) and Martin Gade (Universitat Amburgo) (eds) Remote sensing of European Seas. Springer, Heidelberg/Germany, pp 295–306. doi:10.1007/978-1-4020-6772-3_22

Cipollini P, Benveniste J, Bouffard J et al (2009) The role of altimetry in coastal observing systems. Community White Paper contributed to the OceanObs'09 conference, 21–25 September 2009, Venice, Italy

Dube SK, Rao AD, Sinha PC, Murty TS, Bahulayan N (1997) Storm surge in the Bay of Bengal and Arabian Sea: the problem and its prediction. Mausam 48:283–304

FEMA (2006) Reconstruction guidance using hurricane Katrina surge inundation and advisory base flood elevation maps. Federal Emergency Management Agency. http://www.fema.gov/hazard/flood/recoverydata/katrina/katrina_about.shtm

Flather RA (1988) Storm surge modelling. Lecture notes during the Course on ocean waves and tides at the International Centre for Theoretical Physics, Trieste, 26 September–28 October 1988

Flather RA (2000) Existing operational oceanography. Coast Eng 41:13–40

Flather RA, Smith JA, Richards JD et al (1998) Direct estimates of extreme storm surge elevations from a 40-year numerical model simulations and from observations. Glob Atmos Ocean Syst 6:165–176

Fu LL, Cazenave A (eds) (2001) Satellite altimetry and earth sciences. A handbook of techniques and applications. Academic, San Diego, CA

Gonnert G, Dube SK, Murty TS et al (2001) Global storm surges: theory, observations and applications. German Coastal Engineering Research Council, 623pp

Greenberg DA (1979) A numerical model investigation of tidal phenomena in the Bay of Fundy and Gulf of Maine. Mar Geod 2:161–187

Heaps NS (1983) Storm surges, 1967–1982. Geophys J R Astron Soc 74:331–376

Hirose N, Fukumori I, Zlotnicki V et al (2001) Modeling the high-frequency barotropic response of the ocean to atmospheric disturbances: sensitivity to forcing, topography, and friction. J Geophys Res 106(C12):30987–30995

Horsburgh K (2009) Regional forecast scenarios. Chapter 7 in WMO (2009), ibid

Horsburgh KJ, Williams JA, Flowerdew J et al (2008) Aspects of operational model predictive skill during an extreme storm surge event. J Flood Risk Manag 1:213–221

Hughes CW, Stepanov VN (2004) Ocean dynamics associated with rapid J2 fluctuations: Importance of circumpolar modes and identification of a coherent Arctic mode. J Geophys Res 109:C06002. doi:10.1029/2003JC002176

JCOMM (2009) JCOMM expert team on waves and surges (ETWS) survey on storm surge data sources and storm surge forecasting systems operated by the national meteorological/oceanographical services. To be included within WMO (2009), ibid

Jones JE, Davies AM (2006) Application of a finite element model (TELEMAC) to computing the wind induced response of the Irish Sea. Cont Shelf Res 26:1519–1541

Kistler R, Collins C, Saha S et al (2001) The NCEP-NCAR 50 year reanalysis: monthly means CD-ROM and documentation. Bull Am Meteorol Soc 82:247–267

Lowe JA, Woodworth PL, Knutson T et al (2010) Past and future changes in extreme sea levels and waves. In: Church JA, Woodworth PL, Aarup T, Wilson S (eds) Understanding sea-level rise and variability. Blackwell, London

Mathers EL, Woodworth PL (2001) Departures from the local inverse barometer model observed in altimeter and tide gauge data and in a global barotropic numerical model. J Geophys Res 106(C4):6957–6972. doi:10.1029/2000JC000241

Murty TS, Flather RA, Henry RF (1986) The storm surge problem in the Bay of Bengal. Prog Oceanogr 16:195–233

Pascual A, Marcos M, Gomis D (2008) Comparing the sea level response to pressure and wind forcing of two barotropic models: validation with tide gauge and altimetry data. J Geophys Res 113:C07011. doi:10.1029/2007JC004459

Ponte RM (1994) Understanding the relation between wind- and pressure-driven sea level variability. J Geophys Res 99(C4):8033–8040

Ponte RM, Salstein DA, Rosen RD (1991) Sea level response to pressure forcing in a barotropic numerical model. J Phys Oceanogr 21:1043–1057

Prandle D, Wolf J (1978) Surge-tide interaction in the southern North Sea. In: Nihoul JCJ (ed.) Hydrodynamics of estuaries and fjords: proceedings of the 9th international Liège colloquium on ocean hydrodynamics, 1977. Elsevier Oceanography Series, 23, pp 161–185

Pugh DT (1987) Tides, surges and mean sea-level: a handbook for engineers and scientists. Wiley, Chichester, 472pp

Stepanov VN, Hughes CW (2004) The parameterization of ocean self-attraction and loading in numerical models of the ocean circulation. J Geophys Res 109:C03037. doi:10.1029/2003JC002034

Uppala S, Källberg P, Hernandez A et al (2004) ERA-40: ECMWF 45-year reanalysis of the global atmosphere and surface conditions 1957-2002. ECMWF Newsletter, No. 101 (Summer/Autumn 2004), pp 2–21. European Centre for Medium Range Weather Forecasts, Reading

Webb DJ, Coward AC, de Cuevas BA et al (1997) A multiprocessor ocean general circulation model using message passing. J Atmos Ocean Technol 14:175–182

WMO (1996) In: Ryabinin VE, Zilberstein OI, Seifert W. (eds) Storm surges. WMO marine meteorology and related oceanographic activities report no. 33, WMO/TD-No.779, 121pp. World Meteorological Organisation, Geneva, Switzerland

WMO (2009) Joint Technical Commission on Oceanography and Marine Meteorology (JCOMM) guide to storm surge forecasting. World Meteorological Organisation, Geneva, Switzerland (in press)

Wolf J, Flather RA (2005) Modelling waves and surges during the 1953 storm. Phil Trans R Soc 363:1359–1375. doi:10.1098/rsta.2005.1572

Woodworth PL, Moore P, Dong X et al (2004) Absolute calibration of the Jason-1 altimeter using UK tide gauges. Mar Geod 27(1–2):95–106

Wortley S, Crompton E, Orrell R et al (2007) Storm tide forecasting service: operational report to the environment agency for the period 1st June 2006 to 31st May 2007. Met Office, Exeter, UK, 67pp

Wunsch C, Stammer D (1997) Atmospheric loading and the oceanic inverted barometer effect. Rev Geophys 35

Tide Predictions in Shelf and Coastal Waters: Status and Prospects

8

R.D. Ray, G.D. Egbert, and S.Y. Erofeeva

Contents

8.1	Introduction	191
8.2	Tide Prediction in Shallow Water	193
	8.2.1 Nonlinearity	194
	8.2.2 Temporal Variability	196
	8.2.3 Summary	199
8.3	Assessment of Selected Global Models	199
	8.3.1 Example Regional Assessment: Gulf of Maine	203
8.4	Along-track Tide Corrections	204
8.5	Data Assimilation: Northern Australia	206
8.6	Prospects	211
8.7	Appendix	212
References		213

Abbreviations

ADCIRC	ADvanced CIRCulation and Storm Surge Model
ADCP	Acoustic Doppler Current Profiler
EOT	Empirical Ocean Tide model
FES	Finite Element Solution
GFO	Geosat Follow-On
GLOUP	GLObal Undersea Pressure
GOT	Global Ocean Tide
GRACE	Gravity Recovery and Climate Experiment
IAPSO	International Association for the Physical Sciences of the Oceans
INVTP	INVerse solution from the original T/P ground track

R.D. Ray (✉)
NASA Goddard Space Flight Center, Greenbelt, Maryland
e-mail: richard.ray@nasa.gov

G.D. Egbert and S.Y. Erofeeva
COAS, Oregon State University, Corvallis, Oregon

INVTP2	INVerse solution 2 from the original and interleaved T/P ground track
OSU	Ohio State University
OTIS	Oregon state university Tidal Inversion Software
POL	Proudman Oceanography Laboratory
RMS	Root Mean Square
RSS	Root Sum of Squares
SWOT	Surface Water Ocean Topography
T/P	TOPEX/Poseidon
TPXO	TOPEX/Poseidon Ocean tide model

8.1
Introduction

Seventeen years after the launch of the TOPEX/Poseidon (T/P) satellite, users of satellite altimetry have grown accustomed to having global tide models capable of providing high-quality corrections. Tidal predictions in the open ocean are now generally accurate to about 2–3 cm or better. Predictions in the polar seas, or specifically in those latitudes polewards of the T/P turning latitude of 66°, are much less reliable, owing to a general lack of high-quality observational constraints; satellite missions such as Envisat, ICESat, and GRACE are impacted by tide prediction errors in these regions (Ray et al. 2003; Peacock and Laxon 2004; Kwok et al. 2006).

Similarly problematic are tidal predictions in shallow seas and near-coastal waters. In some shallow-water regions, tide corrections to altimetry can be nearly as accurate as in the open ocean, but in other regions errors in the present generation of global models can exceed tens of cm, and even several meters in isolated spots (e.g., in the White Sea (Lebedev et al. 2008)). This chapter discusses the status of our current abilities to provide near-global tidal predictions in shelf and near-coastal waters. It highlights some of the difficulties that must be overcome and attempts to divine a path toward some degree of progress.

There are, of course, many groups worldwide who model tides over fairly localized shallow-water regions, with many varied applications ranging from navigation to pollution dispersal to sediment transport. Such work can be (after appropriate testing) extremely valuable to any altimeter study limited to those regions. In fact, we highly recommend use of local tide models to any user of altimetry who is focusing on a particular shallow-water region. Some examples are (Foreman et al. 1998; Han et al. 2002; Cherniawsky et al. 2004; Vignudelli et al. 2005; Lebedev et al. 2008). This chapter, however, concentrates on the more global models necessary for the general user. There have indeed been efforts to patch local and global models together, but such work is very difficult to maintain over many model updates (e.g., any update to one model requires remerging over its transition zones with all its neighbors) and can often encounter problems of a proprietary or political nature. Such a path, however, may prove the most fruitful one, especially when local models build on local expertise. The use of nested global models is likely to be more common in the future.

Sect. 8.2 briefly reviews some of the major problems encountered when attempting tide predictions in shelf and coastal waters, with special reference to satellite altimetry. A useful and more comprehensive review, but without explicit discussion of altimetry, is by Parker et al. (1999). In Sect. 8.3 we attempt to evaluate existing altimeter tide corrections in shallow regions, using existing global tide models that were developed primarily for deep-ocean altimetry; this is done mainly by comparison to a new set of validation data. Sect. 8.4 considers the special case of along-track tide corrections. Sect. 8.5 describes some recent work on modeling tides of the northern Australian shelf seas; it illustrates in concrete fashion some of the previously described problems and examines some possible approaches toward progress.

Historically, the applications of satellite altimetry have rightly focused on the deep ocean, where the data provide a unique and global perspective. Current global tide models were developed for such applications. As altimetry is pushed closer and closer towards the coastlines, tidal prediction becomes ever more challenging. In fact, as the reader will see, the subject of this chapter is focused primarily on shelf tides, as opposed to coastal tides. Tidal corrections in shelf and shallow seas are sufficiently problematic for our current suite of global tide models (and corresponding tide-prediction software packages) that we are well justified in devoting our attention to these regions. Truly coastal tide prediction on global scales is, in our opinion, too crude at the present time to warrant serious discussion; it is a subject we defer to the future – perhaps to a time when several years of wide-swath satellite altimeter data (Alsdorf et al. 2007b) are in hand.

8.2
Tide Prediction in Shallow Water

Relative to tides in the deep ocean, tides in shallow waters tend to be larger, of shorter wavelength, and possibly nonlinear (Parker et al. 1999). The first two aspects are highlighted in Fig. 8.1, an M_2 cotidal chart of the Yellow Sea (according to Lyard et al. 2006). Within the confines of such a small region exist three complete amphidromic systems. Amplitudes along the western boundary of the Korean Peninsula, especially off Inchŏn, are large, exceeding 2 m. Short wavelengths in such regions reflect, of course, the decreased phase-propagation speed of shallow-water waves as water depths decrease. Large amplitudes reflect, in grossest terms, the prevalence in the ocean of coastal Kelvin waves and quarter-wavelength shelf resonances at these frequencies.

Fig. 8.1 also displays the ground tracks of the long T/P-Jason altimeter time series. One sees that most of the large-amplitude features of the Yellow Sea are completely missed by these tracks. In our experience, regions of large amplitude are where tide models most benefit from measurement constraints. For altimetry to provide such constraints in the Yellow Sea, or any similar shallow sea, it seems necessary that far denser coverage is required. Coastal-tide gauges can, of course, provide invaluable constraints, and some recent global models have profited from them (Lyard et al. 2006). Coastal gauges, however, are of variable quality, are lacking from many parts of the world, and are often prone to very localized tidal effects that are unrepresentative of the tide even a few km away. For shelf seas such as in Fig. 8.1, denser coverage by satellite altimetry would be of great benefit.

Most recent tide models have in fact utilized multiple satellite altimeter datasets, at least in regions such as Fig. 8.1, and sometimes in deeper water as well (Lyard et al. 2006; Savcenko and Bosch 2008). In addition to the 17-year T/P-Jason time series, data from the

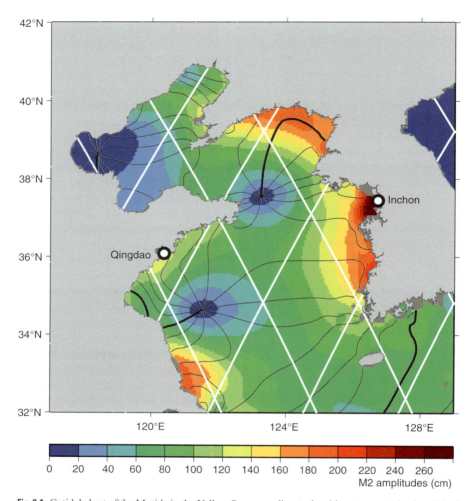

Fig. 8.1 Cotidal chart of the M_2 tide in the Yellow Sea, according to the altimeter-constrained model of Lyard et al. (2006). Phase lags are drawn every 30°, with heavy dark lines denoting Greenwich phase lag of 0°. Heavy white lines mark the ground-tracks of the TOPEX/Poseidon and Jason satellites

TOPEX interleaved (or extended) mission (3 years) as well as the Geosat Follow-On (GFO, 6+ years), and ERS-1, ERS-2, and Envisat series (17+ years) have proven valuable. Some of these missions have somewhat unfavorable aliasing characteristics (e.g., GFO) or pathological aliasing (sun-synchronous in the case of ERS and Envisat), so the T/P-Jason series remains the backbone of any well-determined altimeter constraints on tidal models.

8.2.1
Nonlinearity

Aside from the problem of insufficient measurements to constrain models, tidal prediction in shallow water is complicated by nonlinearities in the dynamical equations of tidal motion.

Weak nonlinearities are responsible for the appearance of overtides and compound tides, which occur at frequencies of the sums and differences of frequencies of interacting linear tides (Le Provost 1991). The number of nonlinearly generated constituents required to produce acceptably accurate tide predictions can be large at some locations; e.g., Andersen et al. (2006) tabulate 22 nonlinear constituents at Dover, U.K., with amplitudes exceeding 15 mm. Altimeter tide-prediction packages have only recently begun including a single one of these many constituents: the quarterdiurnal M_4. Further nonlinear effects can be induced by tide and storm surge interactions (Prandle and Wolf 1981; Bernier and Thompson 2007), as noted in the chapter by Woodworth and Horsburgh, this volume. In extreme cases, strong nonlinearities can culminate in tidal bores (Pugh 1987); although these are relatively isolated phenomena, they will certainly corrupt altimetric monitoring of any region prone to them, such as the Amazon delta (Birkett et al. 2002). Properly disentangling even weak nonlinear effects with their multiplicity of additional constituents calls for considerably more data than in linear regimes, so the lack of data at the required small scales compounds the problem of developing accurate shallow-water tidal models for altimetry.

As an illustration, consider the sea level spectra from two very different sites: "Myrtle," a seafloor pressure recorder from the deep South Atlantic (Spencer and Vassie 1997), a place with predominantly linear tides, and the tide-gauge at Calais, France, on the English Channel, with significant nonlinear tides – see Fig. 8.2. As is common with bottom-pressure data in the deep ocean, the overall noise level at Myrtle is extremely low. This low noise does allow the detection of many terdiurnal and higher frequency tidal lines, which are of interest in their own right, but they are quite small and are seen to be below the background continuum at Calais. In contrast, the quarter-diurnal tides at Calais are larger than the diurnal tides at both sites. The presence of many very large, high-frequency spectral peaks at Calais neatly summarizes the complexity of the tide-prediction problem that coastal altimetry faces.

As is well known, the frequencies of many compound tides coincide with frequencies of linear tides. This is especially common in the semidiurnal band, a result of third-order interactions between typically strong semidiurnal tides. For example, the compound $2MS_2$, denoting the interaction $M_2 + M_2 - S_2$, has a frequency equal to that of the linear tide μ_2.

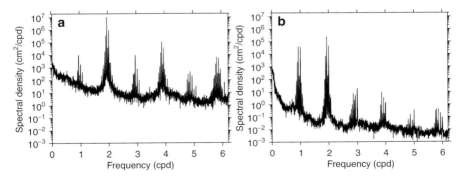

Fig. 8.2 Sea level spectra at (**a**) Calais, France and (**b**) site "Myrtle," on the seafloor in the South Atlantic (Spencer and Vassie 1997). Both spectra are based on 4 years of continuous hourly sea-level measurements. Scales are identical in both diagrams. Most of the nonlinear tides at Myrtle are below the background continuum at Calais

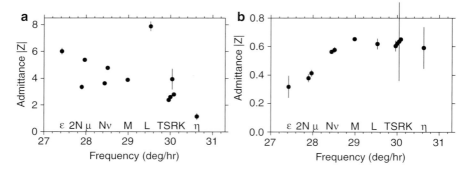

Fig. 8.3 Semidiurnal tidal admittances at (**a**) Calais, France and (**b**) site "Myrtle," on the seafloor of the South Atlantic (Spencer and Vassie 1997). Myrtle is a linear regime, with fairly smooth admittance as function of frequency. Calais is nonlinear; its erratic admittance is caused by nonlinear interactions occurring at the same frequencies as several linear constituents. Myrtle error bars are larger because its time series is shorter (4 years versus 44 years for Calais). Small jitter surrounding T_2, S_2, R_2 (frequencies near 30°/h) in both diagrams is caused mainly by effects of the radiational (i.e., atmospheric) tide (Cartwright and Ray 1994)

When such compound tides have significant amplitudes, observed tidal admittances[1] are no longer smooth functions of frequency. Fig. 8.3 gives a striking example, again comparing Calais and Myrtle. For altimeter applications, a discontinuous admittance has two important undesirable consequences: (1) The tide-prediction packages currently in common use by most altimeter users (including the FES, Global ocean tide (GOT), and TPXO packages) account for about two dozen minor constituents by inferring them from the admittances of major constituents. A nonsmooth admittance, as at Calais, invalidates these inference calculations. (2) The use of the response method (Munk and Cartwright 1966) to determine tides from altimetry, which has proven useful because of its power to extract tides reliably from short time series (Cartwright and Ray 1990), is unsound, because it assumes smooth admittances across each tidal band. Cartwright (1968) investigated extending the response method to nonlinear tidal regimes, but his approach is rarely used for tide-gauge data and has never been attempted for the irregular and relatively sparse temporal sampling of satellite altimetry.

Another somewhat subtler nonlinear effect arises from quadratic bottom friction, which acts to dampen weaker constituents in the presence of the tidal currents of stronger constituents. The effect can be especially noticeable for the nodal sidelines of large constituents (Amin 1976; Ku et al. 1985). For example, Ku et al. (1985) report that the nodal modulation of the M_2 amplitude in the Gulf of Maine is ±2.4% rather than the astronomical value of ±3.7%. To our knowledge, all general-purpose tide prediction packages use the astronomical values. Unfortunately, because this effect arises when a constituent like M_2 is

---

[1]An "admittance" is the complex ratio of the observed tide to the astronomical potential. In the deep ocean, it tends to be a smooth function of frequency, which reflects the ocean's predominantly linear response to tidal forcing and its lack of very sharp resonances (Munk and Cartwright 1966).

very large, a small percentage error in the constituent's modulation results in relatively large prediction errors. It is therefore worth considering how to adjust nodal corrections in those marginal seas where the astronomical values are in error. It is possible, but difficult, to extract nodal information from satellite altimetry (Cherniawsky et al. 2010). The most promising approach is probably to run a well-tuned numerical model with its forcing amplified by the astronomical modulation and then map the resulting tidal changes. Ku et al. (1985) explored such an approach in the Gulf of Maine.

8.2.2
Temporal Variability

While the astronomical forcing of tides is for all essential purposes a line spectrum, the ocean's response is not – the ocean is a physically complex and noisy filter. In consequence, tidal harmonic constants are not strictly constant. Temporal variability of tides can occur even in the deep ocean; the primary mechanism is presumably stratification changes affecting the generation and propagation of internal tides (Radok et al. 1967; Mitchum and Chiswell 2000). Variability can be far more pronounced in shallow water, where seasonally varying stratification is common, and varying tidal fronts, river outflow, buoyant plumes, upwelling, water level, sea ice, etc., in addition to storm surges, can all noticeably perturb tidal currents and elevations (Prinsenberg 1988; Lwiza et al. 1991; Souza and Simpson 1996; Howarth 1998; Byun and Wang 2005; St-Laurent et al. 2008). In this section, we concentrate on two prominent frequencies – seasonal and secular – which can be incorporated into standard tide-prediction methods with relatively minor modifications.

Given the multitude of possible environmental influences that have strong seasonal changes, it is not surprising to observe at some locations seasonal variations in tidal constants. Such variations have been studied at least since Corkan (1934). More recent works are by Kang et al. (1995), Huess and Andersen (2001), and St-Laurent et al. (2008). Outside of polar regions where annual changes in ice can profoundly affect tides, the largest seasonal tidal modulation probably occurs in the Yellow Sea. Fig. 8.4a shows the difference in M_2 amplitude between winter and summer as determined by Kang et al. (2002) using a two-layer numerical model. Differences between summer (strong stratification) and winter (well mixed) conditions give rise to differences in current shear and bottom frictional dissipation, and hence differences in tidal elevations. Very large M_2 amplitude changes are seen to occur both in the interior and along the boundaries of the Yellow Sea, but especially along sections of the coast of China where changes exceed 30 cm. The model satisfactorily simulates large (>10 cm) M_2 changes, which we have derived from 21 years of hourly sea-level data from the tide gauge at Lusi, China, shown in Fig. 8.4b. Kang et al. (2002) modeled only M_2, but interestingly enough, the tide gauge data show a significant semiannual as well as annual variation in the O_1 constituent (Fig. 8.4c).

Annual variations in tidal constituents have long been observed in British waters (Cartwright 1968; Amin 1985), and researchers there routinely account for M_2 variations by employing two artificial constituents at frequencies ±1 cpy away from the M_2 frequency. These lines are sometimes labeled MA_2 and MB_2. There is no need to introduce similar

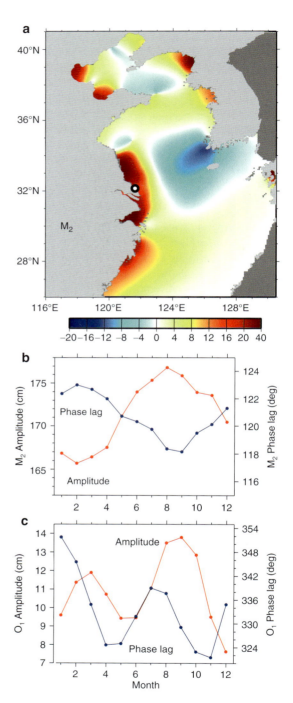

Fig. 8.4 (a) Seasonal differences (cm) in M_2 amplitudes in the Yellow Sea, in the sense summer minus winter, as simulated by a two-layer numerical model (Kang et al. 2002). Data provided courtesy of S. K. Kang. (b) Mean monthly amplitude and phase lag of the M_2 constituent, deduced from 21 years of hourly data from the tide gauge at Lusi, China, which is marked by the white dot in (a). (c) Similar for the O_1 constituent

new lines to model seasonal changes in solar tides, since standard constituents at ±1 cpy already exist – for example, S_2 is surrounded by T_2 and R_2. These seasonal effects, however, do introduce further jitter into the admittance curves around S_2. There is no standard nomenclature for new lines around O_1, probably because that tide is small around Britain. But note that the additional semiannual variation seen in Fig. 8.4c necessitates the introduction of lines at ±1 cpy and ±2 cpy to account for the observed variations; the τ_1 constituent exists at one of these frequencies.

Secular trends in tidal constituents have been observed at a number of coastal-tide gauges (Woodworth et al. 1991). It is sometimes difficult to know if an observed trend is localized just to the vicinity of the gauge or is perhaps even caused by instrumental problems (e.g., slow fouling of the stilling well by marine growth). In a recent analysis of United States gauges by Flick et al. (2003), the largest change in the U.S. (excluding Alaska) is at Wilmington, North Carolina, where mean tidal range has increased by 542 mm/century. The gauge, however, sits well up into the Cape Fear River and is undoubtedly reflecting local environmental changes. Consistent trends from neighboring gauges may well reflect more regional changes that would concern users of altimetry. For example, all southern Alaskan gauges between Kodiak and Cordova are experiencing an increase in mean diurnal tidal range (Flick et al. 2003). Jay (2009) notes consistent increases in the amplitudes of both M_2 and K_1 tides for a series of gauges along the northeast Pacific Ocean, with a mean increase of 0.6 mm/year. The cause is presently unknown. Some of the largest consistent tidal changes are occurring in the Gulf of Maine (Ray 2006) where a series of gauges between Boston and St. John, New Brunswick, shows increasing M_2 amplitudes throughout the twentieth century. M_2 at Eastport, Maine, has increased roughly 13 cm over the century, although it may now be leveling off. The tides of the Gulf of Maine, of course, are close to resonance (Ku et al. 1985), so presumably, the changes there reflect small changes in basin shape or depth induced by the combination of sea level rise and glacial rebound. The Eastport trend of 1.3 mm/year may still seem too small to concern altimeter users, but 30 years have now passed between the launches of Seasat and Jason-2, and our altimeter series continues to lengthen.

8.2.3 Summary

In summary, tides in shallow water present a number of challenges to users of satellite altimetry. Experience suggests that the development of accurate tide models requires constraints from high-quality observations. These are lacking in most shelf regions and even along many coastlines, so the constraints must often be provided by the altimeter data themselves. But the short spatial scales of shallow-water tides require data with denser sampling than for the deep ocean. In addition, disentangling the multitude of additional compound constituents, including those occurring at frequencies of linear constituents, requires considerably longer time series at any one location. Tidal analysis techniques based on exploiting smooth admittances must be modified. Currently employed prediction packages that assume smooth admittances must also be modified. In some locations this should include the addition of sidelines that incorporate seasonal modulations of larger constituents.

8.3 Assessment of Selected Global Models

In an effort to help assess existing and future global tide models in shallow-water regions, we have compiled (and continue to augment) a test set of independent tide-gauge harmonic constants. Comparisons against gauge data have previously proven invaluable for testing (Cartwright and Ray 1990; Shum et al. 1997) and even improving (Lyard et al. 2006) open-ocean tide models. Our current shallow-water test dataset focuses primarily on shelf tides, as opposed to truly coastal tides, since these are more directly relevant to altimeter applications. We have therefore tended to select bottom-pressure stations whenever possible, which unfortunately overemphasizes the Northwest European Shelf. We have also selected coastal-tide gauges from small, offshore islands; these form the entirety of our stations surrounding Australia. Our current dataset, comprising 179 stations, still lacks adequate global coverage (Fig. 8.5) and should be considered a work-in-progress. Both linear and nonlinear tidal constituents with frequencies between 1 and 8 cpd are included as availability dictates. (We have neglected, however, linear and compound long-period constituents. High-quality gauge estimates of these require very long-time series as well as corrections for nontidal oceanographic signals (Luther 1980). Inclusion of these constituents is surely warranted, but it is a task we have not attempted). Our main data sources are described in the appendix.

In this section, we examine some recent global models (or atlases) that have been commonly adopted for tide corrections to altimetry. The selection of models is by no means comprehensive, but our goals are simply to give a general sense of the current quality of shallow-water tide corrections via global models and to document some level of improvement over the past few years. The models examined are: GOT00.2, GOT4.1, and GOT4.7 (all successive updates of Ray 1999), FES2002 and FES2004 (Lyard et al. 2006), and EOT08a (Savcenko and Bosch 2008).

All these examined models have been strongly constrained by satellite altimeter data. In fact, any present-day global model not so constrained will be simply uncompetitive (Arbic et al. 2004; Egbert et al. 2004), its predictions too inaccurate for the severe demands of most satellite altimeter applications. So all models are constrained by T/P and (for the later ones) Jason-1 measurements; the GOT models also use ERS-1, ERS-2, and GFO altimetry in shallow water and in regions polewards of 66°; the FES models use ERS data mainly (but not

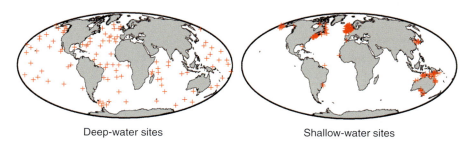

Deep-water sites　　　　　　　　　　Shallow-water sites

Fig. 8.5 Locations of "ground truth" tide gauge sites used in Tables 8.1 and 8.2. The 102 deep-water sites are identical to the dataset used by Shum et al. (1997). The 179 shallow-water sites are primarily clumped on the Northwest European Shelf and off the eastern coasts of the United States and Canada

exclusively) in shallow water, and they also assimilate data from 671 tide gauges; EOT08a uses data from all these altimeters plus Envisat and uses FES2004 as a prior solution. The FES solutions are initially computed on a dense finite-element grid with over one million nodes, but the released solutions are resampled to an equal latitude-longitude grid – FES2002 on a 0.25° grid and FES2004 on a 0.125° grid. EOT08a keeps the same FES2004 grid. The GOT models are on a 0.5° global grid, which is hardly adequate for most shallow seas. The fact that the FES solutions have assimilated tide-gauge data somewhat invalidates the independence of our tests, because there is undoubtedly some overlap between their and our dataset.

Table 8.1 presents rms differences between the examined models and our 179 test stations for the major diurnal, semidiurnal, and quarter-diurnal constituents. The evaluation of model values at station positions is done by bilinear interpolation. The root-sum-of-squares over the nine listed constituents are tabulated in the right-most column and are all of order 10 cm. More recent models are seen to represent improvements over earlier models. (A slight increase in RSS for FES2004 relative to FES2002 is caused by the former's anomalous behavior for M_4 and K_2; FES2004 is otherwise a clear advancement.)

It is of interest to compare these numbers with those for deep-ocean tides. Table 8.2 displays rms gauge differences for model GOT00.2 for both shallow and deep-ocean test stations; the latter consist of the same 102 stations used by Shum et al. (1997). The table clearly emphasizes the inferior quality of present-day tide-prediction capability in shallow water, in sharp contrast to the relatively accurate deep-ocean predictions possible in standard open-ocean altimeter applications. (In fact, the deep-ocean rms differences are here so small that some significant fraction of them must owe to errors in the gauge data, so the model is likely more accurate than the final RSS of 2.4 cm suggests.)

Table 8.1 RMS differences (cm) with validation gauge data: 179 shallow-water sites[a]

	Q1	O1	P1	K1	N2	M2	S2	K2	M_4[b]	RSS
GOT00.2	0.89	1.50	0.99	2.02	2.48	7.89	5.80	2.48	–	11.40
GOT4.1	1.03	1.48	0.99	1.86	2.17	7.25	4.14	1.78	–	9.95
GOT4.7	**0.84**	1.30	0.94	1.64	**2.05**	**6.02**	3.37	1.64	2.32	8.11
FES2002	1.06	1.44	1.04	1.96	2.80	7.70	4.89	2.19	–	10.85
FES2004	0.93	1.30	1.08	1.87	2.67	7.53	4.82	2.94	4.21	10.99
EOT08a	0.87	**1.23**	**0.80**	**1.62**	2.18	6.80	4.11	**1.54**	**2.27**	9.00

[a]Bold font marks smallest rms for each constituent.
[b]The rms signal of M_4 at the 179 stations is 3.70 cm, so this value is used to compute RSS whenever the M_4 constituent is missing from the given model.

Table 8.2 RMS differences (cm) with validation gauge data: GOT00.2

	Q1	O1	P1	K1	N2	M2	S2	K2	Inferred minors	Error of emission	Total RSS
Deep water	0.28	0.86	0.37	1.02	0.63	1.45	0.93	0.42	0.28	0.53	2.43
Shallow water	0.89	1.50	0.99	2.02	2.48	7.89	5.80	2.48	4.71	6.36	13.38

Table 8.2 includes two additional columns that together attempt to account for prediction error across the complete tidal spectrum (excepting, as noted above, the long-period band). The "inferred minor" tides refer to those relatively small tidal constituents not explicitly listed in the table, which, at least in many distributed tidal prediction packages, are usually handled by inference from major constituents. There are typically about two dozen of these constituents. (Note that the FES package includes the constituent $2N_2$ as a distributed, not inferred, constituent.) For the deep-ocean sites, we estimate the Table 8.2 rms differences for these inferred constituents by simply assuming the Q_1 admittance rms applies to each of the two dozen constituents linearly. This cannot be done for the shallow-water sites because of the presence of nonlinear perturbations (Fig. 8.3), but our new test dataset does already include some of these minor tides. We therefore test directly constituents μ_2, ν_2, and L_2 and infer the remainder from Q_1, to arrive at the value 4.71 cm. The Table 8.2 column of "omitted" constituents accounts for compound tides and overtides, mostly in the quarter-diurnal band (now including M_4) but also a few in the semidiurnal band like $2SM_2$, which do not coincide with linear tides and are therefore not included in the "inferred" column. For the shallow-water sites, we can again use our 179-station test dataset to sum these rms differences directly. For the deep-ocean sites, we adopt the value 0.34 cm for M_4 corresponding to 40 sites from a recent study of the Atlantic Ocean (Ray 2007) and, as but a rough guide, scale this value by 1.5 to account for other compound tides (the same ratio implied by our shallow-water sites).

Table 8.2 further highlights the importance of nonlinear tides – both unmodeled ones and wrongly inferred ones – in causing tide-prediction errors in shallow water. Their importance is second only to errors in the major M_2 and S_2 tides in the overall error budget.

Our shallow-water assessments given in Tables 8.1 and 8.2 should be viewed with a great degree of caution. The statistics depend crucially on the distribution of test stations, and this distribution is clearly inadequate. The large number of northeast Atlantic Ocean sites is problematic because it tends to overestimate semidiurnal errors (whose amplitudes are large on the European Shelf) and underestimate diurnal errors (whose amplitudes are small).

Moreover, the error at many sites can be considerably less than 13.4 cm. Fig. 8.6 shows the histogram of M_2 vector differences for all 179 stations, and clearly, the majority of stations have differences well below 5 cm. It would appear that our tabulated rms statistics are dominated by a few stations with large model errors. In fact, the four worst stations in Fig. 8.6 are located in the northern edge of Hudson Bay, the Strait of Gibraltar, the North Channel of the U.K. between Scotland and Ireland, and Torres Strait between Queensland

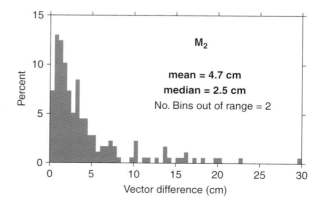

Fig. 8.6 Histogram of vector differences between 179 shallow-water test stations and the GOT4.7 global model, for constituent M_2. Two differences (of 40.6 and 64.1 cm) are out of range. Twenty percent of the stations agree with the model to better than 1 cm; fifty percent to better than 2.5 cm

and New Guinea. One of the largest discrepancies for the EOT08a model is Penrith Island, off the northwest coast of Australia. All these locations experience large tides, and all have very irregular coastlines nearby, demanding models with high spatial resolution. It is not surprising that large model errors occur in these places.

We conclude that at many shallow-water locations, the error in present-generation global-tide models may be only marginally worse than typical deep-ocean locations, but at many other sites, the error is substantial, tens of centimeters. The statistics of Table 8.1 are but a gross description of shallow-water prediction errors. Any shallow-water error statistics are highly location-dependent.

8.3.1
Example Regional Assessment: Gulf of Maine

It is of interest to follow up our global tests with at least one example of a more local nature and to extend the analysis completely to the coastlines. For this, we have selected the region shown in Fig. 8.7, covering the Gulf of Maine, the lower Bay of Fundy, and surrounding shelf

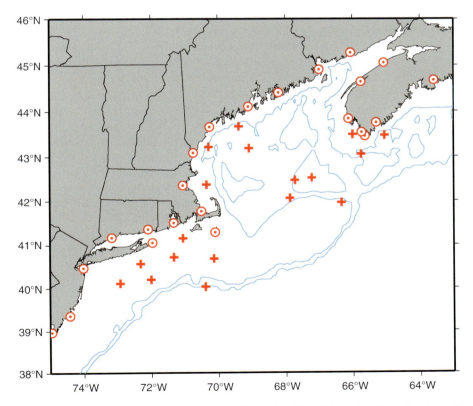

Fig. 8.7 "Ground truth" test sites in the Gulf of Maine region. The offshore sites (19 stations) are all based on bottom-pressure measurements; the coastal sites (24 stations) are standard coastal tide gauges. Blue contours denote 100-m and 200-m isobaths

Table 8.3 RMS differences (cm) with validation gauge data: Gulf of Maine and surrounds[a]

Offshore stations						Coastal stations				
	O1	K1	N2	M2	S2	O1	K1	N2	M2	S2
GOT00	0.90	1.39	1.35	3.40	1.86	1.39	2.03	9.46	31.7	6.10
GOT4.7	0.66	**0.73**	1.06	**2.40**	1.58	1.42	**1.49**	7.16	18.9	4.39
FES2004	0.77	0.95	1.30	3.00	3.36	1.52	1.66	6.85	15.6	5.13
EOT08a	**0.62**	0.75	**0.94**	2.76	1.74	**1.31**	1.56	6.74	15.2	4.27
ADCIRC.2e	1.08	1.54	1.10	4.62	**1.32**	1.50	2.22	**6.25**	**10.2**	**3.82**

[a] Bold font marks smallest rms for each constituent.

seas off Canada and the eastern United States. For such a restricted region, we can now include comparisons for regional tide models as well as the previous global ones. We have selected the widely used ADCIRC model (Westerink et al. 1994; Mukai et al. 2002), which has been applied to a relatively large region of the western North Atlantic Ocean between the 60°W meridian and the North American continent. It is a finite-element model comprising over 250,000 nodes. We used version ec2001_v2e, which includes interpolation software for the harmonic constants.

RMS differences for the 19 offshore and 24 coastal stations of Fig. 8.7 are tabulated in Table 8.3. The coastal test stations are especially challenging for the GOT models with their relatively coarse 0.5° resolution. It is thus not surprising that the AD-CIRC model, with its high-resolution finite-element grid, has the best comparisons with the coastal stations (an exception being the diurnal tides, which are small in this part of the Atlantic). The global models, however, appear superior for the offshore stations. This again is understandable given the use of satellite altimeter constraints in all four global models. We conclude that for many shallow-water altimeter applications our existing global models can be competitive with high-quality regional or local models, excepting perhaps very near the coastlines. The tests in Sect. 8.5 below shed further light on this point.

For the offshore stations in Table 8.3 the ADCIRC model does show the best agreement for the solar S_2 constituent. It is interesting to speculate whether the altimeter-constrained global models have slightly corrupted S_2 fields caused by reliance on the sun-synchronous ERS and Envisat data.

8.4 Along-track Tide Corrections

For the special case of altimeter data collected along the original T/P-Jason ground-track, another approach to tide prediction and altimeter corrections is conceivable. One can envision applying tide corrections point-by-point along the track, based simply on tidal analysis of the long-time series now available at those exact (within 1 km) locations. Aside from the benefit of easily accommodating short spatial scales, corrections based on along-track tide estimates automatically include the surface elevations associated with internal tides (Carrère et al. 2004), whereas most other tide corrections are barotropic only.

As of this writing over 600 repeat cycles of high-precision altimetry have been collected along the T/P-Jason ground-track. Along-track tidal analyses of these data have been undertaken previously (Tierney et al. 1998; Carrère et al. 2004), and the outputs successfully used for assimilation into numerical tide models (Matsumoto et al. 2000; Egbert and Erofeeva 2002), as well as for studies of internal tides and other short-wavelength phenomena. The question of employing along-track tide estimates directly as corrections depends primarily on whether some aliasing issues involving (mostly) minor tides can be adequately resolved. Fortunately, as the time series continues to lengthen the reliability of the along-track estimates strengthens.

Along-track estimates of major constituents like M_2 are now fairly well determined. The mean standard error for T/P-Jason along-track estimates of the M_2 constituent over the global ocean is 0.79 cm, and this number can be somewhat reduced by along-track smoothing. This error is consistent with, and hence validated by, an analysis of the along-track tide estimates at cross-over intersections of independent ascending and descending arcs; the global rms cross-over difference for M_2 is 1.08 cm (roughly $\sqrt{2}$ greater than the mean along-track error, as expected for differences of independent variates). The largest cross-over discrepancies occur in the vicinity of boundary currents, where intense mesoscale variability tends to corrupt tidal estimation. Statistics for constituents O_1, N_2, and S_2 are comparable. Of the major constituents only K_1 appears anomalous, with rms cross-over difference of 2.11 cm, more than twice the M_2 rms, presumably a result of cross-correlation between K_1 (alias period 173 days) and the semi-annual cycle of sea level (Ray 1998).

The along-track tide estimates are slightly degraded in shallow water, probably for a number of reasons, many of them highlighted in other chapters of this book. The errors should not be considered excessive, however, in light of generally larger tidal amplitudes in shallow water. For example, consider the cross-overs in the Yellow Sea (Fig. 8.1). The rms cross-over differences of several estimated along-track constituents are listed in Table 8.4. (The numbers are from five of the six cross-over locations north of 33°N shown in Fig. 8.1; our present altimeter database has insufficient valid data for the cross-over at 34.8°N, 126.1°E, offshore Korea, to allow analysis.) The rms differences for the major tides are not unduly large; K_1 is again the largest, although actually smaller than its global rms. The compound tides MS_4 and MN_4, however, have rms differences larger than their mean amplitudes, so the along-track estimates are unacceptable in this case. Both these constituents have long alias periods in the T/P sampling (1,083 and 244 days, respectively) and hence are more likely susceptible to noise from the red, non-tidal, background continuum. Using along-track estimates of these compound constituents would be warranted only in those locations where their amplitudes exceed several centimeters.

Table 8.4 RMS cross-over differences (cm): Yellow Sea

Tide	Mean Amp.	RMS Xover difference	Tide	Mean Amp.	RMS Xover difference	Tide	Mean Amp.	RMS Xover difference
Q_1	3.5	1.20	N_2	15.8	0.88	M_4	3.2	1.12
O_1	17.5	0.70	M_2	83.8	1.69	MS_4	2.2	2.39
K_1	23.0	1.90	S_2	28.1	1.76	MN_4	1.3	1.75

Along-track tide corrections, strictly applicable to data collected on the T/P-J ground-track, thus appear feasible. Further study is warranted to refine methods for noise reduction (including along-track smoothing, based possibly on independent data such as bathymetry) and to determine if these corrections are superior to standard corrections based on interpolations of gridded fields. Note that along-track corrections are unlikely to be satisfactory for other satellite ground-tracks, either because of limited amounts of data or because of poor aliasing characteristics (e.g., solar tides in ERS).

8.5
Data Assimilation: Northern Australia

Apart from the special case described in Sect. 8.4, the prospects for improved tidal corrections for shallow-water altimetry rest with the development of more accurate high-resolution models, which will probably be local or regional by necessity, but will need to blend smoothly to global models if they are to garner widespread use by the altimeter community. Owing to the lack of sufficient data, high-resolution models must rely heavily on hydrodynamics, so the need for more accurate bathymetric information will be critical. This section illustrates these points with results from a recent assimilation study of tides off the coast of northern Australia.

For this study the OSU Tidal Inversion Software (OTIS; Egbert and Erofeeva 2002) was used. OTIS uses a rigorous, and relatively sophisticated, variational scheme for data assimilation, but comparatively simple shallow-water dynamics, implemented numerically with a finite-difference approach. The OTIS assimilation scheme, based on a representer approach (Bennett 2002), was originally designed for global scale inverse modeling of tides (Egbert et al. 1994). It was subsequently developed into a relocatable system, for both global and regional tidal modeling (Egbert and Erofeeva 2002). Forward tidal solutions can be computed either by time-stepping the nonlinear shallow-water equations or by solving the linear frequency-domain time-separated equations for each constituent; for data assimilation the latter approach is used. OTIS has been applied in a number of regions at a range of resolutions, fitting altimetry, tide gauge, current mooring, and ship mounted ADCP data (Erofeeva et al. 2003, 2005; Andersen et al. 2006; Zu et al. 2008).

Our study area is shown in Fig. 8.8. In addition to the full domain, inverse solutions were constructed for two smaller, higher-resolution domains closer to the coast, marked off by rectangles in Fig. 8.8. The full domain is discretized with a resolution of 1/24° (roughly 4 km). Bathymetry for most of the domain was obtained by averaging of the 1-km Australian Bathymetry and Topography Grid (Geoscience Australia, 2008; http://www.ga.gov.au/meta/ANZCW0703004403.html). For the first subdomain (surrounding the Cape York Peninsula) the grid resolution is 1 km; for the second (King Sound, offshore Derby) 0.5 km. For the higher resolution domain, the land mask was set using an accurate coastline, and bathymetry for the 0.5-km grid was interpolated from the 1-km database.

Our validation tide-gauge dataset here consists of 148 stations. They include all the Australian sites used in Sect. 8.3 along with additional gauges on the main Australian and New Guinea coasts, but excluding for the most part those in small bays or estuaries. In this

8 Tide Predictions in Shelf and Coastal Waters: Status and Prospects

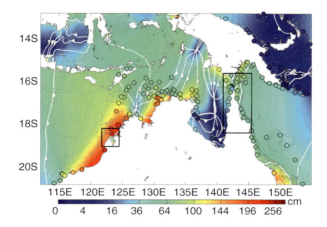

Fig. 8.8 M$_2$ cotidal chart of the seas off northern Australia, obtained by assimilating T/P and Jason altimeter data into a regional hydrodynamic model of horizontal resolution approximately 4 km. This solution is denoted 'InvTPT2' in the main text. Circles show locations of 148 validation tidegauge data, with colors denoting gauge-derived M$_2$ amplitudes. Rectangles mark off areas of local inverse solutions for King Sound (*left*) and Cape York Peninsula (*right*)

area we unfortunately have no access to bottom-pressure measurements, which are generally preferable for testing shelf tide models. Tide gauge locations are shown in Fig. 8.8.

Prior hydrodynamic solutions for the regional 4-km domain were obtained for the two largest semidiurnal (M$_2$, S$_2$) and diurnal (K$_1$, O$_1$) constituents by solving the linear shallow-water equations in the frequency domain. A global altimetric inverse solution was used to provide elevations on open boundaries and to compute the self-attraction and loading forces. Dissipation was parameterized linearly in the current velocity **u** as $\mathbf{F} = c\mathbf{u}u_f/H$, where H is water depth, c is a friction coefficient (taken as a constant 0.0025), and u_f is an effective friction speed, tuned, by minimizing disagreement with validation tide gauges, to 0.6 and 0.3 ms^{-1} for semidiurnal and diurnal waves, respectively. (This friction speed is much greater than would be expected for typical tidal current speeds in this area, presumably because other dissipative processes such as internal wave drag need to be accounted for.) RMS tide-gauge differences for the prior solutions are given in Table 8.5, along with comparable results for two global solutions, GOT4.7 and TPXO.7.1 (the latter is an update to Egbert and Erofeeva 2002). For the semidiurnal constituents regional-scale prior solutions which assimilate no data (except indirectly through the open boundary conditions) are already superior to the global inverse solution TPXO.7.1, and have only slightly larger tide-gauge differences than GOT4.7. For the two diurnal constituents prior gauge differences are significantly larger than for either of the altimetrically constrained global solutions.

Inverse solutions for the four constituents were computed using the standard OTIS model error covariance (Egbert and Erofeeva 2002) with a decorrelation length scale of 50 km. Along-track harmonic constants from T/P and Jason were used as data, with error bars determined formally from the harmonic analysis. A regularization parameter, varied to trade off between dynamical and data misfit, was chosen to minimize the rms differences with the validation tide gauges. Two inverse solutions were computed: InvTP, which was fit to both

Table 8.5 RMS tide-gauge differences (cm) for global and regional Australian solutions: 148 gauges

	M2	S2	K1	O1	M4
RMS signal	56.05	30.25	26.08	16.44	2.35
Global TPXO7.1	19.93	12.42	6.45	3.41	2.31
Global GOT4.7	14.10	8.31	4.58	2.75	2.36
Regional hydrodynamic prior	15.74	8.71	10.11	7.39	2.24
Regional inverse solution InvTP	12.60	–	–	–	–
Regional inverse solution InvTPT2	9.59	4.16	3.54	2.07	1.71

T/P and Jason data from the original T/P ground track (hereinafter TP data); and InvTPT2, which also incorporated the 3 years of T/P data from its interleaved ground track (hereafter "T2 data"). Including the T2 data has a very substantial impact on the solution for M_2, reducing rms tide-gauge differences from 12.60 to 9.59 cm. The M_2 constituent for InvTPT2 is displayed in Fig. 8.8 with tide-gauge amplitudes overlain for comparison. Relative to the large amplitudes of M_2 in this area (up to 3 m) agreement with the gauges appears excellent. Differences with the tide gauges are quantified for the two cases more precisely in Table 8.5. Assimilation of the T/P-Jason altimetry results in significant reductions in rms tide-gauge differences relative to the prior, from 15.74 cm to 9.59 cm for M_2. The other constituents considered show even more dramatic reductions. For all five constituents InvTPT2 agrees with the validation gauges significantly better than the global GOT4.7 solution.

One of the key points we wish to stress in this chapter is the dependence of shallow-water tide models on available bathymetric information and the need to improve both the accuracy and the resolution of those data. Sensitivity of our Australian modeling results to bathymetry was tested by recomputing all our solutions at the same 4-km resolution as before but with bathymetric grids derived from two versions of the Smith and Sandwell (1997) global bathymetric database: version 8.2, which was released in 2000, and version 11.1 from 2008. These two versions have nominal resolutions of roughly 1/30° and 1/60°, but are certainly not uniformly accurate at these scales. Differences between the 4-km grids derived from the Australian Bathymetry and Topography and the S&S v11.1 databases are of the order of 30% over the shallow (~50–100 m) shelf off northwest Australia. Significantly greater relative differences are seen in the shallowest water. Relative differences are also greater for the grid derived from the older version of the S&S bathymetry – in this case roughly 45% over the shelf. Tides computed with the two S&S grids (with all run parameters set as for our original Australian 4-km grid) provide a significantly poorer fit to the validation tide gauges for all constituents (Table 8.6). The newer version of the S&S bathymetry reduces rms differences, especially for the semidiurnal constituents, but the local bathymetric database still results in the best grid for tidal modeling. Inverse solutions for M_2 reveal a similar story, with the best results obtained with the Australian grid, and the worst with the S&S v8.2 grid. Now, however, differences are less dramatic – with the inclusion of nearby altimeter data to constrain the solution, accurate implementation of the dynamics becomes less critical.

Table 8.6 RMS tide-gauge differences (cm) for regional models: Dependence on bathymetry[a]

Bathymetry dataset	M_2	S_2	K_1	O_1	M_2 Inverse
Australian-4 km	15.74	8.34	9.28	6.65	9.59
Smith-Sandwell v8.2	27.68	14.83	12.71	8.88	13.80
Smith-Sandwell v11.1	21.85	11.26	12.27	8.32	10.44

[a]All three bathymetry dataset modified to approximate 4-km horizontal resolution.

The regional assimilation solution is a significant improvement over any of the available global solutions, but rms differences with the (coastal) validation gauges are still fairly large – approximately 10 cm for M_2, with much larger differences at many coastal locations where tidal amplitudes are large. How much improvement can be expected from higher resolution modeling? To explore this we consider prior and inverse solutions for the M_2 constituent in the two smaller coastal domains. The rms difference of the M_2 InvTPT2 solution for the validation gauges within the Cape York domain is 10.02 cm. The 1-km resolution prior solution, with open boundary conditions taken from InvTPT2, is already noticeably superior, with the rms difference for M_2 reduced to 5.81 cm (Table 8.7). This clearly demonstrates the value of high-resolution (accurate) bathymetry for modeling coastal tides. Assimilation of TP and T2 data results in virtually no further reduction in tide gauge differences; the optimal regularization parameter gives almost no weight to the data in this case. However, assimilating ERS data in addition reduces the tide gauge differences for M_2 to 4.7 cm.

The second high-resolution domain, King Sound, is too small to be resolved by the global solutions (Fig. 8.9a and b). There are four tide gauges within the domain, including two near the entrance to the Sound, and one at the head. Only one of these four gauges, which have an rms signal of nearly 2 m, was included in the validation dataset used for the regional model. Although King Sound is marginally resolved in the 4-km grid, tide-gauge comparisons of both the regional prior and inverse solutions to these gauges are poor, with differences exceeding 80 cm rms (Table 8.7; Fig. 8.9c and d). Solutions at higher resolution (0.5 km) result in significant improvements, with the rms reduced to 46 cm for the prior (Fig. 8.9e), and to 26.27 cm for a local high-resolution solution assimilating TP and T2 data (Fig. 8.9f). Incorporating ERS data as well further improves agreement with the tide gauges – at least for this lunar M_2 constituent – reducing the rms differences to 23.28 cm. This is about 10% of the rms M_2 signal in this area and represents a significant

Table 8.7 M_2 RMS tide-gauge differences (cm) for regional[a] and local[b] solutions

	RMS signal	Regional Prior Hydrodynamic	Local Prior Hydrodynamic	Regional Inversion InvTPT2	Local Inversion InvTPT2	Local Inversion InvTPT2+ERS
Cape York	43.61	11.91	5.81	10.02	5.83	4.68
King Sound	195.64	84.62	46.04	82.89	26.27	23.28

[a]Regional model (entire domain depicted in Fig. 8.8) is based on 4-km Australian bathymetry.
[b]Local models (rectangular domains depicted in Fig. 8.8) are based on 1-km (Cape York) and 0.5-km (King Sound) bathymetry.

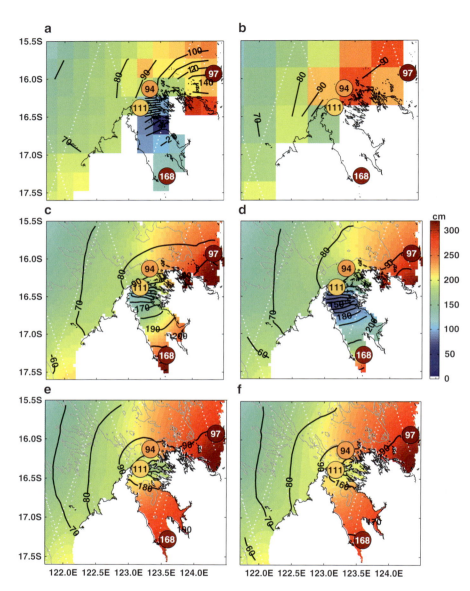

Fig. 8.9 M_2 cotidal charts of King Sound and surrounding shelf off northwest Australia. Four circles denote locations of validation tide gauges, with color denoting M_2 amplitude and interior numbers denoting phase lag. Various charts are: (**a**) global TPXO.7.1 model, (**b**) global GOT4.7 model, (**c**) regional hydrodynamic solution based on 4-km bathymetry, (**d**) regional inversion (called 'InvTPT2' in main text), (**e–f**) similar to (**c–d**) but for the 0.5-km bathymetry. Chart (**f**) gives best agreement with tide gauges (see Table 8.7)

improvement over the coarser-resolution models. Yet such a discrepancy is still very large if it is indicative of the error in an altimeter tide correction.

The above North Australian modeling effort has been described in some detail, because it seems highly illustrative of the kinds of problems that are encountered as we push satellite altimeter applications into shallow and near-coastal environments. For the required tide corrections, more and denser data and better, higher resolution bathymetry are seen to be necessary components to significant progress.

8.6
Prospects

In our efforts to improve tide corrections for satellite altimetry, the altimeter data themselves – as usual – are likely to supply the bulk of high-quality constraints. In shallow water, however, there is a significant mismatch between the short spatial scales of tides and the relatively widely spaced ground-tracks of present-day altimetry. Tide-model improvements must therefore rely heavily on data assimilation methods, such as those described in Sect. 8.5.

Of course, many alternatives to the simple OTIS approach employed in Sect. 8.5 can and have been applied to modeling of tides, and some of these may have significant advantages in shallow seas. For example, finite elements provide for more efficient discretization of a coastal domain, allowing higher resolution where needed and better conformation to a complex coastline (Greenberg et al. 1998). Some phenomena (e.g., interactions of tides and storm surge; wetting and drying) can only realistically be modeled with a time-domain approach, allowing more explicitly for nonlinearities. And in some cases stratification effects may be important enough (e.g., in estuaries) that fully three-dimensional modeling is required. Similarly, other (mostly simpler) assimilation approaches have been applied to mapping the tides, including nudging of the time-stepped shallow-water equations (Matsumoto et al. 2000) and ensemble methods. Given the need for detailed local-scale modeling in diverse shallow water environments around the globe, it is likely that many of these different modeling and assimilation techniques will be applied and compared over the coming years.

In our experience with OTIS over the past 15 years we have not found it possible to provide uniform high-accuracy solutions in shallow and deep water. Increasing data density only in the shallow areas where needed results in degradation of inverse solutions in deep water. Furthermore, providing sufficient degrees of freedom (e.g., by increasing the number of representers used in the inversion (Egbert and Erofeeva 2002)) to accommodate the complexity of tides in all shallow areas results in a computationally intractable problem when the variational assimilation approach is applied globally. Empirical mapping schemes, such as used for GOT4.7 and EOT08a, depend on more local analysis, and can more easily provide the requisite degrees of freedom and fit larger datasets. However, at short scales they depend entirely on their underlying prior hydrodynamic models – e.g., EOT08a is in essence a correction of FES2004 over horizontal scales supported by present-day altimeter coverage. At short scales a reliance on modeling and assimilation cannot be avoided.

The advantage of data assimilation, of course, derives from using the dynamics to interpolate, smooth, and extrapolate the data, and this requires bathymetry of sufficient accuracy. Although global databases, such as Smith-Sandwell (Smith and Sandwell 1997) are improving, there are still clear advantages to using higher resolution local products. Zooming in to very local areas to take full advantage of high-resolution, accurate, local bathymetry can further improve accuracy near the coast. Our example in Sect. 8.5 from King Sound, Australia, is particularly encouraging: with sufficiently accurate bathymetry, and good constraints on open boundary forcing provided by accurate regional shelf models, forward modeling with very sparse altimetric data constraints can provide decent maps of tidal elevations in a complex coastal environment. Providing a complete atlas of such maps for routine applications will, however, require a major community-wide effort to compute and validate model solutions for a huge number of individual domains. Accurate bathymetry for each would, of course, be a prerequisite. Currently available global bathymetric databases are certainly not up to the task.

Reliance on hydrodynamics could be relaxed if sufficient sea level time series measurements were to exist. "Sufficient" refers to both space and time – dense in space to support short tidal scales in shallow water, and long in time to support estimation of the many shallow-water tidal constituents. There is hope that the next generation of satellite altimeters could provide such data. Multi-beam altimeters, scanning altimeters, and swath-mapping interferometric radars have been proposed for future missions, and one or more of these appear likely to fly sometime in the next decade. The SWOT mission (Alsdorf et al. 2007a) envisions mapping the near-global ocean topography with 1-km resolution every 22 days (or better). If history is any guide, several years of such data are likely to result in significant improvements in shallow-water tide models; a decade of such data could result in coastal tide models with accuracies comparable to those now possible in the open ocean.

Acknowledgments This work was supported by the U.S. National Aeronautical and Space Administration's Ocean Surface Topography project. We thank Philip Woodworth for useful discussions. Comments from anonymous reviewers proved greatly helpful.

8.7
Appendix

The tide-gauge validation dataset used in Sect. 8.3 consists of tidal harmonic constants at 179 sites. This appendix briefly describes the sources of these data. Our present compilation should be considered preliminary owing to its inadequate spatial distribution (Fig. 8.5) and an incomplete catalog of constituents at many of the sites. For these reasons we are endeavoring to build on this initial dataset.

For reasons discussed above our dataset focuses on shelf tides as opposed to truly coastal tides. Most of our data are therefore tidal constants from bottom-pressure recorders sitting in shelf waters well away from coastlines. We also use data from small offshore islands.

Of the 179 stations, 50 were extracted from the Global Undersea Pressure (GLOUP) archives, housed at the Proudman Oceanographic Laboratory (POL), U.K. In addition to the pressure data, the GLOUP site stores outputs of tidal harmonic analyzes, which we have adopted for selected sites. We have selected only those GLOUP stations with at least 1 month of bottom-pressure data and have rejected a number of stations with evident errors. The most common errors appear to be time-tag problems. GLOUP analyses having poor agreement with the Cartwright-Zetler IAPSO (Cartwright and Zetler 1985) compilation of tidal constants were also rejected (or corrected if possible). An additional 32 bottom-pressure stations were extracted from the IAPSO compilations (Cartwright and Zetler 1985; Smithson 1992), although these sources usually list only the eight major constituents, but sometimes include M_4. Again only tidal constants based on at least 1 month of measurements are included.

Although the GLOUP database covers well the northwest European Shelf, we have included an additional 26 stations in the North Sea and offshore Belgium and The Netherlands, originally analyzed and compiled by Mariene Informatie Service, Rijswijk, The Netherlands, and distributed as "North Sea Tidal Data 1984–1993." Some of these stations are bottom gauges, others are from North Sea oil platforms. Nine time series are from 1 to 5 years duration; all the remaining are at least 1 month and thus probably of comparable quality to the majority of our GLOUP stations.

Another large set of stations (41 sites) were extracted from the Australian National Tide Tables (Anonymous 2007). For the tests of Sect. 8.3 we tried to select only offshore gauges that appear to be open-ocean sites, although it is possible that a few selected stations too close to large barrier reefs have been inadvertently included. For Sect. 8.5 we also added a large number of stations on the main northern Australian coast, although we attempted to avoid small bays and estuaries or stations located in large rivers. The Australian tables include a number of minor tides often susceptible to nonlinear perturbations (e.g., μ_2), which are useful to our Table 8.2 tests, and they also include compound tides M_4, MS_4, and $2MS_6$.

Additional bottom-pressure sites from the western North Atlantic shelf were extracted from the compilation of Moody et al. (1984) and from one Woods Hole report (Irish and Signell 1992). One station from the latter (located in Wilkinson Basin) appears highly suspicious of a timing problem, but we have kept the site for now. Unfortunately, the Moody report tabulates only the five constituents M_2, N_2, S_2, K_1, and O_1. To obtain a more complete tidal spectrum at these important sites, a reanalysis of the original pressure time series would be highly desirable.

The remaining stations were taken from miscellaneous sources, e.g. (Teague et al. 1998, 2001).

References

Alsdorf DE, Fu LL, Mognard N, Cazenave A, Rodriguez E, Chelton D, Lettenmaier DP (2007a) Measuring global oceans and terrestrial freshwater from space. Eos 88(24):253, 257

Alsdorf DE, Rodriguez E, Lettenmaier DP (2007b) Measuring surface water from space. Rev Geophys 45:RG2002

Amin M (1976) The fine resolution of tidal harmonics. Geophys J R Astron Soc 44:293–310

Amin M (1985) Temporal variations of tides on the west coast of Great Britain. Geophys J R Astron Soc 82:279–299

Andersen OB, Egbert GD, Erofeeva SY, Ray RD (2006) Mapping nonlinear shallow-water tides: a look at the past and future. Ocean Dyn 56:416–429

Anonymous (2007) Australian national tide tables 2007, Austr Hydrogr Pub 11, 404pp

Arbic BK, Garner ST, Hallberg RW, Simmons HL (2004) The accuracy of surface elevations in forward global barotropic and baroclinic models. Deep Sea Res II 51:3069–3101

Bennett AF (2002) Inverse modeling of the ocean and atmosphere. Cambridge University Press, 234pp

Bernier NB, Thompson KR (2007) Tide-surge interaction off the east coast of Canada and northeastern United States. J Geophys Res 112:C06008

Birkett CM, Mertes LAK, Dunne T, Costa MH, Jasinski MJ (2002) Surface water dynamics in the Amazon Basin: application of satellite radar altimetry. J Geophys Res 107. doi:10.1029/2001JD000609

Byun DS, Wang XH (2005) The effect of sediment stratification on tidal dynamics and sediment transport patterns. J Geophys Res 110:C03011

Carrère L, Le Provost C, Lyard F (2004) On the statistical stability of the M_2 barotropic and baroclinic tidal characteristics from along-track TOPEX/Poseidon satellite altimetry analysis. J Geophys Res 109:C03033

Cartwright DE (1968) A unified analysis of tides and surges round north and east Britain. Phil Trans R Soc Lond A263:1–55

Cartwright DE, Ray RD (1990) Oceanic tides from Geosat altimetry. J Geophys Res 95(C3): 3069–3090

Cartwright DE, Ray RD (1994) On the radiational anomaly in the global ocean tide, with reference to satellite altimetry. Oceanol Acta 17:453–459

Cartwright DE, Zetler BD (1985) Pelagic tidal constants, Publ Sci No 33. IAPSO, Paris

Cherniawsky JY, Foreman MGG, Crawford WR, Beckley BD (2004) Altimeter observations of sea-level variability off the west coast of North America. Int J Remote Sensing 25:1303–1306

Cherniawsky JY, Foreman MGG, Kang SK, Scharroo R, Eert AJ (2010) 18.6-year lunar nodal tides from altimeter data. Cont Shelf Res 30:575–587)

Corkan RH (1934) An annual perturbation in the range of tide. Proc R Soc A144:537–559

Egbert GD, Erofeeva SY (2002) Efficient inverse modeling of barotropic ocean tides. J Atmos Ocean Technol 19:183–204

Egbert GD, Bennett AF, Foreman MGG (1994) TOPEX/Poseidon tides estimated using a global inverse model. J Geophys Res 99:24821–24852

Egbert GD, Ray RD, Bills BG (2004) Numerical modeling of the global semidiurnal tide in the present day and in the last glacial maximum. J Geophys Res 109:C03003

Erofeeva SY, Egbert GD, Kosro PM (2003) Tidal currents on the central Oregon shelf. J Geophys Res 108. doi:10.1029/2002JC001615

Erofeeva SY, Padman L, Egbert GD (2005) Assimilation of ship-mounted ADCP data for barotropic tides: application to the Ross Sea. J Atmos Ocean Technol 22:721–734

Flick RE, Murray JF, Ewing LC (2003) Trends in United States tidal datum statistics and tide range. J Waterway Port Coast Eng 129:155–164

Foreman MGG, Crawford WR, Cherniawsky JY, Gower JFR, Cuypers L, Ballantyne VA (1998) Tidal correction of TOPEX/Poseidon altimetry for seasonal sea surface elevation and current determination off the Pacific coast of Canada. J Geophys Res 103:27979–27998

Greenberg DA, Werner FE, Lynch DR (1998) A diagnostic finite-element ocean circulation model in spherical coordinates. J Atmos Ocean Technol 15:942–958

Han G, Tang CL, Smith PC (2002) Annual variations of sea surface elevation and currents over the Scotian shelf and slope. J Phys Oceanogr 32:1794–1810

Howarth MJ (1998) The effect of stratification on tidal current profiles. Cont Shelf Res 18:1235–1254

Huess V, Andersen OB (2001) Seasonal variation in the main tidal constituent from altimetry. Geophys Res Lett 28:567–570

Irish JD, Signell RP (1992) Tides of Massachusetts and Cape Cod Bays. WHOI-92-35, Woods Hole, 62pp

Jay DA (2009) Evolution of tidal amplitudes in the eastern Pacific Ocean. Geophys Res Lett 36:L04603

Kang SK, Chung JY, Lee SR, Yum KD (1995) Seasonal variability of the M_2 tide in the seas adjacent to Korea. Cont Shelf Res 15:1087–1113

Kang SK, Foreman MGG, Lie HJ, Lee JH, Cherniawsky J, Yum KD (2002) Two-layer modeling of the Yellow and East China Seas with application to seasonal variability of the M_2 tide. J Geophys Res 107(C3):3020

Ku LF, Greenberg DA, Garrett CJR, Dobson FW (1985) Nodal modulation of the lunar semidiurnal tide in the Bay of Fundy and Gulf of Maine. Science 230:69–71

Kwok R, Cunningham GF, Zwally HJ, Yi D (2006) ICESat over Arctic sea ice: interpretation of altimetric and reflectivity profiles. J Geophys Res 111:C06006

Le Provost C (1991) Generation of overtides and compound tides. In: Parker BB (ed) Tidal hydrodynamics. Wiley, New York

Lebedev S et al (2008) Exploiting satellite altimetry in coastal ocean through the Alticore project. Russ J Earth Sci 10:ES1002

Luther DS (1980) Observations of long period waves in the tropical oceans and atmosphere. PhD thesis, MIT/WHOI, 210pp

Lwiza KMM, Bowers DG, Simpson JH (1991) Residual and tidal flow at a tidal mixing front in the North Sea. Cont Shelf Res 11:1379–1395

Lyard F, Lefevre F, Letellier T, Francis O (2006) Modelling the global ocean tides: modern insights from FES2004. Ocean Dyn 56:394–415

Matsumoto K, Takanezawa T, Ooe M (2000) Ocean tide models developed by assimilating TOPEX/Poseidon altimeter data into hydrodynamic model: a global model and a regional model around Japan. J Oceanogr 56:567–581

Mitchum GT, Chiswell SM (2000) Coherence of internal tide modulations along the Hawaiian Ridge. J Geophys Res 105:28653–28661

Moody JA et al. (1984) Atlas of tidal elevation and current observations on the northeast American continental shelf and slope. U.S. Geological Survey Bull 1611:122

Mukai AY, Westerink JJ, Luettich RA, Mark D (2002) Eastcoast 2001: A tidal constituent database for the western North Atlantic, Gulf of Mexico, and Caribbean Sea. U.S. Army ERDC/CHL TR-02-24, 194pp

Munk WH, Cartwright DE (1966) Tidal spectroscopy and prediction. Phil Trans R Soc Lond A259:533–583

Parker BB, Davies AM, Xing J (1999) Tidal height and current prediction. In: Mooers CNK (ed) Coastal ocean prediction. American Geophysical Union, Washington

Peacock NR, Laxon SW (2004) Sea surface height determination in the Arctic Ocean from ERS altimetry. J Geophys Res 109:C07001

Prandle D, Wolf J (1981) The interaction of surge and tide in the North Sea and River Thames. Geophys J R Astron Soc 55:203–216

Prinsenberg SJ (1988) Damping and phase advance of the tide in western Hudson Bay by annual ice cover. J Phys Oceanogr 18:1744–1751

Pugh DT (1987) Tides, surges, and mean sea-level. Wiley, Chichester, 473pp

Radok R, Munk W, Isaacs J (1967) A note on mid-ocean internal tides. Deep-Sea Res 14:121–124

Ray RD (1998) Spectral analysis of highly aliased sea-level signals. J Geophys Res 103:24991–25003

Ray RD (1999) A global ocean tide model from TOPEX/Poseidon altimetry: GOT99, NASA Tech Memo 209478. Goddard Space Flight Center, Maryland, 58pp

Ray RD (2006) Secular changes of the M_2 tide in the Gulf of Maine. Cont Shelf Res 26:422–427

Ray RD (2007) Propagation of the overtide M_4 through the deep Atlantic Ocean. Geophys Res Lett 34:L21602

Ray RD, Rowlands DD, Egbert GD (2003) Tidal models in new era of satellite gravimetry. Space Sci Rev 108:271–282

Savcenko R, Bosch W (2008) EOT08a – empirical ocean tide model from multi-mission satellite altimetry. DGFI Rep 81, München, 37pp

Shum CK, Woodworth PL, Andersen OB, Egbert GD, Francis O, King C, Klosko SM, Le Provost C, Li X, Molines JM, Parke M, Ray RD, Schlax M, Stammer D, Tierney C, Vincent P, Wunch C (1997) Accuracy assessment of recent ocean tide models. J Geophys Res 102 (C11):25173–25194

Smith WHF, Sandwell DT (1997) Global sea floor topography from satellite altimetry and ship depth soundings. Science 277:1956–1962

Smithson MJ (1992) Pelagic tidal constants 3, Publ Sci No 35. IAPSO, Paris

Souza AJ, Simpson JH (1996) The modification of tidal ellipses by stratification in the Rhine ROFI. Cont Shelf Res 16:997–1009

Spencer R, Vassie JM (1997) The evolution of deep ocean pressure measurements in the UK. Prog Oceanogr 40:423–435

St-Laurent P, Saucier FJ, Dumais JF (2008) On the modification of tides in a seasonally icecovered sea. J Geophys Res 113:C11014

Teague WJ, Perkins HT, Hallock ZR, Jacobs GA (1998) Current and tide observations in the southern Yellow Sea. J Geophys Res 105:3401–3411

Teague WJ, Perkins HT, Jacobs GA, Book JW (2001) Tide observations inthe Korean-Tsushima Strait. Cont Shelf Res 21:545–561

Tierney C, Parke ME, Born GH (1998) An investigation of ocean tides derived from alongtrack altimetry. J Geophys Res 103:10273–10287

Vignudelli S, Cipollini P, Roblou L, Lyard F, Gasparini GP, Manzella G, Astraldi M (2005) Improved satellite altimetry in coastal systems: case study of the Corsica Channel. Geophys Res Lett 32:L07608

Westerink JJ, Luettich RA, Muccino JC (1994) Modeling tides in the western North Atlantic using unstructured graded grids. Tellus 46A:178–199

Woodworth PL, Shaw SM, Blackman DB (1991) Secular trends in mean tidal range around the British Isles and along the adjacent European coastline. Geophys J Int 104:593–610

Zu T, Gan J, Erofeeva SY (2008) Numerical study of the tide and tidal dynamics in the South China Sea. Deep-Sea Res 55:137–154

Post-processing Altimeter Data Towards Coastal Applications and Integration into Coastal Models

9

L. Roblou, J. Lamouroux, J. Bouffard, F. Lyard, M. Le Hénaff, A. Lombard, P. Marsaleix, P. De Mey, and F. Birol

Contents

9.1 Introduction 219
9.2 Post-processing Altimetry in Coastal Zones: Early Developments 220
    9.2.1 Redefining Editing Strategies 221
    9.2.2 Improving De-aliasing Corrections 222
    9.2.3 Improving the Vertical Reference Frame 225
9.3 Latest Upgrades of the X-TRACK Processor 227
    9.3.1 Orbit and Large-Scale Error Reduction 228
    9.3.2 High Rate Data Stream 229
9.4 Matching Satellite Altimetry with Coastal Circulation Models 230
    9.4.1 Rationale 230
    9.4.2 Consistency Between Altimeter Measurements and Model Estimates 232
    9.4.3 Methodologies and Case Studies 234
9.5 Concluding Remarks and Perspectives 241
References 243

L. Roblou (✉), F. Lyard, M. Le Hénaff, P. Marsaleix, P. De Mey and F. Birol
Laboratoire d'Etudes en Géophysique et Océanographie Spatiales (LEGOS), Université de Toulouse; UPS (OMP-PCA), 14 Avenue Edouard Belin, F-31400, Toulouse, France and CNRS, LEGOS, F-31400, Toulouse, France
e-mail: laurent.roblou@legos.obs-mip.fr

P. Marsaleix
CNRS, Laboratoire d'Aérologie (LA), F-31400Toulouse, France

J. Lamouroux
NOVELTIS SA, 2 Avenue de l'Europe, F-31520, Ramonville-Saint-Agne, France

J. Bouffard
Institut Mediterrani d'Estudis Avançats, C/Miquel Marquès, 21, 07190 Esporles, Mallorca, Illes Balears, Spain

A. Lombard
Centre National d'Études Spatiales (CNES), 18 Avenue Edouard Belin, F-31400, Toulouse, France

Keywords Coastal altimetry · Data correction retrieval · Data editing · Post-processing · Regional de-aliasing · Synergy with coastal models

Abbreviations

ALBICOCCA	ALtimeter-Based Investigations in COrsica, Capraia and Contiguous Areas
ASI	Agenzia Spaziale Italiana
AVISO	Archivage, Validation et Interprétation des données des Satellites Océanographiques
CLS	Collecte Localisation Satellites
CNES	Centre National d'Études Spatiales
COASTALT	ESA development of COASTal ALTimetry
CTOH	Centre de Topographie des Océans et de l'Hydrosphère
DUACS	Data Unification and Altimeter Combination System
ENVISAT	ENVIronmental SATellite
ESA	European Space Agency
EU	European Union
FES	Finite Element Solution
GDR	Geophysical data record
GEOSAT	GEodetic & Oceanographic SATellite
GFO	GEOSAT Follow-On
GMES	Global Monitoring for Environment and Security
GOCE	Gravity field and steady-state Ocean Circulation Explorer
GODAE	Global Ocean Data Assimilation Experiment
GOT	Global Ocean Tide
IB	Inverted Barometer
IMBER	Integrated Marine Biogeochemistry and Ecosystem Research
LA	Laboratoire d'Aérologie
LEGOS	Laboratoire d'Études en Géophysique et Océanographie Spatiales
LPC	Liguro-Provençal-Catalan
LSER	Large-scale error reduction
MARINA	MARgin INtegrated Approach
MDT	Mean dynamic topography
MERSEA	Marine Environment and Security for the European Area
MFSTEP	Mediterranean Forecasting System Toward Environmental Prediction
Mog2D	Modèle d'Ondes de Gravité 2D
MSS	Mean Sea Surface
NWMED	North-Western MEDiterranean
OGCM	Ocean General Circulation Model
OSTST	Ocean Surface Topography Science Team
PISTACH	Prototype Innovant de Système de Traitement pour l'Altimétrie Côtière et l'Hydrologie
SARAL	Satellite with ARgos and ALtika

SLA	Sea Level Anomaly
SSALTO	Segment Sol multimissions d'ALTimétrie, d'Orbitographie et de localisation précise
SSH	Sea Surface Height
SST	Sea Surface Temperature
SVD	Singular Value Decomposition
SWOT	Surface Water and Ocean Topography
TOPEX	TOPography EXperiment
TOSCA	Terre, Ocean, Surfaces Continentales, Atmosphère
T-UGOm	Toulouse Unstructured Grid Ocean Model
UNESCO	United Nations Educational, Scientific and Cultural Organization

9.1
Introduction

The raw data provided by radar altimeters onboard satellite missions do not come ready for use. They must be processed to remove unwanted instrumental effects. In the case of SSH estimates, additional information about orbits and auxiliary corrections due to atmosphere and ocean effects need to be applied (Fu and Cazenave 2001). Official altimetry products generally contain sensor measurements, orbit estimations and a full set of corrections (AVISO 1996).

In coastal systems, shorter spatial and temporal scales make ocean dynamics particularly complex, and the temporal and spatial sampling of current altimeter missions is not fine enough to capture such variability (Chelton 2001). Moreover, the error budget of SSH inferred from satellite radar altimetry measurements in coastal regions is increased by intrinsic difficulties that can be separated into two categories.

Firstly, close to the coastline, the altimeter footprint is contaminated by land (and also inland water) surface reflections (Mantripp 1996; Andersen and Knudsen 2000). These surfaces thus contribute to the backscattered power received by the altimeter and significantly affect the shape of the waveform and consequently its tracking by onboard algorithms based on the Brown model (Brown 1977). Returned echoes from the radar altimeter are also often noisy in the coastal zone and the assumption of the Brown model, which is based on specular reflection from the open sea surface, is questionable.

Secondly, environmental and geophysical corrections are less reliable in coastal regions than in open oceans. Environmental corrections derived from onboard instruments (altimeter and radiometer) are contaminated by non-ocean surface reflections (e.g., Ruf et al. 1994; Chelton et al. 2001; Fernandes et al. 2003). Sea state bias correction (i.e. the path delay due to sea surface state) as well as tides and the ocean response to atmospheric forcing have been tailored to deep ocean conditions (e.g., Anzenhofer et al. 1999; Chelton et al. 2001; Le Provost 2001; Carrère and Lyard 2003; Labroue et al. 2004). Finally, the standard altimeter data validity checks are not well suited for coastal regions and often reject many of the measurements acquired close to the coasts (Vignudelli et al. 2000).

These issues limit the effective use of altimeter-derived products in coastal areas (Vignudelli et al. 2005). As a consequence, particular care has to be taken to improve past and current altimeter data reprocessing and corrections.

Since the pioneering effort constituted by the ALBICOCCA (ALtimeter-Based Investigations in COrsica, Capraia and Contiguous Areas) project (Lyard et al. 2003; Vignudelli et al. 2005), various groups are currently working to correct the known weaknesses in the overall processing phase that prevent the use of altimetry in coastal and shelf seas. For example, regional modelling is used to de-aliase the tidal and the short-term meteorologically induced signal instead of using global models lacking horizontal resolution (e.g., Jan et al. 2004; Pascual et al. 2006; Bouffard et al. 2008a, b). Other groups are working on the removal of land emissivity in the radiometer footprint for wet tropospheric path delay correction (e.g., Coppens et al. 2007; Desportes et al. 2007; Obligis et al., this volume) or using new waveform models for the retracking of coastal waveforms (e.g., Berry et al. 2005; Deng and Featherstone 2006; Quesney et al. 2007; Gommenginger et al., this volume). More details can be found in the various chapters of this volume. Some of these algorithms (e.g., new retracking algorithms, new radiometer wet tropospheric correction, new regional de-aliasing corrections) are currently being implemented by the French space agency (CNES) and the European Space Agency (ESA) for the definition of new coastal products, within the framework of the PISTACH (Lambin et al. 2008), COMAPI (Lyard et al. 2009), and COASTALT (Cipollini et al. 2008a) project.

This paper discusses the ongoing work at CTOH/LEGOS to reprocess the standard altimeter records by applying adapted strategies for all of the corrective terms of the post-processing stage (the benefit of pre-processing strategies are discussed in detail in the chapter by Gommenginger et al., this volume). Sect. 9.3 presents the latest technical upgrades for higher resolution and multi-satellite monitoring of coastal processes. The physically consistent integration of satellite altimetry into models for coastal ocean analysis is discussed in Sect. 9.4. Finally, Sect. 9.5 presents conclusions and perspectives.

9.2
Post-processing Altimetry in Coastal Zones: Early Developments

The innovative data processing presented hereinafter was originally developed within the framework of the ALBICOCCA project in the North-Western Mediterranean (NWMED) Sea and coordinated by Florent Lyard for the CTOH group at LEGOS, Toulouse, France (Lyard et al. 2003). One of the key tasks of this French-Italian project was to design and implement an altimeter data processing tool adapted to coastal ocean applications (e.g., absolute calibration and drift monitoring of the altimeter, estimation of water transport and validation of numerical models). Additional developments led to the X-TRACK altimeter data processor (Roblou et al. 2007). The objective of this processor is to improve both the quantity and quality of altimeter sea surface height measurements in coastal regions, by tackling all components in the altimeter measurement error budget as recommended by earlier studies (Manzella et al. 1997; Anzenhofer et al. 1999; Vignudelli et al. 2000). This means giving priority to redefining the data editing strategy to minimize the loss of data during the correction phase, by using improved local modelling of tides and short-period ocean response to wind and atmospheric pressure forcing where possible and by using an optimal, high resolution, along-track vertical reference surface.

9.2.1
Redefining Editing Strategies

The ALBICOCCA project revealed a number of important issues about the use of altimetry for coastal applications. One of them was that current products reject a significant amount of data in the coastal zone not only because of radar echo interferences from the surrounding land but also due to suboptimal editing criteria. The standard validity checks for altimeter data editing have been designed for open-ocean regions. However, the relevance of these checks needs to be re-investigated in coastal systems where particular care is required to select altimeter data of sufficient quality and quantity. Typically, the usual data editing is excessively restrictive close to the coastline, in particular because of the poor spatial resolution of the land flag from the radiometer swath, which limits the amount of exploitable altimeter data in an almost 50 km-width coastal strip. The standard editing criteria are basically built on radiometer or altimeter measurements and/or derived parameters, which are often degraded in accuracy as their footprints are contaminated by the presence of land.

To make the most of coastal altimetry however, we need to go beyond the single point data editing approach. Editing criteria based on thresholds and applied to individual ground-point data are not enough: many data are incorrectly flagged as unreliable because of standard, conservative open ocean flags and editing criteria. Inversely, erroneous data could also be qualified as "good." The methodology applied for the X-TRACK processor therefore adopts a new data-screening strategy and along-track filtering techniques which allow more data to be recovered in the coastal zone (Lyard et al. 2003).

The guiding principle of this methodology is to screen multiple along-track altimeter data together, rather than individual ground-points. The first step is to maintain all data points where possible in the coastal zone. This implies the use of a high-resolution land mask based on up-to-date coastline data bases and that standard open-ocean editing criteria are not applied. After carefully revisiting the source of the flags, more lenient editing criteria and coastally-dedicated statistical thresholds have been defined (see Table 9.1).

Then, in a second step, avoiding any subjective human procedure, for each single environmental and sea state correction (e.g., wet and dry tropospheric path delay, ionospheric path delay, sea state bias), erroneous corrective terms are detected but not rejected.

Table 9.1 X-TRACK redefined thresholds for data editing in the coastal zone

Parameter	Min. threshold	Max. threshold
Backscatter coefficient (Ku band)	1.0 dB	30.0 dB
Backscatter coefficient (C, S band)	7.0 dB	30.0 dB
Wet tropospheric path delay	−0.5 m	0.0 m
Sea state bias	N/A	0.0 m
Ionospheric path delay	N/A	0.0 m

This second along-track screening flags as "non valid" contiguous measurements of 0-value for sea state bias, wet tropospheric and ionospheric path delays corrections. All previous corrections and the range are also flagged as "non valid" if they are more than three along-track standard deviation from the along-track mean computed from each single valid datasets. Finally, once these valid datasets (meaning altimeter measurements plus associated corrections) are selected, the along-track corrective terms are then filtered and interpolated at the time of the valid range measurements using high order polynomial and minimal curvature interpolation (e.g., Bézier surfaces). As a consequence, this procedure allows invalid correction terms to be retrieved everywhere, where the altimetric range is valid.

Fig. 9.1 shows an example of the data recovery along the TOPEX/Poseidon (T/P) pass 137 for cycle 200 (February 22, 1998). The original GDR wet tropospheric correction from the onboard radiometer is affected by several erroneous values (black curve) in the vicinity of land (i.e. approaching the Portuguese coast, coming-off Spain, approaching and leaving Brittany, crossing the English Channel and overflying the Danish peninsula) which were flagged as bad in the X-TRACK editing strategy. The remaining valid wet tropospheric corrective terms provided by the radiometer are indicated (red curve). De-flagging and re-interpolation of the wet tropospheric correction yields a reconstructed level profile (purple curve). In this example, the large oscillations close to the coastlines induced by land effects are smoothed and the smaller ones over the ocean, assumed to be noise effects, are filtered. Moreover, a corrective term is provided if the range measurement is available and valid (e.g., extrapolation approaching/leaving the coasts, or over a Danish lake).

Birol et al. (2010), Durand et al. (2008) and Vignudelli et al. (2005) outlined the benefits of such a strategy to the overall improvement of altimeter product quality. With respect to standard AVISO products, this innovative methodology allows more altimeter data to be acquired in the coastal zone and ensures better quality for the derived products. For instance, Bouffard et al. (see their chapter in this book) improved the consistency of altimeter sea level anomalies (SLA) estimates with tide gauge sea level by 7% over the NWMED. This concept of working along the satellite track is similarly applied by Tournadre (2006) in his rainfall retrieval studies or by Ollivier et al. (2005) in the framework of their pre-processing of the altimeter waveforms.

9.2.2
Improving De-aliasing Corrections

A second obstacle to tackle when studying coastal ocean processes with satellite altimetry is the aliasing of the tides and short-period ocean response to wind and atmospheric pressure forcing in the altimetry time series. The high frequency energetic variability clearly cannot be resolved by the 10-day, 17-day or 35-day repeat periods of the T/P, Jason-1/2, GFO, ERS-1/2 and Envisat orbits. This high frequency variability appears as noise which is then aliased into lower frequencies. This is a major problem when estimating the seasonal or longer time scale of oceanic circulation in altimeter data. The impact of this aliasing phenomenon is worse for coastal regions because the high frequency signal has larger amplitude in shallow waters (Andersen 1999). The classical solution to this issue is to remove this high frequency signal by using realistic ocean models.

9 Post-processing Altimeter Data Towards Coastal Applications and Integration into Coastal Models 223

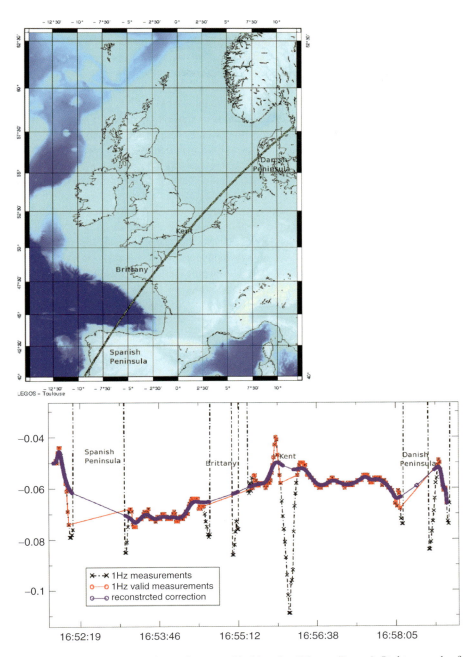

Fig. 9.1 Correction reconstruction. *Left*: geographical location (Western Europe). *Right*: example of wet tropospheric path delay correction (T/P pass 137, cycle 200). Black: raw correction from T/P GDR; red: valid corrective terms as selected by the processing scheme; purple: reconstructed corrective terms for the 10 Hz range measurements. *X*-axis unit is time in hours during February 22, 1998. *Y*-axis unit is meters

However, global ocean models do not provide a precise high frequency, high-resolution correction in terms of both tides and meteorologically forced ocean response in coastal seas and over continental shelves. Indeed, in shallow waters, the propagation of gravity waves is perturbed by numerous factors, e.g., local bathymetry features, the geometry of the shelf and the shorelines, bottom roughness, non-linear interactions, baroclinicity, etc.

Despite their impressive improvement during the past decade and the advent of the precise T/P mission (Le Provost 2001), global tidal models are generally too inaccurate over shallow water areas to fully remove tidal effects from altimeter measurements (see the chapter by Ray et al., this volume). Due to their insufficient spatial resolution, global models cannot resolve rapid changes in tidal features and incorrectly represent the frictional dissipation. Indeed, current global tidal models cannot represent tides in coastal and shelf seas below a decimetre error level. In particular, the omission or the mismodelling of secondary constituents and non-linear tides is largely responsible for the increase in the error budget of global models over continental shelves and along coasts.

In addition to tides, recent results have also shown that atmospherically-forced high frequency signals can be a source of large aliasing in altimeter records at mid and high latitudes (Gaspar and Ponte 1997; Stammer et al. 2000; Hirose et al. 2001; Carrère and Lyard 2003). The atmospheric effects are normally represented by the so-called Inverse Barometer (IB) correction, which is adequate for the open ocean, but known to be unsatisfactory over continental shelves and shallow water seas and at high latitude regions (Carrère and Lyard 2003). In reality, sea surface variation depends both statically and dynamically on meteorological forcing whereas the IB approximation merely formulates the hydrostatic equilibrium between the sea level and the applied atmospheric pressure gradients. Moreover, the IB approximation totally ignores wind-forced sea level variations that can particularly prevail around the 10-day period.

These issues have already been addressed by several teams (Han et al. 2002; Vignudelli et al. 2005; Volkov et al. 2007) and it has been demonstrated that coastal altimetry needs specific, regional models to help reduce the aliasing issue for short-period ocean response to astronomic and atmospheric loading. As recommended by the Ocean Surface Topography Science Team (SALP 2006), the strategy initiated by Lyard et al. at LEGOS is to develop state-of-the-art modelling for ocean response to both astronomic and atmospheric loading. Their approach consists of developing realistic regional models based on the barotropic mode of the T-UGOm code, with additional resolution and locally adapted bathymetry.

The shallow water 2D mode of the T-UGOm code is similar to that previously described in the literature as Mog2D (Carrère and Lyard 2003). Briefly, the code is derived originally from Lynch and Gray (1979) and has since been developed for coastal to global tidal and atmospheric driven applications (Greenberg and Lyard, personal communication). The classical depth-averaged continuity and momentum shallow water equations are solved through a non-linear wave equation with a quasi-elliptic formulation, which improves numerical stability. Currents are derived through the non-conservative momentum equation. A finite element method is applied for converting the problem into a discrete formulation over an unstructured triangular mesh. This spatial discretization method allows a refinement of the model resolution in regions of interest, such as complex coastlines, shallow waters (where most bottom friction dissipation occurs) and areas of strong topographic gradients (showing strong current variability and internal wave generation). It is therefore particularly well adapted to coastal and shelf modelling of gravity wave propagation. To improve the computational efficiency, a multiple reduced time-stepping scheme is introduced for the purpose of smoothing potential instabilities at model nodes. Carrère and Lyard (2003) and Mangiarotti and Lyard (2008) among others have already shown its efficiency in modelling the high frequency barotropic response to meteorological forcing both at a global and regional scale. The quality of the regional tidal solutions has been highlighted by Pairaud et al. (2008) over the European Shelf or Maraldi et al. (2007) over the Kerguelen Shelf (Southern Indian Ocean) for instance. For these two shelves, the root sum square error of the principal constituents is at least two times lower than for global models.

As a consequence, where possible, local modelling of tides is used preferentially at CTOH/LEGOS in the X-TRACK processor, as an alternative to global open-ocean tide models provided in the standard altimeter products, such as FES2004 (Lyard et al. 2006) or GOT00b (or its updated version GOT4.7). Regional models based on the T-UGOm 2D hydrodynamic code are currently available[1] over the European shelf (Pairaud et al. 2008), over the Kerguelen shelf in the Southern Indian Ocean (Maraldi et al. 2007), over the Caspian Sea (see the chapter by Kouraev et al., this volume) or in the Arabo-Persic Gulf (Lyard and Roblou 2009). In addition, composite spectra can also be designed if the regional tidal atlas exhibits weaknesses. For instance, the K1 constituent solution derived from the regional T-UGOm 2D simulations over the Mediterranean Sea suffers large biases in phase lag over the western basin and leads to greater root mean square differences in comparison to tide gauge observations than for global tidal models (Roblou 2004). As a consequence, it has been replaced by the K1 solution taken from the FES2004 atlas in the optimal spectrum used for the Mediterranean Sea in the X-TRACK processor. Finally, although purely hydrodynamic modelling gives a satisfactory performance compared to the standard global databases (e.g., FES2004, GOT00b or its update GOT4.7), Lyard and Roblou (2009) strongly recommend constraining regional tidal solutions by data assimilation techniques to provide the full required accuracy in the tidal corrections for coastal altimetry.

Regional models of short-period ocean response to wind and atmospheric pressure forcing have been developed or are currently under development based on the same regional meshes as for tides. For example, Cipollini et al. (2008b) used the Mediterranean regional model to correct for high frequency atmospheric variability in a tide gauge data set. They found that the averaged improvement for periods of less than 20 days is 46% in comparison with the IB model and 5% in comparison with the Mog2D-G global model (recommended correction for short-term atmospheric effects). Similar improvements are obtained when de-aliasing an altimeter dataset (see the chapter by Bouffard et al., this volume).

9.2.3
Improving the Vertical Reference Frame

In the absence of a geoid model with sufficient accuracy at small scales which are typical of regional and coastal ocean dynamics, most researchers overcome that difficulty by referencing

---

[1]Tidal atlases are freely available on request on the SIROCCO web site: http://sirocco.omp.obs-mip.fr/outils/Tugo/Produits/TugoProduits.htm

the radar altimetry observations to a hydrographic reference surface. The SLA is defined as the difference between the sea levels inferred from satellite altimetry and this reference surface. It is usually approached by the time average of altimeter SSHs, the so-called mean surface height or MSS. However, attention has to be paid to respect the consistency of the corrections applied to the SSH estimates and those applied when building the MSS.

In coastal areas, the resolution of generic MSS products derived from altimetry data is degraded for several reasons. First, altimeter standard data are often missing in the coastal band (from the coastline to 50 km offshore). This implies that the ocean mean sea surface products are merged with a terrestrial geoid model (Hernandez et al. 2000) whose weight becomes predominant in the inverse method applied when building the resulting MSS product. Secondly, the too-sparse altimetry data measurements that are available in coastal and shelf seas are associated with significant errors due to the specific de-aliasing of the short time-space scales of the high-frequency dynamics through the use of global models that are generally too inaccurate in coastal areas. Finally, current satellite gravimeter missions do not provide reliable geoid information for wavelengths lower than 200 km; this results de facto in limited-resolution MSS products.

To overcome some of these problems, the X-TRACK processor allows the possibility of computing a high resolution, along-track mean sea surface consistent with the optimized altimeter dataset and on a high resolution, regular grid following the satellite ground track (Lyard et al. 2003). This along-track MSS is computed on a regular grid (0.01° × 0.01°) following the satellite ground track from an inverse method (optimal interpolation):

$$J(x) = (y - Hx)^T R^{-1}(y - Hx) + (x - x_0)^T B^{-1}(x - x_0) \tag{9.1}$$

where x represents the mean sea surface model vector, y the altimeter data vector and H the observation matrix. The observation matrix uses the coastal-improved altimeter SSHs described in the previous sections. The covariance operator R describes uncertainties on both the instrumental measurements and on the representativeness of the observation operator H. It is assumed to be diagonal (which means that errors are uncorrelated) with observation errors variances set to 20 cm². Covariance operator B describes the Gaussian-shaped uncertainties for the mean sea surface model. Vector x_0, representing the prior model, is the interpolation of the MSS CLS 01 product onto the optimal model grid. Error variances are set to 0.1 m² for the prior model (x_0) and its correlation length is set to the grid resolution, e.g., correlation is 0.5 at 0.01° distance. The resulting MSS thus takes into account the across-track gradient of SSH measurements. The time period over which this average is computed is also an important parameter, as it should take into account typical physical time scales (especially the annual cycle).

Finally, the resulting MSS contains the geoid, the MDT with regard to this geoid and the averaged ocean variability for the period over which the MSS has been computed.

Fig. 9.2 illustrates the mean sea surface obtained following the X-TRACK methodology along the T/P ground track 009. It represents T/P sea level anomalies obtained by computing the mean sea surface using the X-TRACK optimal mean sea surface model (red dots) versus the original GDR correction from the mean sea surface model MSS CLS01 (green dots) for a particular ground track in a near-shore region. Black dots illustrate the mean sea surface difference (i.e. optimal model versus its prior model MSS CLS01) as a function of latitude. The

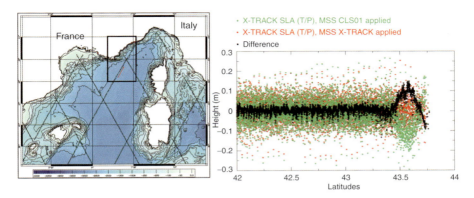

Fig. 9.2 Mean sea surface computed for TOPEX/Poseidon ground track 009 in the NWMED Sea. *Left panel*: location of altimeter passes; *right panel*: T/P SLA including X-TRACK MSS in red dots, including MSS-CLS01 product in green dots and mean sea surfaces differences versus latitude in black dots. (*Y*-axis unit in metres) (Adapted from Lyard et al. 2003)

difference between the X-TRACK optimal MSS and the MSS CLS01 is mainly centred on zero, with oscillations apparently related to MSS CLS01 grid resolution and a bump in the coastal zone. Similar discrepancies have been found for other ground tracks (Lyard et al. 2003). This is a first indication that the X-TRACK optimal mean sea surface model performs better in the near-coast regions, since this higher spatial resolution mean sea surface permits better modelling of the local geodetic variability. These discrepancies could arise from a lack of near-coast altimeter data in the dataset used to derive MSS CLS01, resulting in an imprecise extrapolation between the satellite MSS and the EGM96 geoid in the coastal band.

After such processing, X-TRACK SLAs can be provided[2] both on their original along-track position (meaning for each individual SSH measurement) and projected onto a mean reference track (meaning that time series are available on a fixed spatial grid).

9.3 Latest Upgrades of the X-TRACK Processor

Based on the X-TRACK along-track altimetry profiles, Durand et al. (2008) have set up a novel methodology for monitoring the East India Coastal current. Their method relies on geophysical parameters such as the length of the first baroclinic Rossby radius of deformation or the maximal acceptable magnitude of the surface current. However, expanding this methodology to other coastal seas like the European seas will need a higher resolution because the first baroclinic Rossby radius of deformation is much smaller there than in the tropics (Chelton et al. 1998). Consequently, observing the complex dynamics of coastal

---

[2]Coastal-oriented data sets are computed over several coastal regions and made freely available through the CTOH/LEGOS website: http://ctoh.legos.obs-mip.fr/products/coastal-products

Kelvin waves in the mid to high latitudes or western currents along coasts and continental shelves requires ongoing improvement of the space and time sampling of the altimeter products using several satellites (Pascual et al. 2006) and a higher rate for data streams (10 and 20 Hz, e.g., Bouffard et al. 2008a; Le Hénaff et al. 2010).

9.3.1
Orbit and Large-Scale Error Reduction

Taking advantage of multiple satellite missions increases the coverage of the coastal area (Bouffard et al. 2008a), but this does not necessarily improve the quality of the final data product because of the inhomogeneous error budgets between the different satellite altimetry missions. One of the most important sources of error is large-scale orbit error. A Large-Scale Error Reduction (LSER) method designed for regional application has been developed and implemented in the X-TRACK processor. This method, which is not expensive in terms of computational cost, keeps each satellite dataset independent as opposed to a standard dual-satellite crossover minimization (Le Traon and Ogor 1998). Over short distances (arc lengths smaller than 1,500 km), which is the case for most shelf seas, the orbit error can be modelled by a polynomial approximation; it is generally a first order bias and tilt fit (Tai 1989, 1991). The LSER method is based on this previous assumption.

This new algorithm functions as follows. The first step is to compute the time evolution of the along-track-averaged SLA in order to identify large-scale errors. The SLAs used in this algorithm are smoothed along the arc-orbit with a 20 km low-pass Loess filter to reduce the influence of mesoscale dynamics in the calculation of the large-scale error. Only SLAs that are less than two standard deviations from the along-track median value are used. This eliminates spurious data in the coastal zone and allows having consistent geographical coverage when averaging along-track SLAs for each cycle. The black curve on Fig. 9.3 (step 1) shows, such a time series for the GFO track 74 between 2001 and 2005. As shown in Fig. 9.3 (step 1) some large offsets appear in a random way: the averaged along-track SLA suddenly jumps by 10–80 cm within the time span of two cycles. These physically impossible jumps are the consequences of large-scale errors, essentially due to bad orbit determination, and have to be removed. In a second step, the annual and inter-annual signal is removed using a Loess filter in order to eliminate the low frequency signal due to steric effects (see Fig. 9.3, step 2). Then time-averaged statistics (mean and standard deviation) are computed for the spatially-averaged SLA time series. Outliers greater than three standard deviation from the spatial temporal mean of the along-track SLA identify cycles that have to be corrected. The correction consists in removing a bias for the overall ground track of the rejected cycles. The biases are determined with respect to a smoothed version of the along-track averages. The smoothing consists of a combination of Bézier surfaces for smoothing purposes and linear interpolation (see dashed curve in Fig. 9.3, step 3). The previous three steps are integrated into an iterative process in order to select better, discriminate and correct the outliers.

This method is especially well adapted for regional studies. Although it is a very basic orbit and large-scale adjustment, it is sufficient for a regional study, especially when other errors dominate (e.g., data coverage, de-aliasing corrections, land contamination, etc.).

9 Post-processing Altimeter Data Towards Coastal Applications and Integration into Coastal Models

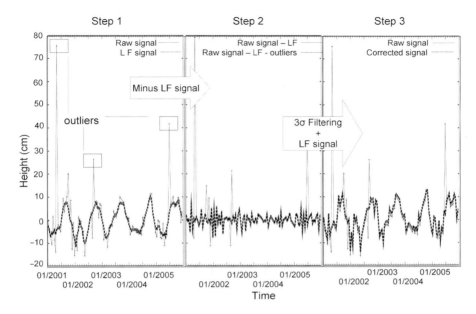

Fig. 9.3 Time evolution of the along-track averaged SLA, GFO track 074. *Step 1*: Raw signal versus low frequency (LF) signal. *Step 2*: Detection and removal of the erroneous cycles by 3 σ-filtering of the high frequency signal. *Step 3*: Reconstruction of the corrected signal by a linear regression over a four-cycle time window

9.3.2
High Rate Data Stream

Observing small mesoscale and coastal current sea level variability is a challenging issue which requires intensive spatial sampling. Energetic small-scale ocean dynamics are found along the continental slope and they are very difficult to observe with standard altimetry. The first baroclinic Rossby radius of deformation is generally only a few kilometres and despite a low signal-to-noise ratio, characteristic length scales of energetic disturbances of the coastal currents could be of sufficient size and magnitude to be detected with high-resolution altimetry. The availability of 10/20 Hz values (350–700 m along-track resolution) can provide denser spatial coverage of sea level variability and also enables the estimation of an MSS at a higher resolution along the satellite ground track. The 20 Hz data (respectively the 10 Hz) also allows a closer approach to the shoreline before the single measurement encounters land in its footprint, by reconstructing the 10 (respectively 5) points between the last 1 Hz value available and the coastline (AVISO 1996).

In order to assess the ability of high frequency sampling to observe small coherent dynamic signals rather than noise, Bouffard et al. (2006) computed along-track SLA spatial spectra with all of the T/P ground tracks over the NWMED Sea. For convenience, they performed a time average over all available cycles. The analysis of the spectra shows that the 10 Hz T/P data exhibit the onset of white noise below 3 km, a value much lower than the standard 1 Hz signal (for which the smallest resolved scale is about 13 km due to the Nyquist-Shannon theorem[3]). Up to 3 km the signal is dominated by white noise (due to the onboard tracking mode), whereas the oceanographic signal emerges from the noise level at scales greater than 3 km. Similar results have been found for the Jason-1 satellite where the power spectrum shape is slightly different in the small scales due to a different onboard tracking mode but a coherent signal exists for spatial scales greater than 3 km. As a consequence, a 1.5 km high-pass Loess filter is applied to differentiate between high rate orbit and range parameters for the processing of high rate SLAs in the X-TRACK processor.

As demonstrated in Bouffard et al. (2008b) over the Corsica channel area, such a simple post-processing strategy is sufficient to increase the quality and availability of coastal X-TRACK data compared to 1 Hz AVISO standard products. Fig. 9.4 (adapted from Le Hénaff et al. 2010) illustrates the ability of the adopted methodology to retrieve data closer to the coast compared to AVISO regional datasets (e.g., SSALTO/DUACS MERSEA products DT-SLAext), here along the northern coast of Spain. The use of the high rate data stream provides altimeter data with finer spatial resolution and improved temporal sampling. These improvements allow them to study processes associated with the shelf break, such as the slope current, which would not have been possible with classical 1 Hz altimeter products.

However, despite a finer spatial resolution than 1 Hz data, caution is required: this post-processing technique still cannot provide reliable SLAs either at real high frequency sampling (e.g., at length scales of 0.3 or 0.6 km) or right up to the coastline, where the Brown model (used for retrieving ranges to surface from altimeter waveforms) is not automatically well adapted and the altimeter and radiometer footprints are contaminated by land reflections.

9.4
Matching Satellite Altimetry with Coastal Circulation Models

9.4.1
Rationale

There is increasing consensus that coastal management requires a holistic view, based on better quality and more integrated geospatial and environmental information on which a scientifically-sound policy can be built, instead of a sector by sector approach. In the future, an important, international research effort will be made on coastal zones (within the GMES and IMBER initiatives) and will aim at developing operational oceanography and services

---

[3] In essence, the Nyquist-Shannon sampling theorem asserts that an analog signal that has been sampled can be reconstructed from the samples if the sampling rate exceeds 2N samples per second, where N is the highest frequency in the original signal. In the case of the T/P satellite, 1 Hz sampling along the satellite ground track represents 6–7 km spatial sampling (depending on the latitude of the satellite), meaning that only scales greater than about 13 km can be reconstructed from the samples.

9 Post-processing Altimeter Data Towards Coastal Applications and Integration into Coastal Models 231

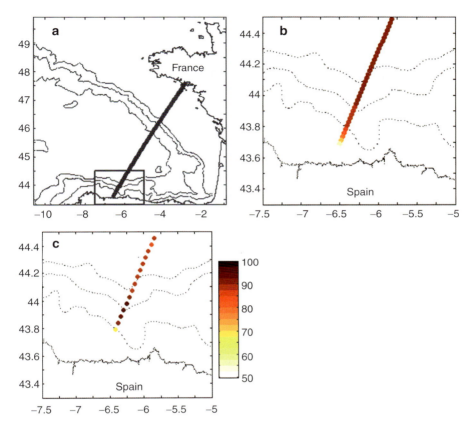

Fig. 9.4 Zone of study (**a**) and altimeter data coverage close to the northern Spanish coast for TOPEX/Poseidon track 137 (November 1992–August 2002), for (**b**) X-TRACK dataset – 2 Hz data interpolated from the high rate 10 Hz data stream – and (**c**) AVISO 1 Hz dataset. In grey scale is the temporal coverage (% of the total number of cycles available). In dotted lines are the 200, 1,500, and 4,000 m bathymetric contours

at regional and coastal scales. As for the deep ocean, regional operational oceanography will be made possible only in an integrated approach, merging process-oriented studies, remote-sensed and *in situ* observation systems, ocean modelling and data assimilation.

Ocean modelling is a precious tool for studying physical processes because all state variables of the system can be accessed. However, models only partially represent the true state of the ocean. Data assimilation of satellite altimetry can thus help in reducing model errors. Data assimilation of coastally-improved satellite altimetry (e.g., CTOH/LEGOS SLA products based on X-TRACK processor, forthcoming COASTALT and PISTACH datasets, etc.) could be highly valuable for regional circulation and even for deep convection studies. At a more local level, a realistic representation of regional circulation is required for the study of the current-shelf/slope interactions or more coastal processes. Although satellite altimetry is not capable of direct monitoring of such small-scale coastal phenomena as wind-induced processes or upwelling systems, the observations it provides of regional oceanic features will still enhance the realism of coastal models through improved monitoring of the offshore circulation by data assimilation.

The objective of the following section is to address the issue of an optimal matching of improved, coastal-oriented altimeter observations with regional hydrodynamic model sea surface elevation estimates. Methodologies of matching are proposed and the results of case studies are discussed.

9.4.2
Consistency Between Altimeter Measurements and Model Estimates

Once a reference level is removed and the effects of tides and short-period ocean response to wind and atmospheric pressure forcing are corrected, the most important component of satellite-observed sea level variability on a seasonal timescale is due to the so-called steric effect. This can be defined as the variation of the sea level due to the expansion/contraction of the ocean water volume in response to density changes in the whole water column (see Lombard et al. 2006). This steric expansion/contraction is an ocean process of first importance when it comes to local sea level variations in the seasonal and inter-annual time scales (e.g., Pattullo et al. 1955; Church et al. 1991; Gregory 1993), and its surface signature is added to the seasonal and inter-annual surface variations due to ocean dynamics.

9.4.2.1
Steric Effect Surface Signature

From a physics standpoint, the sea level change caused by steric effect, h, can be estimated by integrating the density anomalies over the whole water column (Tomczak and Godfrey 1994):

$$h(\lambda,\varphi,t) = \int_{-H}^{0} \frac{\rho(\lambda,\varphi,z,t) - \rho_{\text{ref}}(\lambda,\varphi,z)}{\rho_{\text{ref}}(\lambda,\varphi,z)} dz, \qquad (9.2)$$

where the density ρ is a function of the longitude λ, the latitude φ, the depth z and the time t, ρ_{ref} is a characteristic density over the area depending only on the geographical location and depth, and H is the total water depth. Considering the UNESCO state equation that links the sea water density to its temperature and salinity (UNESCO 1981), the sea level change caused by steric effect only depends on temperature and salinity variations in the water column.

Even if the thermal-steric contribution of the ocean upper layers can account for most of the sea level variations, the thermal-steric contribution of the deeper ocean layers, as well as the halo-steric contribution, should not be neglected at the regional scale. In particular, recent works have demonstrated that the halo-steric contribution (coming from salinity freshening due to fresh water input by river discharge, evaporation, precipitation, ice melting, etc.) can locally balance the thermo-steric effects, in particular in the Northern Atlantic ocean (Antonov et al. 2002; Ishii et al. 2006; Wunsch et al. 2007; Köhl and Stammer 2008; Lombard et al. 2009). In the same way, the thermal-steric contribution of the deeper layers can explain the sea level variations in the Southern Ocean (Morrow et al. 2008; Lombard et al. 2009).

Recent studies (Carton et al. 2005; Lombard et al. 2009) have shown that the regional variations of the steric effect signature mostly result from the general ocean circulation in response to wind and atmospheric pressure forcing (through the buoyancy forcing term). Heat fluxes and fresh water inputs could also account for a significant part, in particular in the Southern Ocean and in the Northern hemisphere (Köhl and Stammer 2008). In addition, these seasonal low frequency phenomena are also enhanced by the sea level trend caused by global warming. It has been shown that its impact on coastal seas and its regional discrepancies can be significant (e.g., Lombard et al. 2009).

9.4.2.2
Steric Height in Coastal Ocean Models

Regional and coastal circulation models mainly feature a solution of the free surface which satisfies the primitive equations. In addition, these models generally formulate the hydrostatic and the Boussinesq hypotheses, respectively assuming negligible vertical velocities and volume conservation (rather than mass-conservation). For the sake of convenience, the model assuming the volume conservation will be referred to in the following lines as the "Boussinesq model."

Compared with the basic situation described in the left side of Fig. 9.5, a uniform temperature flux entering the upper layer of the left column generates a decrease of the density in this column. The upper layer water column then expands and its thickness increases from $h2$ to $h2 + dh$ (in the middle part of Fig. 9.5). Due to the fact that the total mass of

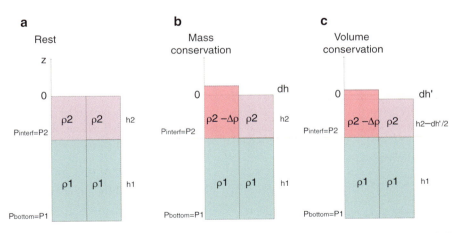

Fig. 9.5 Steric effect modelling in mass-conserving and volume-conserving models – simplified scheme. (**a**) Representation of two water volumes at rest and subjected to a uniform heat flux coming into the upper layer of the left-hand volume. (**b**) Free-surface adjustment of a mass-conserving model. (**c**) Free-surface adjustment of an incompressible volume-conserving model

each column is unchanged, the hydrostatic pressure at the interface also remains unmodified in both columns.

The conservation of mass in the water columns presupposes that:

$$(h2 + dh) \times (\rho2 - \Delta\rho) = h2 \times \rho2 \qquad (9.3)$$

where $h2$ is the height of the upper layer of density $\rho2$, dh is the rise of sea level induced by the variation of density $\Delta\rho$ (see left side and centre of Fig. 9.5).

Consequently,

$$dh = h2 \times \Delta\rho / (\rho2 - \Delta\rho) \qquad (9.4)$$

Assuming, that the variations of density are much smaller than the density itself (generally with a ratio of 1 for 1,000), the sea surface elevation due to the expansion of the left column in Fig. 9.5 can be approximated by:

$$dh = h2 \times \Delta\rho / \rho2 \qquad (9.5)$$

In the Boussinesq model (see right side of Fig. 9.5), it is assumed that the fluid is incompressible, which implies that the expansion of the total volume (i.e. the volume of both columns) is null. On the other hand, one can assume that, at a regional scale, the dynamic response of the model should imply a modification of the column height in order to keep the hydrostatic pressure at the interface constant, i.e.

$$(h2 + dh'/2) \times (\rho2 - \Delta\rho) = (h2 - dh'/2) \times \rho2 \qquad (9.6)$$

where $h2$ is the height of the upper layer of density $\rho2$, dh' is the rise of sea level induced by the variation of density $\Delta\rho$ (see right side of Fig. 9.5).

Finally,

$$dh' = h2 \times \Delta\rho / (\rho2 - \Delta\rho / 2) \qquad (9.7)$$

In the same way, since the variation of density $\Delta\rho$ is much smaller than the density $\rho2$:

$$dh' = h2 \times \Delta\rho / \rho2 \qquad (9.8)$$

As a result, the incompressible Boussinesq models are able to reproduce the realistic difference of height between the heated column and the cold one. On the other hand, these models do not reproduce the mean sea surface rise, i.e. they do not reproduce the mean steric effect over the modelling area. In other words, in these models, the spatial variability of the steric effect is taken into account but its mean time variation is zero, due to the Boussinesq approximation. This, for example, removes part of the seasonal steric effect and any mean sea level rise.

9.4.3
Methodologies and Case Studies

Once the correction of tides, short-period wind and atmospheric pressure effects and a reference level has been applied, combining altimetry observations and coastal circulation models can be reconsidered as a combination of altimetric and modelled sea level anomalies.

The sea level anomaly observed in the altimetry measurements is mainly dominated by the surface signature of the seasonal steric effect, which is generally not reproduced by coastal circulation models. To account for this, either the steric effect contribution needs to be removed from altimetry or added to the model. In the following section, we will consider both of these approaches.

However, the following discussion is a first step of an ongoing effort at CTOH/LEGOS to address the issue of matching altimeter-derived estimates and hydrodynamic models in the coastal systems. The methodologies are here assessed in a pilot system, the NWMED Sea. They need to be implemented and tested in other coastal and shelf seas before providing the community with any guideline, and in particular for an operational correction.

9.4.3.1
Modelling Approach

Strategy

Greatbatch (1994) showed that it was possible to correct the modelled free-surface elevation computed with models taking into account the Boussinesq approximation by adjusting a specific constant. This adjustment constant is spatially uniform over the modelling area, but has a time variability taking into account the net expansion/contraction of the ocean. Furthermore, this correction, being spatially uniform, has no dynamic impact on the current velocities computed by models. The methodology proposed here consists of preferentially removing the steric effect contribution from the altimeter measurements.

Based on the sea water state equation, the density field can be obtained from the model salinity and potential temperature fields at the time of the altimeter measurements. Then, in order to avoid local differences, the resulting steric sea level elevation field needs to be smoothed over a large enough area (e.g., by averaging over the whole regional domain). Finally, this mean steric change would preferably be averaged over a window of several days, since only its low frequency time variability is required in the case of regional and coastal models.

In order to exclude all steric change processes with non-zero average contribution over the run period, a second corrective term should be added, taking into account the mean model sea surface elevation and the mean steric height differential over the study period. In practice, the choice of the averaging period is initially made from the start and end dates of the model run, so that in a second step altimeter SLAs are consistently averaged over the same period.

As a consequence, a steric correction from model simulation outputs can result as follows:

$$\delta = \left(\overline{h\text{steric}(x,y,t)}^{\text{domain}} \right)_{LF} - \left(\overline{\eta(x,y,t)}^{\text{period}} + \overline{h\text{steric}}^{\text{domain, period}} \right) \quad (9.9)$$

where δ is the adjustment constant to apply, h_{steric} is the steric surface signature on the model grid points (x,y) and at time t of observations, η is the model sea surface elevation at grid points (x,y) and at time t of observations; the overbar $\overline{X}^P$ denotes the average of X over period P and $(X)_{LF}$ denotes the low-passed filtering of X.

Model Data

The hydrodynamic model used for this case study is the three-dimensional, primitive equation coastal ocean model SYMPHONIE (Estournel et al. 1997; Marsaleix et al. 2006). It is in widespread use for shelf circulation, sediment transport and coupled bio-physical applications (e.g., Estournel et al. 2003; Dufau-Julliand et al. 2004; Petrenko et al. 2005; Ulses et al. 2005; Gatti et al. 2006; Guizien et al. 2006; Jorda et al. 2007; Hermann et al. 2008). This model is based on the hydrostatic assumption and the Boussinesq approximation. It is eddy-resolving. The three components of the current, the free-surface elevation, the temperature and the salinity are computed on an Arakawa C grid (Arakawa and Suarez 1983), using an energy-conserving finite difference method (Marsaleix et al. 2008). A hybrid sigma-step coordinate system is used for the vertical discretization and a curvilinear coordinate for the horizontal discretization. More details on the model characteristics can be found in Marsaleix et al. (2008). For this study, the model was run over the NWMED Sea from September 2, 2004 to February 28, 2005, on a 3 km horizontal grid spacing which is about three times smaller than the first baroclinic Rossby radius of deformation in this area, of about 10 km. The model is one-way nested on a daily basis into the MFSTEP project OGCM (Estournel et al. 2007). Fig. 9.6 displays the bathymetry of the regional model simulation.

Coastal Altimeter Data

SLA estimates were computed with the X-TRACK processor for the four altimetry missions available over the study period, namely TOPEX/Poseidon, Jason-1, Envisat

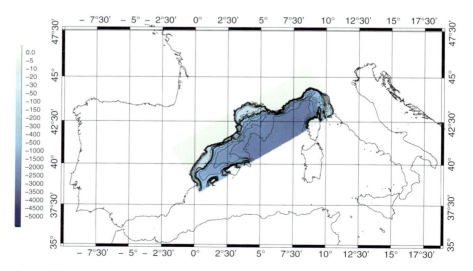

Fig. 9.6 SYMPHONIE model bathymetry for the NWMED Sea configuration

and GFO. The altimeter data were corrected for environmental and sea state perturbations, applying the edition/rebuilding procedure as described in previous sections; the tidal correction was computed from a composite spectrum based on the T-UGOm 2D for all constituents except K1 which was taken from FES2004; atmospheric loading effects have been corrected based on the T-UGOm 2D regional model simulations and the MSS was computed over the model run period (02/09/2004–28/02/2005) as recommended previously. The steric effect surface signature was corrected using the model itself, applying Eq. 9.9 1 Hz SLA estimates were provided along the ground track.

Results and Analysis

Model elevations are first compared with altimeter estimates. Fig. 9.7 displays examples of comparisons between the SLA derived from both model and altimeter along different satellite tracks. These comparisons between model outputs and observations globally show good agreement in the typical synoptic scale and mesoscale. The slopes and large-scale structures represented in the model simulation are close to those captured in the altimeter observations (a low-pass filter would highlight this feature). Moreover, altimeter data seems to provide additional information at scales shorter than 50–60 km that are not represented in the model simulation.

In addition, Fig. 9.8 represents the spatial power spectra of model elevation (blue curve) and corrected SLAs from the T/P altimeter (red line). It can be noticed that short scales (less than 10 km) are not present in the model, which is not surprising owing to the 3 km resolution of its horizontal grid. For wavelengths between 10 and 60 km, the model and altimeter data represent the same level of energy, implying an overall, statistical agreement for these scales.

In this example (Fig. 9.7b), an offset of nearly 10 cm can be seen between the model elevation and the altimeter measurement (along GFO track 401). This bias is also observed for a coincident Jason-1 pass. As the two satellites are independent, it cannot be residual orbit error but is more probably due to an internal error in the computation of the steric correction at that time. The time variation of the sea surface elevation spatial average in a Boussinesq incompressible model depends on the net flux of water at its lateral boundaries. In the present case, it thus depends on the sea surface elevation provided by the OGCM. If the sea surface elevation is erroneous in the OGCM, it introduces an error on the coastal model sea surface elevation through the open boundary forcing and then in the computation of the steric correction. Hopefully, this kind of error does not affect regional circulation modelling, which primarily depends on the sea surface elevation gradients rather than on the absolute value of the sea surface elevation. However, when assimilated, such an outlier will inevitably have an impact on the analysis solution and smart techniques have to be implemented in the assimilation schemes to prevent this spurious model adjustment (e.g., by placing an upper limit on the difference between model proxy and observation).

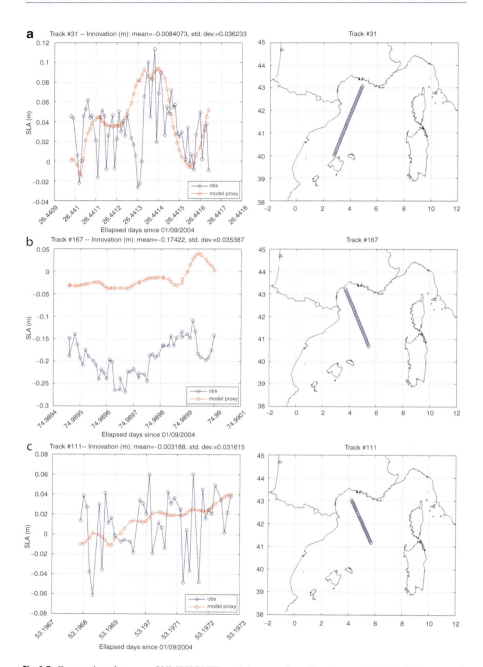

Fig. 9.7 Comparison between SYMPHONIE model sea surface elevation and X-TRACK corrected SLAs along (**a**) GFO track 031, (**b**) GFO track 401, (**c**) Jason-1 track 146, and (**d**) TOPEX/Poseidon track 009 in the NWMED Sea. Track locations are given on the associated right panels

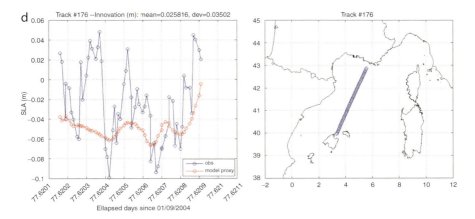

Fig. 9.7 (continued)

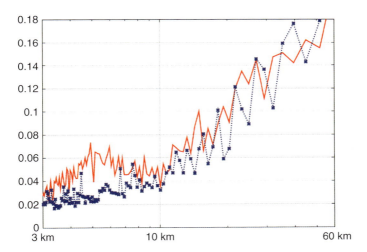

Fig. 9.8 Mean spatial power spectra: SYMPHONIE model sea surface elevation power spectrum (blue line) and T/P SLA estimates power spectrum (red line). Power amplitude is given in meters

9.4.3.2
Statistical Approach

Methodology

Another approach is to use a statistical analysis of the altimeter dataset itself to compute a steric correction. This consists in performing along-track singular value decomposition

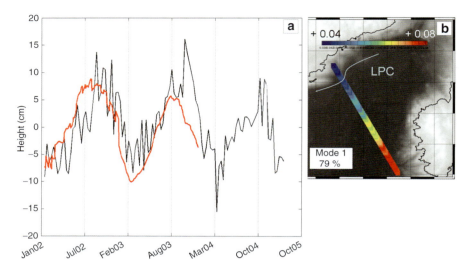

Fig. 9.9 Singular value decomposition of T/P SLAs along ground track 222. (**a**) First mode time component (rescaled by the singular value) and steric effect surface signature as computed from the SYMPHONIE model outputs. Unit is centimetres. (**b**) First mode spatial component (dimensionless)

(SVD) of the altimeter SLA time series. At seasonal time scales, sea surface variability is generally dominated by ocean currents and the annual large-scale steric signal. In this respect, in a statistics point of view, the first mode of the SVD would therefore represent this annual steric signal and could then be removed from the altimeter SLA. Unlike a high-pass digital filter with a cut-off period of 1 year, this method allows the altimetry to retain the variability of the annual dynamics (that is simulated in Boussinesq models) since the SVD method allows discrimination between coherent spatio-temporal dominant processes (from a statistical standpoint).

Case Study

This approach is illustrated through a case study over the Ligurian Sea. In this part of the NWMED Sea, the coastal section of the Jason-1 pass 222 intercepts the Liguro-Provençal-Catalan (LPC) Current quasi-perpendicularly (Fig. 9.9b). This coastal current flows seasonally and counter-clockwise along the continental slopes, from the Ligurian Sea to the Catalan Sea (e.g., Millot 1987, 1991). SLA estimates were computed with the X-TRACK processor for the Jason-1 pass 222 from January 2002 to December 2004.

Whereas modes greater than 1 represent less than 5% each, mode 1 is strongly dominant and represents 79% of the total energy. The associated time series (Fig. 9.9a, black curve) show a marked annual signal with negative and positive maxima respectively at the end of winter (March–April) and summer (September–October). The along-track structure of the mode 1 spatial component (Fig. 9.9b) shows a very small across-shore slope that would generate an insignificant across-track geostrophic flow. Compared to the strong across-track gradient featuring the LPC geostrophic zonal component, and noticing that the whole mode

1 spatial component is of same sign, this mode can be seen as a large-scale translation at the annual period. This suggests that this mode corresponds to the large-scale steric signals.

In order to validate this assumption, the time component has been compared with the steric time-dependent spatial constant built from the SYMPHONIE temperature and salinity fields following Eq. 9.9 (Fig. 9.9a, red curve). Fig. 9.9 shows that, although the modelled signal is smoother, the two signals agree very well in terms of magnitude and phasing which confirm that mode 1 represents the large-scale annual steric signal.

As a consequence, the mode 1 time evolution can be used for steric correction if the altimeter records are long enough (greater than 2 years) to be analysed using the SVD method. Unfortunately, this implies that this methodology is not applicable in operational conditions. Moreover, this methodology assumes that the annual steric effect surface signature is sufficiently marked over the area of interest to be isolated from the dynamic annual signal. This assumption would not be automatically verified in all shelf and coastal seas.

9.5
Concluding Remarks and Perspectives

The work initiated by the ALBICOCCA project and carried out at the CTOH/LEGOS in the NWMED has paved the way for improved post-processing of official altimeter products for coastal applications. Thanks to the X-TRACK processor, it is now possible to get more data close to the coasts (and even in open sea regions), with a reduced error budget (see also the chapter by Bouffard et al., this volume). In this coastal-oriented post-processing, in addition to more lenient editing criteria, land effects on the altimeter and radiometer measurements are minimized during the correction phase and missing or corrupted corrective terms are replaced by interpolation/extrapolation curves based on valid measurements. The X-TRACK processor benefits from state-of-the-art local modelling of tides and ocean response to wind and atmospheric pressure forcing as provided by the T-UGOm 2D (Mog2D follow-on) code for the de-aliasing of short-term barotropic signals and next generation models are expected to provide an even better accuracy. The high-resolution vertical reference frame needed to compute SLAs is locally adjusted using a consistent and coastal-tuned altimeter dataset.

However, features and processes of interest in the coastal systems are those which drive the local spatial variability on short time scales, constrained by the seabed and interaction with the mean flow, tides, seasonal cycles, inflow and outflow. In these conditions, additional developments are required within the X-TRACK processor in order to use several satellites (large-scale error reduction) with their higher acquisition rate leading to increased space and time sampling of altimeter observations.

Unquestionably, the X-TRACK processor will benefit in the future from advances in the new generation altimeters (Ka-band, Delay-Doppler, interferometry, etc.) and synergies with other sensors (tide gauges, SST, ocean colour, etc.). But the next major input in the short term should come from the pre-processing phase and improvements in wet tropospheric path delay correction. The retrieval of information from radar altimeter returned echoes and the decontamination of land effects in the radiometer measurements are expected to be the source of a considerable reduction in the error budget of altimeter products in the future. Finally, major improvements are also expected from the update of regional tidal atlases.

This chapter also focused on matching coastal-oriented altimetry and regional circulation models. In the prospect of integrating altimeter data into models, the present study has pointed out the specific problem of taking into account the steric effect surface signature in regional ocean models that generally do not represent it well. Where possible, an SVD decomposition of altimeter SLA estimates would need to be performed in a delayed mode so that the steric and dynamic annual surface signatures could be statistically dissociable. In order to create physically consistent datasets, an alternate strategy, illustrated here for the NWMED, consists in removing from the altimeter SLA a steric height computed from a diagnostic equation and based on temperature and salinity fields computed by the model itself. However, such a strategy highly depends on the model used and forming and applying an operational correction term to altimetry observations computed from temperature and salinity (provided by forecasting systems for instance) remains a challenging issue.

This methodological work constitutes a first step toward assimilation of coastal altimetry into regional hydrodynamic models and opens up the way for an operational survey of coastal environments. However, numerous investigations still need to be made before the routine monitoring of coastal dynamics can become possible. First of all, small- and mesoscale dynamics are poorly or not reproduced in coastal models, except from a statistical point of view, whereas they seem to be captured in altimeter observations. Consequently, investigations need to be made in terms of hydrodynamic modelling in order to fully represent these scales. Comparisons with independent in situ measurements would also provide an idea of which scales are truly captured by satellite altimetry systems. The capabilities of data assimilation have been clearly demonstrated within the Global Ocean Data Assimilation Experiment (GODAE) for deep ocean or basin-scale systems and it now needs to be downscaled to regional level in coastal and shelf seas. This also requires substantial investigation in terms of data assimilation scheme definition and error settings, before running realistic experiments and finally real-time applications.

Acknowledgements The authors would like to thank the entire staff at the Centre de Topographie des Océans et de l'Hydrosphère (CTOH) at LEGOS for access to the altimeter data bases (http://ctoh.legos.obs-mip.fr/) and its expertise. This paper was significantly improved by comments from Rosemary Morrow. Very special thanks are due to Stefano Vignudelli and Paolo Cipollini for strong interactions on the coastal altimetry issue since the ALBICOCCA project. The ALBICOCCA project was funded by CNES in the framework of TOSCA programme and ASI. The SYMPHONIE model data come from the outcome of the MFSTEP project, funded by the European Commission 5th Framework Programme on Energy, Environment and Sustainable Development (EU contract number EVK3-CT-2002-00075). This work has been partially carried out in the framework of the "GOCE Gravity Improvement of Continental Slope and Shelf Ocean Circulation Modelling" study, funded by the European Space Agency (ESA contract number 19740/06/NL/HE). This work was carried out as a contribution to the MARINA (MARgin INtegrated Approach) project, funded by CNES in the framework of the Ocean Surface Topography Science Team investigation plan (CNES/EUMETSAT/NASA/NOAA).

In addition, the authors would like to honour the warmth and kind guidance of a pioneer, colleague and friend, Dr. Yves Ménard. This chapter is dedicated to his memory.

References

Andersen OB (1999) Shallow water tides on the northwest European shelf from TOPEX/POSEIDON altimetry. J Geophys Res 104:7729–7741

Andersen OB, Knudsen P (2000) The role of satellite altimetry in gravity field modelling in coastal areas. Phys Chem Earth, Part A: Solid Earth Geodesy 25(1): pp 17–24

Antonov JI, Levitus S, Boyer TP (2002) Steric sea level variations during 1957–1994: importance of salinity. J Geophys Res 107(C12):8013

Anzenhofer M, Shum CK, Rentsh M (1999) Coastal altimetry and applications, Tech. Rep. n. 464, Geodetic Science and Surveying, The Ohio State University, Columbus, USA

Arakawa A, Suarez MJ (1983) Vertical differencing of the primitive equations in sigma coordinates. Mon Wea Rev 111:34–45

Archiving, Validation, and Interpretation of Satellite Oceanographic Data (AVISO) (1996) In: Aviso Handbook for Merged TOPEX/Poseidon products, 3rd edn. Toulouse, France

Berry PAM, Garlick JD, Freeman JA, Mathers EL (2005) Global inland water monitoring from multi-mission altimetry. Geophys Res Lett 32(16):1–4

Birol F, Cancet M, Estournel C (2010) Aspects of the seasonal variability of the Northern Current (NW Mediterranean Sea) observed by altimetry. J Mar Syst 81(4):297–311

Bouffard J, Roblou L, Ménard Y, Marsaleix P, De Mey P (2006) Observing the ocean variability in the western Mediterranean Sea by using coastal multi-satellite data and models. In: Proceedings of the Symposium on 15 years of Progress on Radar Altimetry, March 13–18, 2006, ESA SP-614, July 2006

Bouffard J, Vignudelli S, Herrmann M, Lyard F, Marsaleix P, Ménard Y, Cipollini P (2008a) Comparison of ocean dynamics with a regional circulation model and improved altimetry in the Northwestern Mediterranean. Terr Atmos Ocean Sci 19:117–133

Bouffard J, Vignudelli S, Cipollini P, Ménard Y (2008b) Exploiting the potential of an improved multimission altimetric data set over the coastal ocean. Geophys Res Lett 35:L10601

Brown GS (1977) The average impulse response of a rough surface and its applications. IEEE Trans Antennas Propag 25:67–74

Carrère L, Lyard F (2003) Modeling the barotropic response of the global ocean to atmospheric wind and pressure forcing – comparisons with observations. Geophys Res Lett 30(6):1275. doi:10.1029/2002GL016473

Carton JA, Giese BS, Grodsky SA (2005) Sea level rise and the warming of the oceans in the Simple Ocean Data Assimilation (SODA) ocean reanalysis. J Geophys Res 110:C09006

Chelton DB (2001) Report of the High-Resolution Ocean Topography Science Working Group Meeting, Oregon State

Chelton DB, deSzoeke RA, Schlax MG, Naggar KE, Siwertz N (1998) Geographical variability of the first baroclinic Rossby Radius of deformation. J Phys Oceanogr 28:433–460

Chelton DB, Ries JC, Haines BJ, Fu LL, Callahan PS (2001) Satellite altimetry. In: Fu L, Cazenave A (eds) Satellite altimetry and earth sciences, Int. Geophys. Ser., vol. 69. Elsevier, New York, pp 1–131

Church JA, Godfrey JS, Jackett DR, Mc-Dougall TJ (1991) A model of sea level rise caused by ocean thermal expansion. J Clim 4:438–444

Cipollini P, Gomez-Enri J, Gommenginger C, Martin-Puig C, Vignudelli S, Woodworth P, Benveniste J (2008a) Developing radar altimetry in the oceanic coastal zone: the COASTALT project. In: paper presented at General Assembly 2008, European Geoscience Union, Vienna, April 17–18

Cipollini P, Vignudelli S, Lyard F, Roblou L (2008b) 15 years of altimetry at various scales over the Mediterranean. In: Vittorio Barale (JRC Ispra) and Martin Gade (Universitat Amburgo) (eds) Remote sensing of European Seas. Springer, Heidelberg/Germany, pp 295–306

Coppens D, Bijac S, Prunet P, Jeansou E (2007) Improvement of sea surface brightness temperature retrieval in coastal areas for the estimation of the wet tropospheric path delay. In: Paper presented at the Ocean Surface Topography Science Team meeting, Hobart, Australia

Deng XL, Featherstone WE (2006) A coastal retracking system for satellite radar altimeter waveforms: application to ERS-2 around Australia. J Geophys Res. doi:10.1029/2005JC003039

Desportes C, Obligis E, Eymard L (2007) On the wet tropospheric correction for altimetry in coastal regions. IEEE Trans Geosci Remote Sensing 45(7):2139–2149

Dufau-Julliand C, Marsaleix P, Petrenko A, Dekeyser I (2004) Three-dimensional modelling of the Gulf of Lion's hydrodynamics (northwest Mediterranean) during January 1999 (MOOGLI3 Experiment) and late winter 1999: Western Mediterranean Intermediate Water's (WIW's) formation and its cascading over the shelf break. J Geophys Res. doi:10.1029/2003JC002019

Durand F, Shankar D, Birol F, Shenoi SSC (2008) Estimating boundary currents from satellite altimetry: a case study for the east coast of India. J Oceanogr 64:831–845

Estournel C, Kondrachoff V, Marsaleix P, Vehil R (1997) The plume of the Rhône: numerical simulation and remote sensing. Cont Shelf Res 17:899–924

Estournel C, Durrieu de Madron X, Marsaleix P, Auclair F, Julliand C, Vehil R (2003) Observation and modelisation of the winter coastal oceanic circulation in the Gulf of Lions under wind conditions influenced by the continental orography (FETCH experiment). J Geophys Res 108(C3):1–18

Estournel C, Auclair F, Lux M, Nguyen C, Marsaleix P (2007) "Scale Oriented" embedded modelling of the North-Western Mediterranean in the frame of MFSTEP. Ocean Sci Discuss 4:145–187

Fernandes MJ, Bastos L, Antunes M (2003) Coastal satellite altimetry – methods for data recovery and validation. In: Tziavo IN (ed) Proceedings of the 3rd meeting of the International Gravity & Geoid Commission (GG2002), Editions ZITI, pp 302–307

Fu LL, Cazenave A (2001) Satellite altimetry and earth sciences: a handbook of techniques and applications. Int Geophys Ser, vol 69. Elsevier, New York

Gaspar P, Ponte RM (1997) Relation between sea level and barometric pressure determined from altimeter data and model simulations. J Geophys Res 102(C1):961–971

Gatti J, Petrenko A, Devenon JL, Leredde Y, Ulses C (2006) The Rhone river dilution zone present in the northeastern shelf of the Gulf of Lion in December 2003. Cont Shelf Res 26: 1794–1805

Gregory JM (1993) Sea level changes under increasing atmospheric CO_2 in a transient coupled ocean-atmosphere GCM experiment. J Clim 6(12):2247–2262

Greatbatch RJ (1994) A note on the representation of steric sea levels in models that conserve volume rather than mass. J Geophys Res 99(12):767–771

Guizien K, Brochier T, Duchène JC, Koh BS, Marsaleix P (2006) Dispersal of owenia fusiformis larvae by wind-driven currents: turbulence, swimming behaviour and mortality in a three-dimensional stochastic model. Mar Ecol Prog Ser 311 47–66

Han G, Tang CL, Smith PC (2002) Annual variations of sea surface elevation and currents over the Scotian Shelf and Slope. J Phys Ocean 32:1794–1819

Herrmann M, Somot S, Sevault F, Estournel C, Déqué M (2008) Modelling the deep convection in the northwestern Mediterranean Sea using an eddy-permitting and an eddy-resolving model: case study of winter 1986–1987. J Geophys Res. doi:10.1029/2006JC003991

Hernandez F, Calvez MH, Dorandeu J, Faugère Y, Mertz F, Schaeffer P (2000) Surface Moyenne Océanique: Support scientifique à la mission altimétrique Jason-1, et à une mission micro-satellite altimétrique. Contrat Ssalto 2945 – Lot 2 – A.1. Rapport d'avancement. Rapport n° CLS/DOS/NT/00.313, published by CLS, Ramonville St Agne, pp 40

Hirose N, Fukumori I, Zlotnicki V (2001) Modelling the high frequency barotropic response of the ocean to atmospheric disturbances: sensitivity to forcing, topography and friction. J Geophys Res 106(C12):30987–30995

Ishii M, Kimoto M, Sakamoto K, Iwasaki SI (2006) Steric sea level changes estimated from historical ocean subsurface temperature and salinity analyses. J Oceanogr 62(2):155–170

Jan G, Ménard Y, Faillot M, Lyard F, Jeansou E, Bonnefond P (2004) Offshore absolute calibration of space borne radar altimeters. Mar Geod. Special issue on Jason-1 calibration/validation, Part 3, 27:3–4, 615–629

Jorda G, Comerma E, Bolaños R, Espino M (2007) Impact of forcing errors in the CAMCAT oil spill forecasting system. A sensitivity study. J Mar Syst 65:134–157

Köhl A, Stammer D (2008) Variability of the meridional overturning in the North Atlantic from the 50-year GECCO state estimation. J Phys Oceanogr 38:1913–1930

Labroue S, Gaspar P, Dorandeu J, Zanifé OZ, Mertz F, Vincent P, Choquet P (2004) Non parametric of the sea states bias for Jason 1 radar altimeter. Mar Geod 27:453–481

Lambin J, Lombard A, Picot N (2008) CNES initiative for altimeter processing in coastal zone: PISTACH. In: Paper presented at the First altimeter workshop, Silver Spring, MD, February 5–7

Le Hénaff M, Roblou L, Bouffard J (2010) Characterizing the Navidad Current interannual variability using coastal altimetry. Ocean Dyn (in press)

Le Provost C (2001) Ocean tides. In: Fu LL, Cazenave A (eds) Satellite altimetry and earth sciences: a handbook of techniques and applications. Int Geophys Ser, vol 69. Academic, San Diego, CA, pp 267–303

Le Traon PY, Ogor F (1998) ERS-1/2 orbit improvement using TOPEX/POSEIDON: the 2cm challenge. J Geophys Res 103:8045–8057

Lombard A, Cazenave A, Le Traon PY, Guinehut S, Cabanes C (2006) Perspectives on present day sea level change: a tribute to Christian Le Provost. Ocean Dyn 56:445–451

Lombard A, Garric G, Penduff T (2009) Regional patterns of observed sea level change: insights from a ¼° global ocean/sea-ice hindcast. Ocean Dyn. doi:10.1007/s10236-008-0161-6

Lyard F, Lefevre F, Letellier T, Francis O (2006) Modelling the global ocean tides: modern insights from FES2004. Ocean Dyn. doi:10.1007/s10236-006-0086-x

Lyard F, Roblou L, Mangiarotti S, Marsaleix P (2003) En route to coastal oceanography from altimetric data: some ALBICOCCA project insight. In: Paper presented at the Ocean surface topography science team meeting, Arles, France

Lyard F, Roblou L (2009) Robust methods for high accuracy tidal modelling in coastal and shelf seas. In: Paper presented at the Ocean Surface Topography Science Team meeting, Seattle, Washington, DC, USA

Lyard F, Roblou L, De Mey P, Lamouroux J (2009) Ocean high frequency dynamics at global and regional scales. In: Paper presented at the first SARAL/AltiKa scientific workshop, Ahmedabad, India

Lynch DR, Gray WG (1979) A wave equation model for finite element tidal computations. Comput Fluids 7:207–228

Mangiarotti S, Lyard F (2008) Surface pressure and wind stress effects on sea level change estimations from TOPEX/Poseidon satellite altimetry in the Mediterranean Sea. J Atmos Ocean Technol 25:464–474

Mantripp D (1996) Radar altimetry. In: Fancey NE, Gardiner ID, Vaughan RA (eds) The determination of geophysical parameters from space. Scottish Universities Summer School in Physics and Institute of Physics Publishing, Bristol/Philadelphia

Manzella GMR, Borzelli GL, Cipollini P, Guymer TH, Snaith HM, Vignudelli S (1997) Potential use of satellite data to infer the circulation dynamics in a marginal area of the Mediterranean Sea. In: Proceedings of 3rd ERS symposium – space at the service of our environment, Florence (Italy), 17–21 March 1997, vol 3, pp 1461–1466, European Space Agency Special Publication ESA SP-414

Maraldi C, Galton-Fenzi B, Lyard F, Testut L, Coleman R (2007) Barotropic tides of the Southern Indian Ocean and the Amery Ice Shelf cavity. Geophys Res Lett. doi:10.1029/2007GL030900

Marsaleix P, Auclair F, Estournel C (2006) Considerations on open boundary conditions for regional and coastal ocean models. J Atmos Ocean Technol 23:1604–1613

Marsaleix P, Auclair F, Floor JW, Herrmann MJ, Estournel C, Pairaud I, Ulses C (2008) Energy conservation issues in sigma-coordinate free-surface ocean models. Ocean Model 20:61–89

Millot C (1987) Mesoscale and seasonal variabilities of circulation in western Mediterranean. Dyn Atmos Ocean 15:179–214

Millot C (1991) Circulation in the western Mediterranean. Oceanol Acta 10:143–149

Morrow R, Valladeau G, Sallee JB (2008) Observed subsurface signature of Southern Ocean sea level rise. Prog Oceanogr 77(4):351–366. A new view of water masses after WOCE. A special edition for Professor Matthias Tomczak

Ollivier A, Le Bihan N, Lacoume JL, Zanife OZ (2005) Improving speckle filtering with SVD to extract ocean parameters from altimetry radar echos. In: Proceeding of PSIP 2005, Toulouse

Pairaud I, Lyard L, Auclair F, Letellier T, Marsaleix P (2008) Dynamics of the semi-diurnal and quarter-diurnal internal tides in the Bay of Biscay. Part 1: barotropic tides. Cont Shelf Res. doi:10.1016/j.csr.2008.03.004

Pascual A, Faugère Y, Larnicol G, Le Traon PY (2006) Improved description of the mesoscale variability by combining four satellite altimeters. J Geophys Res. doi:10.1029/2005GL024633

Pattullo J, Munk W, Revelle R, Strong E (1955) The seasonal oscillation in sea level. J Mar Res 14:88–156

Petrenko A, Leredde Y, Marsaleix P (2005) Circulation in a stratified and wind-forced Gulf of Lions, NW Mediterranean Sea: in-situ and modelling data. Cont Shelf Res 25:7–27

Quesney A, Jeansou E, Lambin J, Picot N (2007) A new altimeter waveform retracking algorithm based on neural networks. In: Paper presented at the Ocean surface topography science team meeting, Hobart, Australia

Roblou L (2004) Validation des solutions de marée en Mer Méditerranée: comparaisons aux observations. Technical Report, POC-TR-02-04, 67pp, Pôle d'Océanographie Côtière, Toulouse, France

Roblou L, Lyard F, Le Hénaff M, Maraldi C (2007) X-TRACK, a new processing tool for altimetry in coastal ocean. In: ESA ENVISAT Symposium, Montreux, Switzerland, April 23–27, 2007, ESA SP-636, July 2007

Ruf CS, Keihm SJ, Subramanya B, Janssen MA (1994) TOPEX/Poseidon microwave radiometer performance and in-flight calibration. J Geophys Res 99(C12):24915–24926

Service d'Altimétrie et de Localisation Précise (SALP) (1996) Minutes of the ocean surface topography science team meeting, Venice, Italy, March 16–18, 2006, SALP-CR-MA-EA-15640-CNES, Y. Ménard (ed), CNES, p 90

Stammer D, Wunsch C, Ponte RM (2000) De-aliasing of global high frequency barotropic motions in altimeter observations. Geophys Res Lett 27(8):1175–1178

Tai CK (1989) Accuracy assessment of widely used orbit error approximations in satellite altimetry. J Atmos Ocean Technol 6:147–150

Tai CK (1991) How to observe the gyre to global-scale variability in satellite altimetry: signal attenuation by orbit error removal. J Atmos Ocean Technol 8:271–288

Tournadre J (2006) Improved level-3 oceanic rainfall retrieval from dual-frequency spaceborne radar altimeter systems. J Atmos Ocean Technol. doi:10.1175/JTECH1897.1

Tomczak M, Godfrey JS (1994) Regional oceanography: an introduction. Pergamon, New York

Ulses C, Grenz C, Marsaleix P, Schaaff E, Estournel C, Meulé S, Pinazo C (2005) Circulation in a semi enclosed bay under the influence of strong fresh water input. J Mar Syst 56:113–132

UNESCO, ICES, SCOR, IAPSO (1981) Background papers and supporting data on the international equation of state of seawater, 1980, UNESCO technical papers in Mar Science. Nr. 38, UNESCO

Vignudelli S, Cipollini P, Astraldi M, Gasparini GP, Manzella GMR (2000) Integrated use of altimeter and in situ data for understanding the water exchanges between the Tyrrhenian and Ligurian Seas. J Geophys Res 105(19):649–663

Vignudelli S, Cipollini P, Roblou L, Lyard F, Gasparini GP, Manzella G, Astraldi M (2005) Improved satellite altimetry in coastal systems: case study of the Corsica Channel (Mediterranean Sea). Geophys Res Lett. doi:10.1029/2005GL022602

Volkov DL, Larnicol G, Dorandeu J (2007) Improving the quality of satellite altimetry data over continental shelves. J Geophys Res. doi:10.1029/2006JC003765

Wunsch C, Ponte R, Heimbach P (2007) Decadal trends in sea level patterns: 1993–2004. J Clim 20:5889–5911

Coastal Challenges for Altimeter Data Dissemination and Services

10

H.M. Snaith and R. Scharroo

Contents

10.1	Introduction	248
10.2	Generating Custom Products	250
10.3	Coastal Altimetry Integration in Data Systems	252
10.4	Implications for Data Products and Systems	253
	10.4.1 Standard Formats	253
	10.4.2 Metadata Tracking	253
	10.4.3 Data Distribution Tools	254
	10.4.4 Data Service Solutions	254
10.5	A Vision for a Coastal Data Service	255
References		257

Keywords Altimetry • Data services

Abbreviations

ACCESS07	Web-based Altimeter Service and Tools
ALTICORE	ALTImetry for COastal REgions
CEOS	Committee for Earth Observation Satellites
COASTALT	ESA development of COASTal ALTimetry
GDR	Geophysical Data Record
GFO	Geosat Follow-On
GIS	Geographic Information System
HDF	Hierarchical Data Format (http://www.hdfgroup.org)

H.M. Snaith (✉)
National Oceanography Centre, Southampton, European Way, Southampton SO14 3ZH, UK
e-mail: h.snaith@noc.soton.ac.uk

R. Scharroo
Altimetrics LLC, Cornish, New Hampshire, USA

S. Vignudelli et al. (eds.), *Coastal Altimetry*,
DOI: 10.1007/978-3-642-12796-0_10, © Springer-Verlag Berlin Heidelberg 2011

JPL	Jet Propulsion Laboratory
MATLAB	MATrix LABoratory (http://www.matlab.com)
netCDF	network Common Data Form (http://www.unidata.ucar.edu/software/netcdf/)
netCDF CF	netCDF Climate Forecast Conventions (http://cf-pcmdi.llnl.gov/)
OGSA	Open Grid Services Architecture (http://www.globus.org/ogsa/)
OPeNDAP	Open-source Project for a Network Data Access Protocol (http://opendap.org/)
OSTST	Ocean Surface Topography Science Team
PISTACH	Prototype Innovant de Système de Traitement pour l'Altimétrie Côtière et l'Hydrologie (ftp://ftpsedr.cls.fr/pub/oceano/pistach/)
RADS	Radar Altimeter Database System (http://rads.tudelft.nl/)
RDF	Rads Data File
RMF	Rads Meta data File
SWT	TOPEX Science Working Team

10.1
Introduction

Existing data dissemination systems and services have been designed, largely, for the open ocean use of satellite altimetry and optimized for the study of ocean currents or high precision sea level. A few more specialized, higher-level products have been made available, for example, for monitoring lake and river levels, or digital elevation models incorporating land altimetry measurements. However, the wider distribution of pure altimetry products has focused on the core user community in open ocean research. As the altimeter data themselves are now being improved and adapted to allow data to be retrieved closer to the coast, the data are becoming useful to a wider potential community, and this will have an impact on the data products and services required (Fig. 10.1).

By itself, the use of satellite altimetry in the coastal zone does not create new issues for data services. However, the nature of the data usage does highlight some fundamental issues in the provision of satellite data that have not been significant for the majority of existing users. Key to these issues, there are two driving requirements:

- Generation of custom products
- Use of altimeter data within data assimilation or regional observation systems

These drivers are already beginning to influence the range of products available and the way in which data, with their associated metadata, are disseminated to users. In the remainder of this chapter, we will discuss the changes already occurring in the area of data dissemination and propose a vision for an altimeter service that could fulfill the future requirements of coastal users.

10 Coastal Challenges for Altimeter Data Dissemination and Services   249

	Mission	Media	Data Format
1985	**Geosat (1985-90)** Originally delivered on reel-to-reel tapes. Re-release on CD set in 1990.		Introduction of GDRs Flat files, fixed length records, no headers, no description, one file per day.
1990	**ERS-1 (1991-96)** Fast-delivery product on Exabyte (Video 8) tapes. Final product on CD. **T/P (1992-2005)** Delivery via FTP and on CD.		ERS-1 adopts CEOS (image format). Impractical for altimeter data. One file per pass simplifies data search. T/P data gets header records that describe pass, not content.
1995	**ERS-2 (1995-present)** Data available through web interface (time/area selection) and CD. **GFO (1999-2008)** FTP and DVD.		ERS-2 reverts fixed-length files. Headers only on CD archive (not through web server). GFO mimics Geosat product, plus headers.
2000	**Jason-1 (2000-present)** Primarily FTP (DVD option). **Envisat (2001-present)** FTP and DVD.		Jason-1 mimics T/P format. Envisat adopts yet another ESA format. Headers surround embedded products. More flexible for expansion, but not descriptive.
2005	**Jason-2 (2008-present)** FTP and web server.		Jason-2 adopts netCDF as delivery vehicle. Contains product and data descriptors, units and scales. Fully expandable. Easy data selection. RADS adopts similar format.

Fig. 10.1 Evolution of altimeter data products

10.2 Generating Custom Products

Since the launch of Geosat, the fundamental data product for satellite altimetry has been based on an archive format – the geophysical data record (GDR). The GDR format has been massively constrained by the available means of storage and distribution. When the first GDRs of Geosat came out in 1986, they were distributed on reel-to-reel half-inch wide nine-track tapes with a capacity of about 110 megabytes each (as seen in Fig. 10.2). At the time, a high-class desktop computer could contain twice as much disk space. At that time, when both archival and work spaces were at a premium, it was essential that the GDR product was as concise as possible. As a consequence it contained no meta data and no information on the data content, not even on such vital characteristics as cycle and pass numbers or processing software versions. All knowledge of the products had to be obtained from off-line documentation.

The Geosat GDR data structure was what we call *flat files*: a stream of records of a fixed length (78 bytes at the time), each record containing the measurement and correction values of one data point (at left in Fig. 10.2). This was practical when computer memory was scarce, as each measurement could be easily processed individually. But it had significant downsides:

- Fixed length left no space for extension. So the product was basically set in stone with little or no possibility of improvement over time.
- The lack of metadata meant that the users each had to carefully transcribe the available documentation into code; not a job for the non-expert. And once hardcoded, again, it left no flexibility for changes.

At the time, computing resources were actually still so limited that users were not even able to process all these GDR format data from hard disk, so users built their own condensed versions, retaining only the parameters essential to their own research, which did not promote data sharing.

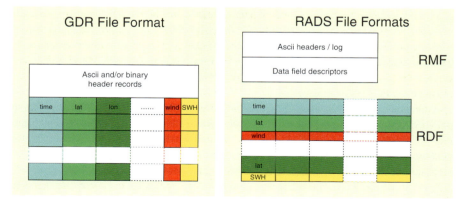

Fig. 10.2 Flat file and netCDF file formats

Despite these downsides, *flat files* remained the common data format until very recently. Although header records, to include the data period, pass and cycle number, processing version, and the like, were added to the TOPEX GDRs and later products, and record lengths grew to 440 bytes (including numerous spares), by the time of Jason-1, not much had changed in over 20 years. By this time, users had to carefully read and interpret the off-line documentation with the emergence of every new product or change in the processing version, which could occur several times in the course of a single satellite mission.

An early attempt to improve data access was to adopt the CEOS (Committee for Earth Observation Satellites) data standard for ERS-1 data. This back-fired as the format was far too complicated, being more suitable for two-dimensional, multi-spectral image data than for one-dimensional altimeter data and also being designed for data distributed on unlabelled tapes rather than random access media. With the GDR *flat file* format so engrained into the user community, simplifications were sought in the content, rather than the format.

As the altimetry measurement system began to be better understood, the user community was able to define *standard* processing systems, to simplify the use of the data for non-specialists. For many years, the TOPEX Science Working Team (SWT) and, more recently, the Ocean Surface Topography Science Team (OSTST), have made significant efforts to synthesize much of the research using satellite altimetry, to generate recommendations for the use of satellite data, including all the required corrections, for open ocean use. This has been possible largely because there is consistency in the recommendations for all the open ocean areas. The development of new products has concentrated on preapplication of these recommendations to the *basic* source data to provide consistent, higher-level data.

Once there was consensus on the suite of corrections to apply for the majority of users, products could be generated that preapplied these corrections to give corrected anomaly products. The generation of gridded products, incorporating more sophisticated interpolation procedures, better mean sea surface and dynamic topography models, and using data assimilation schemes, now gives us a wide range of products, tailored to the majority users: those interested in open ocean applications. As a result of these efforts, there has been extensive use of the open ocean products by the oceanographic community for a wide range of research, as highlighted in numerous publications (e.g., Cheney 1995; Fu and Cazenave 2001). These products, including mapped sea level anomalies, mean sea surfaces, and derived geostrophic velocity maps have provided an invaluable resource for the non-specialist user of altimetry data.

For coastal applications there is, as yet, no similar consensus on corrections and processing. Indeed, the wide range of potential users and region-specific auxiliary data and corrections are expected to result in a wide range of regional-specific corrections or recommendations. For example: each region under consideration may have a different region-specific tidal model available; or users may prefer a tropospheric model tailored to particular atmospheric conditions, incorporating in situ observations; or end products may require all heights to be related to a regional datum. As a result, the new coastal altimeter data products are being designed with the flexibility to allow different correction sets and algorithms for different regions, in accordance with user preferences.

The latest product from the Jason-2 mission is directly aimed at the coastal user. The PISTACH (Prototype Innovant de Système de Traitement pour l'Altimétrie Côtière et l'Hydrologie) project has generated a product that includes all possible corrections and source data, including values from a number of different retracking scenarios, for the global ocean. This approach provides a wealth of detailed information for the technical specialist, but results in very large products, which are the domain of the expert. The COASTALT project, initially investigating Envisat data, has a similar approach, generating a product with a range of retracking scenarios and possible corrections, although data will only be generated for coastal areas. Only those corrections that can be calculated globally will be included in the standard product and users will be able to add regional-specific corrections to provide a more tailored solution. Although initially targeted at Jason-2 and Envisat, respectively, these products could be generated for other satellite missions.

While these products will allow the flexibility necessary for provision of regional-specific altimetry solutions, the trade-off is that they are complex, including a wide range of data options that require specialist knowledge to use effectively.

10.3
Coastal Altimetry Integration in Data Systems

The second key driver for coastal data systems is the potential use of the altimetry data in integrated data systems. Over the open ocean, there is a scarcity of data sources. A prime reason for the massive success of satellite altimetry for research into oceanographic circulation is that it is the only source of (almost) globally available data that are directly representative of a geophysical parameter – sea surface height. As a result, there is a significant body of research carried out with satellite altimetry alone, or with alternative data sources used for validation. This is not the case for coastal areas.

While there are areas of the global coastal ocean where altimetry may offer the only long-term data source, as a generality, coastal areas have a wider potential number of data sources, as they are easier to reach, easier to instrument for in situ measurements, and also covered by national monitoring bodies. Hence, many coastal areas have denser data available and there are a wider number of data sources available. In addition, the scales of the oceanographic processes, both spatial and temporal, in the coastal zone, mean that even multi-mission altimetry does not, yet, provide optimal, or even sufficient, sampling in many regions; for example, see the chapters by Kouraev et al., Ginzburg et al., Lebedev et al., Emery et al., and Deng et al., this volume.

As a result, it is highly likely that many coastal applications will be far less reliant on altimetry alone than their open ocean counterparts (e.g., see chapter by Emery et al., this volume). Altimetry will probably become an input to wider observational systems, with many users anticipating use of altimetry in data assimilation systems (see chapter by Dufau et al., this volume), together with multiple data sources, or incorporated into GIS data systems. Researchers, and particularly operational agencies, will not be able to invest the same level of time in "learning" how to use coastal altimetry data as their open ocean counterparts, for whom altimetry is a very high priority dataset.

So we have a conundrum: altimetry processing becomes inherently more complex in the coastal zone, requiring more complex data streams to incorporate the wider range of potential corrections. At the same time, if we wish to increase the application of coastal altimetry, the *use* of the products needs to be much simpler, allowing data selection and processing ready for ingestion along with multiple data streams, with minimal investment of time. This gap can be addressed by the development of tools and services to manipulate the coastal products.

10.4 Implications for Data Products and Systems

10.4.1 Standard Formats

As described previously, the formats of traditional altimetry data products have been dominated, not by user needs or applications, but by the requirement for a compact data storage format. As the number of missions increased and space was no longer a major constraint, there was a major incentive to provide multi-mission data in consistent formats. Higher-level products, and more recent products, under pressure and guidance from the open-data, open-source community, have moved towards using more standardized data formats across the geosciences. These formats are not generally as compact, but allow for greater flexibility in content and increased levels of included metadata. These formats are normally self-describing, meaning that software can be designed to interrogate the data themselves to determine the content, so the software tools need less configuration for different altimetry missions and fewer modifications to allow for data processing versions. The two most commonly used of these formats are HDF (Hierarchical Data Format) and netCDF (network Common Data Form) and of these, netCDF is widely used in the climate community and has become a de facto standard for altimetry.

10.4.2 Metadata Tracking

One result of the need for more customizable data products is a requirement for traceability of the data sources and metadata, which is far more stringent than for the open ocean. While best practice is to maintain the full suite of metadata on data sources throughout the data stream, the use of global recommendations has made this less important and the metadata for many products have existed only in the data handbooks. Standard data formats and reduced space restriction mean that it is now feasible to provide more metadata, including algorithms and calibration applied, processing streams, and links to more auxiliary data, within the data products themselves, allowing access to this information without searching external handbooks or websites.

Across the geosciences, there are also an increasing number of initiatives to generate standards in metadata content and naming. Among these standards, the netCDF Climate Forecast (CF) Conventions are increasingly gaining acceptance and have been adopted by a number of projects and groups as a primary standard. The conventions define metadata that provide a definitive description of what the data in each variable represents, and the spatial and temporal properties of the data. This enables users of data from different sources, and users from different communities, to decide which quantities are comparable. Perhaps even more importantly these conventions facilitate building applications with powerful extraction, regridding, and display capabilities. Allowing computer code to determine the content of data files *without* external information allows for advanced automation of the inclusion of data into integrated systems, even into systems that were not originally designed to handle altimeter data.

10.4.3
Data Distribution Tools

One stage further in the development of data services is to move away from the generation of fixed, predetermined products and towards the generation of tailor-made products containing specific data items, corrections, or data collection in response to user requests. Products incorporating this concept are already in existence, and further systems are under development. The simplest level, as demonstrated by the RADS (Radar Altimeter Database System) web delivery system, will generate a sea level anomaly according to user specified flags and corrections. The drive for this approach will continue from operational systems: data assimilation systems, observing systems, and disaster monitoring applications are likely to move towards ingestion of products generated with parameters defined by the user, rather than by the data provider.

As the push for customized products, targeted at complex coastal environments, increases, there will also be an increasing need for tools to manipulate these data by non-specialists. The coastal user, with a wide range of potential data sources, incorporating altimetry into an observation system or assimilation scheme, needs support to use the data without investing a disproportionate amount of time in understanding how to use them. These tools should incorporate the coastal equivalent of the OSTST recommendations for coastal environments, which may well need to be modified by region and by users' specialist knowledge of a region.

10.4.4
Data Service Solutions

In the Open Grid Services Architecture (OGSA) terminology, a data service is "a service that provides interfaces to the capabilities and data of one or more data resources within a service-oriented architecture." And here, a data resource is specifically "an entity (and its associated framework) that provides a data access mechanism or can act as a data source or data sink." The provision of coastal altimetry data on demand by users is moving towards providing data as a resource through a data service. Such a service can allow access to a range of different data sources, and can be accessed by a number of means, including directly from software, such as a GIS system, or through provision of a web interface to interrogate the service. A single service is able to provide data via a range of customizable interfaces.

OPeNDAP (Open-source Project for a Network Data Access Protocol) is a simple data service and visualization tool that has been in use for several years, and incorporates the essential coastal functionality of regional subsetting. The primary data sources are held in appropriate data centers. This makes updating data much simpler, with no necessity for distribution of updated products or corrections, as they can be applied at source. The data center then runs a service, which can be accessed, using standard protocols. OPeNDAP can also be configured to allow access directly from within applications such as MATLAB, allowing direct access to data from remote servers without the need for local copies. This approach was successfully trialed within the ALTICORE project, allowing access by collaborative sites to regional archives of altimetry.

10.5
A Vision for a Coastal Data Service

Many coastal regions have been extensively monitored and researched, and there is a wealth of auxiliary information, from distributed sources, that can be used to generate customized products for many regions. This information is often already produced by regional data centers, or national institutes, and may only be relevant to relatively small areas. Attempting to duplicate this information in its entirety at a central altimetry data center may no longer be the most appropriate way to manage and distribute these data. An alternative approach has been proposed previously, and is now being trialed as part of the ACCESS07 project at the Jet Propulsion Laboratory (JPL). The regional centers and specialist data providers generate the required corrections, models, or algorithms and provide them to the primary data center, where they are then accessible to users when they request products. These corrections do not need to be available globally, but their existence can be flagged and recommended if a user requests data for that region. The service is required to provide a range of functions, including spatial and temporal data selection, selection of either complete or limited parameters, processing of data on demand, and retrieval in the preferred format, including visualization of results.

This service is the first step towards achieving the possible complexity that could be managed by a more complete, integrated data service. Fig. 10.3 illustrates how such a range of services could be connected. At the input side of the service, data centers generating specialist corrections, algorithms, or even retracked data for specific locations can be accessed via a central data service, with data requests being "passed off" to the most appropriate secondary center as appropriate. On the output side of the service, users can access the data directly from software applications, potentially including modelling assimilation schemes. Web portals can be generated to provide access to specialized subsets of the data, by region or scientific interest, restricting the complexity of offered solutions to those most appropriate to that specialty. Also providing input to the service are specialists in the field, who are able to provide recommendations on the most appropriate use of data, correction sets, and quality control that can be utilized by the services in determining appropriate sources and solutions for users.

Using this service, we could envisage a scenario where an end user wishes to provide oil spill spread prediction in an emergency response situation. The user's model can be

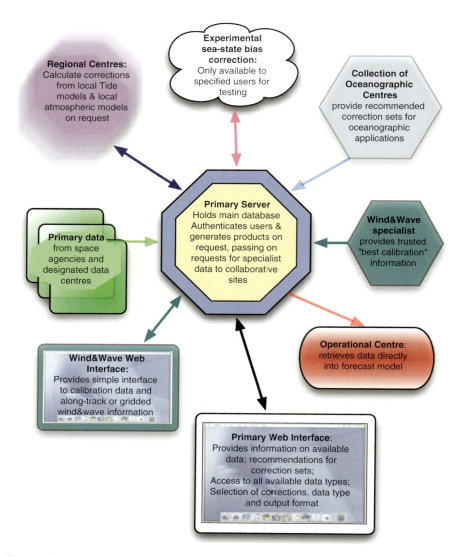

Fig. 10.3 Complexity of altimeter data services

configured to operate in any region, and in order to provide the most accurate prediction, it will require the widest possible range of local forcing and boundary data. The model could access a range of services, including the satellite altimetry service, for data. An initial request to the server from the model is configured for "the most accurate climate state sea surface height data" in order to initialize the model, and the service responds by selecting averaged off-line products, processed using a regional tide model and regional calibrated atmospheric corrections. The model then requests further data; this time the requirement is for "the most up to date sea surface height, wind, and wave data," to constrain and force the model. In this case, the service will respond by selecting near-real-time products, but will substitute the same regional tide model and atmospheric corrections, and include the wind and wave parameters.

In order for this vision to be achieved, and efficiently provide coastal altimetry users with the data they require, the underlying data providers must continue to develop their systems towards the increased use of standard formats and metadata, and traceability of sources that has already begun. An altimetry service configured to meet the demands of the coastal community would ultimately benefit the wider altimetric community, with potential benefits for open ocean users, as well as ice and land specialists.

References

Cheney RE (ed) (1995) TOPEX/Poseidon: scientific results, reprinted from the J Geophys Res 100(C12), AGU

Fu LL, Cazenave A (eds) (2001) Satellite altimetry and earth sciences: a handbook of techniques and applications. Academic, San Diego, CA

In situ Absolute Calibration and Validation: A Link from Coastal to Open-Ocean Altimetry

11

P. Bonnefond, B.J. Haines, and C. Watson

Contents

11.1	Introduction	261
	11.1.1 Overview	261
	11.1.2 A Review of Absolute Calibration Experiments	263
	11.1.3 Absolute Versus Relative Calibration	264
	11.1.4 Outline	265
11.2	Calibration Geometry and Measurements	265
	11.2.1 Absolute Calibration Techniques	265
	11.2.2 Formation Flying and Tandem Mission Verification	268
	11.2.3 In situ Measurements	268
	11.2.4 Geophysical Corrections	273
	11.2.5 Reference Frames	274
	11.2.6 Orbit	275
11.3	Dedicated Calibration Sites	275
	11.3.1 Overview	275
	11.3.2 Harvest	276
	11.3.3 Corsica	278
	11.3.4 Bass Strait	281
	11.3.5 Data Analysis Comparison	283
11.4	Salient Results	283
	11.4.1 Detection of Instruments/Model Problems	283
	11.4.2 Recent Results	289
	11.4.3 Revised Error Budget in Altimetry	290
11.5	Conclusion	292
References		293

P. Bonnefond (✉)
Observatoire de la Côte d'Azur – GeoAzur, Avenue Nicolas Copernic, 06130 Grasse, France
Pascal.Bonnefond@obs-azur.fr

B.J. Haines
Jet Propulsion Laboratory, California Institute of Technology, 4800 Oak Grove Drive, MS 238-600, Pasadena, CA 91109, USA

C. Watson
Surveying and Spatial Science Group, School of Geography and Environmental Studies, University of Tasmania, Hobart, TAS, Australia, 7001

S. Vignudelli et al. (eds.), *Coastal Altimetry*,
DOI: 10.1007/978-3-642-12796-0_11, © Springer-Verlag Berlin Heidelberg 2011

Keywords Altimetry • GPS • In situ instrumentation • Reference frame • Sea level

Abbreviations

ALT-A or B	Side A or B of the TOPEX altimeter
AMR	Advanced Microwave Radiometer (OSTM/Jason-2)
Cal/val	Calibration/validation
CCAR	Colorado Centre for Astrodynamics Research
CERGA	Centre d'Etudes et de Recherches Geodynamiques et Astronomiques
CGPS	Continuous Global Positioning System
CNES	Centre National d'Étude Spatiales
CU	Colorado University
DORIS	Détermination d'Orbite et Radiopositionnement Intégrés par Satellite
ECMWF	European Centre for Medium-Range Weather Forecasts
EM	Electro Magnetic
Envisat	Environmental Satellite
ERS (1 & 2)	European Remote Sensing
EUMETSAT	European Organisation for the Exploitation of Meteorological Satellites
FTLRS	French Transportable Laser Ranging System
GCE	Geographically Correlated Errors
GDR	Geophysical Data Records (prefix O = Operational, I = Interim)
GFO	Geosat Follow-On
GLONASS	GLObal NAvigation Satellite System
GOES	Geostationary Operational Environmental Satellites
GNSS	Global Navigation Satellite Systems
GPS	Global Positioning System
GRACE	Gravity Recovery and Climate Experiment
GSFC	Goddard Space Flight Centre
IERS	International Earth Rotation Service
IGDR	Interim Geophysical Data Record
IGN	Institut Géographique National
IGS	International GPS Service for geodynamics
ISDGM/CNR	Istituto per lo Studio della Dinamica delle Grandi Masse / Consiglio Nazionale delle Ricerche
ITRF	International Terrestrial Reference Frame
JMR	Jason-1 Microwave Radiometer
JPL	Jet Propulsion Laboratory
MGDR	Merged Geophysical Data Record
MLE	Maximum Likelihood Estimator
NASA	National Aeronautics and Space Administration
NOAA	National Oceanic and Atmospheric Administration
OGDR	Operational Geophysical Data Record
OMR	Orbit Minus Range
OSTM	Ocean Surface Topography Mission
OSTST	Ocean Surface Topography Science Team
POD	Precision Orbit Determination

POE	Precision Orbit Ephemerides
RMS	Root Mean Square
RSS	Root Sum of Squares
SHOM	Service Hydrographique et Océanographique de la Marine
SLR	Satellite Laser Ranging
SPTR	Scanning Point Target Response
SSB	Sea State Bias
SSH	Sea Surface Height
SWH	Significant Wave Height
T/P	TOPEX/Poseidon
TMR	TOPEX Microwave Radiometer
TCA	Time of Closest Approach
USO	Ultra Stable Oscillator
VLBI	Very Long Baseline Interferometry
VSAT	Very Small Aperture Terminal
WVR	Water Vapor Radiometer

11.1 Introduction

11.1.1 Overview

The observation of rising mean sea levels over the Earth's oceans using satellite altimetry has helped inform and shape the debate on global climate change that has emerged over the last 2 decades (Cazenave and Nerem 2004; Bindoff et al. 2007). The ongoing monitoring of secular changes in global mean sea level with accuracy tolerances of better than 1 mm/year remains a fundamental science goal within satellite altimetry and represents one of the most challenging objectives in space geodesy. Central to this objective has been the recognition that calibration and validation (cal/val) are vital components of the altimeter measurement system technique. Cal/val defines a multidisciplinary problem in and of itself, which pushes the limits of available terrestrial, oceanographic, and space-based observational techniques.

The primary aim of the calibration process is to characterize the performance of each component of the measurement system, providing early warnings about changing instrumental behaviors and sources of possible spurious changes in derived sea-surface heights over time. Notable drifts discovered relatively well into the TOPEX/Poseidon mission (Nerem et al. 1997) and absolute sea-surface height biases in the Jason-1 and Jason-2 missions that remain unexplained, underscore the complexity of the measurement system and intrinsic importance of calibration and validation activities. The requirement of a multi-technique and geographically diverse calibration approach has elicited the emergence of two main complementary techniques in the calibration effort. First, as emphasized in this chapter, dedicated absolute in situ calibration sites have enabled the calibration of SSH and its component measurements in a truly absolute sense, thus providing insight into a range of instrument behaviors, geographically dependent effects and absolute biases. Second, calibration based on the utilization of the global tide gauge network provides a global relative calibration capable of estimating SSH bias drift to within the accuracy tolerances that are required (Mitchum 1998).

Regardless of the technique, the calibration problem is complex given the array of components forming the altimeter measurement system, the lack of truly independent and more accurate measurement techniques and the temporal and spatial scale at which systematic error manifests in satellite altimetry. These different components – orbit and reference frame realization, range determination, path length corrections and, geophysical corrections and associated calibrations – form the basis of the error budgets published in the framework of the TOPEX/Poseidon (T/P) and Jason missions during both geophysical evaluation and scientific results phases (Fu et al. 1994; Ménard et al. 2003). They have permitted: (1) the constraint of results from the mission providing limits to the geophysical interpretation of altimetric derived signals, (2) the identification of possible sources of future improvement and development, and (3) the development of new and refined evaluation, validation and calibration methods for the future.

During the last years, complementary altimetric missions operating concurrently have provided the unprecedented ability to compare measurement systems by undertaking relative calibrations. Studies have shown, through global statistics and results, the power of such a technique (e.g., Vincent et al. 2003). Given the technical challenges of operating spacecraft for many years in orbit, however, the ability to cross-calibrate multiple missions in this manner cannot be assured. Problems have been discovered both in the algorithms and the instruments on various missions: the Ultra Stable Oscillator (USO) drift and Scanning Point Target Response (SPTR) correction for ERS-1 (Benveniste 1997), the T/P oscillator drift error discovered by Zanife and others in 1996 (see also Nerem et al. 1997) and more recently in the JMR wet path delay correction for Jason-1 (see Sect. 11.4.1.1). These issues reinforce the need for a range of complementary calibration methodologies – including a geographically diverse array of in situ absolute calibration sites that can assess changes to instrument behavior in near real time.

Beyond the calibration of the sea surface height (SSH), the calibration sites also are very useful for assessing the various components of the altimetric measurement systems. Absolute in situ calibration sites are often equipped with a complete system of in situ instruments which have the capability of measuring very accurately the environmental parameters that bear on the altimetric measurement of SSH: sea level, sea state, atmospheric effects (troposphere and ionosphere), reference frame stability, etc. (Christensen et al. 1994; Ménard et al. 1994).

Despite advances in the technique, absolute calibration of radar altimeters at the centimeter level or better remains one of the most difficult challenges in space geodesy. The solution of the so-called closure equation that is fundamental to absolute calibration activities requires the direct comparison of terrestrial sea level measurements with sea-surface heights deduced from satellite altimetry. This exercise requires that the terrestrial estimate of sea level is precisely tied to the same reference frame used for the satellite orbit. The terrestrial measurement must also be sufficiently out in the open-ocean to avoid land contamination, due principally to the large footprint of the space borne water vapor radiometers used to derive the wet path delay. This constraint introduces considerable logistical and operational difficulties, further exacerbated by the need to observe terrestrial observations simultaneously at each overflight of the altimeter satellite. Finally, the terrestrial measurement of sea-surface height in a dynamic open sea environment represents a demanding geodetic problem in its own right. These problems and the attendant systematic errors make it very difficult to fully realize the calibration site error budgets, which are driven by the substantial demands of sea level monitoring from space. The earliest absolute calibration experiments (Francis 1992; Christensen et al. 1994; Ménard et al. 1994; White et al. 1994) showed these difficulties clearly. As a consequence of the improved precision over the last 10 years of each component within the satellite altimetry system (radar instrumentation, path delay observation, orbit determination, and corrections) (Barlier and Lefebvre 2001), requirements are now at the centimeter level. Against this backdrop, absolute calibration has emerged as a continuous, high performance and multifaceted field campaign. Despite the inherent challenges, the calibration efforts are vital components of the succession of high-accuracy altimeter missions.

11.1.2
A Review of Absolute Calibration Experiments

The Bermuda experiments for GEOS-3 (1976) and Seasat (1978) provided the earliest examples of the overhead altimeter calibration concept (Martin and Kolenkiewicz 1981; Kolenkiewicz and Martin 1982). Sea-surface-height data from the in situ and satellite measurement systems were compared at overflight times to infer the overall bias in the GEOS-3 and Seasat measurement systems. Laser data from the NASA station at Bermuda were the most important source of tracking data available for determining the satellite orbit height as it passed overhead.

For the ERS-1 mission, the *Acqua Alta* platform of the Istituto per lo Studio della Dinamica delle Grandi Masse of the Consiglio Nazionale delle Richerche (ISDGM/CNR) served as the calibration site (Francis 1992). Located in the Adriatic Sea 16 km off the coast off Venice, *Acqua Alta* was equipped with a tide gauge and various meteorological instruments.

For TOPEX/Poseidon, calibration sites were maintained at the Harvest oil platform (Christensen et al. 1994) and Lampedusa Island, although the latter was used only during the first 6 months of the mission (Ménard et al. 1994). Although satellite laser ranging systems remain valuable for determining accurate estimates of the orbit heights over experiment sites (Murphy et al. 1996; Exertier et al. 2004), the DORIS and GPS systems have emerged as important tools for both satellite and tide gauge positioning in calibration experiments. For example, Christensen et al. (1994) used GPS-based estimates for determining the heights of both the Harvest tide gauge system and the satellite in their validation of the TOPEX/Poseidon data.

A calibration of the TOPEX/Poseidon and ERS-1 altimeters was undertaken using altimetry across the English Channel (Murphy et al. 1996; Woodworth et al. 2004). Data from the tide gauge installed at Newhaven and the satellite laser ranger at Herstmonceux were used in conjunction with local models for the geoid, ocean tide, and storm surge to generate bias values at each sub satellite altimetry point.

Dedicated calibration sites continue to serve a vital function in modern altimeter missions. Jason-1 and its successor Jason-2 are served by three primary, dedicated sites: the Harvest platform (Christensen et al. 1994; Haines et al. 2003), Corsica (Bonnefond et al. 1997, 2003a) and Bass Strait (White et al. 1994; Watson et al. 2004). The longest, continuous calibration time series (e.g., the Harvest series depicted in Fig. 11.1), dating back to T/P, have emerged from these locations (Fig. 11.6). Due to their long observing record and vital role in monitoring the T/P and Jason series of missions, we devote separate sections (see Sect. 11.3 and 11.4) to the discussion of the experiments and results.

A permanent satellite facility on the island of Gavdos, Crete, Greece has been established since 2001 to: (1) carry out calibration of satellite altimeters, and (2) monitor absolute sea level changes. The European Union, NASA-GSFC and the Swiss Government funded the initial development of the infrastructure in Gavdos. The calibration facility is under a crossing point of the ground tracks of the Earth-observing satellite altimetry missions of Jason-1 (and thus

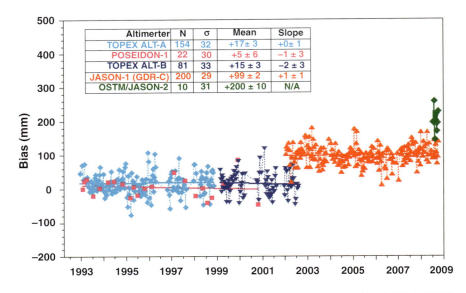

Fig. 11.1 Altimeters biases time series from Harvest platform for TOPEX/Poseidon (ALT-A, ALT-B, Poseidon-1), Jason-1 (Poseidon-2), and Jason-2 (Poseidon-3)

Jason-2), and adjacent to an Envisat pass (Pavlis et al. 2004). The infrastructure of the permanent facility on Gavdos includes tide gauges, permanent GPS satellite receivers, meteorological and oceanographic instruments, a DORIS satellite beacon, an electronic transponder (Cristea and Moore 2007), a EUMETSAT antenna, communications systems for the transmission of data, etc. Given the importance of the Gavdos installations, both from the scientific point of view and as a model satellite research center, the continuation of its operation for the collection, analysis and distribution of data and its extension to other satellites are necessary.

Three experiments were conducted near Begur, Spain: March 16–19, 1999, which was the first altimeter calibration ever developed in Spain; July 4–7, 2000, and August 25–28, 2002. Direct absolute altimeter calibration, estimating the TOPEX Alt-B bias, was made from direct overflights using GPS buoys. In the framework of the Jason-1 mission, a campaign was conducted on June 9–17, 2003, at the absolute calibration site on the Island of Ibiza. The objective was to determine the local marine geoid slope under the ascending (187) and descending (248) Jason-1 ground tracks, in order to allow a better extrapolation of the open-ocean altimetric data to on-shore tide gauge locations, and thereby improve the overall precision of the calibration process (Martinez-Benjamin et al. 2004).

11.1.3
Absolute Versus Relative Calibration

The ability to sample varied geographically correlated errors and characterize them in an absolute sense is a significant benefit of a well-distributed set of calibration sites. There is no doubt, however, that the "calibration task" requires a multifaceted approach, including both in situ calibration sites and global studies using the tide gauge network. The two techniques are therefore considered complementary and fundamental to altimeter missions.

Current estimates of regional and global change in mean sea level are only possible with careful and ongoing calibration of altimeter missions. Cross-calibration of future altimeter missions will remain essential for continued sea level studies. Calibration of Jason-1 with respect to T/P has been partially simplified by the simultaneous operation of both spacecraft in a formation flight configuration. The complexities associated with operating satellite equipment dictates this may not be the case for future follow-on missions. In this eventuality, relative calibration will rely heavily on absolute calibration sites. In situ calibration sites will also remain imperative for the verification of geographically correlated errors. Estimates of altimeter drift will require continued operation of global tide gauge networks and complementary calibration techniques, underscoring the need to maintain tide gauge networks into the future, and to monitor the vertical position of each tide gauge site.

11.1.4
Outline

This chapter focuses on absolute in situ calibration of the altimetric measurement system – the available methodologies, instrumentation, dedicated sites and salient results. Following the overview previously presented, Sect. 11.2 provides a review of the geometry associated with the calibration task and discusses the available instrumentation that can be utilized to monitor the different parameters involved in the derivation of the altimeter sea-surface height (SSH). Section 11.3 presents a detailed review of the three dedicated absolute calibration sites at Harvest, Corsica and Bass Strait respectively. We highlight on one hand the benefits of adopting common analysis standards, and on the other hand, the importance of adapting a site-specific methodology to better respond to local conditions and situations that are often unique to each calibration site. The final section (Sect. 11.4) discusses many of the salient results that have emerged from in situ calibration experiments, highlighting the importance of the technique to detect time-dependent changes in the altimetric measurement system, and changes in the absolute bias from one mission to the next.

11.2
Calibration Geometry and Measurements

11.2.1
Absolute Calibration Techniques

The traditional "overhead" concept of in situ altimeter calibration involves the direct satellite overflight of a thoroughly instrumented experiment site (Fig. 11.2). It is essential that such a calibration site has some means of observing sea level in situ (using, for example, a conventional tide gauge, ocean mooring or GPS-equipped buoy) and subsequently tying the sea level estimates to a terrestrial reference frame comparable to the satellite altimeter. In an ideal situation, the experiment site is located on a repeating ground track (or better still a crossover of an ascending and descending altimeter pass), sufficiently out in the open-ocean to avoid contamination of either the altimeter or radiometer footprints by the land.

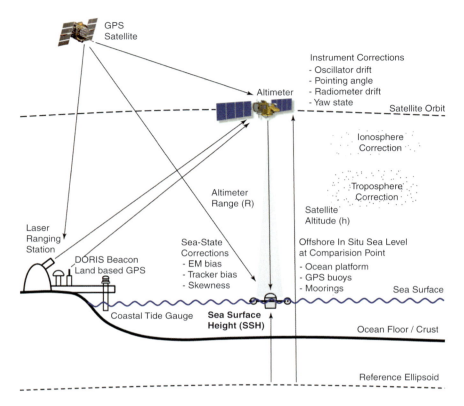

Fig. 11.2 Schematic diagram showing geometry of altimeter point calibration. At the instant the satellite passes overhead, geocentric sea-surface height (SSH) is observed independently by the altimeter and in situ measurement systems. The difference, assuming the in situ systems are properly calibrated, represents the "absolute altimeter bias"

The potential for a number of geographically correlated errors within the altimeter system underscores the need to undertake calibration experiments at a range of well-distributed calibration sites. This also ensures diverse sampling of ocean and atmospheric conditions, and promotes the use of different methodologies and processing software to help isolate systematic errors in any of the geodetic techniques involved.

The determination of absolute altimeter bias requires the in situ measurement of SSH in a comparable terrestrial reference frame at some chosen comparison point ($SSH_{ComparisonPoint}$). The absolute bias of the altimeter ($Bias_{Alt}$) may be determined using the simple relationship:

$$Bias_{Alt} = SSH_{Alt} - SSH_{ComparisonPoint} \tag{11.1}$$

where SSH_{Alt} is the altimeter derived SSH estimate:

$$SSH_{Alt} = h - (R + \Delta R_{DRY} + \Delta R_{WET} + \Delta R_{ION} + \Delta R_{SSB}) \tag{11.2}$$

where:

h is the height of the center of mass of the satellite above the reference ellipsoid, determined from the precise orbit determination (POD) analysis.

R is the nadir range from the center of mass of the satellite to the sea-surface, corrected for instrument effects.

ΔR_{DRY} is the atmospheric refraction range delay caused by the dry gas component of the troposphere.

ΔR_{WET} is the atmospheric refraction range delay caused by the water vapor and cloud liquid water content of the troposphere.

ΔR_{ION} is the atmospheric refraction range delay caused by the free electron content of the ionosphere.

ΔR_{SSB} is the range correction caused by the interaction of the large radar footprint and the sea-surface. The "sea state bias" (SSB) is a combined correction referring collectively to the electromagnetic (EM), skewness and tracker biases.

A negative bias is therefore indicative of SSH being measured too low by the altimeter (i.e., the altimeter range is too long or the orbit biased downwards).

Two distinct methodologies exist for the measurement of in situ SSH at the comparison point (i.e., $SSH_{ComparisonPoint}$); the techniques and underlying algorithms are quite disparate depending on the particular application and will not be developed here, but must generally consider – either directly or indirectly – geophysical, oceanographic and atmospheric phenomena that cause the variation in sea level over time.

- *Direct measurement*: In this case, SSH is physically observed at the comparison point (see "Offshore in situ sea level" on Fig. 11.2). In the case of the NASA calibration site at Harvest (Haines et al. 2003), the platform itself (with associated geodetic and sea level instrumentation) is located at the comparison point, allowing the direct estimation of SSH for each overflight. Studies utilizing solely GPS-equipped buoys are other examples of the direct calibration methodology.
- *Indirect measurement*: In this case, the SSH measurement involves the observation of sea level away from the comparison point, typically using a tide gauge at nearby (typically coastal) location (Fig. 11.2). The remote SSH is then "transferred" or "extrapolated" offshore through the use of precise regional geoid models, and in many cases, numerical tide models. Examples include the CNES calibration site (Bonnefond et al. 2003a), the United Kingdom project (Woodworth et al. 2004), and the Greek GAVDOS project (Pavlis et al. 2004). The indirect technique provides logistical advantages whilst maintaining the ability to determine cycle-by-cycle estimates of absolute bias. The accuracy of the SSH transfer technique (i.e., the accuracy of the geoid and tidal models) is the limiting factor for this methodology. One must also take into account differential effects of tides and atmospheric pressure in the error budget (see the chapter by Woodworth et al., this volume). The magnitude of these effects depends not only on local conditions (e.g., shape of the coast, bathymetry), but also the distance between the location of the in situ sea level observation and the comparison point. The Australian calibration site (Watson et al. 2004) uses a combined approach employing both GPS buoys and an offshore oceanographic mooring to provide cycle-by-cycle estimates directly at the comparison point. Outside of the intensive calibration phases of the respective missions (i.e., when the mooring is deployed), a coastal tide gauge is used "indirectly," combining it with an observed tidal difference (that may be predicted at any time using standard tidal prediction routines) to extrapolate to the sea surface height to the comparison point. The various

in situ measurements introduced above for both calibration methodologies are discussed in the sections below.

11.2.2
Formation Flying and Tandem Mission Verification

The tandem verification mission (also named the Formation Flight Phase) between Jason-1 and TOPEX/Poseidon provided an important and unique opportunity to make quasi-direct comparisons of the whole altimetric measurement system thanks to a very short time separation between both satellites (~1 min). This method allows comparison of all geophysical corrections and corrected sea-surface height. It also enables estimation of relative bias and drift, along with a characterization of the specific contributions of all underlying parameters. During T/P–Jason-1 verification phase, unexpected differences between Jason-1 and T/P were discovered. These differences were intensively investigated and led to substantial improvements of the altimeter data products of the two missions. In particular, the altimeter data processing has been improved (retracking, see the chapter by Gommenginger et al., this volume) for both missions and large improvements have also been brought to the precise orbit determination (including gravity fields from GRACE and new standards) and to geophysical corrections. Particular studies have been dedicated to the reduction of the geographically correlated differences between T/P and Jason-1 notably to better understand differences between global and regional scales, which can help coastal studies. The same configuration has been chosen for Jason-1 and Jason-2 and some preliminary results are presented in Sect. 11.4.2.

11.2.3
In situ Measurements

11.2.3.1
Sea Surface Heights

A range of techniques and instrumentation is available for the observation of in situ sea-surface height. Each technique has its advantages and disadvantages with respect to sampling, cost and operational duration.

Tide Gauges. Usually placed on piers within coastal harbors or working port facilities, they measure the sea level relative to a nearby geodetic benchmark. They are typically divided into two categories (Pugh 1987): those measuring the water level (1) above the sensor (for example, a submerged pressure sensor), and (2) below the sensor (for example, gauges that use a float or the propagation of an acoustic, radar or laser signal to measure water level). Given that the gauge is fixed to the crust, in some cases on a potentially dynamic structure, it is very important to continuously (or at least periodically) monitor their vertical reference. Moreover, as discussed in the Sect. 11.2.5, the benchmark must be tied to a reference frame which is homogeneous with that used in the orbit determination process of the satellite altimeter. This monitoring must also be extended to observe any instrumental drift within the tide gauge (often caused by temporal changes to the performance of the sensor) as well as any localized movement between the tide gauge and other inland geodetic monuments. Regular instrumental calibration or comparison with independent measurements is therefore often important (e.g., Bonnefond et al. 2003a; and Watson et al. 2008). As an example, the accuracy of the AANDERAA pressure tide gauges installed at Senetosa Cape (see Sect. 11.3.3) has been estimated from the cycle-by-cycle independent determination of the altimeter bias (over more than 200 cycles for the 3 historical tide gauges): results show that the error on the absolute sea level height is close to 2 mm with a scatter of 8 mm. Finally, the integration times, as well as the sampling times, have to be defined carefully to take into account local sea state dependencies (e.g., Haines et al. 2003).

GPS Buoys. A promising alternative technique that is used for point location calibration studies involves instrumenting a floating platform with a global positioning system (GPS) receiver. Such "GPS buoys" enable the measurement of sea level in an absolute reference frame on an epoch-by-epoch basis, with the buoy fixed or left to drift freely. By deploying a buoy along an altimeter ground track, direct comparison can be made to determine the bias in altimeter estimates of sea-surface height. Using data from a GPS-equipped spar buoy near Harvest, Born et al. (1994) first computed a TOPEX single-pass bias estimate that agreed with the corresponding figure from the platform tide gauge at the 1-cm level. More recently, a number of investigators have utilized this technique for calibration studies, employing both small "wave-rider" designs in addition to larger more permanent oceanic buoy structures (see Watson 2005, for review). The use of GPS buoys offer the advantage of portability and flexible deployment locations, limited in terms of accuracy by proximity to land based reference stations. Ideally, GPS buoy deployments should be carried out within 20–30 km of a coastal fiducial site that is directly part of or processed within the terrestrial reference frame. In these cases, filtered estimates of sea surface height can be achieved at the 1–2 cm level with respect to the defined reference frame (e.g., Watson et al. 2004). Advances in GPS technology have also enabled routine, accurate and timely positioning even for isolated, roving GPS receivers (e.g., Zumberge et al. 1998; Bar-Sever 2008).

At the Corsica calibration site, a wave-rider GPS buoy (originally designed by CCAR at the University of Colorado) is deployed for approximately 1 h centered on the time of satellite overflight, 10 km offshore on the satellite ground track. The epoch-by-epoch (1 Hz) estimates of sea-surface height deduced from the kinematic processing of the GPS data are filtered using a low-pass filter (Vondrak 1977) with a cutoff period of 5 min and compared to the altimeter data. The cutoff period was chosen to minimize the impact of waves and to be comparable to the tide gauge time sampling. The GPS buoy is also placed at each tide gauge location for a period of at least 30 min before and after the overflight. This ensures there will be enough data to support a statistically meaningful comparison between the two in situ data types (tide gauge and buoy). The cross-comparison between instruments provides an important validation of the in situ measurement systems: which has often resulted in the discovery of tide gauge related errors (Bonnefond et al. 2003a). Apart from the use of GPS buoys for direct sea level comparisons, other experiments have been conducted to map the sea surface (see Sect. 11.3.3 for more details).

GPS buoys are also used frequently at the Bass Strait calibration site, for both cycle-by-cycle comparison with the altimeter but more importantly, to define the absolute datum of oceanographic moorings (see below). Wave-rider buoys used at this site have undergone various developments since first used in 2001 following the launch of Jason-1. The latest design raises the antenna above water level whilst minimizing tilt in order to prevent loss

Fig. 11.3 Mk II (*left*) and Mk III (*right*) GPS buoys from the Bass Strait calibration site

of lock caused by breaking waves as experienced with the earlier design (Fig. 11.3). The Mk III buoys have been utilized since 2008 and throughout the Jason-2 calibration phase, undergoing repeated deployments for durations of up to 10 h. The extended deployment duration assists in reducing the uncertainty associated with the solution of the mooring datum with respect to the terrestrial reference frame.

Moorings. The relatively long time series of precise SSH derived from the GPS buoys at the Bass Strait calibration site are subsequently used to constrain the datum of a pressure gauge, which forms part of an oceanographic mooring deployed at the same location. Mooring deployments offer continuous observation of relative sea level and hence require this connection to an absolute datum provided by the GPS buoys. Variations in the density of the water column must be observed as part of the mooring array in order to rigorously convert the bottom pressure observations into sea-surface heights (see Fig. 11.4). Placing additional moorings along track, Watson et al. (2004) were able to quantify very small along and cross-track changes in dynamic height of the ocean which ensures the accuracy of the extrapolation of coastal tide gauge data out to the comparison point outside the times the moorings were deployed. A similar strategy has been adopted for the intensive calibration phase of the Jason-2 mission at this site.

Testut et al. (2005) provides another example of a mooring installed for cal/val activities, this time near the entrance of the Bay of Kerguelen Island, directly under a Jason-1 track. An additional mooring will be installed south of the Island in order to estimate local oceanographic differences between Port-aux-Français (recorded by tide gauges) and the entrance of the bay and the open sea.

GNSS Reflectometry. The use of GNSS (Global Navigation Satellite Systems) signals reflecting off the sea surface can also be considered a potential new technique for altimeter calibration but its merits or drawbacks have still to be precisely defined. In fact, with the development of the future European constellation GALILEO in complement to the current GPS or GLONASS constellations, the potential of this technique will increase.

The principle is the use of the differenced time delay between direct and reflected GNSS signals tracked by a dedicated receiver in order to determine information about the sea-surface height.

Several experiments have been performed with a receiver located on the ground (Treuhaft et al. 2001), in an airplane (Lowe et al. 2002), in a balloon, and even on the Harvest platform (Treuhaft et al. 2003). However, this technique remains experimental and will need improved accuracies, especially over the open ocean, to be viable for most cal/val activities.

11 In situ Absolute Calibration and Validation: A Link from Coastal to Open-Ocean Altimetry

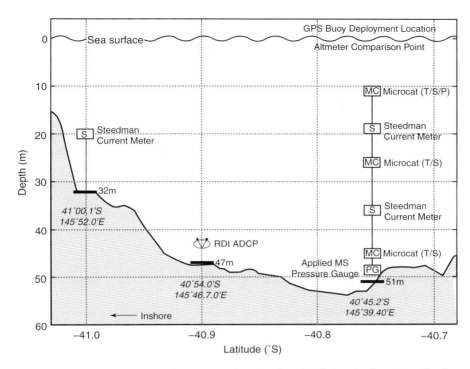

Fig. 11.4 Mooring arrays used at the Bass Strait calibration site during the Jason-1 calibration phase. The moorings are placed along the altimeter ground track from Burnie (tide gauge location) to the offshore comparison point, assisting in the determination of sea surface height at the comparison point, in addition to the along and cross-track currents inshore to Burnie. These currents were required to assist in the extrapolation of sea surface height from Burnie to the comparison point, outside of the period of operation of the mooring. (Adapted from Watson et al. 2004)

11.2.3.2
Significant Wave Height

The high frequency GPS buoy measurements (1 Hz) also provide the sea height variations at the buoy location caused by waves (Hein et al. 1990; Bonnefond et al. 2003a, b). The standard deviation on the GPS buoy sea height residuals (σ_{shr}) is the root square sum of σ_{gps} and σ_{wave} (where σ_{wave} is the standard deviation of GPS buoy measurements due to waves and σ_{gps} the internal error of GPS buoy measurements):

$$\sigma_{shr}^2 = \sigma_{gps}^2 + \sigma_{wave}^2 \; ; \; so, \sigma_{wave} = \sqrt{(\sigma_{shr}^2 - \sigma_{gps}^2)}.$$ SWH (or H1/3) is then deduced from the formula below:

$$SWH_{buoy} = 4 \cdot \sigma_{wave}$$

where SWH_{buoy} is the resulting significant wave height deduced from GPS buoy measurements that is compared to altimetric data.

In addition to GPS buoys, wave buoys and wave recorders (such as those used at Gavdos) can provide measurements of SWH. The scatter of high-rate data from tide gauges also reveals SWH.

11.2.3.3
Range Comparison

The calibration of the altimeter satellites can also be undertaken using transponders (Powell 1986; MacDoran and Born 1991; Cristea and Moore 2007). Instead of measuring an in situ sea level that will be compared to an altimetric measurement of the same quantity, this technique uses a transponder on ground that will receive a signal from the altimeter, amplify it and then send it back to the satellite. The transponder backscatter echo will then appear as a strong echo in the satellite received waveforms and the distance between the radar altimeter and the transponder can then be estimated.

The error budget relative to the altimetric part has to be considered for measurements on land. Moreover, additional error sources have to be taken into account:

- GPS positioning of the transponder.
- Precise leveling of the reflectors position.
- Signal processing: Techniques for the detection of the transponder echo in raw data.
- Estimation of the path delays caused by the propagation in the atmosphere: the most critical parameter is the water vapor. Correction can be measured or modeled but corrections on land are not optimal for the moment.
- The quality of the backscattered signal depending of the amplification factor of the transponder.

Even if the transponder technique introduces other potential errors, it eliminates other error sources such as tides, sea state bias and geoid extrapolation. If the potential of the technique is fully realized, the distance between the radar altimeter and the transponder can be measured with a resolution better than 1 cm (Cristea and Moore 2007).

11.2.3.4
Wet Tropospheric Correction

The troposphere causes a delay in the propagation time of the altimetric radar wave due to the presence of water vapor and cloud liquid water in the atmosphere. Corrections for this wet component of the troposphere can be determined using passive measurements from onboard radiometer measurements or by the using meteorological parameters from models. The wet troposphere correction is an important source of uncertainty for the coastal studies, because of the sensitivity of the correction to land contamination (see the chapter by Obligis et al., this volume). The wet tropospheric path delay can also be measured on the ground (using GPS or upward looking radiometers) and then compared to the one derived from the onboard radiometer.

Using GPS data from a geodetic reference point close to the ground track (e.g., the CGPS site at Senetosa for the Corsica experiment, or GPS receiver on the Harvest platform), the total zenith troposphere path delay is computed (e.g., Haines and Bar-Sever 1998). This radiometric technique relies on the simultaneous observations from a single station to multiple GPS satellites. The geometrically diverse paths traversed by the signals through the atmosphere enable estimation of a total delay in the zenith direction. An a priori mapping function (e.g., Niell 1996) is needed to provide a model for the variation of the signal as a function mainly of elevation. By using the meteorological parameters from a local weather station, or from a weather model, the dry component can be fixed a priori, or subtracted from the total troposphere path delay derived from GPS. The resulting wet troposphere path delay is then computed by interpolation at the time of closest approach and compared to the reading from the satellites radiometer. A significant advantage of this technique is that the large (and increasing) numbers of geodetic-quality GPS stations at coastal and island location can serve as ad hoc calibration sites for spaceborne radiometers (Desai and Haines 2004).

Ground truth measurements for wet troposphere can also be obtained from dedicated uplooking Water Vapor Radiometers (Ruf et al. 1994; Brown et al. 2007). To support the T/P cal/val effort, WVRs were deployed on Lampedusa Island (Italy), Chichi Jima (Japan), Norfolk Island (Australia) and on the Harvest platform. Intercomparisons with the space borne radiometers can be performed using both path delay and brightness temperature readings. Harvest continues to host a WVR on a periodic basis to support calibration of the radiometers of Jason-1 and Jason-2 (JMR and AMR respectively).

11.2.3.5
Local Weather Conditions

Local weather stations provide an important contribution to in situ calibration experiments. Not only is atmospheric pressure required in some tide gauge reductions, wind speed and direction can also be used to characterize localized ocean conditions. Used in conjunction with regional meteorological models or reanalysis products, observed pressure and wind observations can assist with the prediction of dynamic sea surface changes, such as wind set-up relative to an offshore comparison point (see Watson et al. 2004).

11.2.4
Geophysical Corrections

A common geodetic reference surface is required in order to make reliable and meaningful comparison between in situ and altimeter observations of SSH. Thus various geographical and instrumental corrections are required to compensate for various well-known geophysical phenomena. Corrections to the range can be divided into two categories (except the purely instrumental ones): (1) those physically affecting the sea level (ocean loading, solid/permanent tides, pole tide), and (2) those affecting the distance measured by the altimeter (ionospheric path delay, dry tropospheric path delay, wet tropospheric path delay and sea state bias). Corrections for the former affects are computed from models generally based on IERS conventions (McCarthy and Petit 2004). The latter corrections are more specific to the spacecraft payload (dual-frequency altimeter, onboard radiometer) or range measurement process (retracking). Except for the wet tropospheric correction, for which we will provide in situ comparisons, we will not further discuss these corrections throughout this chapter while it is the object of the following other chapters in this book:

- "Range corrections near coasts" by Andersen et al.
- "Tropospheric corrections for coastal altimetry" by Obligis et al.
- "Retracking altimeter waveforms near coasts" by Gommenginger et al.

11.2.5
Reference Frames

The absolute calibration process involves two techniques applied within the same reference frame to measure the instantaneous mean sea-surface height: (1) radar altimetry and orbit determination; and (2) terrestrial and space geodesy for tide gauges and/or GPS buoys positioning.

Over the last 40 years, different techniques have been developed to measure the position, velocity, and acceleration of objects orbiting the Earth: Doppler and laser tracking of satellites, very long baseline interferometry (VLBI) astrometry, and later, Global Positioning System (GPS) tracking. These techniques have been implemented by major space geodetic projects such as space oceanography and gravity missions, ensuring today, high levels of accuracy and repeated spatial and temporal coverage of oceans, continents and atmosphere. As a result, the International Terrestrial Reference Frame (ITRF) provides a set of station positions and linear velocities. It is produced by the combination of the four global geodetic technique results, DORIS, GPS, SLR, and VLBI. This reference frame – the latest being ITRF 2005 (Altamimi et al. 2007) – is used for orbit determination on many of the altimetric missions. It is also very important to tie the local instruments (tide gauges, GPS buoys, oceanographic moorings, transponders, etc.) to the same frame. This is often realized by using local GPS receivers to define a reference geodetic marker from which other instruments are tied using leveling or relative GPS positioning. However, using other space geodesy techniques such as Satellite Laser Ranging can be very useful to detect systematic errors between observational systems. To that end, the French Transportable laser Ranging System has been used during dedicated campaigns at three different calibration sites: Corsica in 2002, 2005, and 2008 (Gourine et al. 2008), Gavdos in 2004 (Pavlis et al. 2004), and Bass Strait in 2007. Results of these different campaigns show consistency at the 1 cm level or better in the up component between GPS-based and SLR-based coordinates. However, absolute positioning remains a significant challenge given the presence of many unresolved systematic errors within GPS in the measurement system (for example, near field multipath, aliasing of sub-daily signals and biases in antenna phase center location) and also SLR (e.g., range biases) (Rutkowska and Noomen 1998; Exertier et al. 2006).

Despite these challenges, collocation of each of these techniques provides one possible means to identify and mitigate systematic error. With respect to altimeter cal/val, it is often impossible to set-up a permanent SLR site close to the calibration experiment, and GPS remains the preferred solution (in term of cost and precision) to insure a continuous monitoring of the vertical reference. To estimate measurement system drift at the 1 mm/year level requires a minimum of 5 years of data, as well as tight control over potential systematic errors, between the GPS and sea level instrumentation, such as unmodeled subsidence or uplift of the tide gauges. Harvest is one of the longest operating calibration sites with the potential to support this level of accuracy. One of the challenges at Harvest is measuring the platform subsidence (Fig. 11.5) – estimated at close to 6 mm/year – with an accuracy of 1 mm/year from ongoing GPS measurements (Haines et al. 2003).

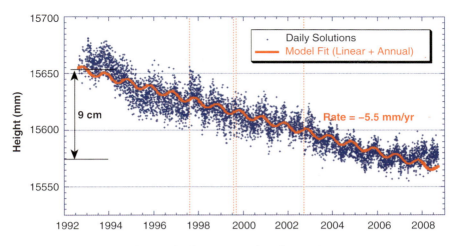

Fig. 11.5 Harvest platform height monitoring 16 years of continuous GPS

11.2.6
Orbit

Significant improvements to orbit determination have been realized due in large measure to the intensive efforts dedicated to T/P and Jason-1 (e.g., Chelton et al. 2001; Luthcke et al. 2003; Haines et al. 2004). Other missions are in the process of minimizing the orbit error part in the global altimetric error budget. However, from a local point of view (as afforded by absolute calibration sites), Geographically Correlated Errors (GCE) in the orbit estimate continue to be a non negligible error source: achieving 1 cm accuracy is still a challenge. Some improvements realized by processing the tracking data using specialized techniques (e.g., laser-based short-arc approach (Bonnefond et al. 1995, 1999) or GPS-reduced-dynamic tracking (Haines et al. 2004) can be exploited to better separate the error sources in the closure equation (Bonnefond et al. 2003a). These techniques can also help detect potential problems with the orbit processing (see Sect. 11.4.1.1) and monitor the orbit quality during the whole satellite mission.

11.3
Dedicated Calibration Sites

11.3.1
Overview

Three dedicated absolute in situ calibration sites have operated in various capacities since the launch of TOPEX/Poseidon in 1992. In conjunction with, a number of regional and episodic sites operated by other research groups, the facilities at Harvest, Corsica and Bass Strait (Fig. 11.6) continue to monitor the absolute biases of the Jason-2 mission, offering new insights into the performance of satellite altimetry. Throughout this section, we

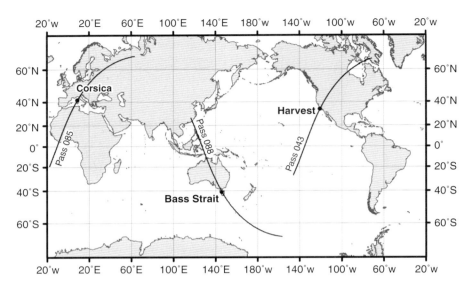

Fig. 11.6 Locations of the three dedicated cal/val sites

provide a brief description of the three dedicated sites followed by a summation of the similarities and differences between sites. Readers are referred to Haines et al. (2003), Bonnefond et al. (2003a), and Watson et al. (2004) for specific details on each calibration site.

Harvest, Corsica and the Bass Strait sites provide a range of calibration metrics computed using a diverse range of techniques but with a common set of analysis procedures. The Harvest site uses a purely direct approach as described in Sect. 11.2.1. Corsica and Bass Strait combine direct and indirect methodologies each with specific advantages and disadvantages as discussed throughout the following sections.

11.3.2
Harvest

The Harvest Oil Platform is located in the Pacific Ocean about 10 km off the coast of central California near Pt. Conception (Fig. 11.7a). Currently owned and operated by Plains Exploration and Production (PXP), the 30,000 t platform is anchored to the sea floor and sits in about 200 m of water near the western entrance to the Santa Barbara Channel (Fig. 11.7b). Harvest was installed by Texaco in 1985 and has been in production since 1991, withdrawing oil and gas from the underlying Pt Arguello Field.

In the search for a dedicated calibration site to support the T/P mission, the potential of Harvest was recognized early on (Christensen et al. 1994; Morris et al. 1995). The platform offers a number of important advantages as a spaceborne altimeter calibration site (e.g., Chelton et al. 2001). First, Harvest is located sufficiently far offshore so that the area illuminated by the altimeter is covered entirely by ocean when the satellite is directly

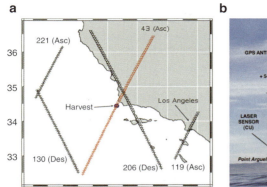 

Fig. 11.7 (**a**) Location of PXP Harvest platform and (**b**) PXP Harvest platform with locations of in situ calibration systems

overhead, offering the ability to observe altimeter biases directly (Sect. 11.2.1). At the same time, the platform itself is too small to significantly influence the reflected radar signal. In addition, conditions at Harvest are typical of the open ocean, with significant wave heights averaging 2–3 m. This implies that the altimeter measurement systems are monitored for ocean conditions under which they are designed to best operate. Finally, the platform is located in the vicinity of important tracking stations in California and the western U.S, data from which contribute to measuring the geocentric positions of the altimeter satellites and platform through space-based surveying techniques. Its location off California's central coast is also appealing from a logistical standpoint, since the experiment is managed from JPL in Pasadena, California. Helicopter flights are scheduled from Santa Maria, California, airport to Harvest (and neighboring platforms) each day, providing ready access to the platform on a space-available basis.

In August 1992, the T/P mission was launched into a 10-d repeating orbit that took it directly over Harvest and a companion site, managed by CNES, on Lampione (Lampedusa) Island. Data from Harvest were used to continually monitor T/P measurements for 10 years until the satellite was moved to an interleaving orbit in 2002. Following in the footsteps of T/P, the Jason-1 (2001–) and OSTM/Jason-2 (2008–) missions have continued to pass over the platform, implying that the Harvest calibration record has grown to over 16 years. Due to its long, continuous, homogenous and accurate record of monitoring, Harvest has emerged as a crucial international resource for the study of sea level from space, and it will continue to serve a vital role in validating data from precise space borne radar altimeter systems.

The Harvest experiment features carefully designed collocations of in situ measurement systems to support the absolute calibration of the altimetric SSH and ancillary measurements. Operated by NOAA, the primary water-level sensors are dual redundant Nitrogen Bubbler (Digibub) systems, each of them equipped with Paroscientific pressure transducers (Gill et al. 1995). The University of Colorado (CU) maintains an experimental water-level measurement system, which uses optical (lidar) ranging technology. The lidar system has been particularly valuable for exposing sea state dependencies in the primary

Digibub time series. Harvest also hosts one of the oldest, continuously operating GPS stations in the global geodetic (IGS) network. GPS data have been collected continuously since 1992, and show that the platform is subsiding by ~6 mm/year due to the extraction of oil and other fluids from the underlying Pt. Arguello field (Fig. 11.5). The GPS data are also invaluable for computing and calibrating the wet path delay measurements (Haines and Bar-Sever 1998). Additional systems at Harvest include a meteorological package, local server, and satellite communication systems (VSAT and GOES). A water vapor radiometer has also been deployed episodically to support more comprehensive calibrations of the space borne radiometer data (e.g., Ruf et al. 1994).

While the most conspicuous results from the Harvest experiment are the SSH bias determinations, data from the long-term occupation of the platform have lent insight on many signals of geodetic, oceanographic and environmental interest (Haines et al. 2003). The wealth of information from the Harvest experiment underscores the unique contributions of a dedicated, well instrumented and continuously maintained calibration site.

11.3.3
Corsica

The Corsica site that refers collectively to instrumentation and facilities located at Ajaccio-Aspretto, Senetosa Cape, and Capraia (Italy) in the western Mediterranean was selected to maximize the ability to provide absolute calibration of a range of satellite altimeters (see Fig. 11.8 respective sites and satellite ground tracks). Thanks to the French Transportable Laser Ranging System (FTLRS) for accurate orbit determination, and to various geodetic measurements of the land and sea level, the primary objective is to estimate altimeter biases and associated drifts. At the same time, the site offers an opportunity to contribute to the orbit tracking of oceanographic and geodetic satellites and to enable the analysis of the different error sources that affect altimetry and space geodesy. From a geophysical perspective, data from the site also contributes to the decorrelation of possible vertical displacements of the site (near and far field tectonic processes related to the Earth's crust) and variations in mean sea level.

The primary facilities for the Corsica calibration site have been operational at Senetosa since 1998 to monitor the TOPEX/Poseidon and Jason-1 missions. The site was expanded to facilitate comparison at two additional comparison points, Ajaccio and Capraia, also adding the ability to monitor the Envisat satellite. This provides a unique opportunity to cross-calibrate all these disparate missions with common processes and standards. For example, the wet tropospheric path delays for comparison to the satellite radiometer estimates can be determined from GPS and ground-based met stations located at both Ajaccio and Senetosa. The close proximity of each site also provides economic and logistical advantages, such as the ability to use to the same GPS buoy to regularly perform independent calibrations at the various comparison points. An evolution of the "overhead" calibration methods (Sect. 11.2.1) to a regional approach has been also developed based on the extended Corsica site capabilities (Jan et al. 2004).

Since 1998, the Senetosa site has been equipped with 3 tide gauges, located on each side of T/P and Jason-1 ground track (M_3 on the east side and $M_{4/5}$ on the west side),

Fig. 11.8 General configuration of the Corsica calibration site. The coloured surfaces correspond to local geoids determined by using the Catamaran-GPS method described in Bonnefond et al. (2003b)

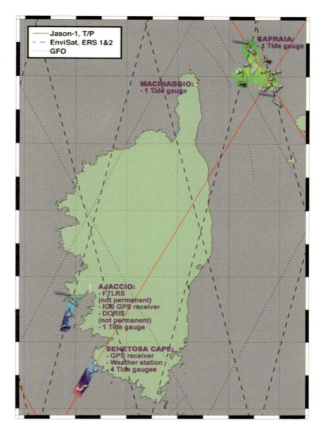

dedicated to the altimeter calibration process. Surveys of the geodetic markers and tide gauge locations have been undertaken regularly since 1998 and the repeatability of the GPS solutions and the optical leveling are below 1 cm and 5 mm, respectively. In June 2005, a fourth tide gauge (M_6) was installed close to M_3 in order to monitor long-term behavior of the other tide gauges. The sample rate at M_6 is set to 30 min, in order to enable longer autonomous operations: more than 1 year compared to few months with the 5 min sampling used for the others gauges. Data from M_6 are recovered by a diver every 6 months, and at different times than for the 3 tide gauges nominally used for calibration. These staggered servicing intervals allow better characterization of possible spurious effects of recovering and redeploying the gauges (as manifest, for example, in different time durations required for the pressure sensors to stabilize).

A weather station has been installed at the lighthouse since 2000, near the GPS reference point. The main goal of this station is to provide atmospheric pressure to correct the tide gauge measurements and to derive the dry component of the tropospheric correction. Atmospheric pressures from Ajaccio (~40 km North) and Figari (40 km East) provided by Météo-France are also used as back-up in case of local station outages. Other environmental parameters, such as wind speed, are used to compare against corresponding measurements from the satellites. For example, a study to compare wind speed was conducted during the T/P & Jason-1 Formation Flight Phase (2002) and showed good correlation (~80%) between the altimeter and in situ data. The differences, however, are difficult to interpret because the dominant winds at the weather station can be very different than those offshore, due mainly to the configuration of the coast configuration and the difference in altitude.

The most recent extension of this calibration experiment includes Capraia Island and features a tide gauge at Macinaggio. A dedicated GPS campaign was realized here in 2004 to measure the geoid slopes under the Jason-1, Envisat and GFO tracks around the island. In October 2005, a similar campaign was realized for the Envisat ground track close to Ajaccio.

The double site configuration (Senetosa and Ajaccio) of the Corsica calibration methodology has been derived to maximize logistics and multi-platform monitoring. The closest nominal Jason (1 & 2) and TOPEX/Poseidon ground track (Fig. 11.8) is ascending (number 85) and located at the southwest of the island (Cape Senetosa). Conversely, an installation at Aspretto's air and sea military base, located ~2 km from Ajaccio, provides an improved location for deploying the French Transportable Laser Ranging Station (FTLRS). This system requires greater logistical support for its episodic deployments. The FTLRS (CNES-IGN-CERGA) was operated in 2002, 2005, and 2008 on the Aspretto hill by OCA teams for a period of 6 months for each campaign. Being an exact and altimetry-independent tracking system, the primary goal of these deployments has been to install this highly transportable station near the calibration pass (within ~25 km of nadir) in order to assess the accuracy of the radial component of the orbit. The port at the Ajaccio site is relatively small with very little shipping traffic. A permanent GPS station (IGN) and an automatic tide gauge (SHOM) has been installed since 1999. This site is close to a descending ERS and Envisat ground track (number 130) and has been used to calibrate the Envisat altimeter since 2005. The various facilities that comprise the Corsica calibration site therefore minimize the challenging logistical and financial demands of operating a cycle-by-cycle calibration facility in a remote location.

Since 2000, a GPS buoy has also been used in the calibration process at Corsica: the GPS buoy is deployed for 1 h surrounding overflights (~10 km offshore) whenever sea state conditions are not too harsh to ensure safe navigation. Bonnefond et al. (2003a) provide an example of where both the direct and indirect methodologies are combined. GPS-equipped buoy deployments are made on an episodic basis allowing direct calibration. At the same time, the estimates of $SSH_{ComparisonPoint}$ are compared with SSH estimates taken at coastal tide gauge sites. Analysis of differences yields information on both the relative geoid and tidal phase and amplitude differences. In the case of the Senetosa site, the impact of the differential tides is generally negligible (also confirmed by model), but the geoid slope is strong (6 cm/km) and has to be applied even for relatively close distances (a few hundred meters). This information can then be applied using the indirect methodology to transfer the tide gauge SSH estimates to the comparison point, without the need to deploy the GPS buoy. Bonnefond et al. (2003b) include an additional step to compute a "map" of the sea surface topography (see Fig. 11.8) between the coastal tide gauges and surrounding the offshore altimeter comparison point by using a "Catamaran-GPS," which can then be used in the calibration process (for the Ajaccio and Capraia locations too).

11.3.4
Bass Strait

The Bass Strait calibration site (40° 39′S, 145° 36′E) has provided the first southern hemisphere contribution to the absolute calibration effort of T/P, Jason-1 and Jason-2. Unlike the Harvest and Corsica sites, the Bass Strait facility is located on a descending altimeter pass (pass 088), and utilizes a novel combination of the direct and indirect calibration methodologies. Bass Strait itself separates Tasmania from the Australian mainland, with the calibration site located on the south west side, near the city of Burnie (Fig. 11.9). Bass Strait is a shallow body of water with typical depths between 60 and 80 m (~51 m at the calibration site).

The first study undertaken at the site coincided with the cal/val objectives of the T/P altimeter in 1992 (see White et al. 1994). This initial study adopted an indirect methodology using data from a single coastal tide gauge augmented with land based GPS observations and a marine geoid computed from local gravity and altimeter data. This strategy allowed the estimation of absolute bias for T/P following its launch, but was limited by a lack of in situ data observed at or around the offshore comparison point.

In preparation for the launch of Jason-1, the Bass Strait site was revisited utilizing an updated and enhanced calibration methodology. To remove the uncertainty associated with estimating a marine geoid, an innovative approach to both the direct and indirect methodologies was developed (Watson et al. 2003). This approach subsequently demonstrated the value of the Bass Strait site by providing independent estimates of altimeter absolute bias, computed using a different methodology and processing strategy to the other dedicated calibration sites (Watson et al. 2004).

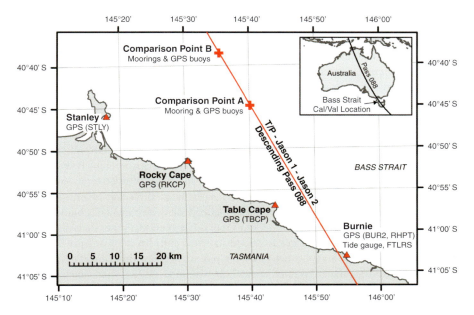

Fig. 11.9 The Bass Strait calibration site

To achieve a direct and truly geometric calibration on a cycle-by-cycle basis, in situ observations are clearly required at the offshore comparison point (Sect. 11.2.1). In Bass Strait, an oceanographic mooring was deployed at the comparison point (see point A, Fig. 11.9) to observe for the full calibration phase of the Jason-1 mission. The absolute datum of the relative sea level record obtained from the mooring (corrected for ocean density changes in the water column) was defined using several episodic GPS buoy deployments (see Watson 2005 for further detail). The reference frame for the GPS buoy observations was in turn realized through the use of a regional network of continuously operating GPS reference stations. This defined an absolute SSH time series at the comparison point ($SSH_{Comparison Point}$) suitable for direct comparison to the SSH_{Alt} in order to compute the altimeter bias (Eq. 11.1, Sect. 11.2.1).

To facilitate further comparison outside of the period of mooring operation, concurrent estimates of sea level observed from a coastal tide gauge were compared to the absolute mooring time series. The difference between these time series includes a mean offset to account for the datum separation between sensors (note the tide gauge data in this case has a local chart datum and is not connected to a geocentric reference frame). The remaining difference between the two series is dominated by a tidal signal caused by the ~40 km separation between instruments. Tidal analyses of this difference provides a means of predicting the sea surface height at the offshore comparison point at any given time when tide gauge data is available, and mooring data is not. As this tidal prediction includes the mean offset (i.e., datum offset), adding the predicted differences to the raw tide gauge data effectively transforms the tide gauge observations to the offshore comparison point without the need to estimate a geoid, and is limited only by any (small) non-tidal differences in SSH between the two locations. This refined indirect methodology has allowed the determination of absolute bias since the first cycles of T/P.

The use of episodic GPS buoy deployments to solve for the datum of the offshore mooring and pressure gauge, which is then used to compute the geometrical and tidal differences from coastal tide gauge data, provides an elegant geometrical solution to the calibration problem. The approach has removed the need to estimate a precise relative geoid, the source of greatest uncertainty in indirect calibration studies (including the original study at the Bass Strait site).

In the lead up to the launch of Jason-2, a further period of development and intensive observations were initiated at the site. The primary offshore comparison point used for previous studies was moved northwest along the descending pass to help eliminate land contamination within the satellite radiometer data (see comparison point B, Fig. 11.9). The calibration methodology remains centered on use of oceanographic moorings with high-accuracy pressure gauges and associated instrumentation for density measurement. The absolute datum of the moorings has been defined using repeated GPS buoy deployments with an extended 8–10 h duration. The French Transportable Laser Ranging System (FTLRS) was also located in Burnie for the period November 30, 2007 to April 30, 2008, allowing us to directly assess orbit quality in this region. For this campaign, oceanographic instrumentation was installed to coincide with the FTLRS period, and extends to cover the duration of the Jason-1/Jason-2 inter-calibration period. Outside of these times, the refined indirect technique is again utilized to extend the cycle-by-cycle estimates of altimeter bias.

As with the Harvest and Corsica sites, a number of additional and significant findings have emerged from the analyses undertaken with data from the Bass Strait facility. When comparing the solid Earth tide model used in the reduction of satellite altimeter data with that used in the GAMIT/GLOBK GPS analysis suite (used to define the reference frame for all Bass Strait data), a superseded model was found and corrected in the GPS software. This seemingly small difference was investigated by Watson et al. (2006) providing one of the first confirmations using real data of mismodeled or residual sub-daily geophysical signals aliasing or propagating into long-period signals as predicted by Penna and Stewart (2003). This work is now part of a body of literature (see also, for example, Penna et al. 2007; King et al. 2008) that helps us understand the complex nature of systematic error sources present within GPS analyses, and affecting all geophysical interpretations of such data, including absolute studies such as altimeter calibration. Given the multi-technique and absolute nature of the altimetry calibration process, it is clear that such investigations are central to refining not only satellite altimetry, but also the various other space geodetic techniques that are used for Earth observation.

11.3.5
Data Analysis Comparison

The analysis strategies adopted by each of the primary calibration sites (Harvest, Corsica and Bass Strait) are highlighted in Table 11.1. To the extent practical, uniform strategies are adopted and form the basis for estimating and comparing various calibration parameters. However, certain aspects of the strategies are unique to each site, reflecting particular aspects of the local environment, the geometry of the overflight and nearby lands, and the bias computation technique itself (e.g., direct or indirect).

Combination techniques (based on direct and indirect measurements, see Sect. 11.2.1), as described by Bonnefond et al. (2003b), improve the extrapolation technique and hence the accuracy of the $SSH_{ComparisonPoint}$ estimates. Techniques such as these are still limited by the short time series of measurements used to determine the extrapolation parameters of interest (geoid slope, tidal differences, etc.). Effects which may remain unresolved include aliasing from seasonal oceanographic signals and meteorological effects.

11.4
Salient Results

11.4.1
Detection of Instruments/Model Problems

The absolute calibration experiments have yielded valuable insights into various instrument performance and model accuracy issues that combine to manifest as absolute bias in altimeter SSH. Issues related to the microwave radiometer (both instrument specific and cause by land contamination) and orbit accuracy are discussed as examples below.

Table 11.1 Processing strategies for Corsica, Harvest, and Bass Strait calibration sites

Correction	Harvest	Corsica	Bass Strait
Iono.	Mean over −21 s to −1 s around the TCA (corresponds to the 140 km smoothing requirement)	Mean over −21 s to −1 s around the TCA (corresponds to the 140 km smoothing requirement)	Mean of all values between 39°48′S and 40°48′S, equivalent to the 140 km smoothing requirement
Dry tropo.	Linear fit over −5 s to +2 s around the TCA interpolated at the TCA	Linear fit (over −5 s to +2 s around the TCA) used for 1 and 20 Hz overflight data	Linear fit of 1 Hz data either side of the comparison point
Wet tropo.	Linear fit over −15 s to −5 s around the TCA interpolated at the TCA-5 s (avoid land contamination, 30 km)	Linear fit over −15 s to −5 s from TCA (avoid land contamination, 30 km). Last fit value used for all overflight data	Linear fit to all values between 39°54′S and 40°27′S, interpolated to the TCA
Sea State Bias	Cubic fit over −10 s to +1.1 s around the TCA interpolated at the TCA	Cubic fit over −10 s to −1 s from TCA used for all overflight data till TCA −1 s, TCA −1 s fit value after	Linear fit of 1 Hz data either side of the comparison point
Range Ku	Fifth-order polynomial (high-rate data) over −10 s to +1 s, evaluated at TCA	Averaged on the 20 km surface w/geoid correction from Catamaran-GPS surface	Linear fit of 1 Hz data either side of the comparison point
Geophysical corrections	Own corrections: Ocean loading[a] Solid/permanent tides[a] Pole tide[a] + (atmospheric loading, effect of temperature on the platform, ground water loading, ice loading, non-tidal ocean loading)	GDR corrections for: Ocean loading Solid/permanent tides Pole tide	Correction for the altimeter as supplied in the GDR IERS 2003 Solid Earth and pole tides, FES 2004 OTL, and atmospheric loading (non-tidal plus tidal) for GPS reference network
Tide gauge data	Linear fit from −1,100 to +1,100 s (~30 min of data) evaluated at TCA (6 min sampling)	Linear fit over 30 min centered on the TCA (5 min sampling)	Tide gauge: Linear fit of 6 min data to TCA Mooring: Linear fit of 5 min data to TCA
GPS Buoys		Sea level heights filtered with a cutoff period of 5 min	Weighted moving average with 25 min duration

TCA: time of closest approach
[a]Fully compliant with GDR ones

11.4.1.1
JMR and Orbit impact on Jason-1 GDR-A products

Early Jason-1 results from Harvest, Corsica and Bass Strait calibration sites helped identify a spurious drift in the absolute bias of the Jason-1 GDR-A data (Bonnefond et al., Haines et al., and Watson et al. in OSTST minutes 2004). The drift was linked to instabilities in the noise diodes of the JMR (Brown et al. 2007) and to the preliminary orbit solution. Provided in the following Table 11.2 are the results from four different strategies used:

- 1: Fully compliant with GDR-A
- 2: The POE orbit (CNES) is replaced by the GPS-reduced-dynamic solution (JPL)
- 3: Same as 2, but using the wet troposphere correction from the ECMWF model instead of the JMR
- 4: Same as 2, but using the wet troposphere correction from the local GPS data

The POE orbits (CNES) were compared to the GPS-based reduced-dynamic orbits (JPL, see Haines et al. 2004) from repeat cycles 1–88. The radial orbit differences were then mapped at 1°×1° resolution for each cycle and a drift was computed at each node for the whole period (Fig. 11.10). While the pass used for the Corsica calibration site (#085) is located in an area where the drift is close to zero, it is not the case for Harvest (#043) and Bass Strait (#088) were the effect reaches respectively –7 and –5 mm/year.

The wet path delays from JMR were compared to those derived from GPS data from Harvest and Corsica sites. Fig. 11.11 clearly shows that results from both sites give evidence of a common drift (compared to GPS) at the level of 4-5 mm/year. When using the wet troposphere correction from the local GPS data in the calibration exercise, the drift in the altimeter bias time series is reduced by this amount (strategy 4 compared to 2 in Table 11.2).

The results presented in Table 11.2 clearly show the impact of geographically correlated drifts in the GDR orbit (POE), and in the wet path delay due to instabilities in the JMR measurements. After replacing the POE height measurements and JMR path delay corrections with improved values, the drift estimates from the individual sites are not statistically distinguishable from zero (strategy 3 and 4 in Table 11.2). This reinforces the importance of having dedicated instruments and processes to monitor all the parameters in the altimetric measurement system. It also underscores the benefit of a well-distributed set of calibration sites.

Table 11.2 Differences in terms of Jason-1 bias and drift using different strategies (described in the text) for Corsica, Harvest, and Bass Strait calibration sites

S[a]	Corsica		Harvest		Bass Strait	
	Drift (mm/year)	Mean bias (mm)	Drift (mm/year)	Mean bias (mm)	Drift (mm/year)	Mean bias (mm)
1	–13	99	–16	119	–8	144
2	–11	103	–9	122	–4	128
3	–5	122	0	120	–	–
4	–6	124	–4	120	–	–

[a]Strategies

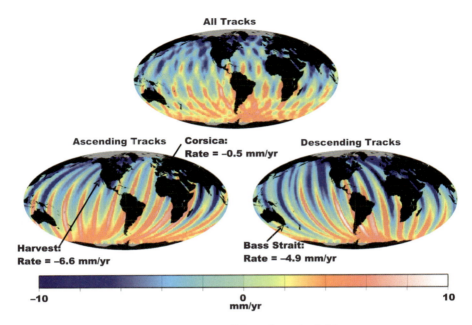

Fig. 11.10 Jason-1 GDR-A Orbit – JPL GPS: radial rate for cycles 1–90

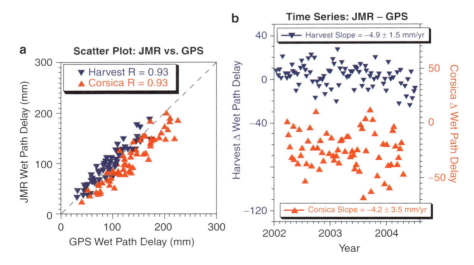

Fig. 11.11 Estimation of the wet path delay drift due to JMR calibration coefficient instabilities (in GDR-A products) using comparison with GPS for the Harvest (*blue triangles*) and Corsica (*red triangles*) calibration sites: (**a**) scatter plot and (**b**) time series with the estimated drifts

11.4.1.2
Radiometer Behavior Close to the Coast

The wet tropospheric correction (path delay with a negative sign) is an important source of geographically correlated bias. Indeed, it is mainly linked to radiometer land contamination, with differences existing between calibration sites depending on the distance from the coast and the orientation of the satellite approach to the land. For example, at the Corsica site for the calibration pass #85 (Fig. 11.8) the satellite first overflies Sardinia and then enters over a channel where the maximum distance to the coast is about 40 km. Fig. 11.12 and 11.13 show that the radiometer behavior is different for T/P (TMR) and Jason-1 (JMR) when flying over Sardinia and crossing the Sardinia-Corsica channel. The TMR contamination seems to not really affect the wet path delay corrections, but most of the data are missing over Sardinia. For Jason-1, the JMR behavior is more complex: the correction is decreasing (in absolute value) when approaching the coast (Sardinia and then Corsica) and increasing when moving away from coast. This effect implies that in the area where the correction is interpolated, its value is stable for T/P (Fig. 11.12, green line) while for Jason-1 (Fig. 11.12, blue line) it will be overestimated by 10–15 mm (even using classical criteria that suggest maintaining data no nearer than 30 km from the coast). On Fig. 11.13, the (c) and (d) maps show the differences between wet path delay data from both radiometers with respect to the ECMWF model

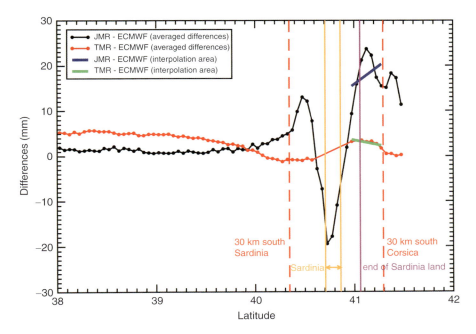

Fig. 11.12 Comparison of averaged differences (every 5 km) between wet tropospheric correction from radiometers and ECMWF model for Jason-1 (JMR, in *black*, cycle 1–144) and T/P (TMR, in *red*, cycle 221–364) at Senetosa (Corsica calibration site). "End of Sardinia land line" corresponds to the end of the small Asinara Island

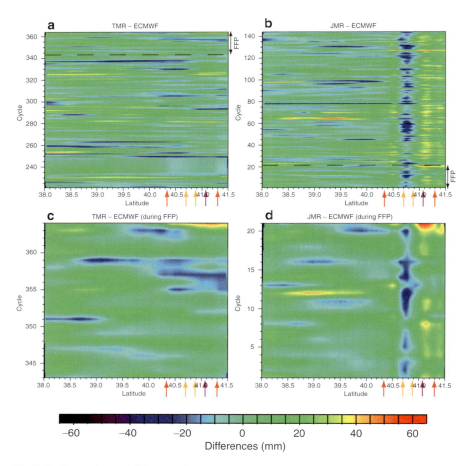

Fig. 11.13 Comparison of differences between wet tropospheric correction from radiometers and ECMWF model at Senetosa (Corsica calibration site); over 4 years of data for (**a**) T/P (cycle 221–364) and (**b**) Jason-1 (cycle 1–144); over the formation flight phase (FFP) for (**c**) T/P (cycle 343–364) and (**d**) Jason-1 (cycle 1–21). The coloured *arrows* on the latitude axis correspond to the lines defined in Fig. 11.12

when T/P and Jason-1 were in formation flight (January to August 2002): before Sardinia (open ocean) there's a relatively good correlation in the patterns showing a good consistency between both radiometers and revealing some coherent signatures over few cycles which are probably due to model inconsistencies for some meteorological conditions.

One possibility for estimating the bias induced by the land contamination is to make comparisons with the wet tropospheric correction determined from in situ GPS data. Recent results (Table 11.3) suggest that AMR (onboard Jason-2) is less sensitive to land contamination than JMR: indeed the 1 m, unblocked reflector of the AMR improves coastal resolution and reduces geographically correlated errors, compared to partially blocked 0.6 m reflector for JMR/TMR; spatial resolution is nearly doubled from TMR (T/P radiometer)

Table 11.3 Mean differences between radiometer and GPS wet tropospheric correction (in mm)

Differences	Harvest	Corsica
AMR − GPS[a]	+1 ± 4 (9)	−0.5 ± 4 (10)
JMR − GPS[a]	+7 ± 1 (234)	+12 ±3 (60)

Number of samples in bracket
[a]Opposite sign if working directly on path delay

and JMR (~26 km for AMR compared to ~50 km for JMR/TMR). For Corsica, this confirms the observed discrepancy which was not also seen on TMR (Bonnefond et al. 2003a): this 12 mm bias partly explains the low value of the altimeter bias determined at the Corsica site in comparison with the corresponding values from Harvest and Bass Strait.

The long series of JMR vs. GPS comparisons also permit monitoring of drift: it has been found to be negligible (below 0.3 mm/year for both sites). The wet tropospheric path delay derived from GPS data demonstrates that it is a very powerful and accurate method (Haines and Bar-Sever 1998) to monitor onboard radiometers (see also Fig. 11.11 in Sect. 11.4.1.2).

11.4.2 Recent Results

During the Jason-2 verification phase, the satellite flew in formation with Jason-1, with 58-s lag along track. This enabled a direct comparison of the two altimetric measurement systems under nearly the same environmental conditions. Of particular importance, it allowed a direct comparison of the "orbit minus range" ($h - R$, in Eq. 11.2, Sect. 11.2.1, hereinafter OMR) without the need for additional corrections. Table 11.4 summarizes the preliminary results of detailed studies on common overflights for Jason-1 and Jason-2 based on either OMR or SSH differences (corrections applied, see also Eq. 11.2, Sect. 11.2.1). It also provides the differences (Jason-1 − Jason-2) of the correction terms used in the SSH computation. On average, the correction differences are close to zero so the ΔOMR and ΔSSH biases are very close (Table 11.4). There are however differences between the sites: Bass Strait and Corsica are very close while Harvest exhibits a relative bias higher by about 25 mm. The ΔOMR and ΔSSH biases are also very close to the corresponding value from global analysis studies (~80 mm for SSH differences, Desai et al. and Ablain et al. in OSTST minutes 2008). Moreover, the differences of the absolute bias computed at each calibration site separately for each satellite are very comparable to the relative biases (Δabsolute biases compared to ΔSSH biases in Table 11.4).

This preliminary result clearly indicates that the primary source of Jason-1 and Jason-2 relative bias is the altimeter range. The mean effect of the orbit, ionosphere, and wet troposphere (AMR-JMR in Table 11.4) errors is at the 1-cm level or smaller.

The dominant and coherent source of difference in terms of correction is the ionosphere (~10 mm) for which the origin has been linked to the relative bias differences between C and Ku bands (respectively 84 and 132 mm from global analysis, Desai et al. in OSTST minutes 2008).

Recently, updated products from both Jason-1 and T/P have been used to reprocess data from their Formation Flight Phase (January to August 2002). For T/P, the update consists

Table 11.4 Corrections differences in mm (Jason-1 – Jason-2)

Correction	Corsica	Harvest	Bass Strait
Dry Tropo.	0	+1	0
AMR-JMR	−10	−5	−9
Iono.	+12	+10	+9
SSB	−4	−6	−3
Δ MSS Grad	Corrected for[a]	−2	−3
Total	−2	−2	−6
ΔOMR	+70	+95	+77
ΔSSH biases	+72	+97	+71
Δabsolute biases	+73	+90	+77

Standard deviations of the correction differences are always below 10 mm

Results are based on the first 4 months of the Formation Flight Phase, July to October 2008 (cycle 0–12)

[a]See Sect. 11.3.3

of replacing the MGDR data with a new wet tropospheric correction (TMR replacement product from JPL; Brown et al. 2010) and the latest orbit computed at the Goddard Space Flight Center (TVG orbits based on ITRF 2005; Lemoine et al. 2007). For Jason-1, the study is based on the latest available products (GDR-C). The differences (Jason-1 − T/P) from the three calibration sites are very coherent at the level of a few mm.

Bass Strait: +87 mm (17 common overflights)

Corsica: +85 mm (11 common overflights)

Harvest: +78 mm (16 common overflights)

When comparing the results of the T/P and Jason-1 Formation Flight Phase to those of Jason-1 and Jason-2, it appears that the SSH bias for each follow-on mission is increased by ~80 mm. Whether this is purely coincidence is not presently known. Because ground calibration of the altimeter cannot completely simulate the in-flight behavior of the instrument, it is technologically impossible to predict prior to launch the altimeter range bias with the necessary accuracy. The increased probability of non-overlapping missions in the future reinforces the importance of cal/val activities to detect and monitor such effects, either in the altimeter data or other corrections involved in the SSH determination.

11.4.3
Revised Error Budget in Altimetry

The recent improvements (models, corrections, orbit, etc.) realized in the frame of Jason-1 and T/P missions as well as the technological improvements for the OSTM/Jason-2 mission (altimeter, radiometer, etc.) have lead to revise the error budget during the redaction of the OSTM/Jason-2 cal/val plan and to publish it in OSTM/Jason-2 Products Handbook

Table 11.5 OSTM/Jason-2 error budget (in cm), extracted from the OSTM/Jason-2 Products Handbook (Issue: 1 rev 3, January 20, 2009)

	OGDR Delay: 3 h	IGDR Delay: 1–1.5 days	GDR Delay: 40 days	GOALS
Altimeter noise	2.5 (a)(c)(d)	1.7 (b)(c)(d)	1.7 (b)(c)(d)	1.5 (b)(c)(d)
Ionosphere	1 (e)(d)	0.5 (e)(d)	0.5 (e)(d)	0.5 (e)(d)
Sea state bias	3.5	2	2	1
Dry troposphere	1	0.7	0.7	0.7
Wet troposphere	1.2	1.2	1.2	1
Altimeter range RSS	5	3	3	2.25
RMS orbit (radial component)	10 (h)	2.5	1.5	1
Total RSS sea surface height	11.2	3.9	3.4	2.5
Significant wave height	10% or 0.5 m (i)	10% or 0.4 m (i)	10% or 0.4 m (i)	5% or 0.25 m (i)
Wind speed	1.6 m/s	1.5 m/s	1.5 m/s	1.5 m/s
Sigma naught (absolute)	0.7 dB	0.7 dB	0.7 dB	0.5 dB
System drift	–	–	–	1 mm/year (j)

All numbers are for 1 s average, 2 m SWH, 11 dB sigma naught
(a) Combined Ku + C measurement
(b) Ku band after ground retracking
(c) Averaged over 1 s
(d) Assuming 320 MHz C bandwidth
(e) Filtered over 100 km
(f) Can also be expressed as 1% of H1/3
(g) After ground retracking
(h) Real time DORIS onboard ephemeris
(i) Which ever is greater
(j) On global mean sea level, after calibration

(Table 11.5). However, this error budget is mainly dedicated to altimetric measurements in open-ocean conditions: for example, the precision of the altimeter and some corrections (e.g., SSB) are estimated for 2 m SWH which do not often correspond to typical waves in coastal areas. Moreover, the ionospheric correction being relatively noisy it should be filtered (or averaged) over about 100 km leading to difficulties for complex shapes of the coast and presence of islands.

In consequence such an error budget cannot be used as it is for specific studies, notably in coastal areas, where authors should conduct specific studies to derive their own error budget. Indeed, we have demonstrated (Sect. 11.4.1.2) that some coastal conditions can lead not only to a degradation of the precision but can also introduce some systematisms which can make difficult the link between open-ocean and coastal measurements.

Finally, the altimetric correction that represents the most significant current challenge is the sea state bias (SSB). Different models for the sea state bias can impact the computed altimeter (SSH) bias at the few-cm level. The correction itself is commonly determined empirically (Labroue et al. 2004) from the sea state conditions (wind and waves) and SSH information (anomalies or crossover differences) and it is highly dependent on the altimeter retracking algorithm used. As an example, the evolution of the SSB from MLE3 to MLE4 retracking algorithm (maximum likelihood estimator) has changed the relative and absolute altimeter bias for T/P and Jason-1 by several centimeters (2–5 cm). This was mainly constant at the global scale, but also revealed some geographical patterns at the centimeter level.

11.5
Conclusion

While comparisons of sea-surface heights derived from overlapping missions (e.g., T/P, Jason-1, Envisat) at the global scale are able to yield exceptionally precise estimates of relative bias due to the large sample size, absolute calibration experiments at point locations provide a unique opportunity to further investigate component wise biases and geographical dependencies in the altimeter measurement system. Further, absolute calibration sites provide an important source of continuous observations to assist in the assimilation of multi mission data sets in the event of a gap between missions. Moreover, thanks to a diverse array of in situ instrumentation, such experiments provide insights into the origin of such biases (orbit, range, or corrections for example), and often provide valuable insight into systematic errors in other observational systems (SLR and the GPS for example). On the other hand, some of the errors affecting the altimetric measurement system (Table 11.5) can be attributed to conditions local to the calibration sites (e.g., wet tropospheric correction, sea state bias, geographically correlated orbit errors, etc.). Moreover, some of these errors can be amplified in coastal areas (e.g., the wet troposphere), not only in terms of scatter but also for the systematic component. The desired approach to absolute calibration must aspire to ensure continuity among past, present and future missions, but also ensure a proper reconciliation of offshore (open ocean) and coastal altimetric measurements (see the chapter by Bouffard et al., this volume).

Due to the increasing need for altimetry to provide monitoring of inland waters, and in preparation for new missions (e.g., AltiKa, CryoSat-2, SWOT), calibration activities directed at river and lake data are also underway (Calmant and Seyler 2006; Crétaux et al. 2008; see also the chapter by Crétaux et al., this volume). Cal/val activities focused in the oceanic domain have a long history and protocols are well established. Cal/val activities focused on rivers and lakes are more recent developments, but in turn enable mitigation of spurious results from sea state bias (SSB) and ocean tides. They also enable new perspectives on other problems, such as the performance of the various tracking/retracking algorithms, and on the quality of the geophysical corrections under more extreme conditions.

Acknowledgments This chapter is dedicated to the memory of Dr. Yves Ménard for his deep involvement in the calibration and validation activities. The authors collectively acknowledge other members of their teams that have contributed to the design, operation and analyses of the calibration sites. Bass Strait: Neil White, Richard Coleman, John Church, Paul Tregoning, Jason Zhang, and Reed Burgette. Corsica: Olivier Laurain, Pierre Exertier, François Barlier, Gwénaële Jan, Yves Ménard, Claude Gaillemin, and the FTLRS team. Harvest: Ed Christensen (posthumous), George Born, Dan Kubitschek, Steve Gill, Dave Stowers, Plains Exploration and Production. A portion of this work was conducted by the Jet Propulsion Laboratory, California Institute of Technology, under contract with NASA.

References

Altamimi Z, Collilieux X, Legrand J, Garayt B, Boucher C (2007) ITRF2005: a new release of the International Terrestrial Reference Frame based on time series of station positions and Earth Orientation Parameters. J Geophys Res 112:B09401. doi:10.1029/2007JB004949

Barlier F, Lefebvre M (2001) A new look at planet Earth: satellite geodesy and geosciences, The Century of Space Science. Kluwer, Dordrecht, The Netherlands, pp 1623–1651

Bar-Sever Y (2008) APPS: The automatic precise point positioning service from JPL's global differential GPS (GDGPD) system. ION GNSS 2008 meeting, Savannah, USA, September (http://apps.gdgps.net)

Benveniste J (1997) ERS-2 altimetry calibration. In: Proceedings of the 3rd ERS symposium. March 17–21, ESA SP-414, vol III

Bindoff N, Willebrand J, Artale V, Cazenave A, Gregory J, Gulev S, Hanawa K, Le Quéré C, Levitus S, Nojiri Y, Shum CK, Talley L, Unnikrishnan A (2007) Ocean climate change and sea level, In: Climate change 2007: the physical science basis. Contribution of working group I to the fourth assessment report of the Intergovernmental Panel on Climate Change (IPCC) Ed, Cambridge University Press, Cambridge/New York

Bonnefond P, Exertier P, Schaeffer P, Bruinsma S, Barlier F (1995) Satellite altimetry from a short-arc orbit technique: application to the Mediterranean. J Geophys Res 100(C12):25365–25382

Bonnefond P, Exertier P, Ménard Y, Jeansou E, Manzella G, Sparnocchia S, Barlier F (1997) Calibration of radar altimeters and validation of orbit determination in the Corsica-Capraia Area. In: 3rd ERS symposium, Florence, vol III, pp 1525–1528

Bonnefond P, Exertier P, Barlier F (1999) Geographically correlated errors observed from a laser-based short-arc technique. J Geophys Res 104(C7):15885–15893

Bonnefond P, Exertier P, Laurain O, Ménard Y, Orsoni A, Jan G, Jeansou E (2003a) Absolute calibration of Jason-1 and TOPEX/Poseidon altimeters in Corsica. In: Special issue on Jason-1 calibration/validation, Part 1. Mar Geod 26(3–4):261–284

Bonnefond P, Exertier P, Laurain O, Ménard Y, Orsoni A, Jeansou E, Haines B J, Kubitschek D, Born G (2003b) Leveling sea surface using a GPS catamaran. In: Special issue on Jason-1 calibration/validation, Part 1. Mar Geod 26(3–4):319–334

Born GH et al (1994) Calibration of the TOPEX altimeter using a GPS buoy. J Geophys Res 99(C12):24517–24526

Brown ST, Desai S, Wenwen L, Tanner AB (2007) On the long-term stability of microwave radiometers using noise diodes for calibration. IEEE Trans Geosci Remote Sensing 45(7):1908–1920

Brown S, Desai S, Keihm S, Lu W (2010) Microwave radiometer calibration on decadal time scales using On-Earth brightness temperature references: application to the Topex microwave radiometer. J Atmos Ocean Technol 26:2579–2591

Calmant S, Seyler F (2006) Continental surface waters from satellite altimetry. CR Geosci 338: 1113–1122

Cazenave A, Nerem RS (2004) Present-day sea level change: observations and causes. Rev Geophys 42, RG3001. doi:10.1029/2003RG000139

Chelton DB, Ries JC, Haines BJ, Fu LL, Callahan PS (2001) Satellite altimetry in "satellite altimetry and earth sciences. In: Fu LL, Cazenave A (eds) A handbook of techniques and applications. Academic, San Diego, CA

Christensen EJ, Haines BJ, Keihm SJ, Morris CS, Norman RA, Purcell GH, Williams BG, Wilson BD, Born GH, Parke ME, Gill SK, Shum CK, Tapley BD, Kolenkiewicz R, Nerem RS (1994) Calibration of TOPEX/Poseidon at platform harvest. J Geophys Res 99(C12):24465–24485

Crétaux J-F, Calmant S, Romanovski V, Shibuyin A, Lyard F, Bergé-Nguyen M, Cazenave A, Hernandez F, Perosanz F (2008) Implementation of a new absolute calibration site for radar altimeter in the continental area: lake Issykkul in Central Asia. J Geod. doi:10.1007/s00190-008-0289-7

Cristea E, Moore P (2007) Altimeter bias determination using two years of transponder observations. In: Proceedings of the Envisat symposium 2007, Montreux, Switzerland (ESA SP-636, July 2007)

Desai S, Haines BJ (2004) Monitoring measurements from the Jason-1 microwave radiometer and independent validation with GPS. In: Special issue on Jason-1 calibration/validation, Part 2. Mar Geod 27(1–2):221–240

Exertier P, Nicolas J, Berio P, Coulot D, Bonnefond P, Laurain O (2004) The role of laser ranging for calibrating Jason-1: the Corsica tracking campaign. In: Special issue on Jason-1 calibration/validation, Part 2. Mar Geod 27(1–2):333–340

Exertier P, Bonnefond P, Deleflie F, Barlier F, Kasser M, Biancale R, Ménard Y (2006) Contribution of laser ranging to earth sciences. CR Geosci 338:958–967

Francis CR (1992) The height calibration of the ERS-1 radar altimeter. In: Proceedings of the first ERS-1 symposium – space at the service of our environment. ESA Spec Pub ESA SP-359(1): 381–393

Fu LL, Christensen EJ, Yamarone CA Jr, Lefebvre M, Ménard Y, Dorrer M, Escudier P (1994) TOPEX/Poseidon mission overview. J Geophys Res 100(C12):24369–24381

Gill SK, Edwing RF, Jones DF, Mero TN, Moss MK, Samant M, Shih HH, Stoney WM (1995) NOAA/national ocean service platform harvest instrumentation. Mar Geod 18(1–2):49–68

Gourine B, Kahlouche S, Exertier P, Berio P, Coulot D, Bonnefond P (2008) Corsica SLR positioning campaigns (2002 and 2005) for satellite altimeter calibration missions. Mar Geod (31): 103–116

Haines BJ, Bar-Sever Y (1998) Monitoring the TOPEX microwave radiometer with GPS: stability of wet tropospheric path delay measurements. Geophys Res Lett 25(19):3563–3566

Haines BJ, Dong D, Born GH, Gill SK (2003) The harvest experiment: monitoring Jason-1 and TOPEX/Poseidon from a California offshore platform. In: Special issue on Jason-1 calibration/validation, Part 1. Mar Geod 26(3–4):239–259

Haines BJ, Bar-Sever Y, Bertiger W, Desai S, Willis P (2004) One-centimeter orbit determination for Jason-1: new GPS-based strategies. In: Special issue on Jason-1 calibration/validation, Part 2. Mar Geod 27(1–2):299–318

Hein G, Landau H, Blomenhofer H (1990) Determination of instantaneous sea surface, wave heights and ocean currents using satellite observations of the global positioning system. Mar Geod 14:217–224

Jan G, Ménard Y, Faillot M, Lyard F, Jeansou E, Bonnefond P (2004) Offshore absolute calibration of space borne radar altimeters. In: Special issue on Jason-1 calibration/validation, Part 3. Mar Geod 27(3–4):615–629

King MA, Watson CS, Penna NT, Clarke PJ (2008) Subdaily signals in GPS observations and their effect at semiannual and annual periods. Geophys Res Lett 35(3):1–5

Kolenkiewicz R, Martin C (1982) Seasat altimeter height calibration. J Geophys Res 87(C5): 3189–3197

Labroue S, Gaspar P, Dorandeu J, Zanife O Z, Mertz F, Vincent P, Choquet D (2004) Non parametric estimates of the sea state bias for the Jason-1 radar altimeter. In: Special issue on Jason-1 calibration/validation, Part 3. Mar Geod 27(3–4):453–481

Lemoine FG, Zelensky NP, Lutchke SB, Rowlands DD, Williams TA, Chinn D (2007) Improvement of the complete TOPEX/Poseidon and Jason-1 orbit time series: current status (abstract). In: Proceedings of the OST/ST Meeting, Hobart, Australia

Lowe ST, Zuffada C, LaBrecque J, Lough M, Lerma J (2002) Five-cm-precision aircraft ocean altimetry using GPS reflections. Geophys Res Lett 29(10):1–4

Luthcke SB, Zelensky NP, Rowlands DD, Lemoine FG, Williams TA (2003) The 1-centimeter orbit: Jason-1 precision orbit determination using GPS, SLR, DORIS and altimeter data. In: Special issue on Jason-1 calibration/validation, Part 1. Mar Geod 26(3–4):399–421

MacDoran PF, Born GH (1991) Time, frequency and space geodesy: impact on the study of climate and global change. Proc IEEE 79(7):1063–1069

Martin C, Kolenkiewicz R (1981) Calibration validation of the GEOS-3 altimeter. J Geophys Res 86(7):6369–6381

Martinez-Benjamin JJ, Martinez-Garcia M, Gonzales Lopez S et al (2004) Ibiza absolute calibration experiment: survey and preliminary results. In: Special issue on Jason-1 calibration/validation, Part 3. Mar Geod 27(3–4):657–681

McCarthy DD, Petit G (2004) IERS Conventions 2003 (IERS Technical Note; 32). Verlag des Bundesamts für Kartographie und Geodäsie, Frankfurt am Main, 127pp, paperback, ISBN 3-89888-884-3

Ménard Y, Jeansou E, Vincent P (1994) Calibration of the TOPEX/POSEIDON altimeters at Lampedusa: additional results at Harvest. J Geophys Res 99(C12):24487–24504

Ménard Y, Fu L-L, Escudier P, Parisot F, Perbos J, Vincent P, Desai S, Haines B, Kunstmann G (2003) The Jason-1 mission. In: Special issue on Jason-1 calibration/validation, Part 1. Mar Geod 26(3–4):131–146

Mitchum GT (1998) Monitoring the stability of satellite altimeters with tide gauges. J Atmos Ocean Technol 15:721–730

Morris CS, Dinardo SJ, Christensen EJ (1995) Overview of the TOPEX/Poseidon platform harvest verification experiment. Mar Geod 18(1–2):25–38

Murphy CM, Moore P, Woodworth P (1996) Short-arc calibration of the TOPEX/Poseidon and ERS 1 altimeters utilizing in situ data. J Geophys Res 101(C6):14191–14200

Nerem RS, Haines BJ, Hendricks J, Minster JF, Mitchum GT, White WB (1997) Improved determination of global mean sea level variations using TOPEX/Poseidon altimeter data. Geophys Res Lett 24(11):1331–1334

Niell AE (1996) Global mapping functions for the atmospheric delay at radio wavelengths. J Geophys Res 101:3227–3246

OSTST minutes of the Nice meeting (2008) France, November 10–12

OSTST minutes of the St Petersburg meeting (2004) Florida, November 4–6

Pavlis EC, Mertikas SP, GAVDOS Team (2004) The GAVDOS mean sea level and altimeter calibration facility: results for Jason-1. In: Special issue on Jason-1 calibration/validation, Part 3. Mar Geod 27(3–4):631–655

Penna NT, Stewart MP (2003) Aliased tidal signatures in continuous GPS height time series. Geophys Res Lett 30(23):2184. doi:10.1029/2003GL018828

Penna NT, King MA, Stewart MP (2007) GPS height time series: short-period origins of spurious long-period signals. J Geophys Res 112(B2):1–19

Powell RJ (1986) Relative vertical positioning using ground transponders with the ERS-1 altimeter. IEEE Trans Geosci Remote Sensing GE-24(3):421–425

Pugh D (1987) Tides, surges and mean sea level: a handbook for engineers and scientists. Wiley, Chichester, 472p

Ruf CS, Keihn SJ, Subramanya B, Jansen M (1994) TOPEX/Poseidon microwave radiometer performance and in-flight calibration. J Geophys Res 99(C12):24915–24926

Rutkowska M, Noomen R (1998) Range biases and station position analysis for network solutions derived from SLR data in the period 1993–1995. In: Proceeding of the 11th international workshop on laser ranging, Deggendorf, Germany, September 21–25, p 70

Testut L, Wöppelmann G, Simon B, Téchiné P (2005) The sea level at Port-aux-Français, Kerguelen Island, from 1950 to the present. Ocean Dyn. doi:10.1007/s10236-005-0056-8

Treuhaft RN, Lowe ST, Zuffada C, Chao Y (2001) 2-cm GPS altimetry over Crater Lake. Geophys Res Lett 28(23):4343–4346

Treuhaft RN, Chao Y, Lowe ST, Young LE, Zuffada C, Cardellach E (2003) Monitoring coastal eddy evolution with GPS altimetry. In: proceeding of international workshop on GPS meteorology. Tsukuba, Japan

Vincent P, Desai SD, Dorandeu J, Ablain M, Soussi B, Callahan PS, Haines BJ (2003) Jason-1 geophysical performance evaluation. In: Special issue on Jason-1 calibration/validation, Part 1. Mar Geod 26(3–4):167–186

Vondrak J (1977) Problem of smoothing observational data II. Bull Astron Inst Czech 28:84–89

Watson CS (2005) Satellite altimeter calibration and validation using GPS buoy technology. Thesis for Doctor of Philosophy, Centre for Spatial Information Science, University of Tasmania, Australia, 264pp. Available at: http://eprints.utas.edu.au/254/

Watson C, Coleman R, White N, Church J, Govind R (2003) Absolute calibration of TOPEX/Poseidon and Jason-1 using GPS Buoys in Bass Strait, Australia. In: Special issue on Jason-1 calibration/validation, Part 1. Mar Geod 26(3–4):285–304

Watson CS, White N, Coleman R, Church J, Morgan P, Govind R (2004) TOPEX/Poseidon and Jason-1 absolute calibration in Bass Strait, Australia. In: Special issue on Jason-1 calibration/validation, Part 2. Mar Geod 27(1–2):107–131

Watson CS, Tregoning P, Coleman R (2006) The impact of solid earth tide models on GPS time series analysis. Geophys Res Lett 33:L08306. doi:10.1029/2005GL025538

Watson CS, Coleman R, Handsworth R (2008) Coastal tide gauge calibration: a case study at Macquarie Island using GPS buoy techniques. J Coastal Res 24(4):1071–1079

White NJ, Coleman R, Church JA, Morgan PJ, Walker SJ (1994) A southern hemisphere verification for the TOPEX/Poseidon altimeter mission. J Geophys Res 99(C12):24505–24516

Woodworth P, Moore P, Dong X, Bingley R (2004) Absolute calibration of the Jason-1 altimeter using UK tide gauges. In: Special issue on Jason-1 calibration/validation, Part 2. Mar Geod 27(1–2):95–106

Zumberge J, Watkins M, Webb F (1998) Characteristics and applications of precise GPS clock solutions every 30 s. Navigation 44(4):449–456

Introduction and Assessment of Improved Coastal Altimetry Strategies: Case Study over the Northwestern Mediterranean Sea

12

J. Bouffard, L. Roblou, F. Birol, A. Pascual, L. Fenoglio-Marc, M. Cancet, R. Morrow, and Y. Ménard

Contents

12.1	Introduction	299
12.2	Study Area Characteristics	302
12.3	Main Characteristics of Existing Data	304
	12.3.1 Standard Data: AVISO Along-track Data	304
	12.3.2 Improved Pre-processed Data: RECOSETO Data	304
	12.3.3 Improved Post-processed Data: X-TRACK Data	305
12.4	Benefit of Improved Strategies over the Northwestern Mediterranean Sea	306
	12.4.1 Improvements from RECOSETO Project	306
	12.4.2 Improvements from X-TRACK Processor	310

J. Bouffard (✉)
Instituto Mediterráneo de Estudios Avanzados, IMEDEA (CSIC-UIB),
C/Miquel Marquès, 21, 07190, Esporles, Illes Balears, Spain, and
Université de Toulouse; UPS (OMP-PCA), Laboratoire d'Études en Géophysique et Océanographie Spatiales (LEGOS), 14 Av, Edouard Belin, F-31400, Toulouse, France
e-mail: jerome.bouffard@uib.es

L. Roblou, F. Birol, M. Cancet and R. Morrow
Université de Toulouse; UPS (OMP-PCA), Laboratoire d'Études en Géophysique et Océanographie Spatiales (LEGOS), 14 Av, Edouard Belin, F-31400, Toulouse, France

A. Pascual
Instituto Mediterráneo de Estudios Avanzados, IMEDEA (CSIC-UIB),
C/Miquel Marquès, 21, 07190, Esporles, Illes Balears, Spain

L. Fenoglio-Marc
Institute of Physical Geodesy, Technical University of Darmstadt,
L5|01 451 Petersenstr. 13, 64287, Darmstadt, Germany

Y. Ménard
Université de Toulouse; UPS (OMP-PCA), Laboratoire d'Études en Géophysique et Océanographie Spatiales (LEGOS), 14 Av, Edouard Belin, F-31400, Toulouse, France, and Centre National d'Études Spatiales (CNES), 18 Avenue Edouard Belin, F-31400, Toulouse, France

12.5   Intercomparison Experiments: X-TRACK-SLA Versus DT-SLA "Upd" 317
   12.5.1 Basin Scale Statistics 317
   12.5.2 Focus on the Gulf of Lion 320
12.6   Conclusion, Ongoing Work, and Perspectives 325
References 327

Keywords   Coastal altimetry • Coastal current • Mediterranean Sea • Tide gauge • Validation

Abbreviations

ADT	Absolute Dynamic Topography
ALBICOCCA	Altimeter-Based Investigations in COrsica, Capraia and Contiguous Area
ASI	Agenzia Spaziale Italiana
AVHRR	Advanced Very High Resolution Radiometer
AVISO	Archiving, Validation and Interpretation of Satellite Oceanographic data
CTD	Conductivity-Temperature-Depth
CNES	Centre National d'Étude Spatiale
CTOH	Centre de Topographie des Océans et de l'Hydrosphère
DFG	Deutsche Forschungsgemeinschaft
DH	Dynamic Height
DUACS	Data Unification and Altimeter Combination System
ECMWF	European Centre for Medium-Range Weather Forecasts
ESA	European Space Agency
FES	Finite Element Solution
GDR	Geophysical Data Record
GFO	Geosat Follow-On
GOT	Goddard Ocean Tide
GVA	Geostrophic Velocity Anomaly
IB	Inverted Barometer
IMEDEA	Institut Mediterrani d'Estudis Avançats
IMP	IMprovement Percentage
ISPRA	Istituto Superiore per la Protezione e la Ricerca Ambientale
J1	Jason-1
JPL	Jet Propulsion Laboratory
LEGOS	Laboratoire d'Études en Géophysique et Océanographie Spatiales
LPC	Liguro Provençal Catalan
LSER	Large-Scale Error Reduction
MDT	Mean Dynamic Topography
MOG2D	Modèle aux Ondes de Gravité 2-Dimensions
MWR	Microwave Radiometer
MSS	Mean Sea Surface
NASA	National Aeronautics and Space Administration
NCEP	National Centres for Environmental Prediction
NP	Normal Point

NWM	Northwestern Mediterranean
OCOG	Offset Centre of Gravity
OSU	Ohio State University
PODAAC	Physical Oceanography Distributed Active Archive Centre
RADS	Radar Altimeter Database System
RECOSETO	Regional COastal Sea level change and sea surface Topography from altimetry, Oceanography, and tide gauge stations in Europe
RGDR	Retracked Geophysical Data Record
RMS	Root Mean Square
SGDR	Sensor Geophysical Data Record
SLA	Sea Level Anomaly
SONEL	Système d'Observation du Niveau des Eaux Littorales
SSALTO	Segment Sol multimissions d'ALTimétrie, d'Orbitographie et de localisation précise
SSH	Sea Surface Height
SST	Sea Surface Temperature
T/P	TOPEX/Poseidon
TG	Tide Gauge
T-UGOm	Toulouse Unstructured Grid Ocean model

12.1
Introduction

Most urban development and economic activities around the world occur in coastal zones. In 1997, 37% of the global population was concentrated within a 100 km coastal band (Cohen et al. 1997), which represents a density of 80 people/km^2 (i.e., more than twice the global average). In the ocean, 95% of the socioeconomic stakes are localized in coastal areas. This raises crucial issues related to the management of environmental and marine resources. It is essential to better understand the physical processes driving the coastal ocean dynamics, which requires the development of monitoring systems, coupling observations and modelling through data assimilation techniques (Mourre et al. 2004; Li et al. 2008).

Radar altimetry constitutes an essential component for observing the ocean circulation complexity (Le Traon and Morrow 2001). The spatial and temporal sampling of satellite altimetry is mostly suited to detecting sea level variations over the open ocean, where the signals and the corrections to be applied are well known. Multi-satellite configurations have allowed a better resolution of the mesoscale variability in the global ocean (Pascual et al. 2006) or at a regional scale (Pascual et al. 2007). Over the past decade, attention has been paid to basin-scale signals in the Mediterranean Sea (e.g., Larnicol et al. 1995, 2002; Ayoub et al. 1998; Iudicone et al. 1998; Pujol and Larnicol 2005, see Cipollini et al. 2008 for a review of studies using altimetry in the Mediterranean Sea) and in a few specific open sea regions (e.g., Buongiorno Nardelli et al. 1999; Vignudelli et al. 2000). These past studies highlight that satellite altimetry observations represent a privileged tool for monitoring oceanic processes over the Mediterranean open ocean. But compared to the open ocean, it turns out that the coastal ocean dynamics are much more complex and difficult to observe with standard altimetric products.

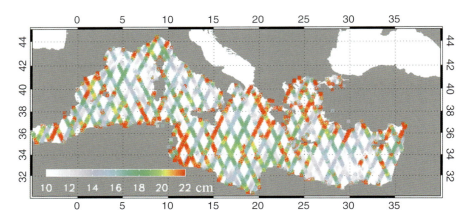

Fig. 12.1 Along-track RMS difference between radiometer and modelled wet tropospheric correction (courtesy R. Dussurget; LEGOS-CTOH). Satellites: GFO+J1; Period: Sept. 2002 to Feb. 2004; Corrections: for GFO: radiometer (model_wet_tropo_corr_mwr) - NCEP model (model_wet_tropo_corr_ncep). For J1: radiometer (rad_wet_tropo_corr) - ECMWF model (model_wet_tropo_corr)

Data distributed by operational centres are not adapted to the studies of the dynamical mechanisms in coastal areas where space scales can be of about 10 km (Chelton 2001). Moreover, the signal-to-noise ratio in the coastal band is rapidly degraded, particularly because the altimeter and radiometer signals are perturbed at a 10 and 50 km distance to the coast, respectively. For instance, Fig. 12.1 shows the differences between a modelled wet tropospheric correction and an onboard-based radiometric correction. The most important disagreements are located in the coastal band. In those regions, the onboard microwave radiometer (MWR) footprint contains land surface causing high contamination of the radiometer measurements (more precisely of the brightness temperature). Therefore, today MWR-derived wet tropospheric corrections are not suitable at distances less than 50 km from the coast (Desportes et al. 2007). Apart from land contamination, data quality in these regions is also reduced because of the inaccuracy of high-frequency corrections (atmospheric and tidal corrections from global models) and the lack of algorithms and quality control procedures dedicated to coastal targets (Vignudelli et al. 2000).

As a first initiative aiming at improving coastal altimetry over the Mediterranean Sea, Manzella et al. (1997) proposed a customised wet tropospheric correction more suited to the coastal zone. In a more general framework, Anzenhofer et al. (1999) identified two main strategies to improve coastal altimetry. The first one consisted of retracking the altimetric waveforms. This strategy proposed reprocessing the single waveforms using a non-standard waveform model in an attempt to recover the ocean surface parameters (sea level anomaly, significant wave height, wind speed). This strategy is usually defined as a "pre-processing" approach. The second strategy lies in a "post-processing" approach, discarding over-conservative flags, using local models for tides and atmospheric effects, filling gaps and filtering out noise in some of the corrections. This strategy was adopted by the ALBICOCCA initiative (ALtimeter-Based Investigations in Corsica, Capraia and Contiguous Area; see Vignudelli et al. 2005), which made a major contribution to the coastal altimetry data post-processing. ALBICOCCA was a joint project funded by ASI (Italia) and CNES (France), aimed at increasing both the quantity and quality of altimetric data in the coastal ocean of the northwestern Mediterranean (NWM). This project led to significant improvements mainly through the reconstruction of data corrections as well as an improved de-aliasing of tides and atmospheric effects through regional modelling. When compared to standard products, these new post-processing techniques reduce the amount of erroneous data and increase the number of observations in the coastal zone. Vignudelli et al. 2005 showed that it is possible to use these improved coastal data sets for scientific applications, but the project success was limited to one satellite. Moreover, the performance of the data set was limited by the relatively poor spatial along-track 1 Hz sampling (about 7 km), the lack of a large-scale error reduction method and of a specific coastal retracking technique.

In this chapter, we will briefly present some techniques for improving the processing of multi-satellite altimetric data and will provide diagnostic evaluations of the different methods used. Two kinds of improved coastal data have been validated over the NWM coastal zone: the X-TRACK-derived data (post-processed data, see Roblou et al. 2007) and RECOSETO-derived data (pre-processed data, see Fenoglio-Marc et al. 2007). The first one is a coastal multi-satellite altimetric data set (from TOPEX/Poseidon, Jason-1, Envisat, Geosat Follow-On raw measurements) mainly derived from routine Geophysical Data Records (GDR). The processing strategy is, however, characterized by the use of local de-aliasing modelling in addition to new methods and quality control procedures dedicated to coastal zone applications (for more details, see the chapter by Roblou et al., this volume). The second data set is computed from Envisat and TOPEX/Poseidon raw measurements, where different kinds of waveform retracking algorithm and instrumental and geophysical corrections have been applied. Here, we assess the X-TRACK and RECOSETO data sets over the NWM by evaluating independently and individually the main aspects associated with these new processing strategies. This evaluation is essentially statistic. The impact of retracking (pre-processing) is shown in terms of standard deviation reductions of:

- The difference between retracked RECOSETO sea surface height (SSH) and the geoid (spatial approach)
- The difference between retracked RECOSETO SSH and tide gauge (TG) time series (temporal approach)

The wet tropospheric correction is analyzed in terms of the number of data rejected in the coastal area by comparing two kinds of corrections applied to the RECOSETO SSHs (one based on modelling and the other one based on on-board radiometric measurements). The impact of local de-aliasing and editing strategies (including rebuilding over flagged corrections) are tested through inter-comparison experiments between X-TRACK data (both with standard and improved corrections) and tide gauge time series. The use of high-frequency sampling, large-scale error reduction and an optimal mean sea surface in the X-track processing will also be evaluated.

Even if the error related to each altimetric correction is often sub-centimetric, the combined effects of the different improved corrections and methods should generate coastal data that are significantly better than standard products for monitoring coastal dynamics. In this respect, an X-TRACK altimetric Sea Level Anomaly (SLA) data set – including all of the improved techniques – will be quantitatively and qualitatively compared to a regional along-track altimetric product distributed by AVISO.

Thus, the main purpose of this chapter is to present and validate improved existing coastal altimetric data, which can be used for coastal ocean monitoring. The NWM represents a case study which is well adapted for such investigations, in virtue of its complex coastal circulation, its wide and shallow continental shelves combined with the availability of long-term TG measurements. As shown in an example over the Gulf of Lion, once the processing data strategy is improved, it is possible to precisely monitor the sea surface variability related to a narrow coastal current.

The chapter is organized as follows: First, we present the study area and its main dynamical characteristics. We then focus on the different datasets used and give details and quantitative evaluations of their benefits to increase the quality and availability of coastal altimetric data. As such, we present intercomparison experiments between TGs, X-TRACK and AVISO SLA. Finally, we discuss the results at a basin scale and over the particular shelf zone of the Gulf of Lion.

12.2
Study Area Characteristics

Our zone of study and validation of the altimetry data sets is the NWM with special emphases on the Ligurian Sea (see blue rectangle in Fig. 12.2 and also refer to Fig. 12.3) and the Gulf of Lion (see pink rectangle in Fig. 12.2).

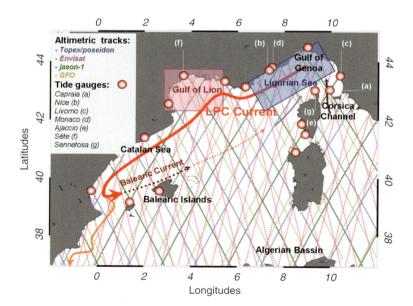

Fig. 12.2 Liguro Provençal Catalan Current multi-satellite ground track and TG locations

Fig. 12.3 Locations of Envisat (circle) and TOPEX/Poseidon (diamond) 1 Hz normal points near Genoa and Imperia with local bathymetry

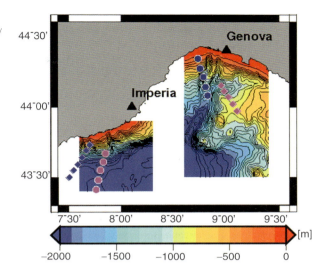

The dynamics of the NWM is dominated by the variability of a coastal slope current, which arises from the merging of the eastern and western Corsica currents (see Fig. 12.1). This current, called the "Liguro Provençal Catalan" (hereafter, LPC) Current or "Mediterranean Northern Current," flows to the southwest along the continental slope until it reaches the Balearic Islands continental shelf. From there, one branch recirculates in the direction of Corsica, forming the so-called Balearic Current, and another branch flows in the direction of the Alboran Sea. The LPC Current exhibits a permanent density front between the light water on the shelf and the denser water over the open ocean. This density front, which has marked seasonal and interannual variability in addition to baroclinic instabilities, is strongly linked to wind forcing and sub-surface topography (for more details on the current characteristics, refer to Millot 1987, 1991; Font et al. 1995). The LPC Current should be observed by altimetric measurements (see Fig. 12.2 for the altimetric track locations), even though this current is only 20–30 km wide and mainly located in a 30 km coastal band. The LPC Current is a complex dynamical system in a micro-tidal zone and a transition area between the continental shelf and the open ocean. It follows a pathway along the continental shelf, which is widest in the Gulf of Lion. Here, intrusions of the LPC Current bring in open ocean water onto the continental shelf (Petrenko et al. 2005). In winter, very strong winds (Tramontana, Mistral) generate upwelling and downwelling, coastal eddies and deep-water convection that can be monitored with altimetry (for details refer to Herrmann et al. 2009).

As mentioned in Send et al. (1999), the complexity of small space and time scales and intense gradients means that understanding the dynamics of the western Mediterranean is quite challenging. Detecting such complex dynamics requires adequate spatial and temporal sampling. Altimetry can be used for this purpose; however, as has been mentioned before, monitoring coastal circulation with altimetry is still a challenging issue because of the numerous limitations. The following sections focus on existing specific coastal data sets and their validations.

12.3
Main Characteristics of Existing Data

12.3.1
Standard Data: AVISO Along-track Data

An along-track multi-satellite SLA product especially dedicated to regional studies over the Mediterranean Sea (Delayed Time-SLA "Update," SSALTO/DUACS User Handbook, 2006[1]) is available and distributed by AVISO. Specific algorithms and corrections have been developed to improve the quality of this regional altimetric data product.

The SSH is corrected for geophysical effects (wet and dry troposphere, ionosphere, etc.; for details, see Le Traon and Ogor 1998). Note two corrections that have been recently updated. The classical inverted barometer (IB) correction, which formulates the static response of the ocean to atmospheric pressure forcing ignoring wind effects, is replaced by a dynamic atmospheric correction based on a barotropic model MOG2D (Modèle aux Ondes de Gravité 2-Dimensions) global simulations. This improves the representation of high-frequency atmospheric forcing, as it takes into account wind and pressure effects (Carrère and Lyard 2003). The dynamic atmospheric correction applied to altimetry combines the high frequencies of the barotropic model MOG2D–G (Carrère and Lyard 2003) forced by pressure and wind (from ECMWF analysis), with the low frequencies of the IB correction. The other modification concerns the tide model, since GOT99.2 is replaced by the updated GOT00.2 (Ray 1999). These two changes, as it is shown in Volkov et al. (2007) and Pascual et al. (2008), constitute an important step forward with respect to previous altimeter products (e.g., those used in Pascual et al. 2006). This is particularly relevant close to the coast (defining here the coastal ocean as the region over the continental shelf and slope, ranging from 0 to 100 km from the coast), where altimetric observations are often of lower accuracy.

The corrected SSH obtained for each mission is then inter-calibrated with a global crossover adjustment of the ERS-2, Geosat Follow-On (hereafter GFO) and Jason-1 (hereafter J1) orbits, using the more precise orbit of TOPEX/Poseidon (hereafter T/P) data as a reference (Le Traon and Ogor 1998). Next, the data are resampled every 7 km along the tracks using cubic splines. A mean profile, <SSH>, is removed from the individual SSH measurement, yielding SLA, for the different missions. The mean profile contains the geoid signal and the mean dynamic topography over the averaging period. For J1 and ERS-2, a mean profile calculated over a 7-year period (1993–1999) is used. In terms of GFO, only several months of data are available, and, for that reason, a specific processing is applied in order to get mean profiles that are consistent with J1 and ERS mean profiles.

12.3.2
Improved Pre-processed Data: RECOSETO Data

RECOSETO is a project founded by DFG (Germany) that aims at an improved estimation of the sea level variability in two specific coastal regions with complementary characteristics, namely, the Gulf of Biscay on the Atlantic Ocean shelf and the NWM Sea.

---

[1] See http://www.aviso.oceanobs.com/en/data/products/sea-surface-height-products/regional/index.html

It consists in an improved pre-processing of altimetry data at 1 Hz and higher frequency because of (a) a combination of improved selection criteria, (b) selection of the most suitable corrections, and (c) dedicated waveform processing.

The 1 Hz T/P Level 2 GDR are extracted from the Radar Altimeter Database System (RADS) (Naeije et al. 2002). The Retracked R-GDR from the T/P mission are provided by NASA-JPL Physical Oceanography Distributed Active Archive Center (PODAAC, P. Callahan, release 2.1). The temporal coverage is from July 28, 2000 to August 11, 2002, corresponding to cycles 290 to 364 (without 362). Two types of retracked data are available, corresponding to two different algorithms (RGDR1 = least squares, RGDR2 = maximum likelihood a posteriori estimation), with both 1 and 10 Hz data.

Envisat SGDR[2] data are provided by ESA in the interval from February 2006 to July 2007 (cycles 44–45, 47, 50, 52–59). Four different retrackers are operationally applied to Envisat waveforms. We further retrack Envisat waveform data in-house with four different retracker procedures: OCOG (offset center of gravity), Beta-5, Threshold (Davis 1995), and improved threshold retracker (Hwang et al. 2006).

12.3.3
Improved Post-processed Data: X-TRACK Data

A detailed description of the methodology used to create this SLA product can be found in the chapter by Roblou et al., this volume. Here, we present a summary of their section dedicated to data processing.

The X-TRACK data processing was originally developed within the framework of the ALBICOCCA project in the NWM Sea. A key task of this French–Italian project was to design and implement an altimeter data processing tool adapted to coastal ocean applications (e.g., altimeter absolute calibration and drift monitoring, water transport estimation, numerical model validation). Additional developments led to the X-TRACK altimeter data processor (Roblou et al. 2007). In order to observe complex dynamics such as coastal Kelvin waves or boundary currents of the coastal and shelf seas, improvements of the space and time sampling of the altimeter products have been added to the X-TRACK processor. An orbit and large-scale error reduction method has been applied in order to reduce the inter-mission biases. The higher along-track sampling (from 10 and 20 Hz, e.g., Bouffard et al. 2006, 2008a, b; Bouffard 2007) is also used.

In summary, the multi-satellite, X-TRACK-SLA data set used for this study includes:

- Higher along-track sampling
- A specific-coastal data edition strategy
- An improved retrieval and sea state corrections
- An optimal, high-resolution tidal correction, as provided by T-UGOm (Toulouse Unstructured Grid Ocean model) 2D (MOG2D follow-up) simulations
- A high-resolution correction over the Mediterranean Sea for the short-period ocean response to atmospheric forcing, also provided by T-UGOm 2D simulations
- A large-scale error reduction
- A high-resolution Mean Sea Surface (MSS) consistent with the improved SSH data

---

[2]Refer to http://earth.esa.int/dataproducts

As a result of the X-TRACK processing, SLAs are interpolated onto an along-track grid for calculating time series.

12.4
Benefit of Improved Strategies over the Northwestern Mediterranean Sea

The main objective of this section is to assess the performance of improved pre- and post-processing strategies integrated in the RECOSETO and X-TRACK processors.

12.4.1
Improvements from RECOSETO Project

12.4.1.1
Methods and Data Used

The SSH derived from T/P and Envisat altimeter data are investigated in the coastal zone of the Mediterranean Sea to: (1) establish to what extent off-line retracked altimeter data are applicable close to the coast and (2) assess the impact of non-uniform local conditions (e.g., sea-land and land-sea along-track direction) on the altimeter retrievals.

For this purpose, the RECOSETO satellite-derived sea level heights have been compared to the sea level variability measured from TGs at selected sites. Ocean and pole tide corrections are not applied and we do not account for the effect of atmospheric pressure on sea level (Fenoglio-Marc et al. 2004). Hourly TG measurements at the stations of Genoa and Imperia (ISPRA, see Fig. 12.3) are directly compared to 1 Hz co-located altimetry at normal points (NP, virtual positions along a reference ground track on which are spatially interpolated altimetric SSH in order to compute along-track time series at fixed locations[3]). The analysis is based on the following parameters: number of usable time-points, correlation and root mean square (RMS) of the differences between altimeter and TG time series. Only time series with a minimum of three observations per NP are considered.

To validate the improvement in sea level height SSHs obtained by retracking the waveform, the improvement percentage (IMP, see Anzenhofer et al. 1999; Hwang et al. 2006) is computed as:

$$IMP = \frac{\delta_{raw} - \delta_{retracked}}{\delta_{raw}} \times 100 \qquad (12.1)$$

where δ_{raw} and $\delta_{retracked}$ are the standard deviations of the differences between raw SSHs, and retracked SSHs and geoid heights (EGM2008), respectively.

---

[3]Refer to http://www.deos.tudelft.nl/ers/operorbs/node10.html for a more detailed definition

12.4.1.2
Wet tropospheric Correction

Firstly, the standard corrections and selection criteria are applied to the altimeter data. We further adjust the standard criteria based on the analysis of data rejection and the distance analysis.

The main cause of data rejections near coast is the editing on the wet tropospheric radiometer correction and on the standard deviation of the 18 Hz retracked ranges that eliminate, respectively, 8 and 4% of the Envisat GDR data (Fig. 12.4). Data eliminated because of the radiometer wet tropospheric correction appear to be usually concentrated on a given track. The statistics of the corrections over the complete Mediterranean show a difference of 1 cm on the average. We therefore use the wet tropospheric model corrections (ECMWF) to avoid data rejections due to the radiometer, while data eliminated because of the big standard deviation of the 18 Hz ranges are marked to be retracked.

We compute SLA relative to the CLS01 mean sea surface at 1 and 18 Hz. Envisat SLA values are calculated using the open ocean retracked range from the SGDRs.

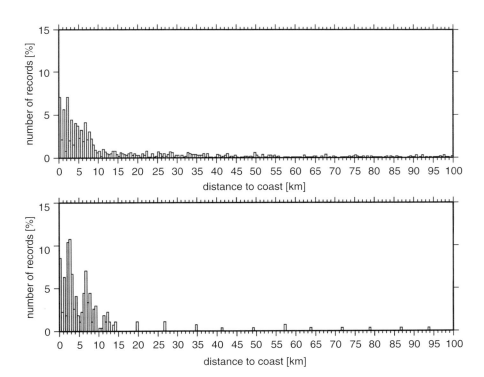

Fig. 12.4 Histogram of Envisat records rejected for the wet tropospheric correction using MWR (*below*) and of records rejected only because of the standard deviation of 18 Hz retracked ranges (*above*). (Adapted from Fenoglio-Marc et al. 2008)

12.4.1.3
Distance Analysis of T/P and Envisat Measurements

Four types of distances from the coast are analysed for each record to assess the performances of the various satellites: (1) distance to TG, (2) distance to the nearest coastline, (3) along-track sea-land (sl) distance, and (4) along-track land-sea (ls) distance. Fig. 12.5 shows for each distance the distribution of the minimum value for each pass over all the analysed cycles for T/P (red) and Envisat (black) at 1 Hz.

We found that when applying analogous selection criteria to both satellites, Envisat data perform better in coastal regions with up to 15% more 1 Hz data available in the last 5 km. As expected, sea-land transitions perform better than land-sea transitions. Envisat provides more usable data near the coast than T/P independently from the land-crossing direction.

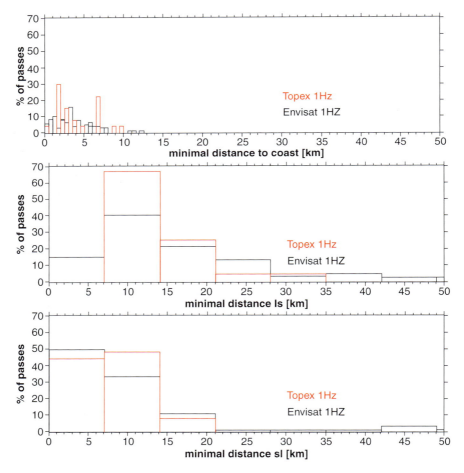

Fig. 12.5 Histogram of the percentage of passes for a given type of minimum distance: distance to coast (*top*), minimum along-track distance land-sea (*centre*) and sea-land (*below*). (Adapted from Fenoglio-Marc et al. 2008)

In the land-sea direction, Envisat provides data up to 7 km offshore, while no T/P data are available in that distance range. In the sea-land direction, 90% of all passes in the Mediterranean Sea of both satellites provide usable data closer than 15 km to the coastline.

12.4.1.4
Improved Retracking

Over coastal regions and land, the altimeteric waveform is contaminated by non-uniform local conditions e.g., the shoreline topography and local ocean tides. Therefore, waveform retracking is implemented to perform waveform data reprocessing and to improve the accuracy of altimeter ranges (Anzenhofer et al. 1999; Deng et al. 2002). See also the Chapter by Gommenginger et al., this volume.

A few subsets of altimeter data have been made available such as the T/P Retracked R-GDR, provided by NASA-JPL (see Sect. 12.3.2). Two types of retracking data are available, corresponding to two different retracking algorithms (least squares and maximum a posteriori).

We have compared the sea level heights for T/P GDR, R-GDR (from both retrackers) for a given cycle and pass (cycle 328 and pass 44) near the Genoa TG station (Fig. 12.6). The comparison with TG records shows a higher correlation and smaller RMS differences when the retracked data are used. Moreover, the closest distance to the coast is obtained with retracked data (17.2 km), where it is 27.3 km for the GDR data. A bias of a few centimeters between the GDR and the R-GDR sea level anomalies is observed.

Within the RECOSETO project, the Envisat waveforms are retracked using the four retracker methods listed in Sect. 12.2. The data and corrections provided in the SGDR Level 2 are used to finally produce the retracked sea level heights.

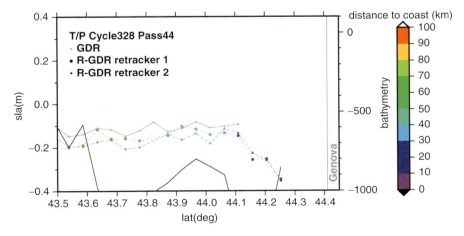

Fig. 12.6 Sea level anomaly relative to the CLS01 mean sea surface along T/P cycle 328, pass 44 in the Ligurian Sea from Geophysical Data Record (GDR) and Retracked Geophysical Data Record (RGDR) retrackers 1 and 2. The bathymetry is given by the blue continuous line (Adapted from Fenoglio-Marc et al. 2008)

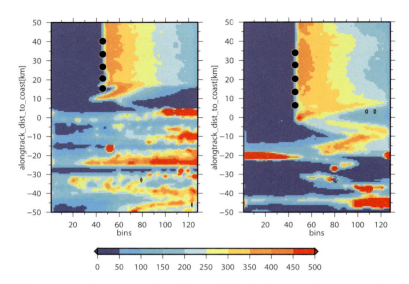

Fig. 12.7 Envisat waveforms at a sea-land (Genoa, *left*) and a land-sea (Imperia, *right*) transition and locations of Envisat 1 Hz normal points (circles)

The power decrease, typical for ocean type waveforms, is found very near to land (almost at the zero crossing) in sea-land transition (see Fig. 12.7, where the power per bin of each waveform is plotted vs. the distance to coast for Envisat).

To quantify the improvement of the retracking process, we compute the standard deviation of the differences between sea surface heights (SSH) calculated from both raw and retracked ranges and geoid heights as well as the improvement percentages (IMP) (Eq. 12.1).

We have retracked the waveforms for cycle 59 track 801 near Genoa for latitudes between 43.3° and 44.3° (307 waveforms). The resulting SSHs are shown in Fig. 12.8. The success rate of the Beta-5 method is 87%, and 100% for the other methods. A negative improvement factor indicates that retracking deteriorates the SSHs which occurs for the OCOG retracker (−2.8). The Beta-5 and the improved threshold methods (both with an IMP of 59.8) turn out to be the most suitable methods with a small improvement with respect to the onboard ocean retracker (58.9). The bias between the retracked SSHs and EGM2008 depends on the retracker used.

12.4.2
Improvements from X-TRACK Processor

12.4.2.1
Methods and Data Used

In order to validate the multi-satellite X-TRACK-SLA data set, we compare the altimetric SLA products with French–Italian TG sea level records available from 1996 to 2004 and provided courtesy of the SONEL and ISPRA networks. The TG locations are shown in Fig. 12.2 (letters "a "to "g" indicate the TG used for T/P), the time series lengths are

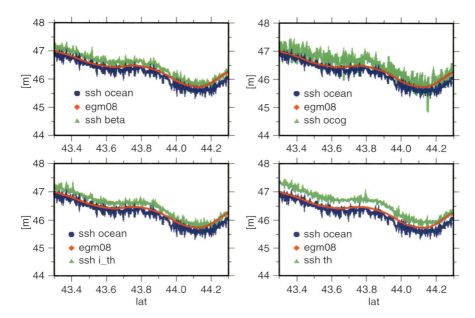

Fig. 12.8 SSH from Beta-5, OCOG, Threshold and Improved Threshold retrackers compared to GDR ocean SSH and EGM2008, cycle 59, pass 801 (Genoa)

between 1 and 8 years. For optimal consistency between the TG sea level and the altimetric SLA, the barotropic sea level response due to the total tidal forcing and the response of the ocean to wind and pressure forcing have been removed from the TGs, the oceanic response to atmospheric forcing is corrected by applying the MOG2D-Medsea model outputs (e.g., Jan et al. 2004; Mangiarotti and Lyard 2008) and an harmonic analysis of this residual signal is used to obtain the tidal constants needed to compute the ocean tidal height. The time average of the sea level over the record length is also removed, as TG records are not attached to an absolute terrestrial reference frame.

The following sections will show some validations of the improved post-processing methods used in the X-TRACK processor.

12.4.2.2
Improved Space-Time Resolution

Observing the small mesoscale and LPC Current sea level variability in the NWM is a challenging issue which requires intensive spatial sampling. Energetic small-scale ocean dynamics are found along the continental slope of the NWM coastal area (see Sect. 12.2) and their observation with standard altimetry is very difficult. The first internal Rossby radius in the NWM Sea is only a few kilometres (about 14 km according to Robinson et al. 2001) and despite the low signal-to-noise ratio, characteristic length scales of energetic disturbances of the LPC Current could be of sufficient size and magnitude to be detected with high-resolution altimetry.

Fig. 12.9 Averaged along-track spatial spectrum at 10 Hz (X-TRACK, red colour), T/P track 146

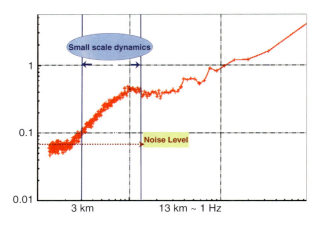

In order to assess the ability of the 10/20 Hz along-track sampling (retracked with the standard ocean1 procedure) to observe small coherent dynamical signals rather than noise, we have computed along-track SLA spatial spectrum with all of the altimetric ground tracks over the NWM. The analysis of the spectra shows that the 10/20 Hz data exhibit the onset of white noise below 3 km, a value much lower than the standard 1Hz signal (for which the smallest resolved scale is about 13 km). Fig. 12.9 shows an example of such a T/P 10 Hz spatial spectrum (average in time over all the available cycles) as a function of the distance along-track. Up to 3 km the signal is dominated by white noise, whereas the oceanographic signal can be distinguished from the noise level at scales greater than 3 km (a second change in the slope can be observed at about 10 km; we do not propose here explanations, which require specific investigations that have not been addressed in the present study). On the basis of these findings, the 10/20 Hz along-track data are thus sub-sampled every 1.5 km within the high rate SLA processing stage of the X-TRACK processor.

As shown in Bouffard et al. (2008b) for the Corsica channel area, such a simple post-processing approach combining high frequency sampling with a coastal-oriented editing strategy is sufficient to increase the quality and availability of coastal X-TRACK data compared to the 1 Hz AVISO products. However, this post-processing technique does not provide reliable SLAs at a high-frequency sampling and right up to the coastline because the Brown waveform model (Brown 1977) is not adapted here. The altimeter waveforms contaminated by land reflections should probably be retracked. Indeed, only a re-processing of the native radar echo data could effectively improve the raw high-frequency range in the 20–30 km coastal band.

12.4.2.3
Improved Data Editing and Correction Rebuilding

Editing GDR corrections is a key point for eliminating spurious data. It has already been demonstrated that edition criteria recommended for deep ocean are not relevant for coastal systems. Roblou et al. (2007) have set up a list of specific coastal criteria that allow them to collect only valid data (meaning range measurements plus associated corrections). Where possible, they rebuild any missing corrections using a polynomial interpolation/extrapolation technique (the method is fully described and illustrated in the chapter by Roblou et al., this volume).

Fig. 12.10 shows the impact of the methodology used in the X-TRACK processor along the T/P track 222, over the Ligurian Sea, in the neighbourhood of the Nice TG. We have calculated the percentage of RMS explained at each altimetric point. The higher the percentage, the higher the consistency between the altimetric observation and the Nice TG measurements. Fig. 12.10 illustrates that with the X-TRACK derived data this consistency not only increases for the measurements close to the coast, but also over the whole pass. At the point of maximum percentage of RMS explained, there is an improvement of 4% due to the improved editing/reconstruction techniques with respect to the standard data. This corresponds to a decrease in the residual standard RMS by about 2 mm.

Such statistical evaluations have also been computed along 12 other altimetric T/P ground tracks located in the neighbourhood of seven other TGs (see Fig. 12.1). Table 12.2 summarizes the averaged scores at the position of maximum percentage of RMS explained by satellite altimetry estimates in comparison to the TG signals.

Table 12.1 shows that the improved editing/reconstruction method allows us to significantly increase the consistency between the altimetry and the sea level stations time series. Indeed, the percentage of RMS explained increases by about 7%, the correlation improves

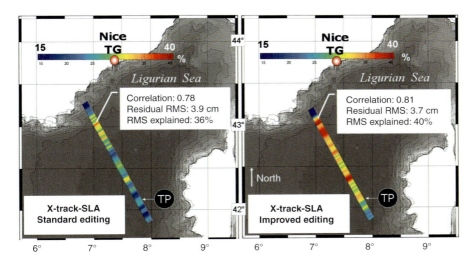

Fig. 12.10 Along-track percentage of Nice TG SLA explained by altimetric SLA time series. *Left side*: standard editing, *right side*: improved editing and correction reconstructions

Table 12.1 Averaged statistical result of intercomparisons between 13 T/P altimetric tracks and 8 TGs at the point of maximum root mean square (RMS) explained

	Number of altimetric data in common with the TG time series	Correlation[4]	Residual RMS	% of RMS explained
Standard editing	104	0.80	4.1 cm	37%
Improved editing	99	0.83	3.7 cm	44%

---

[4]All the correlations shown in this paper are at a 95 % significance level

by 0.03, whereas the residual RMS decreases by 4 mm. The improved editing/reconstruction method also leads to a decrease of about 5% in the number of points in the altimetric time series. This corresponds to the elimination of spurious data because of our stringent selection procedure (also refer to Fig. 12.10, right side: spurious coastal measurements are flagged and eliminated with X-TRACK). If this method is applied to the 1 Hz data (as in Vignudelli et al. 2005), the improvement is less marked with time series reduced by about 10% and an averaged percentage of RMS explained of 39.5%.

12.4.2.4
Improved de-aliasing corrections by using regional modelling

In the multi-satellite X-TRACK-SLA data set used for this study, a regional, high-resolution correction based on T-UGOm 2D (called hereafter, MOG2D-Medsea) simulations is provided as an alternate correction to the lower resolution global corrections for both tides and atmospheric forcing provided in the GDR products (for more details on the model's characteristics, refer to the chapter by Roblou et al., this volume).

In order to quantify the impact of using the MOG2D-Medsea regional model rather than standard global models in our coastal area, we compare the standard deviation of the sea level difference at multi-satellite crossover points over the continental shelves (for depths less than 500 m). The exact crossover point locations are calculated for each cycle. The along-track altimetric SLAs within a 5km neighbourhood are then linearly interpolated onto the theoretical crossover point location for each cycle. Then, the time series at the 68 crossover points for each track are interpolated in time on a common temporal grid. For each crossover point, the satellite with the lower repeat period is chosen as the temporal reference grid.

We then compute the crossover difference between the two X-TRACK-SLA ascending and descending passes time series using (i) global de-aliasing corrections, i.e. FES2004 (Lyard et al. 2006) as the tidal correction and MOG2D-G (Carrère and Lyard 2003) as the atmospheric correction, or (ii) regional de-aliasing correction, i.e. MOG2D-Medsea regional simulation outputs for both tides and atmospheric effects. We assess RMS of these crossover SLA differences using the standard and regional de-aliasing corrections. The average of the residual RMS at all of the crossover points over the continental shelves is about 5.8 cm using the global models. When MOG2D-Medsea is applied both for tidal and atmospheric effects, the RMS values are reduced at most of the crossover points, especially in the Gulf of Lion, in the Ligurian Sea and south of the Balearic Islands. The average residual RMS at the multi-satellite crossover points over the whole domain is 5.1 cm. This illustrates the efficiency of the local model over continental shelves and in coastal seas. We note that further offshore from the continental shelves, de-aliasing correction provided by the MOG2D-Medsea simulations outputs also shows better results than the global corrections in terms of residual RMS at crossover points: respectively, 5.0 and 5.8 cm on average over 258 crossover points.

Fig. 12.11 shows the percentage of the Nice TG RMS explained by each altimeter SLA time series available along the track. The left panel illustrates the residual sea level variance explained (as a percentage) when applying de-aliasing corrections computed from

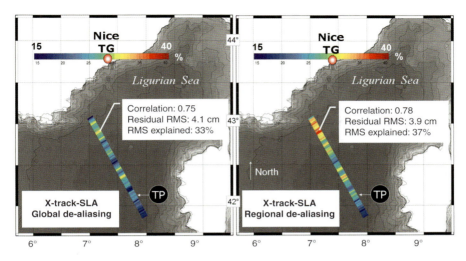

Fig. 12.11 Along-track percentage of Nice TG SLA explained by T/P 222 altimetric SLA time series

global models (e.g., FES2004 plus MOG2D-G). The right panel illustrates the residual sea level variance explained when applying de-aliasing corrections from the regional model (MOG2D-Medsea). According to Fig. 12.11, the percentages are significantly higher when applying the specific, regional de-aliasing model than with the global ones. Again, this is not only true in close proximity of the coast but also along the whole T/P track 222. At the location of the maximum percentage of RMS explained, the gain in explained RMS because of MOG2D-Medsea corrections is about 4%, and the correlation between the TG and altimetry is also increased (0.78 against 0.72).

Similar statistical evaluations have also been computed along 12 other altimetric tracks located in the neighbourhood of seven other TGs. The following table summarizes the averaged statistics at the point of maximum percentage of RMS explained by altimetry in the TG signals.

Table 12.2 shows that using a regional de-aliasing rather than a global one leads to an increase of 4.5% in terms of the RMS variability explained by altimetry SLA estimates at TGs. It also produces higher correlations and a decrease in the residual RMS.

Table 12.2 Averaged statistical results of intercomparisons between 13 T/P altimetric tracks (at 1Hz) and 8 TGs at the point of maximum RMS explained. "Global de-aliasing" stands for the combination of FES2004 + MOG2D-G corrections whereas "Regional de-aliasing" stands for the use of MOG2D-Medsea for both tides and atmospheric effects

	Correlation	Residual RMS (cm)	Percentage of RMS explained (%)
Global de-aliasing	0.78	4.2	35.0
Regional de-aliasing	0.80	4.0	39.5

12.4.2.5
Improved Multi-satellite Data Sets

Taking advantage of multiple satellite missions would increase the coverage of the coastal area (Bouffard 2007), provided the data sets are homogeneous and, in particular, large-scales errors are removed. The following experiments aim at validating the large-scale error reduction (the so-called LSER, fully described in the chapter by Roblou et al., this volume) method used in the X-TRACK processor and quantifying its impact at satellite crossover points.

We compute the RMS SLA differences at each single satellite crossover point. The mean number of data available for the crossover time series computations are between 20 (for Envisat) and 256 for T/P. Table 12.3 shows that the LSER algorithm improves all of the SLA time series in terms of residual RMS reduction. The most significant impact of this large-scale noise reduction concerns the GFO satellite with a decrease of 2.4 cm of the residual RMS at crossovers. T/P and J1 exhibit minimum residual RMS at crossover points, both with or without the LSER algorithm. This is due to the good accuracy in their orbit determination (better than 2.5 cm for T/P, Ries and Tapley 1999). In addition, for T/P and J1 missions, the time difference between the ascending and the descending track is always less than 10 days which reduces the amount of mesoscale variability which can contribute to the crossover SLA differences. The worst results are obtained with the GFO satellite (residual RMS is about 8.7 cm without the LSER algorithm) because of an inaccurate orbit determination. Envisat has a lower RMS than GFO despite its repeat orbit period being twice as long. Its lower residual RMS without LSER is due to its relatively good orbit determination. We have also computed the RMS at dual-satellite crossover points (see "Total" in Table 12.3). The differences of SLAs are only taken into account if the temporal delay between two SSH measurements at crossover point is less than 10 days. Without the LSER algorithm, the averaged residual RMS difference is of 5.2 cm. The LSER algorithm significantly improves this result with a residual RMS difference of 4.8 cm on the average.

12.4.2.6
Improved Mean Sea Level Computation

Another improvement integrated in the X-TRACK processor concerns the MSS calculation. Since the marine geoid is not known with sufficient accuracy, the altimetric MSS is removed from the corrected SSH before computing SLAs. For each satellite mission, the MSS is generated in the same way: from the average of the longest SSH time series of each

Table 12.3 Impact of the LSER algorithm at single and dual satellite crossover points

Satellites		T/P	GFO	JASON	ENVI.	TOTAL
Number of crossover available		11	33	11	75	
mean number of data in the time series		256	47	76	20	
Mean RMS at crossover points (cm)	No LSER	4.3	8.7	3.8	5.5	5.2
	LSER	4.2	6.3	3.6	5.2	4.8

satellite track over an entire number of years. This also allows us to remove the mean bias (because of range and correction biases) between the different missions. In the X-TRACK processor, a high-resolution – consistent with the SSH data sets – MSS is computed from an inverse method (more details in the chapter by Roblou et al., this volume).

To check the validity of the obtained optimal MSS, Vignudelli et al. (2010) have compared TG records with the altimetric signal taken in the closest vicinity of the TGs in the NWM Sea. For instance, the Nice TG is located at the north-eastern extremity of the T/P ground track 009 and its records starts in 1998, meaning that ~100 T/P cycles are usable. T/P observations are located at about 30 ± 5 km from the coast. The sea level difference standard-deviation is 9 cm when using the optimal MSS computed by the X-TRACK processor, and 13 cm when using the MSS CLS01 product. The optimal mean sea surface model performs better in the near-coast regions, because its higher spatial resolution mean sea surface permits better modelling of the local geodetic slope. There are also differences between the X-TRACK optimal MSS and the MSS CLS01, mainly centred on zero, with oscillations apparently related to CLS MSS01 grid resolution and a bump in the coastal strip. These discrepancies could arise from a lack of near-coast altimeter data in the data set used to derived MSS CLS01, and consequently be the result of spurious extrapolation between the satellite MSS and the EGM96 geoid.

12.5
Intercomparison Experiments: X-TRACK-SLA Versus DT-SLA "Upd"

This section compares X-TRACK-SLA (including the whole specific coastal post-processing described in the previous section) and the regional DT-SLA product made available by AVISO (see Sect. 12.3.1). The following comparisons will focus on the J1, T/P and GFO satellites. For Envisat, the common time period between the SLA data set and the TG records is too short to give robust statistics. Moreover, comparisons between altimeter data sets and TG sea level records are only made when the maximum distance between them is less than 150 km. The TG time series are interpolated in time onto the SLA estimates measurement time. The comparisons are only computed if the common period between the altimeter and the TG records is greater than 1 year.

12.5.1
Basin Scale Statistics

12.5.1.1
Data Availability in Coastal Area

To begin with, we consider the availability of the two altimeter SLA data sets in the neighbourhood of the TG locations. Since the TGs are located at the coast, this first statistic gives information about the quantity of altimeter data available in the coastal zones for both data sets.

Table 12.4 shows the number of TGs used for the cross-comparisons between the X-TRACK-SLA and the DT-SLA product, the number of satellite tracks available in the

Table 12.4 Comparisons between the X-TRACK and the DT-SLA data sets in terms of distance to the TGs and number of data available. In bracket, statistics at the minimum distance between the TG and the closest altimetric point

Satellites	Kind of data	Number of TG used	Number of tracks available	Mean distance (km)	Mean number of data per TG
T/P	X-TRACK-SLA	7	6	88 (59)	99 (88)
	DT-SLA		5	111 (73)	90 (87)
GFO	X-TRACK-SLA	14	13	107 (60)	40 (38)
	DT-SLA		11	86 (69)	38 (37)
J1	X-TRACK-SLA	13	7	105 (49)	39 (35)
	DT-SLA		6	110 (60)	39 (37)

neighbourhood of TG locations (measurement distance < 150 km), the mean distance between the altimeter SLA estimates and the TG locations, and the number of common data available for the comparison. These statistics are computed at the minimum residual RMS point (i.e. the points where the RMS of altimeter SLA minus TG SLA is minimal). Also shown in brackets are the statistics at the minimum distance to the TG.

Table 12.4 demonstrates that both at the nearest and at the minimum residual RMS point, the X-TRACK-SLA time series are longer than the DT-SLA ones. This is especially the case for T/P and GFO whereas the score concerning X-TRACK-SLA and DT-SLA products are equivalent for J1. This is essentially due to the fact that for J1 the 20 Hz data is not distributed by the operational centers if there is less than 10 data around the 1 Hz range measurement.

Although both altimetric data products are based on the same original GDR data, the DT-SLA processing systematically eliminates large sections of altimetric tracks, especially in coastal zones and over continental shelves. This is illustrated in Fig. 12.12.

In the Corsica Channel, several tracks are eliminated in the DT-SLA product whereas the X-TRACK-SLA processed in this area exhibit a realistic sea surface RMS and a good consistency with the TG measurements (see Bouffard et al. 2008b). Other large sections of altimetric tracks are also systematically eliminated in the coastal bands of the Ligurian Sea, the Balearic Sea and the Gulf of Lion whereas they exhibit good statistics in the X-TRACK-SLA data set and would thus be considered as valid. These eliminated data in the DT-SLA are more important in the shallow water and coastall zone. In those areas, onboard radar altimeter and the radiometer footprints are contaminated by land and islands, which strongly affect the quality of SSH estimates. Restrictive editing criteria on corrections in the DT-SLA products seem to eliminate valuable coastal measurements.

As a first conclusion, the X-TRACK processor allows us to generate more coastal data, and closer to the coastline than the processing scheme for regional DT-SLA products. The next step consists in the assessment of their quality.

12.5.1.2
Altimetric Data Quality

To evaluate the qualitative improvements due to the X-TRACK processing, we compare the X-TRACK-SLA and DT-SLA at several TGs. Table 12.5 synthesizes the averaged statistical results of the T/P, GFO, and J1 SLA estimates from the two data sets compared

12 Introduction and Assessment of Improved Coastal Altimetry Strategies

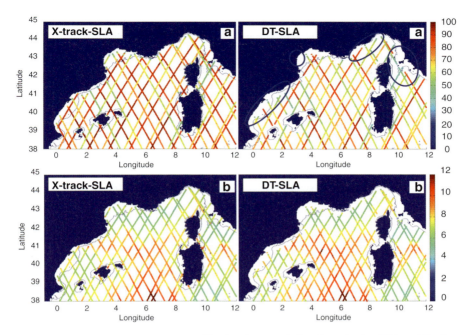

Fig. 12.12 *Top*: percentage of data availability for along-track data from the J1, GFO and Envisat satellite missions. The blue circles highlight regions where AVISO data are greatly reduced in number. *Bottom*: RMS (in cm) of the along-track SLA. Left panels show the improved X-TRACK multi-mission altimetry data set (**a**) and the right panel the regional along-track DT-SLA product (**b**)

to TGs located in their neighbourhood (<150 km). For each satellite, the same TGs are taken into account for the cross-comparisons and the statistics are computed at the minimum residual RMS points (the number of data in each time series and the number of TG used per satellite are shown in Table 12.4).

Table 12.5 shows that the X-TRACK-SLA processing exhibits better statistics in comparison to the TG for both the T/P and the J1 satellites whereas the DT-SLA products show better results for the GFO satellite. Concerning T/P, the mean residual RMS for the X-TRACK-SLA and the DT-SLA are, respectively, 3.7 and 4.4 cm. Moreover, the mean

Table 12.5 Comparisons between the X-TRACK-SLA and the DT-SLA data sets in terms of correlation with the TGs, residual RMS, and percentage of RMS explained

Satellites	Kind of data	Correlation	Residual RMS (in cm)	Percentage of RMS explained (%)
T/P	X-TRACK-SLA	0.89	3.7	44
	DT-SLA	0.85	4.4	34
Jason-1	X-TRACK-SLA	0.82	3.9	51
	DT-SLA	0.75	4.8	46
GFO	X-TRACK-SLA	0.73	4.5	28
	DT-SLA	0.80	4.2	36

correlation between the X-TRACK-SLA and the TG are greater than for the DT-SLA by about 0.04. The X-TRACK-SLA processing applied to the J1 satellite measurements also shows a significant decrease in the residual RMS and the mean correlation is greater by about 8% compared to the DT-SLA.

Given that the orbit accuracies of T/P and J1 are very good, the orbit and large-scale error reduction method has a small influence on the statistics. Concerning GFO, the orbit has lower precision than the T/P and J1 orbits. The correlations and percentage of RMS explained are notably better with the DT-SLA data than the X-TRACK-SLA because of a better large scale and orbit error reduction method applied (Le Traon and Ogor 1998), based on a multi-satellite crossover minimization. Indeed, their method uses the combined T/P and J1 orbit information to improve the GFO orbit. If we adjust the large scales (>1,000 km) of the X-TRACK-SLA to the large scale of the DT-SLA, by applying a simple bias difference, the major impacts due to the residual GFO orbit errors are removed. In that case, the GFO X-TRACK-SLA statistics are improved compared to the DT-SLA GFO product with a mean correlation of 0.88, a residual RMS of 3.9 cm and a percentage of RMS explained of 39%. This method allows us to take advantage of the DT-SLA large wavelength orbit error reduction but also keep the small-scale coastal signature emerging from the noise, because of the customized X-TRACK processing.

12.5.2
Focus on the Gulf of Lion

In this section, we focus on a particular dynamical area of the NWM Sea, the Gulf of Lion, where we compare the Sète TG with the closest T/P along-track data. The Sète TG is located along the coast of the Gulf of Lion at 43.3°N and 3.68°E. The Gulf of Lion is a relatively shallow water zone, characterised by a large continental shelf, and dominated by intense ocean-atmosphere mechanisms, as described by Estournel et al. (2003). The Gulf of Lion is also marked by intrusions of the LPC Current, which usually flows along its continental slope to the southwest.

The aim of this section is to assess the quality of the X-TRACK-SLA and the DT-SLA products and see how well they characterize the different coastal dynamics.

12.5.2.1
Statistical Intercomparisons

Firstly, we have compared the altimetric time series length of the two altimetric data sets. According to Fig. 12.13, the X-TRACK processing allows us to recover more data than for the DT-SLA both close to the coast and offshore. This is particularly useful for sampling the small-scale processes occurring over the continental shelf of the Gulf of Lion and associated with the LPC variability.

Now, we have to evaluate the quality of the recovered data in order to assess whether physical signals emerge from the noise and whether coherent small-scale processes can be monitored with the X-TRACK altimetric data. For this, we consider the statistics of T/P altimetric data from tracks 146 and 187 located in the 150 km neighbourhood of the Sète

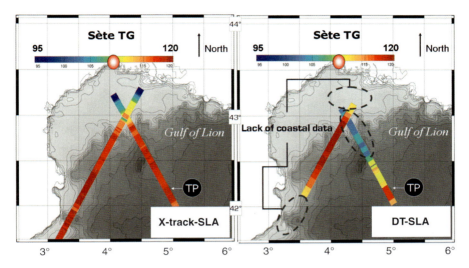

Fig. 12.13 Number of data in the altimetric time series along the track 187 and 146 of T/P (from 1/1996 and 11/1999) and also sampled by the TG measurement

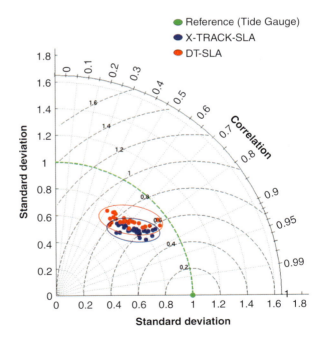

Fig. 12.14 Taylor diagram (The statistics have been normalized by the TG observations)

TG (see Fig. 12.14). The correlation coefficient, the RMS difference and the standard deviations of the two patterns are linked in a simple relationship and can therefore be represented on the same Taylor diagram (Taylor 2001), in order to quantify graphically how closely the data sets resemble each other. In this study, the TG SLA data set is considered as the "reference" field, and the two altimetric along-track SLA data sets (X-TRACK and AVISO) are the "test" fields. The radial distances from the graph origin to the points represent the standard deviation, and the azimuthal positions give the correlation coefficients between the tested data sets and the reference field. Finally, the distances from the reference point (plotted on the horizontal axis, in green) indicate the RMS differences. The more the test points approach the reference point, the closer the data sets are. Thus, from Fig. 12.14, it follows that the consistency between the X-TRACK-SLA data and the Sète TG is better than with DT-SLA data.

On closer consideration, when we look at the along-track percentage of RMS explained (see Fig. 12.15), it is confirmed that the X-TRACK statistics are better, not only at the point of maximum percentage explained (where the improvements are higher than 10%) but also along the whole altimetric track where small-scale structures are clearly distinguished between the continental shelves and the deep ocean, which is not the case with the DT-SLA. The question which remains is whether such small features are physically realistic. As a next step, multi-sensor approaches (ongoing works in several groups at LEGOS, IMEDEA, OSU, etc.) should provide information for resolving this issue.

The LPC Current is characterized by an intrinsic seasonal variability (see Sect. 12.2) whose sea surface signature is nearly out of phase with the large-scale steric signal (see Bouffard 2007; Bouffard et al. 2008a). As a consequence, the seasonal sea surface signal within the LPC Current is very different to the offshore and shelf regions. This constitutes a natural dynamical boundary between the coastal area and the deep ocean. Track 146 of T/P intercepted the LPC Current over the continental slopes of the Gulf of Lion between 42.7° and 42.9° N in latitude. Here, the along-track percentage of RMS explained by altimetry is marked by a minimum (in both X-TRACK and DT-SLA data sets, see Fig. 12.15),

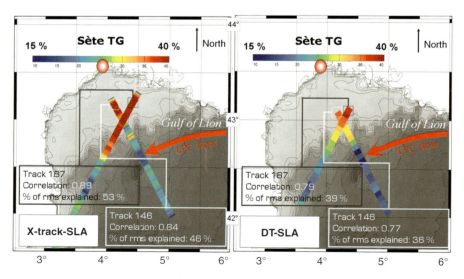

Fig. 12.15 Along-track percentage of root mean square (RMS) explained by altimetry in the Sète TG time series

which is simply due to the fact that the Sète TG and altimetry measurements "see" two different seasonal signals.

12.5.2.2
Application: Monitoring of the LPC Current

In order to evaluate the ability of the two data sets to monitor the spatial and temporal variability of the LPC Current, Hovmuller diagrams (time-position diagram) of across-track geostrophic velocity anomaly (GVA, refer to Bouffard et al. 2008b for the method) have been computed with both X-TRACK and DT-SLA from T/P track 146, during the year 2006. The mean velocity field has been derived from the regional MDT Rio (Rio et al. 2007) and adds to the altimetric GVA.

Some identified patterns are clearly observed in the two data sets (Fig. 12.16). While the geostrophic variability is weak in the open ocean, stronger currents occur over the continental slope of the Gulf of Lion in the area of the LPC Current mean position. As the current circulates southwestward along the shelf break delimiting the Gulf of Lion, it can act as a physical barrier and separates the shelf circulation from the offshore circulation.

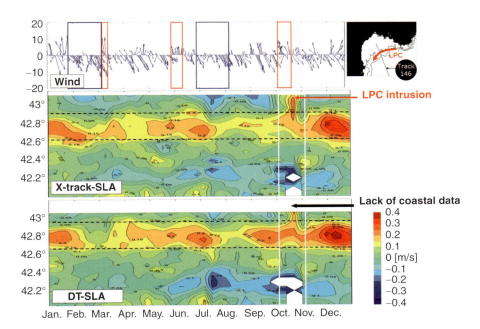

Fig. 12.16 *Top*: wind direction and magnitude over the Gulf of Lion (in m/s, courtesy of meteo France) as a function of time from January 2006 to December 2006 (*left*); track 146 and isobaths 200 and 1,000 m location (*right*). In the blue rectangles (respectively red) are shown northern (respectively southern) wind events. Bottom: Hovmuller of accros track geostrophic velocity (in m/s) of T/P track 146. In dashed lines are the position of isobath 200 and 1,000 m

From Fig. 12.16, it follows that the current is clearly identified both in the X-TRACK data and the DT-SLA data. It is narrow and coherently constrained along the continental slopes between the isobaths 200 and 1,000 m. From January until the end of February, the current accelerates and has velocities greater than 30 cm/s. In spring and summer, the current significantly decreases to about 10–20 cm/s, which is in agreement with its seasonal variability as described earlier (see Sect. 12.2). The velocity and positions seem to be greatly controlled by the wind direction and intensity as noted previously. All during the year, the wind direction changes alternately from north-northwest (the so-called Tramontane and Mistral regimes) to southeast. In the first case, the LPC Current tends to go farther off the coast (in blue on the previous figure) whereas in the second case (in red on the previous picture) the current tends to approach the coast.

As shown in Echevin et al. (2003) and Millot and Wald (1980), the surface currents generated by northwesterly winds do not tend to advect water along the Gulf coast which is in good agreement with our altimetric observations. In contrast, strong southeasterly winds can generate surface currents oriented to the north-northeast direction. The effect of these southerly winds on the LPC Current intrusions has been detected with AVHRR observations (Millot and Wald 1980) but never with altimetry.

Such wind conditions were observed in October 2006 (see Fig. 12.16). During this event, the X-TRACK data set seems to show an intrusion of the LPC Current over the continental shelf, which does not appear as clearly in the DT-SLA product because of a lack of coastal data. The LPC Current thermal signature often appears on SST images (Crépon et al. 1982) as a relatively warm surface water in contrast with the colder and well-mixed waters of the Gulf of Lion. We have used sea surface temperature (SST) data from Medspiration in order to check if the "observed intrusion" is physically realistic (see Fig. 12.16) and whether it had a SST signature. To remove large-scale steric signature in the SST, spatial averages have been subtracted first.

SST images of Fig. 12.17 clearly show that a coastal current progressively advects warm surface waters from the Ligurian Sea to the Gulf of Lion coastal area. Around 23 October, this warm pattern crosses track 146 of J1 over the Gulf of Lion continental shelf. Thus, the penetration of the coastal current onto the shelf is observed in both the temperature and velocity fields (see Fig. 12.17).

Here, we do not propose a precise explanation of the mechanisms driving such a process. This would require a dedicated investigation of the use of regional high-resolution modelling (see Gatti et al. 2005). We do however suggest two potential mechanisms: numerical analysis (see Echevin et al. 2003) has shown that the Ekman transport due to the southeasterly wind forces a downwelling along the east coasts of Provence and the Gulf of Lion, which in turn generates a westward current which flows along the shelf coast. The advection of the surface warm water by the coastal current could be associated to this mechanism. Alternatively, rather than inducing a new current, the southeasterly wind could add extra vorticity locally which pushes the LPC Current off its usual potential vorticity conserving path along a constant depth.

In either event, observing a current intrusion is not surprising given that, under particular wind and stratification conditions, surface waters of the LPC Current tend to penetrate onto the shelf at the eastern entrance of the gulf (Auclair et al. 2003; Petrenko et al. 2005). However, observing such small-scale dynamics both with altimetry and SST, in coastal

12   Introduction and Assessment of Improved Coastal Altimetry Strategies   325

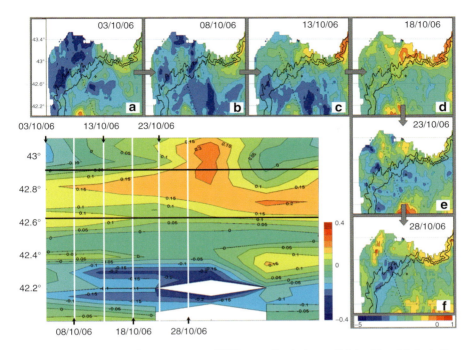

Fig. 12.17 (**a**) to (**f**): sea surface temperature (SST) anomalies in degree Celsius (from Medspiration database) and zoom of the Hovmuller of across track geostrophic velocities (m/s)

zone and over a shallow water area, is a novelty and demonstrates the potential of using a multi-sensor approach.

12.6
Conclusion, Ongoing Work, and Perspectives

Satellite altimetry offers an important measurement source for coastal studies. We have shown that by adopting specific coastal processing strategies, it is possible to increase the quantity and quality of data in those regions. Indeed, studying the complexity of coastal dynamics requires an adapted processing of altimetric data. The comparisons between the X-TRACK-SLA and a standard regional altimetric product confirm this assessment. This is especially the case over the continental shelves where the altimetric signal needs to be de-aliased with a local high-resolution model. Moreover, higher-frequency spatial samplings combined with adapted editing criteria allow us to significantly increase the number of valid data in the coastal zone and improve their quality by about 7%. In addition, this new altimetric processing strategy also allows us to approach closer to the coastline and to better highlight coastal dynamics with small spatial extents, such as coastal jets and fronts.

However, as shown through the example of the RECOSETO data, there are still ways to improve the high-frequency sampling even further, including new retracking of raw waveforms in coastal zones. In the X-TRACK processing scheme, the large-scale error reduction method has still to be improved using a multi-satellite cross-calibration method, especially to reduce the large scale GFO orbit error.

The cross-validation using the improved X-TRACK multi-satellite altimetry and TG records shows encouraging results. The qualitative and quantitative improvements compared to the AVISO regional product demonstrate that a simple data post-processing allows us to build a valuable coastal altimetric data set. Geostrophic velocities derived from the improved coastal data are able to monitor for the first time small-scale coherent coastal processes occurring over the continental shelves, such as the LPC Current seasonal variability and its intrusions over the continental shelf of the Gulf of Lion. This event has also been observed with satellite SST data.

The combined use of coastal altimetry and in situ data is also necessary for understanding coastal processes. For example, glider data, which give information on the oceanic vertical structure, are a good complementary tool to coastal altimetry. An intensive observational program, conducted in the Western Mediterranean, allowed running coastal glider missions along selected altimeter (Envisat, Jason-1/2) tracks. This experiment allowed an assessment of the dynamics of the frontal area, and an evaluation of the limitations of coastal altimetry. Reasonable coherence between absolute dynamic topography (ADT) from altimetry (GDR) and the dynamic height (DH) from the glider CTD data has been found (Ruiz et al. 2009). However, reliable altimetry measurements near the coast (less than 30 km) are not always available (either they are not acquired or are acquired but not available in the 1 Hz products). This reinforces the need for specific altimetric algorithms (retracking, corrections, editing, etc.) for coastal areas.

Owing to the difficulty of deploying and maintaining a dense in situ measurement network, improved coastal altimetric data would form a complementary tool for monitoring the water transport in marginal zones, for understanding oceanic processes that occur and assessing the water budget of the NWM (Astraldi et al. 1999). For this purpose, efforts to improve coastal altimetry have to be maintained. As suggested before, the combined use of improved coastal multi-source data would allow us to efficiently observe the temporal and spatial scales of coastal dynamics. This is within the actual framework of operational and scientific oceanographic applications, such as monitoring the variability of narrow coastal currents.

Acknowledgments This chapter is dedicated to Dr. Yves Ménard memory who was deeply involved in some activities focusing on coastal altimetry. The authors kindly acknowledge SONEL and ISPRA networks for the tide gauge sea level records. We wish to thank JF Bellemare for his revision of English spelling and syntaxes. Many thanks are also due to ESA, CNES, JPL, and NOAA for providing raw altimetric data within the framework of the RECOSETO project (founded by DFG). The X-TRACK altimetric data used in this study were developed, validated, and distributed by the CTOH/LEGOS, France. The DT-SLA product were produced by SSALTO/DUACS and distributed by Aviso, with support from CNES. Two referees made pertinent comments that helped to improve the manuscript.

References

Anzenhofer M, Shum CK, Rentsh M (1999) Coastal altimetry and applications, Tech. Rep. n. 464, Geodetic Science and Surveying, The Ohio State University Columbus, USA

Astraldi M, Balopoulos S, Candela J, Font J, Gacic M, Gasparini GP, Mamap B, Theocharis A, Tintoré J (1999) The role of straits and channels in understanding the characteristics of Mediterranean circulation. Prog Oceanogr 44:65–108

Auclair F, Marsaleix P, De Mey P (2003) Space–time structure and dynamics of the forecast error in a coastal circulation model of the Gulf of Lions. Dyn Atms Ocean 36:309–346

Ayoub N, Le Traon PY, De Mey P (1998) A description of the Mediterranean surface variable circulation from combined ERS-1 and T/P altimetric data. J Mar Syst 18(1–3):3–40

Bouffard J (2007) Amélioration de l'altimétrie côtière appliquée à l'étude de la circulation dans la partie nord du bassin occidental méditerranéen (in French). PhD thesis under the supervision of Y. Ménard and P. De Mey

Bouffard J, Roblou L, Ménard Y, Marsaleix P, De Mey P (2006) Observing the ocean variability in the western Mediterranean sea by using coastal multi-satellite data and Models. In: Proceedings of the symposium on 15 years of progress in radar altimetry, 13–18 March 2006, Venice, Italy (ESA SP-614, July 2006)

Bouffard J, Vignudelli S, Herrmann M, Lyard F, Marsaleix P, Ménard Y, Cipollini P (2008a) Comparison of ocean dynamics with a regional circulation model and improved altimetry in the North-western Mediterranean. Terr Atmos Ocean Sci 19:117–133. doi:10.3319/TAO.2008.19.1-2.117(SA)

Bouffard J, Vignudelli S, Cipollini P, Menard Y (2008b) Exploiting the potential of an improved multimission altimetric data set over the coastal ocean. Geophys Res Lett 35:L10601. doi:10.1029/2008GL033488

Brown GS (1977) Average impulse response of a rough surfaces and its applications. IEEE Trans Antennas Propag 25

Buongiorno Nardelli B, Santoleri R, Marullo S, Iudicone D (1999) Altimetric sea level anomalies and three dimensional structure of the sea in the strait of Sicily. J Geophys Res 104(C9): 20585–20603

Carrère L, Lyard F (2003) Modeling the barotropic response of the global ocean to atmospheric wind and pressure – comparisons with observations. Geophys Res Lett 30(6):1275

Chelton DB (2001) Report of the high-resolution ocean topography science working group meeting, Oregon State

Cipollini P, Vignudelli S, Lyard F, Roblou L (2008) 15 years of altimetry at various scales over the Mediterranean, In: V Barale, JRC (Ispra), M Gade (Universitat Amburgo) Remote Sensing of European Seas, Springer, Heidelberg, Germany, pp 295–306. doi:10.1007/978-1-4020-6772-3_22

Cohen JE, Small C, Mellinger A, Gallup J, Sachs J, Vitousek PM, Mooney HA (1997) Estimates of coastal populations. Sciences 278(5341):1209–1213

Crépon M, Wald L, Longuet JM (1982) Low-frequency waves in the Ligurian Sea during December 1977. J Geophys Res 87(C1):595–600

Davis CH (1995) Growth of the Greenland Ice Sheet: a performance assessment of altimeter retracking algorithms. IEEE Trans Geosci Remote Sensing 33(5):1108–1116

Deng X, Featherstone WE, Hwang C, Berry P (2002) Estimation of Contamination of ERS-2 and Poseidon satellite radar altimetry close to the coasts of Australia. Mar Geod 25:249–271

Desportes C, Obligis E, Eymard L (2007) On the Wet tropospheric correction for altimetry in coastal regions. IEEE Trans, Geosci Remote Sensing 45(7):2139–2149

Echevin V, Crépon M, Mortier L (2003) Interaction of a coastal current with a gulf: application to the shelf circulation of the Gulf of Lions in the Mediterranean Sea. J Phys Oceanogr 33: 188–206

Estournel C, Durrieu de Madron X, Marsaleix P, Auclair F, Julliand C, Vehil R (2003) Observation and modelisation of the winter coastal oceanic circulation in the Gulf of Lions under wind conditions influenced by the continental orography (FETCH experiment). J Geophys Res 108(C3):7–1, 7–18. doi:10.1029/2001JC000825

Fenoglio-Marc L, Groten E, Dietz C (2004) Vertical land motion in the Mediterranean Sea from altimetry and TG stations. Mar Geod 27(3–4):683–701

Fenoglio-Marc L, Vignudelli S, Humbert A, Cipollini P, Fehlau M, Becker M (2007) An assessment of satellite altimetry in proximity of the Mediterranean coastline. In 3rd ENVISAT symposium proceedings, SP-636, ESA Publications Division

Fenoglio-Marc L, Fehlau M, Ferri L, Becker M, Gao Y, Vignudelli S (2008) Coastal sea surface heights from improved altimeter data in the Mediterranean Sea. In: proceeding of international geoscience and remote sensing symposium (IGARSS), vol 3, p 423

Font J, Garcia-Ladona E, Gorriz EG (1995) The seasonality of mesoscale motion in the Nothern Current of the Western Mediterranean: several years of evidence. Oceanol Acta 18(2):207–219

Gatti J, Petrenko A, Devenon JL, Leredde Y (2005) Intrusions of the Mediterranean Northern Current on the eastern side of the Gulf of Lion's continental shelf NW Mediterranean Sea. Geophys Res Abst 7:00986

Herrmann M, Bouffard J, Béranger K (2009), Monitoring open-ocean deep convection from space, Geophys. Res. Lett., 36, L03606. doi:10.1029/2008GL036422

Hwang C, Guo JY, Deng XL, Hsu HY, Liu YT (2006) Coastal gravity anomalies from retracked Geosat/GM altimetry: improvement, limitation and the role of airborne gravity data. J Geod 80:204–216

Iudicone D, Santoleri R, Marullo S, Gerosa P (1998) Sea level variability and surface eddy statistics in the Mediterranean Sea from TOPEX/Poseidon data. J Geophys Res 103(C2):2995–3011

Jan G, Ménard Y, Faillot M, Lyard F, Jeansou E, Bonnefond P (2004) Offshore absolute calibration of space borne radar altimeters, Mar Geod, Special issue on Jason-1 Calibration/Validation Part 3 27:3–4, 615–629

Larnicol G, Le Traon PY, Ayoub N, De Mey P (1995) Mean Sea level and surface circulation variability of the Mediterranean Sea from 2 years of T/P altimetry. J Geophys Res 100(12):25163–25178

Larnicol G, Ayoub N, Le Traon PY (2002) Major changes in Mediterranean Sea level variability from 7 years of T/P and ERS-1/2 data. J Mar Syst 33–34:63–89

Le Traon PY, Morrow RA (2001) Ocean currents and mesoscale eddies. In: Fu L-L, Cazenave A (ed) Satellite altimetry and earth sciences. A handbook of techniques and applications. Academic, San Diego, USA, pp 171–215

Le Traon PY, Ogor F (1998) ERS-1/2 orbit improvement using T/P: the 2 cm challenge. J Geophys Res 103:8045–8057

Li Z, Chao Y, McWilliams JC, Ide K (2008) A three-dimensional variational data assimilation scheme for the regional ocean modeling system: implementation and basic experiments. J Geophys Res C05002. doi:10.1029/2006JC004042

Lyard F, Lefevre F, Letellier T, Francis O (2006) Modelling the global ocean tides: modern insights from FES2004. Ocean Dyn. doi:10.1007/s10236-006-0086-x, ISSN

Mangiarotti S, Lyard F (2008) Surface pressure and wind stress effects on sea level change estimations from T/P satellite altimetry in the Mediterranean Sea. J Atmos Ocean Technol 25: 464–474

Manzella GMR, Borzelli GL, Cipollini P, Guymer TH, Snaith HM, Vignudelli S (1997) Potential use of satellite data to infer the circulation dynamics in a marginal area of the Mediterranean Sea. In: ESA SP-414 Proc. of 3rd ERS Symposium, Florence (Italy), 17–21 March 1997, vol 3, pp 1461–1466

Millot C (1987) Circulation in the western Mediterranean. Oceanol Acta 10:143–149

Millot C (1991) Mesoscale and seasonal variabilities of circulation in western Mediterranean. Dyn Atmos Ocean 15:179–214

Millot C, Wald L (1980) The effect of the Mistral wind on the Ligurian current near Provence. Oceanol Acta 3(4):399–402

Mourre B, De Mey P, Lyard F, Le Provost C (2004) Assimilation of sea level data over continental shelves: an ensemble method for the exploration of model errors due to uncertainties in bathymetry. Dyn Atmos Ocean 38:93–121. doi:10.1016/j.dynatmoce.2004.09.001

Naeije ME, Doornbos L, Mathers R, Scharroo E, Schrama Visser P (2002). Radar altimeter database system: exploitation and extension (RADSxx), Final Report. NUSP-2 report 02-06, NUSP-2 project 6.3/IS-66. Space Research Organisation Netherlands (SRON), Utrecht, the Netherlands.

Pascual A, Faugère Y, Larnicol G, Le Traon PY (2006) Improved description of the mesoscale variability by combining four satellite altimeters. J Geophys Res 33. doi:10.1029/2005GL024633.33

Pascual A, Pujol MI, Larnicol G, Le Traon PY, Rio MH (2007) Mesoscale mapping capabilities of multisatellite altimeter missions: first results with real data in the Mediterranean Sea. J Mar Syst 65:190–211

Pascual A, Marcos M, Gomis D (2008) Comparing the sea level response to pressure and wind forcing of two barotropic models: validation with tide gauge and altimetry data. J Geophys Res Oceans 113(C7)

Petrenko A, Leredde Y, Marsaleix P (2005) Circulation in a stratified and wind forced Gulf of Lions, NW Mediterranean Sea: in-situ and modeling data. Cont Shelf Res 25:7–27

Pujol M-I, Larnicol G (2005) Mediterranean Sea eddy kinetic energy variability from 11 years of altimetric data. J Mar Syst 58:121–142

Ray RD (1999) A global ocean tide model from T/P altimetry: GOT99.2, NASA Tech. Memo. 209478, Goddard Space Flight Center, 58pp

Ries JC, Tapley BD (1999) Centimeter level orbit determination for the T/P altimeter satellite. In: Proceedings of the ASS/AIAA Space Flight Mechanism Meeting, Breckenridge, CO, USA, pp 583–597

Rio MH, Poulain PM, Pascual A, Mauri E, Larnicol G, Santoleri R (2007) A mean dynamic topography of the Mediterranean Sea computed from altimetric data, in-situ measurements and a general circulation model. J Mar Sys 65(14):484–508

Robinson A, Leslie W, Theocharis A, Lascaratos A (2001) Encyclopedia of ocean sciences, chap. Mediterranean sea circulation. Academic, London, pp 1689–1706

Roblou L, Lyard F, Le Hénaff M, Maraldi C (2007) X-TRACK, a new processing tool for altimetry in coastal oceans. In: ESA ENVISAT symposium, Montreux, Switzerland, April 23–27, 2007, ESA SP-636

Ruiz S, Pascual A, Garau B, Faugere Y, Alvarez A, Tintoré J (2009) Mesoscale dynamics of the Balearic front, integrating glider, ship and satellite data. J Mar Syst 78, S3-S16. doi:10.1016/j.jmarsys.2009.01.007

Send U, Font J, Krahmann G, Millot C, Rhein M, Tintoré J (1999) Recent advances in observing the physical oceanography of the western Mediterranean Sea. Prog Oceanogr 44:37–64

SSALTO/DUACS User Handbook (2006) (M)SLA and (M)ADT near-real time and delayed time products. SALP-MU-P-EA-21065-CLS

Taylor KE (2001) Summarizing multiple aspects of model performance in a single diagram. J Geophys Res 106(D7):7183–7192

Vignudelli S, Cipollini P, Astraldi M, Gasparini GP, Manzella GMR (2000) Integrated use of altimeter and in situ data for understanding the water ex-changes between the Tyrrhenian and Ligurian Seas. J Geophys Res 105:19649–19663

Vignudelli S, Cipollini P, Roblou L, Lyard F, Gasparini GP, Manzella GRM, Astraldi M (2005) Improved satellite altimetry in coastal systems: Case study of the Corsica Channel (Mediterranean Sea). Geophys Res Lett 32. doi:1029/2005GL22602

Vignudelli S, Cipollini P, Gommenginger C, Snaith H, Coelho H, Fernandes J, Lazaro C, Nunes A, Gómez-Enri J, Martin-Puig C, Woodworth P, Dinardo S, Benveniste J (2010) Satellite altimetry: sailing closer to the coast. In: Gower J, Levy G, Heron M, Tang D, Katsaros K, Singh R (eds) Remote Sensing of the Changing Oceans, Springer, 2010 (in press)

Volkov DL, Larnicol G, Dorandeu J (2007) Improving the quality of satellite altimetry data over continental shelves. J Geophys Res 112:C06020. doi:10.1029/2006JC003765

Satellite Altimetry Applications in the Caspian Sea

13

A.V. Kouraev, J.-F. Crétaux, S.A. Lebedev, A.G. Kostianoy, A.I. Ginzburg, N.A. Sheremet, R. Mamedov, E.A. Zakharova, L. Roblou, F. Lyard, S. Calmant, and M. Bergé-Nguyen

Contents

13.1	Introduction	333
13.2	Caspian Sea Level Variability and Its Observations by Level Gauges	336
13.3	Application of Satellite Radar Altimetry for the Caspian Sea	338
	13.3.1 Tides and Atmospheric Forcing	339
	13.3.2 Atmospheric Corrections	339
	13.3.3 Mean Sea Surface	339
	13.3.4 Influence of Ice Cover on Altimetric Measurements	341
	13.3.5 Multi-satellite Bias	342
	13.3.6 Calibration/Validation (Cal/Val) Activities	342
	13.3.7 Existing Datasets	345
13.4	Results	346
	13.4.1 Seasonal Variability of the Caspian Sea Level/Volume	346
	13.4.2 Storm Surges in Altimetry	348
	13.4.3 Volga Water Level from Altimetry	351
	13.4.4 Water Budget of the Caspian Sea and the Kara-Bogaz-Gol Bay	352
	13.4.5 Wind and Wave Regime	354
	13.4.6 Ice Cover Regime	357
13.5	Discussion and Conclusions	360
References		362

A.V. Kouraev (✉)
Universite de Toulouse; UPS (OMP-PCA), LEGOS, 14 av. Edouard Belin, F-31400 Toulouse, France and

State Oceanographic Institute, St. Petersburg branch, Vasilyevskiy ostrov, 23 liniya, 2a, St. Petersburg, Russia
e-mail: kouraev@legos.obs-mip.fr

J.-F. Crétaux
Universite de Toulouse; UPS (OMP-PCA), LEGOS, 14 av. Edouard Belin, F-31400 Toulouse, France and

CNES, LEGOS, F-31400, Toulouse, France

S.A. Lebedev
Geophysical Center, Russian Academy of Sciences, 3 Molodezhnaya Str., 119296 Moscow, Russia

A.G. Kostianoy, A.I. Ginzburg and N.A. Sheremet
P.P. Shirshov Institute of Oceanology, Russian Academy of Sciences, 36 Nakhimovsky Pr., 117997 Moscow, Russia

R. Mamedov
Institute of Geography, National Academy of Sciences of Azerbaijan, 31H.Javid Pr., AZ1143 Baku, Azerbaijan

E.A. Zakharova
State Oceanographic Institute, St. Petersburg branch, Vasilyevskiy ostrov, 23 liniya, 2a, St. Petersburg, Russia

L. Roblou and F. Lyard
Universite de Toulouse; UPS (OMP-PCA), LEGOS, 14 av. Edouard Belin, F-31400 Toulouse, France and

CNRS, LEGOS, F-31400 Toulouse, France

S. Calmant
Universite de Toulouse; UPS (OMP-PCA), LEGOS, 14 av. Edouard Belin, F-31400 Toulouse, France and

IRD, LEGOS, F-31400 Toulouse, France

M. Bergé-Nguyen
Universite de Toulouse; UPS (OMP-PCA), LEGOS, 14 av. Edouard Belin, F-31400 Toulouse, France and

CNES, LEGOS, F-31400 Toulouse, France

Keywords Altimetry • Caspian Sea • Ice cover • Kara-Bogaz-Gol Bay • Sea level • Volga River • Wave height • Wind speed

Abbreviations

ALTICORE	ALTImetry for COastal REgions
CERSAT	Center for Satellite Exploitation and Research
COAPS	Center for Ocean Atmospheric Prediction Studies
ECMWF	European Centre for Medium-range Weather Forecasts
FAS	Foreign Agricultural Service
GC RAS	Geophysical Center of Russian Academy of Sciences
GCOS	Global Climate Observing System
GDR	Geophysical Data Record
GFO	Geosat Follow-On
GLOSS	Global Sea Level Observing System
GPS	Global Positioning System
GTOS	Global Terrestrial Observing System
IFREMER	Institut Français de Recherche et d'Exploitation de la MER
INTAS	International Association for the promotion of co-operation with scientists from the New Independent States of the former Soviet Union

IPCC	Intergovernmental Panel on Climate Change
ISADB	Integrated Satellite Altimetry Data Base
LEGOS	Laboratoire d'Études en Géophysique et Océanographie Spatiales
MACE	Multidisciplinary Analysis of the Caspian Sea Ecosystem
MODIS	Moderate Resolution Imaging Spectroradiometer
MSS	Mean Sea Surface
NCEP	National Center for Environmental Prediction
NOAA	National Oceanic and Atmospheric Administration
RADS	Radar Altimeter Database System
RMS	Root Mean Square
RA	Radar Altimeter
SMMR	Scanning Multichannel Microwave Radiometer
SSH	Sea Surface Height
SSM/I	Special Sensor Microwave/Imager
T/P	TOPEX/Poseidon
T-UGO 2D	Toulouse Unstructured Grid Ocean model 2D
USDA	US Department of Agriculture
WGS 84	World Geodetic System 1984

13.1 Introduction

The Caspian Sea is the largest continental water body on the Earth. Descendant of the ancient Thetys Sea that once comprised the actual Black, Caspian, and Aral seas, nowadays the Caspian Sea has no connection with the World Ocean. However, its size, depth, chemical properties, and peculiarities of thermohaline structure, water circulation, and biodiversity make it possible to classify it as a deep inland sea. The Caspian Sea has an unique and precious ecosystem, including the largest sturgeon population in the world that in the 1990s gave 90% of world supply of caviar. At the same time, the sea has important oil deposits that have been exploited for more than a century near Baku and only recently (since the 1990s) in the shelf regions in the Northern Caspian. The sea is shared by five countries – the Russian Federation, Kazakhstan, Turkmenistan, the Islamic Republic of Iran, and the Azerbaijan Republic.

Owing to its size and geographical location, the Caspian Sea is truly a large-scale climatic and ecological indicator. The main peculiarities of the sea are rapid and considerable changes of hydrophysical, hydrochemical, and hydrobiological states under the influence of external factors (Kostianoy and Kosarev 2005).

All large-scale or global changes, such as temperature variability, rapidly affect natural conditions, sea level, and the ecosystem of the Caspian Sea (Kostianoy et al. 2008). A recent example is the introduction of exotic organisms (such as *Mnemiopsis leidyi Agassis*) into the Caspian Sea from other regions of the World Ocean.

One of the most important parameters of the Caspian Sea is its level. Explanation of interannual sea level changes puzzles the scientists of many countries and at the same time stimulates studies of water level changes in other seas of the world in the context of climatic changes. Recent increased importance of the Caspian Sea region and its resources attracts interest of various international organizations for problems of the Caspian Sea. There is a possibility that existing experience for solving various problems (such as potential consequences of global sea level rise and climate change, optimization of wildlife management, etc.) can be applied in the context of large changes in the Caspian Sea.

Actually (average data for 2008), the Caspian Sea lies about 27.1 m below the World Ocean (with respect to the Baltic reference altitude system). At this position, the sea surface is more than 410,000 km^2, and the water volume is about 78,000 km^3. The sea extends for more than 1,100 km from north to south, and its width is 200–400 km in the central and southern parts, and more than 500 km in the northern part. Such a large extent and great variability of climatic conditions over the sea result in large regional differences in thermohaline structure and water dynamics.

The Caspian Sea occupies a vast depression in the Earth's crust and comprises three distinct regions with particular bottom topography and natural conditions: Northern, Middle, and Southern Caspian (Fig. 13.1). The shallow Northern Caspian with the mean depth of 5–6 m and maximal depths of 15–20 m completely refers to the shelf zone. The Mangyshlak Ridge separates the Northern Caspian from the Middle Caspian with the Derbent Basin (maximal depth of 788 m). The Middle Caspian is connected through a one-way strait to the large Kara-Bogaz-Gol Bay. Apsheron Ridge (depths of 160–180 m) acts as a boundary between the Middle Caspian and the deepwater Southern Caspian Depression (maximal depth of 1,025 m) (Kosarev 2005).

Being an enclosed water basin, the Caspian Sea is strongly affected by the river input. Among more than 130 rivers entering the Caspian Sea, the most important is Volga. This river has an annual discharge of 236 km^3 and a watershed area of 1,360,000 km^2 (State water cadaster, 1930–1993), and provides more than 78% of the total river input into the Caspian Sea. Other important rivers are Ural, Terek, Sulak, and Kura. River inflow from Volga, Ural, and Terek (84% of the total river input into the Caspian Sea) is especially important for the Northern Caspian. Here, combination of large river runoff, small depths, and significant surface area results in existence of vast zones of fresh water transit and large-scale processes of mixing fresh and saline waters. Main features of this mixing zone explain the high biological productivity in this region of the Caspian Sea.

Winds play an important role in the formation of currents and mixing processes in the upper layer of the sea. In general, two main types of winds dominate over the whole Caspian Sea: the northerly (from northwesterly to northeasterly) and southeasterly. In annual distribution, northerly winds represent 40% of observations (with maximum of up to 50% in summer) and southeasterly winds −36% (with a maximum of about 40% observed in winter and spring) (Kosarev 2005). However, in each specific region of the Caspian Sea, temporal distribution of the wind field has its own particularities. Wind field significantly influences water dynamics and wave field. For the enclosed Caspian Sea, a combination of strong wind and atmospheric pressure can result in significant short-term (from a few hours to 1–2 days) increase or decrease of the sea level (storm surges) compared to its usual values. The highest sea level changes are typical for coastal regions, bays and bights. While in the Middle and Southern Caspian amplitude of storm surges is about 100–150 cm (Kosarev 2005), in the Northern Caspian extreme storm surges have resulted in 3–4 m increase of sea level, flooding large areas and affecting coastal settlements and

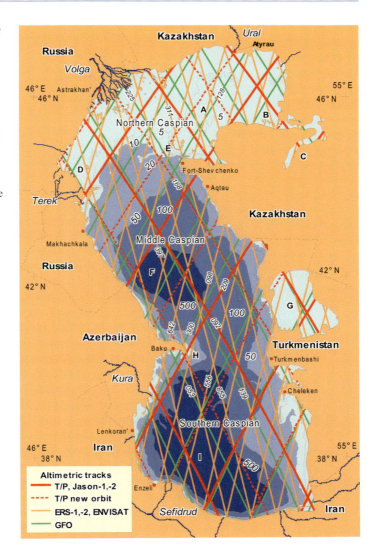

Fig. 13.1 Overview map of the Caspian Sea and coverage of the altimetric ground tracks (selected tracks are numbered). Isobaths are in meters (referring to −28 m). River contours are shown only for the regions near the coast. Coastline position is according to MODIS image of 11 June 2005. A – Ural Furrow, B – Mertviy Kultuk Bay, C – Kaydak Bay, D – Tyuleniy Island, E – Kulaly Island, F – Derbent Depression, G – Kara-Bogaz-Gol Bay, H – Neftyaniye Kamni, I – Southern Caspian Depression

infrastructure. On a smaller scale, seiche and baric waves also affect short-term Caspian Sea level fluctuations (Kosarev and Yablonskaya 1994; Kostianoy and Kosarev 2005).

Every winter, ice forms in the Northern Caspian and stays there for several months, covering large areas. Ice cover is also observed in the Middle Caspian, but on a much smaller scale. The presence of ice negatively affects navigation, fisheries, and other industrial activities in the coastal areas and on the shelf. Of special concern is the impact of sea ice on industrial structures located in the coastal zone, such as Russian and Kazakh oil prospecting rigs operating on the Northern Caspian shelf. Ice conditions make it necessary to maintain an ice-breaker fleet as well as to use oil drilling rigs of arctic-class in the Northern Caspian. Ice cover also affects the accuracy of altimetric measures over the Caspian Sea (see Sect. 13.4).

In Sect. 13.2, we consider variability of the Caspian Sea level and the in situ level gauge network. In Sect. 13.3, specificity of application of satellite radar altimetry for studies of the Caspian Sea is discussed and in Sect. 13.4, various areas of combination of satellite and in situ data for the studies of natural regime of the sea are presented.

13.2
Caspian Sea Level Variability and Its Observations by Level Gauges

The Caspian Sea is characterized by cyclic and high-amplitude water level variations over historical timescales. For the last 2,000 years, fluctuations were around 11 m (Lebedev and Kostianoy 2005) and for the last five centuries around 7 m, with an extreme level of −23 m at the end of first century, mid-seventeenth century, and end of seventeenth century, and −29 m in 1977. Since the start of observations and until the beginning of the twentieth century, sea level was about −25.8 m (Fig. 13.2). Between 1900 and 1929, sea level remained around −26.2 m and its variability was low. In the 1930s, creation and filling of large reservoirs on the Volga River, as well as regional climate change over the watershed, led to level decrease by 1.8 m between 1930 and 1941. Since that time, the study of sea level variability became important not only from the scientific, but also from the applied point of view. After 1941 and until the 1950s, the sea level decrease slowed and in the 1960s level was stable around −28.4 m. Since the 1970s, a rapid sea level decrease started and in 1977 the level reached −29 m: the lowest mark for the last 400 years. Since then, and quite unexpectedly, the Caspian Sea level rapidly began to rise by about 2.5 m (−26.5 m in 1995). After 1995, the level started to decrease again and in winter 2001–2002 reached −27.3 m (Lebedev and Kostianoy 2005). Short-term increase by about 20 cm until 2005 was followed by a decrease of the Caspian Sea level at about the same rate.

Variations of the Caspian Sea level are closely related with components of the sea water balance. However, despite years of study and investigation, the physical mechanism generating long-term sea level changes is not well understood yet, neither are we able to predict the rise or fall of the level in the future. Significant water level fluctuations displace thousands of shoreline inhabitants, bring damage to industrial constructions and infrastructure, and are clearly a topic of public concern for riparian countries.

Sea level measurements of the Caspian Sea started in 1830, when the first level gauge was installed in Baku. The measurements have then been gradually extended to Astrabad Bay (1853), Makhachkala and Kuuly lighthouse (1900), Krasnovodsk (actual Turkmenbashi), Fort-Shevchenko, and Kara-Bogaz-Gol Bay (1921) (Pobedonostsev et al. 2005). After the creation of the Hydrometeorological service in 1929, many more observation points have been opened along the Caspian Sea coast until 1937. The changes of the coastline position led to closing of many stations. Four stations: Baku, Makhachkala, Fort-Shevchenko, and Krasnovodsk (Turkmenbashi) have been considered "century" stations (Terziev et al. 1992).

Sea level measurements were initially made in different height systems (Black Sea, Baltic Sea, local relative systems). In 1961, data for 31 level gauges were catalogued and a new "zero" mark was defined at −28 m (Baltic system). In the 1970s, a new leveling survey was done on the western coast of the sea. As a result, levels of level gauges have been

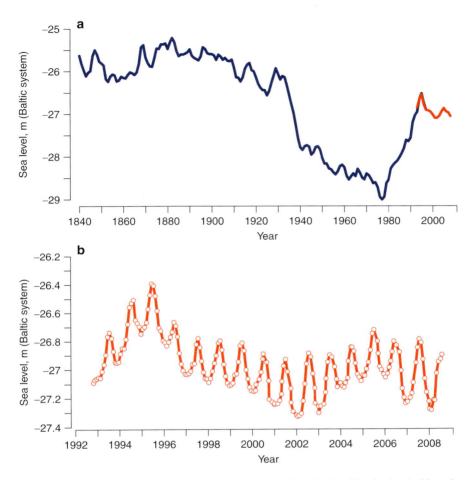

Fig. 13.2 (**a**) Annual Caspian Sea level (m, Baltic system), from in situ (*blue line*) and altimetric (*red line*) observations (Hydroweb data, initial values referred to GGM02C geoid are adjusted to in situ observations by subtracting 46 cm). (**b**) Seasonal variability of the Caspian Sea level (m, Baltic system) from altimetric data (Hydroweb)

recalculated. Because of increased accuracy of leveling and vertical movements of zero marks between the 1950s and 1977 (that are in the order of 2–5 mm/year), new zero marks have shifted (+7 to +11 cm) referring to the 1950s (Pobedonostsev et al. 2005). At the same time, the leveling survey has not been done on the eastern coast of the sea and actually there are two different height systems – 1950 system for the eastern and 1977 system (adopted since 1979) for the western coast. Pobedonostsev et al. (2005) state that as a result, differences between mean annual sea level measurements at the west and east coasts can reach up to 30 cm. To compensate for this, observations at the western coasts often are corrected back to the 1950s system, which creates ambiguities. There is a need for homogenization of sea level measurements at various level gauges and currently a new catalogue of the Caspian

Sea level is being prepared at the State Oceanographic Institute (A.N. Kosarev, personal communication 2010).

13.3
Application of Satellite Radar Altimetry for the Caspian Sea

Continuous weather-independent observations from radar altimetry are a natural choice to complement the existing in situ sea level observations and to provide new information for open sea regions that have been never covered by direct sea level observations. The large Caspian Sea is extremely well sampled by the different radar altimeters' ground tracks (see Fig. 13.1). The existence of the altimetric datasets of decommissioned (TOPEX/Poseidon, or T/P, GFO and potentially ERS-1,-2) or currently in orbit satellites (Envisat, Jason-1 and -2) offers the possibility of continuous high-precision monitoring of the Caspian Sea level since the early 1990s. Future missions (AltiKa Sentinel-3, HY-2) will assure continuity of the altimetric datasets. The Caspian Sea can be considered an intermediate target between the open ocean and closed lakes, and it is an interesting object for evaluating the potential of satellite altimetry for enclosed water bodies. The sea has significant temporal and spatial variability in its sea level; it has been monitored for more than 150 years by in situ level gauges, yet the reasons of the sea level changes are still not fully known and its impact in the coastal zone is considerable.

Numerous studies based on coupled satellite altimetry and in situ gauges measurements have been published in different domains of continental hydrology. Some of them covered broad issues (Morris and Gill 1994a, b; Birkett 1995) while others focused on more specific lake or river basin case studies (Cazenave et al. 1997; Birkett 2000; Mercier et al. 2002; Maheu et al., 2003; Kouraev et al. 2004; Aladin et al. 2005; Coe and Birkett 2005; Crétaux et al. 2005; Frappart et al. 2005; Kouraev et al. 2008, 2009; Crétaux et al. 2009a, b). These studies show that radar altimetry is currently an essential technique for different applications: assessment of hydrological water balance (Cazenave et al. 1997; Crétaux et al. 2005), prediction of lake level variation (Coe and Birkett 2005), studies of anthropogenic impact on water storage in enclosed seas and reservoirs (Aladin et al. 2005; Zakharova et al. 2007), correlation of interannual fluctuations of lake level on a regional scale with ocean–atmosphere interaction (Mercier et al. 2002), study of the impact of climate variability on river basins (Maheu et al. 2003), variability of river level and discharge (Kouraev et al. 2004a; Lebedev and Kostianoy 2005; Zakharova et al. 2006), and analysis of floodplain water storage in the river basin (Frappart et al. 2005), as well as ice cover in enclosed seas and lakes (Kouraev et al. 2008).

Various studies have shown that altimetry data can be successfully used for the studies of sea level variability of the Caspian Sea (Cazenave et al. 1997; Vasiliev et al. 2002; Kostianoy et al. 2004; Lebedev and Kostianoy 2005, 2008b; Crétaux and Birkett 2006).

Prior to proper use of altimetric data, several issues should be addressed. Influence of tides, atmospheric forcing, and dry and wet tropospheric corrections have particularities for the Caspian Sea. Assessment of accuracy of existing geoid fields and definition of an alternative, e.g., a mean sea surface (MSS), need to be done. Potential influence of ice on altimetric measurements and bias between various satellite missions should be estimated, and validation of altimetric results by in situ measurements should be done.

13.3.1
Tides and Atmospheric Forcing

For the correction of both tides and atmospheric forcing, a regional model for high frequency motions for the Caspian Sea based on the T-UGO 2D hydrodynamic code (Mog2D follow-up) has been made. Model horizontal resolution is 7 km offshore, increasing to 1.5 km at the coast. Outputs from the model simulations permit us then to compute tides and short-term atmospheric loading response to be used for dealiasing correction in the X-TRACK processor of altimetric data (see the chapter by Roblou et al., this volume). From the model outputs, it seems that only M2 and S2 tidal constituents (Fig. 13.3) are relevant (with mean amplitudes of 2 and 1 cm and maximal amplitudes of 4 and 1.5 cm, respectively) and other constituents are very small (virtually negligible) in the model. However, there was no hourly level gauge data to validate the model in response to both astronomic and atmospheric loading. There is a strong need for permanent, GLOSS-compliant in situ data for validation and to confirm the regional tides and atmospheric forcing response in the model.

13.3.2
Atmospheric Corrections

Dry tropospheric correction is also one of the problems for using altimetry on enclosed water bodies. The value of this correction is altitude dependant, but T/P and GFO Geophysical Data Records (GDRs) incorrectly compute the correction at sea level, while for other satellites the correction is calculated strictly following the topography – i.e., to the lake bed. There might be large differences between data processing for various satellites, as some of them use the "wrong" pressure for the computation (i.e., they do not account for height of the water body). As for the wet tropospheric correction, it is possible over some enclosed water bodies to recover a radiometer correction, away from land, but sidelobe contamination issues remain (especially for Jason) and there is the need for the best possible models. In that case, one could use a wet tropospheric model based on gridded meteorological data sets at the altitude of the lake such as the European Centre for Medium-Range Weather Forecasts (ECMWF) (Mercier and Zanife 2006).

13.3.3
Mean Sea Surface

For large water bodies, the satellite altimetry data should be averaged over long distances and in order to transform altimetric measures to relative sea level anomalies, a subtraction of geoid is necessary. While the existing geoid models are sufficient for large-scale ocean studies, for inland waters a calculation of a regional mean sea surface represents a better solution. Usually MSS is calculated by averaging altimetric measures over a given region and over a given time period. However, in the case of the Caspian Sea this represents a certain challenge. First of all, interannual changes of sea level here are sometimes much higher than seasonal ones (that are about 30 cm). Formation of ice cover should be

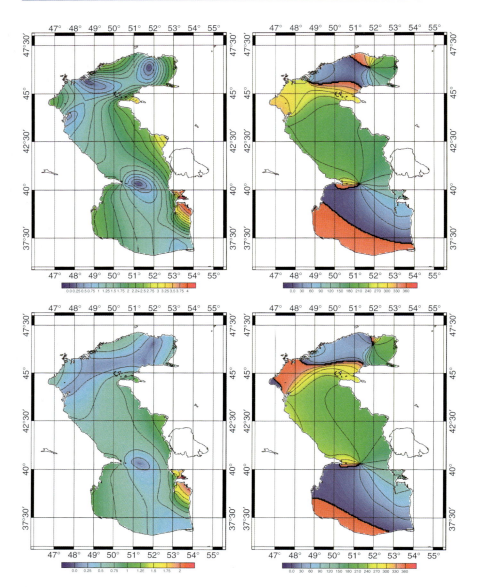

Fig. 13.3 Amplitude (cm) and phase (degrees) of the M2 (*upper panel*) and S2 (*lower panel*) constituents from T-UGO 2D tidal solutions

accounted for in altimetric measures (through exclusion or correction of the data), as well as influence of river input, seiches, and baric waves.

Existing MSS models essentially differ according to the used information or temporal averaging interval. One of the possible solutions implemented in the Integrated Satellite Altimetry Data Base (ISADB) developed in the Geophysical Center of the Russian Academy of Sciences (GC RAS) is to analyze sea level variability at crossover points of ascending and descending passes of T/P and Jason-1 (Lebedev and Kostianoy 2005). In this database, the GCRAS08 MSS model of the Caspian Sea was calculated according to the following scheme. At first from the T/P and Janson-1 Satellite altimetry data, the sea surface height (SSH) synoptic and seasonal variations for all passes of each repeat cycle were eliminated. In the last phase, the GCRAS08 MSS Model was constructed as a function of latitude, longitude, and time with consideration for climatic dynamic topography. This MSS is maximally close to the geoid height, calculated on model EGM96 with decomposition to 360°. In this case, the root-mean-square (RMS) deviation of a difference between SSH and geoid height is minimal at 17 cm, and the correlation coefficient between these two values is maximal at 0.999 (Lebedev and Kostianoy 2005). A possible explanation for this is that this model uses the longest time series available (September 1996–December 2008) of T/P and Jason-1 data. Every year GCRAS08 MSS is recalculated.

13.3.4
Influence of Ice Cover on Altimetric Measurements

Estimates of range between satellite and the echoing surface are obtained using procedures known as altimeter waveform retracking. Retracking retrieves the point of the radar echo that corresponds to the effective satellite-to-ground range. As the primary goal of most altimeters is the study of ocean topography, most of the retracking algorithms used are suited to the open ocean conditions. For example, T/P, Jason-1, and GFO all have only one onboard retracker that is adapted to the ocean surface. For the Caspian Sea, where stable ice cover is present every year for several months in the Northern Caspian, this poses a problem. Presence of ice significantly affects the shape of the returning radar waveform and could result in erroneous range estimates in winter.

In order to assess the degree in which ice presence affects altimeter range measures of the enclosed seas and estimate associated uncertainties, a study for the neighboring Aral Sea has been done using data from Envisat RA-2 altimeter (Kouraev et al. 2009). For this satellite, four different retracking algorithms (one – Ocean – for ocean conditions and three – Ice1, Ice2, and Sea Ice – for ice) are used to process raw RA-2 radar altimeter data. Presence of the four simultaneous range values from these retrackers for each 18 Hz RA-2 measure gives a possibility to precisely quantify the difference between various retrackers. Using the ice discrimination methodology (see Sect. 13.4.6 and Kouraev et al. 2008 for more details), each 18 Hz data has been classed as either open water or ice. Using this dataset, we have calculated the range differences between various retrackers and calculated separate statistics for the open water and ice cases.

It has been found (Kouraev et al. 2009) that because of its algorithm Ice1 retracker provides constantly higher sea level measures (about 25 cm for ice cover and 15–20 cm for open water) than the two other ice retrackers and this should be taken into account. For ice-covered sea, Ocean retracker constantly shows much higher level values than any of ice-adapted retrackers, with differences coming up to 40–45 cm. Thus, for Envisat, it is obviously better to use other retrackers than Ocean when ice cover is present. In the absence of in situ time series of the sea level for the Aral Sea, it was not possible to estimate the absolute difference for each altimetric satellite mission. However, in order to homogenize sea level time series for open water and ice-covered sea, it is suggested to subtract 40–45 cm from sea level measures from T/P, Jason-1, and GFO (all of them use Ocean retracker) when the sea is ice-covered. These findings should also be relevant for the Aral Sea's neighbor and close relative – the Caspian Sea.

13.3.5
Multi-satellite Bias

Same as for ocean studies, merging of time series from different satellite missions requires an accurate assessment of biases and drifts for each parameter contributing to the final estimate of the sea level. To compute the Caspian Sea and Kara-Bogaz-Gol Bay water level in a multi-satellite mode, it is first necessary to estimate biases between satellites. In general, these values are in the range of 10–20 cm. After the MSS is computed (using data from all satellite missions) and subtracted, each satellite data are processed independently and potential radar instrument biases between different satellites are removed using T/P data as reference. Then water levels from the different satellites are merged to provide uniform time series.

13.3.6
Calibration/Validation (Cal/Val) Activities

The calibration technique over enclosed water bodies is however still immature even though important progress has already been achieved. A Cal/Val experiment in the Lake Erie (North American Great Lakes, Shum et al. 2003) has shown interesting limitation on the Sea State bias correction's calculation when algorithms developed over the ocean are used for other types of water surface. They also calculated altimeter biases for Jason-1 quite different from what was observed with ocean calibration sites. This is achievable when measurements of specific and numerous field campaigns and a permanent ground network of level gauges and meteo stations are processed to detect biases, errors in the geophysical corrections, etc.

There is a clearly expressed requirement (cf. Proceedings of the 2nd Space for Hydrology Workshop, WPP-280) calling for a number of regional validation efforts on the basis of in situ data – which most of the time are lacking and may require dedicated validation campaigns.

Shum et al. (2003) pointed out that the variety of calibration sites for altimetry had to be enlarged in order to have more global distribution and more robust assessment of the altimetry system, and to check if specific conditions lead to different estimation of absolute bias of the instruments. Calibration over lakes and enclosed seas surfaces has specific characteristics with respect to the ocean surface: wave and ocean tides are generally low, and dynamic variability is much weaker than in the oceans. On the other hand, while satellite altimeters are capable of providing a detailed picture and regular monitoring of ocean level, they are not optimized for coastal/inland retrievals. This is related to some processing and quality control issues, for example, the fidelity of corrective terms in coastal areas and possible contamination due to the presence of land in the radar footprint. These issues are currently impeding the effective use of altimeter-derived products in coastal areas and continental water bodies.

While the comparison of altimeter-derived Caspian annual mean sea level with level gauge data from different stations generally shows good agreement (maximal discrepancies of 7 cm in 1995), careful and independent validation against in situ data is necessary. This validation should help to assess the uncertainties in the data products and separate instrumental changes from environmental ones. A regional Cal/Val site and experiments in the Caspian Sea would supply invaluable data to formally establish the error budget of altimetry over continental water bodies, in addition to the global mission biases and drift monitoring.

LEGOS (Laboratory of Space Geophysics and Oceanography, Toulouse, France) has taken part in a number of campaigns for absolute calibration of altimetry over lakes and enclosed seas, including one campaign in the Caspian Sea in 2005. In those campaigns, a GPS receiver along the track of the satellite at the time of an overpass has been used (Fig. 13.4). Such an approach helps to reveal specific features of altimetry over continental water bodies, such as the one found at Cal/Val site in Lake Issyk-Kul, in Central Asia, where a big slope of the mean lake surface – up to 5 m from one side to the opposite of the lake – was revealed (Crétaux et al. 2009a, b). Other issues specific to continental water bodies have also been studied during those campaigns to better understand the error budget of the geophysical corrections, and the precision of the different algorithm of retracking (Crétaux et al. 2009a, b).

The first drifter experiment in the Caspian Sea was carried out in October 2006–February 2007 in the framework of the MACE project (Lebedev and Kostianoy 2008a) (Fig. 13.5). Data on atmospheric pressure were used for computing a dry tropospheric correction, and RMS differences in the derived dry tropospheric correction between drifter and closest (in time and space) altimeter overpasses were between 2.25 and 3.86 cm.

Within the framework of the INTAS project ALTICORE (ALTimetry for COastal REgions) (Vignudelli et al. 2007), a pilot sea level station consisting of a bottom pressure recorder has been recently installed and put into operation in June 2008 at Apsheron Port, Azerbaijan (Vignudelli et al. 2008) to produce a high-quality data set of in situ sea level measurements. It is hoped that in 2–3 years this station will be part of the Global Sea Level

Fig. 13.4 GPS measurements of sea level (m) with respect to the ellipsoid of reference (WGS84) along the altimeter tracks (T/P, GFO and Envisat) during the ship cruise in July 2005 near Apsheron Peninsula. It is supposed to match the geoid surface with respect to the ellipsoid. The colour scale is expressed in meters: observe the 4 m slope of the geoid between Baku and Jiloy Island

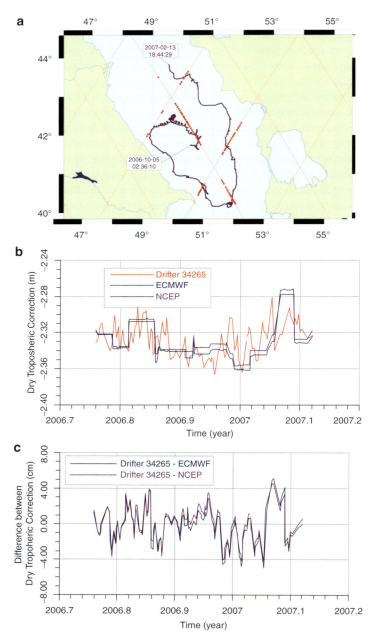

Fig. 13.5 Data from the MACE drifter experiment. (**a**) Red marker shows on pass 092, 133, and 209 altimetry data which are used for comparison with result dry tropospheric corrections calculation on atmospheric sea level pressure 34265 drifter data. Dry tropospheric corrections (in m) on ECMWF and NCEP model and drifter data (**b**) and its difference (drifter minus ECMWF and NCEP, cm) for drifter 34265 from 5 November 2006 to 13 February 2007 (**c**) (Lebedev and Kostianoy 2008a)

Observing System GLOSS. This can be a good Cal/Val site for other missions as it is close to several altimeter ground tracks (including planned AltiKa satellite). The intention is to perform further geodetic leveling campaigns, including GPS measurements at land and sea, and to collect a preliminary accurate record of in situ sea level measurements at high temporal resolution.

It is important to mention that at present there are globally no stations maintained to GLOSS standards and therefore none of the numerous past or current level gauges is really usable for satellite altimetry validation. In addition, the absence of a transnational uniform geodetic leveling network along the coasts does not permit the comparability of in situ measurements of sea level at different sites. As no Caspian Sea country is currently participating in GLOSS programs, a regional example would better convince Caspian Sea countries of the value of sea level measurements and enhance co-operation with international institutions to build up competence in new technologies for sea level observation.

As stated previously, a study and demo project of Cal/Val site in inland waters to be tested using Envisat/Jason is necessary. The Caspian Sea would be the optimal target, being a big water object with perfect location of Envisat and Jason tracks and a station already in operation since June 2008 but with the need to be upgraded to achieve GLOSS-ALT quality requirements. It is important to highlight that no similar existing Cal/Val site in Europe would satisfy these requirements.

Further developing and validating a calibration and drift monitoring system over lake are necessary to reach requirements for planned satellite mission Sentinel-3 and for lake level monitoring with the accuracy required for an Essential Climate Variable (GCOS 2006).

The infrastructure on Apsheron site will include level gauge, permanent GPS satellite receivers, meteorological station, communication systems for data transmission, etc. The place is located close by a crossover point between Envisat and Jason tracks and is well suitable for altimeters calibration tasks.

13.3.7
Existing Datasets

There are several databases of long time series of altimetric sea level observations for the Caspian Sea and its parts. Some are publicly available online (Hydroweb, USDA Reservoirs database), while others not (cf. ISADB). On the basis of the same initial altimetric GDR data, each group of researchers uses different methods to estimate the resulting sea level for the given period.

Hydroweb (Hydroweb web site 2009). This altimetric water level data base at LEGOS, Toulouse, France, contains time series of water levels of over 150 lakes and enclosed seas, and 250 sites over large rivers and wetlands around the world. These time series are mainly on the basis of altimetry data from T/P for rivers, but ERS-1 & -2, Envisat, Jason-1, and GFO data are also used for lakes. For the Caspian Sea, observations since 1992 are used. The altimeter range measurements used for lakes consist of 1 Hz data.

USDA Reservoir Database (USDA Reservoir Database web site 2009). The US Department of Agriculture's Foreign Agricultural Service (USDA-FAS), in co-operation with the National Aeronautics and Space Administration, and the University of Maryland, is monitoring lake and reservoir height variations for about 100 lakes worldwide using T/P and near-real time Jason-1 data. For the Caspian Sea water level, variations are computed with respect to a 10-year mean level derived from T/P altimeter observations and are provided with 10 days resolution without median filtering.

Integrated Satellite Altimetry Data Base (ISADB). This database has been developed at the Geophysical Center of the Russian Academy of Sciences (Medvedev et al. 1997; Lebedev and Kostianoy 2005). For the Caspian Sea, the satellite altimetry data of T/P and Jason-1 are used.

Caspian Sea dataset from the X-TRACK processor. In the framework of the ALTICORE project, a dedicated dataset for the Caspian Sea was developed using the X-TRACK processor (Roblou et al. 2007; Roblou et al., this volume). The objective of this processor is to improve both the quantity and quality of altimeter sea surface height measurements in coastal regions, by improving the correction phase (postprocessing). This X-TRACK-derived dataset for the Caspian Sea is on the basis of the use of 15 years of 1 Hz T/P plus Jason-1 data with a coastal-dedicated data edition strategy, an improved retrieval of environmental and sea state, tidal and atmospheric forcing corrections (provided by T-UGOm 2D simulations outputs), and a high resolution MSS. Currently, the dataset consists of (a) improved sea surface height estimates along each ground track, (b) a high resolution mean sea surface along each ground track, and (c) time series of sea level anomalies along the track or projected onto mean track.

13.4
Results

Besides monitoring interannual changes of the sea level, altimetry provides a valuable tool to monitor level changes at different temporal (from seasonal to short-term) and spatial scales. It can also provide useful information on sea level for storm surge events, as well as on water level of large rivers flowing to the sea, such as Volga. Sea level variability coupled with observations of other constituents of water budget can provide estimations of evaporation over the Caspian Sea or Kara-Bogaz-Gol Bay. Among other less direct applications of altimetry are observation of wind and wave regime over the sea, and study of its ice cover regime using synergy of active (radar altimeter) and passive (radiometer) microwave observations from altimetric satellites.

13.4.1
Seasonal Variability of the Caspian Sea Level/Volume

One of the main drivers of seasonal changes in the Caspian Sea is the river input, especially from the Volga which provides more than 78% of total inflow. Using mean monthly data of river runoff, we can compare (Fig. 13.6) the variability of river discharge and variations of the Caspian Sea volume (recalculated from sea level using a hypsometric curve). There is a marked temporal shift in the maxima – while Volga spring flood is observed in May–June, maximal peak of volume is observed 1–2 months later (June–July). Obviously not all the Caspian Sea volume variation can be explained by the Volga and other rivers discharge: one should also take into account seasonal changes in precipitation and evaporation.

Fig. 13.6 Mean monthly values of Volga inflow (km³, *black*) and the Caspian Sea volume anomalies with respect to mean volume of each year (km³, *dark red*)

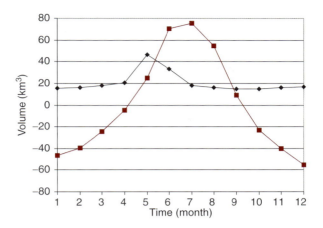

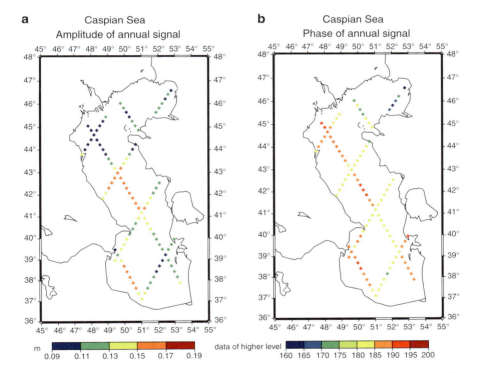

Fig. 13.7 Amplitude (cm) and phase (day of the year when maximum amplitude was observed) for the Caspian Sea level calculated along the T/P tracks

Another interesting aspect is the spatial variability of the sea level that could provide more insight into the influence of Volga discharge on the Caspian Sea level. Spatial variability of amplitude and phase of T/P sea level observations (Fig. 13.7) shows the complexity of sea level changes. There is a marked spatial pattern with more than 15 cm of annual amplitude in the Middle Caspian, while in the Northern or Southern Caspian this value does not exceed 10 cm. The timing of maximal seasonal level has an early (160–175 days, i.e., 9–25 June) peak in the eastern part of the Northern Caspian; 30–35 days later (185–195 days, i.e., 4–14 July) the maximum is observed in the western Northern Caspian, as well as in the Middle and Southern Caspian. It is interesting to note that the latest dates (200 days, i.e., 19 July) are typical for the region near the Volga delta in the Northern Caspian. Apparently, this spatial and temporal variability of sea level can not be attributed to Volga inflow alone – evaporation and precipitation play an important role, as well as influence of wind field on short-term sea level changes.

13.4.2
Storm Surges in Altimetry

The wind-induced storm surges occur along all the shores of the Caspian Sea. The most important and dangerous ones are observed in the large and shallow Northern Caspian. Usually they take place in spring and autumn, when winds from the north–northwest or southeast dominate. The storm surges last from several hours to several days, while the maximal registered duration was 8 days (Terziev et al. 1992).

Storm surge occurrence varies from 2–3 to 15–20 times per year and 60-cm high water level rises, due to the storm, are observed every year in the near-shore zones. Such rises are typically caused by winds with a speed of 10–12 m/s. Winds with a speed of 16–25 m/s can cause the water level rise of 1.5–2 m and for the Northern Caspian with average depth of just 5–7 m this is a dangerous event. The 3–4 m high sea level rise is catastrophic and can lead to an inundation of 30–40 km wide coastal band and to serious damages. Such extreme surges are formed under special wind situations when the strong effective wind is preceded by winds of opposite directions and the sea level decreases. This surge is a combination from wind-induced wave propagation and water level restoration. The catastrophic storm surges took place in 1877, 1910, 1952, and 1995 (Terziev et al. 1992; Kabatchenko 2006). As all the coastal zone of the Northern Caspian represents the low flat plain, the impact and scope of floods from storm surges will increase with the sea level rise.

In spite of some limitations, satellite altimetry is capable of observing the storm surges, as was the case at the beginning of 1999 (in this season sea level is not significantly affected by Volga River discharge). The T/P 10 Hz altimetric data have been processed to derive sea level anomaly above the mean sea surface for cycles 1–360. These values were plotted (Fig. 13.8) and the mean (median) sea level anomalies for the northern part of the ground track 092 (part between the coastline and isobath 20 m) have been calculated. For cycle 235 (02 February 1999) this value was −25.7 cm. The spatial distribution along the track is noisy near the coast and more stable (around −50 mm) seaward. Ten days later, we observe a significant increase of sea level anomaly – mean value rises up to 3.7 cm but extreme values near the Volga delta go up to 100 mm. At cycle 237 (22 February 1999) we observe a rapid decrease – mean values go down −54.7 cm and extreme values near the Volga delta are −500 mm. By the beginning of March (cycle 238), sea level has been restored (mean value is −11.1 cm) and closely resembles the initial sea level observations on 2 February.

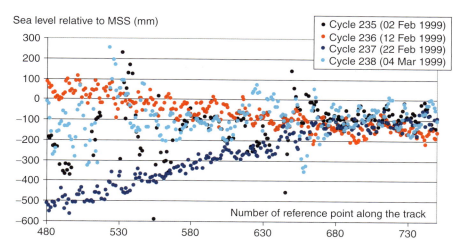

Fig. 13.8 Variability of sea level in the Northern Caspian (mm, relative to the MSS). T/P data for ground track 092, X-axis – number of reference point (each point is separated by about 600 m) along the track (480 – intersection with 5 m isobath, 620 – with 20 m isobath, 745 – 50 m isobath, see also Fig. 13.1)

In order to complement these altimetric values and enhance the analysis, wind fields from AMI-Wind scatterometer have been used (IFREMER CERSAT web page 2009). These are the data from the C-Band active microwave instrument that provides radar backscatter measurements, from which wind speed and direction can be retrieved. Depending on satellite coverage, data for the Caspian Sea are available from two to three times per day to once every several days. We also have used the data on the ice extent and ice edge position using a methodology described in Sect. 13.4.6 (see Kouraev et al. 2008 for more details).

Four wind situations corresponding to the time of the T/P overflight (by ground track 092) have been selected (Fig. 13.9). On 2nd February, the wind was weak, with speed up to 5 m/s in the Middle Caspian and up to 12 m/s for north-westerly winds in the coastal region in the Northern Caspian. In the western Northern Caspian north-westerly winds did not exceed 3–4 m/s. In the next few days, the wind field was unstable but southeasterly winds started to dominate over the large areas, such as for 5 February (not shown). On 12 February we observed steady and uniform southeasterly winds over the whole Middle Caspian with speeds of 10–12 m/s. By entering the Northern Caspian, winds became less strong and their direction became more chaotic. Though in the eastern part of the Northern Caspian north-westerly winds were observed with speeds of 7–10 m/s, this area was ice-covered (see ice edge position on the corresponding image on Fig. 13.9) and the correctness of wind speed retrieval over ice was not known. Ice cover in the eastern part acted as a solid "lid" hindering sea surface increase and apparently the southeasterly winds created a pile-up of water (+30 cm on average for southern part of ground track 092) in the western part of the Northern Caspian.

For 22 February, we observe strong (up to 17 m/s) winds coming from southeast in the Middle Caspian and gradually turning to south and southwesterly winds in the Northern Caspian. In the western part of the Northern Caspian, some wind vectors have direction opposite to the main flow. This could be related to the so-called aliasing of wind solutions

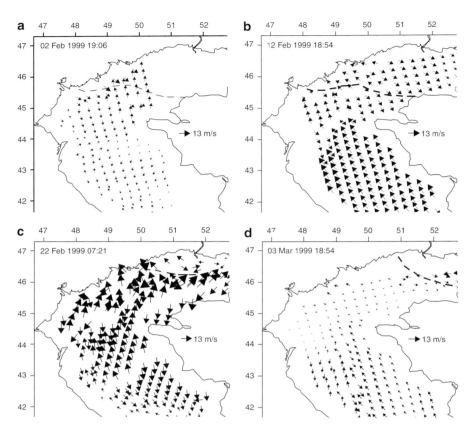

Fig. 13.9 Wind fields (m/s) from AMI-Wind scatterometer onboard ERS-2 for selected dates in February–March 1999. Dashed line – approximate ice edge position (westward from ice edge – open water, eastward – ice cover) obtained using an ice discrimination methodology (Kouraev et al. 2008)

from the scatterometer. The principle of deriving wind direction from scatterometer lies in determining the angle that is most likely to be consistent with the backscatter observed from multiple angles. Then a mathematical function fitting the observed backscatter as a function of the wind direction is applied. This function usually has multiple minima (ambiguities), with the best fit corresponding to the real wind direction and the next best fit being the opposite direction. Noise in the observations can lead to the choice of wrong direction. This problem is typical for weak winds (less than 8 m/s), where signal can be confounded with noise (COAPS web site 2009), and various "dealiasing" algorithms are usually applied by distributing agencies to remove this ambiguity. Apparently in this case for 22 February, dealiasing provided some wrong vectors that should be corrected to the opposite direction (this also applies for some vectors on the boundary between the Middle and Northern Caspian for 12 February). As a result, we would observe a homogeneous wind flow entering the Northern Caspian from the south and then gradually turning northeast. This situation has been confirmed by another satellite overpass later this day (not shown).

Retreat of ice cover to the far end of the eastern Northern Caspian has opened a large area in the central part of the Northern Caspian, and this is where the water has been moving, leading to decrease of water level in the northwestern part (level decrease of 58 cm).

By 3 March 1999, the wind field was weak: less than 6 m/s for mainly southeasterly winds in the Middle Caspian, and no wind in the western part of the Northern Caspian (except several wind vectors up to 6 m/s near the Volga delta). The water level in the northern part of ground track 092 rose by 43 cm, back to similar level anomaly as for 2 February.

The satellite altimetry does not permit to reliably and continuously monitor the storm surges because of inadequate temporal resolution. There is also an issue of the deterioration of the altimetric data quality in the near-shore zone. Often the number of valid observations dramatically decreases during and soon after the storms. As a result, the catastrophic surge of 12–16 March 1995 was not observed by T/P, which flew over on 8 and 18 March. The altimetric measurements of 18 March are scattered near the coastal zone, but show the decrease of sea level that normally takes place soon after the storm surge.

Nevertheless, when altimetric data are available for a specific storm surge situation, they provide information on the surge wave characteristics along the ground track. Such data can successfully complement in situ observations of storm surges and can be used for validation/parameterization of numerical models.

Use of scatterometer-derived data on wind vectors in combination with satellite altimetry can improve our understanding of the water dynamics in the Caspian Sea and can provide detailed information on storm surges, with better spatial coverage as compared to solely in situ observations. However, there are some inherent limitations for this approach. One is that spatial sampling for altimeters is possible only for region covered by the ground tracks. Another problem is limited temporal resolution of both scatterometer and radar altimeter, which does not allow continuous monitoring of storm surges at their typical temporal scale (several hours to 1–2 days). The most appropriate solution will be to combine (a) in situ observations with high temporal sampling from the coastal and island stations, (b) satellite observations from radar altimeters and scatterometers, and (c) numerical modelling of water dynamics response to the wind field.

13.4.3
Volga Water Level from Altimetry

River water level variability has been studied using 1 Hz T/P data for September 1993–August 2000 observations at the cross section of ground track 235 with the river (Lebedev and Kostianoy 2005, 2008b). The obtained time series were compared to water discharge rates at the Volgograd hydroelectric power station. The correlation coefficient was about 0.83 for annual values and 0.71 for monthly mean values. The low spring flood in spring 1996 was well observed in altimetric data. The negative water level trend of 46.7 cm/year in 1993–1996 corresponds to the decrease in average annual water discharge in the Volgograd power station from 10,654 to 5,609 m^3/s. Comparison of satellite water height with daily observations of water level at the three hydrological stations close to the ground track 235 (Enotaevka, Seroglazovka and Verkhnee Lebyazhye) shows a much better correlation of 0.8–0.9 (Lebedev and Kostianoy 2005, 2008b).

13.4.4
Water Budget of the Caspian Sea and the Kara-Bogaz-Gol Bay

Another application of sea level deduced from satellite altimetry is the study of water budget constituents of the Caspian Sea and Kara-Bogaz-Gol Bay. Sea level/volume changes are a dynamical response to the variations in incoming (precipitation, river, and underground inflow) and outgoing (evaporation and, in the case of the Caspian Sea, outflow to the Kara-Bogaz-Gol Bay) components. For the Caspian Sea, the water balance has been computed for the last century thanks to in situ data based on the network of hydrometeorological stations along the coast and on the islands of the Caspian Sea. Traditionally some of these parameters are easily measured (river runoff, precipitation), while for the others their estimation imposes the use of hydrological or climatic modelling (evaporation, underground seepage). For calculating evaporation, information on many parameters pertaining to atmospheric and sea surface conditions (air and sea temperature, air humidity, wind speed, wave heights) is necessary.

Mean values for 1900–1993 (Kouraev 1998) show that river discharge (768 mm/year of water equivalent) is the most dynamically changing constituent of the water budget. Precipitation (191 mm/year) is much lower and while underground discharge is not measured directly, it is estimated around 12 mm/year. Evaporation (969 mm/year) is the highest outgoing constituent but its seasonal changes are low. Mean outflow to the Kara-Bogaz-Gol Bay varied during the twentieth century, but on average it was about 20 mm/year.

Evaporation, which is difficult to measure directly, is usually computed as the residual term of the water balance. Using altimetric time series of sea level and estimating volumetric changes, and by an inversion of the water budget equation, one can estimate the difference between evaporation and precipitation (E−P). Evaporation deduced using this approach for the period 1993–2001 shows an increase of effective (E−P) evaporation of around 22 mm of water/year. At the same time, models of evaporation that use in situ observations of sea level indicate only an increase of evaporation of 17 mm/year during the same period. One of possible reasons for this is the difference between in situ and altimetric observations (see Sect. 13.3.6) and this stresses the need to have very precise measurements of the Caspian Sea level.

Natural conditions of the Kara-Bogaz-Gol Bay are unique and its water budget has been heavily perturbed in the twentieth century by natural and anthropogenic influence (Kosarev and Kostianoy 2005; Kosarev et al. 2009). The bay's depression is a shallow bowl with a flat bottom and bare shorelines. The climate of the region is arid with hot summers and strong winds in the autumn and winter. This is the largest lagoon in the world, separated from the sea by sandbars. A strait 7–9 km long and 400–800 m wide is located between bars. Its depth is about 3–4 m. Seawater races along the strait at speeds of 50–100 cm/s into the bay, where it completely evaporates.

In the first few decades of the 20th century, when sea level was at the high marks, differences between Caspian Sea and bay levels were relatively small. This stipulated hydraulic connection between them and moderated bay level variations. The long-term fall in the Caspian Sea level that began in the 1930s resulted in a decrease in the volume of sea water inflow into the bay. The bay area shrank and the bay became shallower. In order to cut Caspian water losses and stabilize the sea level, in March 1980, the strait was blocked by a solid sandy dam and the Caspian inflow was stopped completely. After its separation the bay began to dry out rapidly and was all but completely dried out by the middle of 1984.

As the sea level had started to rise even before the separation of the Kara-Bogaz-Gol, the opening of the strait was eventually decided upon. In September 1984, tubing was put through the dam and Caspian water once again flowed into the bay. The Kara-Bogaz-Gol began to come back to its usual shape. In the summer of 1992, the dam was destroyed and natural inflow of the Caspian water into the bay took place. Since that time, its level can be followed with the satellite altimetry data (Fig. 13.10). Until the middle of 1996, the Bay has been rapidly filling, and then the level rise stopped and its variations started to reflect seasonal changes well correlated with the seasonal level changes of the Caspian Sea. At present, the level of the bay oscillates near an absolute mark of −27.5...−27.7 m.

Altimetric data can be used to estimate evaporation over the Kara-Bogaz-Gol Bay (Fig. 13.11). E−P over the bay is much higher than that on the Caspian Sea. There is also no increasing trend of E−P, as observed for the Caspian Sea and this is a surprising result. If E−P on the Kara-Bogaz-Gol is mainly driven by variation of temperature, then both curves should follow more or less the same trend. But E−P is also controlled by temporal

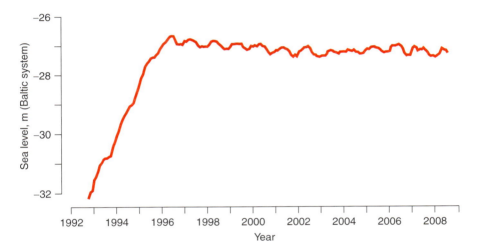

Fig. 13.10 The Kara-Bogaz-Gol water level (m, Hydroweb data, values referred to GGM02C geoid)

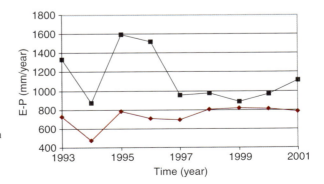

Fig. 13.11 Visible evaporation (E-P) over the Caspian Sea (*Dark red line*) and Kara-Bogaz-Gol Bay (*black line*)

variability of precipitation that is influenced by the presence of the Caspian Sea in the west, in particular under a general eastward wind regime. It is generally known (Krapivin and Phillips 2001) that precipitations over the central Asian region, and in areas close to the Caspian Sea, like the Aral Sea or Kara-Bogaz-Gol, are partially generated by water evaporated from the Caspian Sea. This indicates that an increase of evaporation over the Caspian Sea can be followed by an increase of precipitation over the bay. This may have occurred in 1993 and from 1996 to 2001: when evaporation rose in the Caspian Sea, it was attenuated or even decreasing over the Kara-Bogaz-Gol, which indicates more precipitation. In the years when E–P decreased in the Caspian Sea, it was either amplified on the Kara-Bogaz-Gol (decrease of temperature observed for both lakes, but Kara-Bogaz-Gol still receives some additional precipitation) or reversed (which probably indicates that impact of lesser precipitation than in the previous year prevails on the decreasing evaporation rates because of decreasing of temperature). A better assessment of water balance of the Kara-Bogaz-Gol could be done with precise measurements of lake and air temperature as well as wind on the Kara-Bogaz-Gol and high spatiotemporal resolution precipitation gridded data set. Altimetry data that give monthly-based level of the Caspian Sea and Kara-Bogaz-Gol can then be assimilated within a climatic model of the region, as control or forcing data of the model.

13.4.5
Wind and Wave Regime

Analysis of the relation between wind speed and wave height estimates from altimetric data RADS (Schrama et al. 2000) and in situ measurements was made for a number of places on the Caspian Sea coast in the framework of the ALTICORE Project (Vignudelli et al. 2007). For the methodology of in situ wind speed data selection, processing, analysis, and decomposition of wind vectors in relation to the coastline orientation with the aim to improve the correlation with satellite altimetry data, see the chapter by Ginzburg et al., this volume.

For the comparison of wind speed retrieved from altimeter and in situ data, two places at the Caspian Sea coast were chosen: Baku Airport and Fort-Shevchenko. For Baku Airport, selected ground tracks (see Fig. 13.1 for tracks position) were from ERS/Envisat (642), T/P and Jason-1 (092, 209), and GFO (053, 300). For Fort-Shevchenko, they were from ERS/Envisat (311), GFO (128, 225), and T/P new orbit (168).

For the Jason-1, three estimates of wind speed are provided: one based on the backscatter in Ku band, another based on brightness temperature from JMR radiometer, and a third one from interpolation of ECMWF meteorological fields (called here as "altimeter," "radiometer," and "model" estimates, respectively). For Baku Airport, wind speed values from radiometer are considerably higher (15–25 m/s and higher) than those from model and observations (mostly below 10 m/s). Altimeter and model values are in good agreement, while correlations between altimeter and in situ data (Fig. 13.12a and b) are less. Low values of wind speed (less than 5 m/s in most cases) from in situ database for Baku Airport, which are not typical for this region, raise doubts in the reliability of this database for 2002–2004 taken from Russian Hydrometeorological Center web site (http://meteo.infospace.ru). In Terziev et al. (1992), typical values of wind speed for the Baku region are higher than those on this web site.

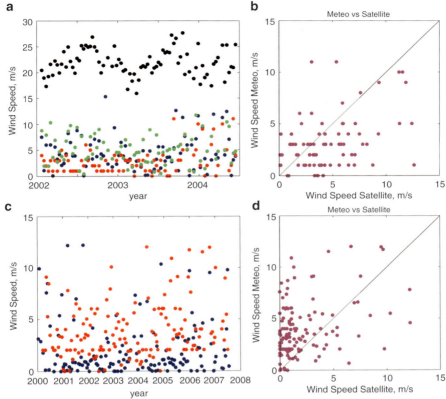

Fig. 13.12 Results of data processing for Baku Airport during 2002–2004: (**a**) wind speed from Jason-1 (track 209) altimeter (*blue circles*), observations (*red*), model (*green*), and radiometer (*black*); (**b**) scatter plot of observed versus altimeter data. The same for Fort-Shevchenko during 2000–2007: (**c**) wind speed from GFO (track 128) altimeter (*blue*) and observations (*red*); (**d**) scatter plot of observed versus altimeter data

For Fort-Shevchenko, GFO altimeter-based estimates of wind speed were compared to in situ observations (Fig. 13.12c and d) presented in meteo databases 2 (2000–2004) and 3 (2004–2007). Wind speed values from in situ measurements are in most cases higher than those from altimeter; correlation coefficient between altimeter and observed data (database 2) is equal to 0.44. The decomposition of wind directions in four quadrants as a function of coastline orientation similar to that described for the Gelendzhik case in the Black Sea (see the chapter by Ginzburg et al., this volume) can increase this correlation. Maximum correlation ($r = 0.87$) is observed for easterly winds, and minimum one for westerly winds (Fig. 13.13).

Satellite-derived estimates of wind speed in general show poor correlation with in situ data. Apparently radiometer-based estimates are significantly affected by land contamination in the radiometer field of view and in the coastal regions they should not be considered as reliable. Possible reasons of such discrepancy, provided altimeter measurements are trustworthy, may be local differences between wind observed at the meteorological station

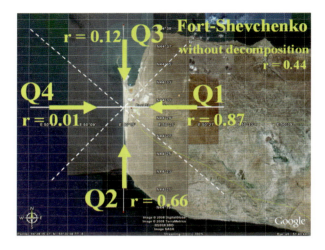

Fig. 13.13 Decomposition of wind direction in four quadrants (Q1–Q4) in relation to the orientation of the coast line for the Fort-Shevchenko case, and the appropriate correlation coefficients between wind speed from altimetry and in situ meteo record. Without decomposition $r = 0.44$. *Arrows* show prevailing wind direction for each quadrant

and along the satellite ground track, influence of land on the backscatter coefficient, as well as temporal difference between in situ and satellite measurements.

To improve the correlation, some additional techniques must be used (e.g., a correct choice of the distance between coordinates of a satellite track point of the measurement and meteo station position in order to exclude a possible influence of land; the decomposition of the winds in the four quadrants in relation to the orientation of the coastal line). It is also possible that some increase of the correlation can be obtained by taking into account the angle at which the satellite track approaches the coast, as well as the direction of the flight – from land to the sea and vice versa.

Comparison of in situ measurements of wave height and corresponding altimeter data was carried out for the Neftyanye Kamni region (outside of Baku), Caspian Sea. Data from satellite tracks of Envisat (098, 397, 556, 855), GFO (053, 139, 300), and Jason-1 (209) were used (see Fig. 13.1). For Envisat (Fig. 13.14a), the total number of observations was low compared to that for Jason-1 (Fig. 13.14b) because of difference in the repeat cycle (35 versus 10 days, respectively). Both datasets show that altimeter underestimates the observed waves. The reason of this discrepancy might be the fact that the current wave algorithms are on the basis of calibration done over the open ocean.

Two other factors – eddies and coastal upwelling – can also contribute to poor correlation between satellite-derived sea level and corresponding data of level gauge stations. The enclosed Caspian Sea, especially its southern part, is rich in mesoscale eddies. In particular, intense cyclonic eddies 10–20 km in diameter are frequently observed during upwelling in the Fort-Shevchenko region (Ginzburg et al. 2006). It is not known whether they are formed during the upwelling events (Fig. 13.15) or cold upwelling water only serves as an indicator for their surface manifestation. It is also still a question under which wind direction they are formed. Physical nature of upwelling along the Caspian eastern coast is also ambiguous – some authors suppose that upwelling in this region may not be wind-driven, but conditioned by along-shore current (see Ginzburg et al. 2006). Therefore, detailed information on local wind speed and direction is needed. Such information is also required for revealing the physical nature of upwelling filaments directed from the coast to the deep sea and propagated over distances up to 150 km in the open sea (see Fig. 13.15).

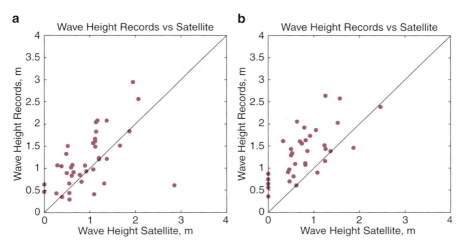

Fig. 13.14 Comparison of altimeter-derived and in situ data of wave height. (**a**) Envisat data from ground track 556 and in situ data for 2003–2006. (**b**) Jason-1 data from ground track 209 and in situ data for 2006

13.4.6
Ice Cover Regime

Studies of the ice cover of the Caspian Sea from the coastal stations began in the second part of the nineteenth century. In the second part of the twentieth century, ice studies became part of the Soviet Union's national hydrometeorological monitoring system and were performed on a regular basis using aerial surveys. However, starting from the late 1970s, the high cost led to a dramatic decrease of aerial ice cover observations. The lack of information was partially compensated for by the use of satellite imagery, mainly in the visible and infrared range (Krasnozhon and Lyubomirova 1987; Buharizin et al. 1992). However, because of frequent cloud cover in the winter, satellite observations have been used only occasionally. Actually all historical published time series of various ice cover parameters for the Caspian Sea stopped in 1984–1985.

To fill in the information gap on ice regime since the mid-1980s, an evident solution is the use of microwave satellite observations that provide reliable, regular, and weather-independent data on ice cover. A promising source of information is presented by radar altimeter satellites that provide continuous and long time series of both active (radar altimeter) and passive (radiometer used to correct altimetric observations) simultaneous observations, from the same platform and with the same incidence angle (nadir-looking).

An ice discrimination methodology that uses more than 15 years long simultaneous active and passive data from several radar altimeters (TOPEX/Poseidon, Jason-1, Envisat and Geosat Follow-On) and complemented by passive microwave observations (SMMR and SSM/I) has been developed. This methodology has been initially tested and applied for the Caspian and Aral seas (Kouraev et al. 2003, 2004a, b). Later on, this approach has been extended to (a) complement the T/P observations by Jason-1, GFO and Envisat data, and

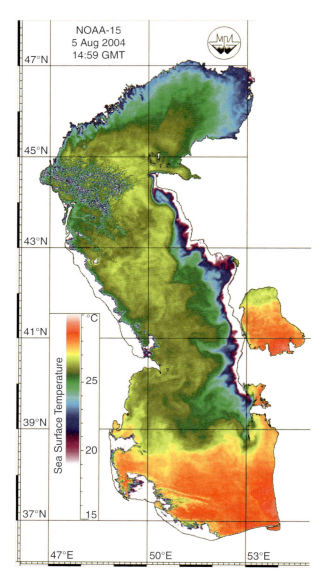

Fig. 13.15 Upwelling along the eastern coast of the Caspian Sea. (NOAA IR image for 5 August 2004) (Courtesy of Dmitry Soloviev, Marine Hydrophysical Institute, Sevastopol, Ukraine)

(b) provide better spatial and temporal resolution using an improved ice discrimination approach that combines all altimetric and SSM/I data. This approach has been implemented for the Lake Baikal (Kouraev et al. 2007a, b) and Large and Small Aral seas (Kouraev et al. 2009). A comprehensive description of the existing state of the art of this methodology, with discussion of drawbacks and benefits of each type of sensor, influence of sensor footprint size, ice roughness and snow cover on satellite measurements, and validation is given in (Kouraev et al. 2008). There, the particularities of the application of the methodology have been demonstrated for the two salt water (Caspian and Aral seas) and three freshwater (lakes Baikal, Onega and Ladoga) water bodies.

This methodology has been successfully validated for the Caspian Sea using independent in situ and visible satellite data in November–December 2001 (Kouraev et al. 2003). In situ observations of sea ice cover (Buharizin 2003) were made on the ships "*Khongay*" and "*Nord,*" operating in the Northern Caspian. "*Nord*" was in the vicinity of the Kazakh oil rig "*Sunkar.*" Analysis of the spatial distribution of ice cover is further extended using satellite image of 3 December 2001 from MODIS sensor aboard the Terra (EOS AM-1) satellite.

The first appearance of close ice in the Northern Caspian in winter 2001 was observed on 30 November from the "*Khongay*" in the region of the "*Sunkar*" oil rig (Fig. 13.16). Decrease of near-surface air temperature (down to −9°C) during the next few days stipulated ice formation, its spreading over the sea, and increase of ice thickness. Under the influence of strong easterly winds, ice drifted rapidly westward, reaching as far west as the region of the Kulaly Island (see Fig. 13.1), where it was observed on 2 December from the "*Khongay.*" The winds also caused a decrease in sea level in the eastern part of the Northern Caspian. Comparison of these in situ observations with T/P data for pass 244 shows that the onset of ice formation is well determined: the surface type in the region of the oil rig was characterized as open water on 26 November 2001 and as ice on 6 December 2001. Ice edge position derived from data for T/P track 168 agrees well with its position on the same day on the MODIS image. In the eastern coastal region, surface type for ground tracks 133 (1 December) and 244 (6 December) was classified as "ice", which corresponded to the ice distribution seen on the MODIS image. The validation described above shows that the proposed method provides robust results for discriminating between open water and ice.

Using this approach, new improved time series of ice events (ice formation, break-up, and duration of ice period) for the eastern and western parts of the Northern Caspian and estimates of mean and maximum ice extent for the whole Northern Caspian have been obtained (Fig. 13.17). Overall variability shows the complicated structure and existence of various cycles in ice conditions. A warming trend seen in the second part of the 1980s (seen both in ice duration and maximal ice development) has been followed by a cooling trend until 1993/94. Since 1995, a steady decrease of ice duration is observed both for

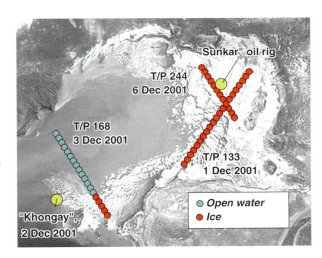

Fig. 13.16 Sea ice distribution from MODIS image for 3 December 2001, surface types derived from T/P data and locations of in situ ship observations (After Kouraev et al. 2003)

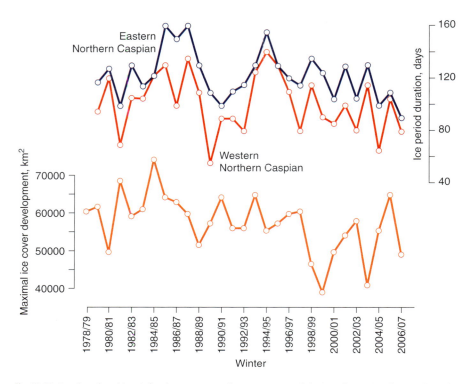

Fig. 13.17 Ice duration (days) for the eastern and western part of the Northern Caspian, and maximal area of ice cover development (km^2, within the EASE-Grid coverage of the Northern Caspian)

western and eastern parts, while for maximal ice cover development we observe general decrease but increase of amplitude associated likely with alternation of warm and cold winters. The same character was observed in the Caspian Sea sea surface temperature interannual variability (Ginzburg et al. 2005). There is a need to continue monitoring of ice conditions, for practical applications for ship routing and protection of industrial objects, for scientific studies of environmental conditions and climate variability, as well as for forcing and verification of general circulation models.

13.5
Discussion and Conclusions

We have seen that the spatial and temporal coverage of the Caspian Sea by different radar altimetry missions offers the possibility of continuous high-precision monitoring of the Caspian Sea level since the early 1990s. Being an intermediate target between open ocean and closed lakes, this sea is an interesting natural object to evaluate the potential of satellite altimetry for enclosed water bodies.

A specificity of application of satellite radar altimetry for the studies of the Caspian Sea suggests several issues to be addressed in order to obtain high-quality and reliable datasets. Influence of tides and atmospheric forcing could be successfully estimated using numerical modelling, such as has been done using the regional model for high frequency motions for the Caspian Sea based on the T-UGO 2D. However, as there are no hourly level gauge data to validate the model, there is a strong need for permanent, GLOSS-compliant in situ data to confirm the regional tides and atmospheric forcing response in the model.

Dry atmospheric correction is altitude dependant, but T/P and GFO GDRs incorrectly compute the correction at sea level, and this needs to be accounted for. Though it is possible over some enclosed water bodies to retrieve a wet tropospheric correction from radiometer, issue of land contamination in the radiometer field of view remain and there is the need for the best possible models. In that case, one could use wet tropospheric model based on gridded meteorological data sets at the altitude of the lake such as ECMWF.

While the existing geoid models are sufficient for large-scale ocean studies, for the Caspian Sea, a calculation of a regional mean sea surface represents a much better solution and this is implemented by various research groups.

Ice cover that is present every year in the Northern Caspian significantly affects altimetric measures. Using Envisat data for the Aral Sea, it has been found that ocean retracker (T/P, Jason-1, and GFO use only this kind of retracker) shows much higher values than any of ice-adapted retrackers. It is suggested to subtract 40–45 cm from sea level measures from T/P, Jason-1, and GFO when the sea is ice-covered.

Cross-calibration of various satellite missions and calibration/validation of altimetric measurements demand dedicated calibration/validation campaigns. GPS measurements from the ship, drifter experiments, as well as installation of a permanent GLOSS-compliant tide-gauge in Baku are all extremely important to ensure high-quality estimates of the Caspian Sea level from radar altimetry.

Currently several altimetric databases for the Caspian Sea exist online or offline and these are sources that need to be promoted for and exploited by a large research community.

Application of radar altimetry and other satellite techniques in combination with in situ observations show various promising results for the Caspian Sea. Temporal and spatial variability of the sea level and its relation with the Volga discharge can be investigated. When altimetric data are available for a specific storm surge situation, they, in combination with wind fields from scatterometers, provide unique information on the surge wave characteristics along the ground track. Such data can successfully complement in situ observations of storm surges and can be used for validation/parameterization of numerical models.

Altimeters also give us a tool to study the water level variability of Volga River levels. Altimeter-derived sea level measures coupled with observations of other constituents of water budget can provide estimations of evaporation over the Caspian Sea or Kara-Bogaz-Gol Bay. There is an increasing trend of visible (E−P) evaporation over the Caspian Sea. For the Kara-Bogaz-Gol Bay, this trend has not been observed, which could be linked to a dynamical relation between both water bodies: increased evaporation over the Caspian Sea can be followed by an increase of precipitation over the bay. A better assessment of water balance of the Kara-Bogaz-Gol could be done with precise measurements of water and air temperature and wind on the Kara-Bogaz-Gol Bay and high spatio-temporal resolution precipitation gridded data set. Altimetry data that give monthly based level of the Caspian

Sea and Kara-Bogaz-Gol can then be assimilated within a climatic model of the region, as control or forcing data.

Less direct applications of altimetry are observations of wind, wave, and ice regime of the Caspian Sea. For the wind field, radiometer-based estimates are significantly affected by land contamination and in the coastal regions should not be considered as reliable. Radar altimeter-derived estimates of wind speed show in general poor correlation with in situ data. Possible reasons of such discrepancy could be local differences between wind observed at the Meteorological Station and along the satellite ground track, influence of land on the backscatter coefficient, as well as temporal difference between in situ and satellite measurements. This could be improved by better geoselection of the altimetric data, and the decomposition of the winds in four quadrants in relation to the orientation of the coast line. For the wave height, comparison of in situ measurements shows that altimeters generally underestimate the observed waves. The reason of this discrepancy might be the fact that the current wave algorithms are on the basis of calibration done over the open ocean. Two other factors – eddies and upwelling – can also contribute to poor correlation between satellite-derived data and corresponding in situ observations.

Study of the Caspian Sea ice cover regime using synergy of active (radar altimeter) and passive (radiometer) microwave observations onboard altimetric satellites in combination with radiometric data from SMMR and SSM/I is a promising tool to provide reliable, regular, and weather-independent data on ice cover. This well developed and validated approach provides new improved time series of ice events (ice formation, break-up, and duration of ice period) for the eastern and western parts of the Northern Caspian and estimates of mean and maximum ice extent for the whole Northern Caspian. Since 1993/94, a steady decrease of ice duration is observed both for western and eastern parts, while for maximal ice cover development we observe general decrease but increase of amplitude.

On the basis of these interesting results, there is a clear need to continue monitoring of various environmental parameters of the Caspian Sea, for practical applications as well as for scientific studies of environmental conditions and climate variability. The most appropriate solution would be the combination of dedicated high-quality in situ observations, satellite observations from radar altimeters and other Earth Observation data, and numerical modelling of the Caspian Sea water dynamics.

Acknowledgments We are grateful to the Centre de Topographie des Océans et de l'Hydrosphère (CTOH, www.legos.obs-mip.fr/observations/ctoh) at LEGOS (Toulouse, France) for providing the initial altimetric data. MODIS images of 11 June 2005 (used for the definition of coastline position) and 3 December 2001 are courtesy of MODIS Rapid Response Project at NASA/GSFC. This work was partially supported by the Russian Foundation for Basic Research grants No. 10-05-00097-a, 08-05-97016, 10-01-00806, 10-05-01123.

References

Aladin N, Crétaux J-F, Plotnikov IS, Kouraev AV, Smurov AO, Cazenave A, Egorov AN, Papa F (2005) Modern hydro-biological state of the Small Aral Sea. Environmetrics 16(4):375–392

Birkett CM (1995) The contribution of TOPEX/POSEIDON to the global monitoring of climatically sensitive lakes. J Geophys Res 100:25179–25204. doi:10.1029/95JC02125

Birkett CM (2000) Synergistic remote sensing of lake Chad: variability of basin inundation. Remote Sens Environ 72:218–236

Buharizin PI (2003) Obzor ledovyh usloviy v severnoy chasti Kaspiyskogo morya i v nizovyah Volgi v zimniy period 2001/2002 gg. (Review of ice conditions in the northern part of the Caspian Sea and in the lower Volga in winter 2001/02). In: Materials for the state report on environmental conditions of the Russian Federation for the Astrakhan region. OOO TsNTEP, Astrakhan (in Russian)

Buharizin PI, Vasyanin MF, Kalinichenko LA (1992) Metod kratkosrochnogo prognoza polozheniya kromki splochennyh l dov na Severnom Kaspii. (A method for short-term forecasting of the pack ice boundary in the Northern Caspian.) Meteorologiya i gidrologiya (Meteorol Hydrol) 4:74–81, Moscow (in Russian)

Cazenave A, Bonnefond P, Dominh K (1997) Caspian Sea level from Topex/Poseidon altimetry: level now falling. Geophys Res Lett 24:881–884

COAPS web site on Scatterometry and ocean vector winds (accessed January 2009). http://www.coaps.fsu.edu/scatterometry/about/overview.php

Coe MT, Birkett CM (2005) Water resources in the lake Chad basin: prediction of river discharge and lake height from satellite radar altimetry. Water Resour Res 40(10). doi:10.1029/2003WR002543

Crétaux J-F, Kouraev AV, Papa F, Bergé-Nguyen M, Cazenave A, Aladin NV, Plotnikov IS (2005) Water balance of the Big Aral Sea from satellite remote sensing and *in situ* observations. J Great Lakes Res 31(4):520–534

Crétaux J-F, Birkett C (2006) Lake studies from satellite altimetry. Geosciences CR. doi:10.1016/j.crte.2006.08.002

Crétaux J-F, Letolle R, Calmant S (2009a) Investigations on Aral Sea regressions from Mirabilite deposits and remote sensing. Aquat Geochem. doi:10.1007/s10498-008-9051-2, 2009

Crétaux J-F, Calmant S, Romanovski V, Shibuyin A, Lyard F, Bergé-Nguyen M, Cazenave A, Hernandez F, Perosanz F (2009b) An absolute calibration site for radar altimeter in the continental area: lake Issykkul in Central Asia. J Geodesy. doi:10.1007/s00190-008-0289-7, 2009

Frappart F, Seyler F, Martinez JM, Leon J, Cazenave A (2005) Determination of the water volume in the Negro River sub basin by combination of satellite and *in situ* data. Remote Sens Environ 99:387–399

Ginzburg AI, Kostianoy AG, Sheremet NA (2005) Sea surface temperature. In: Kostianoy AG, Kosarev AN (eds) The Caspian Sea environment. Springer-Verlag, Berlin/Heidelberg/New York, pp 59–81

Ginzburg AI, Kostianoy AG, Soloviev DM, Sheremet NA (2006) Frontal zone of upwelling near the eastern coast of the Caspian Sea as inferred from satellite data. Issled Zemli iz kosmosa 4:3–12 (in Russian)

GCOS 2006, Systematic observation requirements for satellite-based products for climate – supplemental details to the GCOS implementation plan, GCOS 107 (WMO/TD No. 1338)

Hydroweb web site (accessed January 2009): http://www.legos.obs-mip.fr/en/soa/hydrologie/hydroweb/

IFREMER CERSAT web page (accessed January 2009). http://cersat.ifremer.fr/fr/data/discovery/by_product_type/swath_products/ers_wnf

Kabatchenko IM (2006) Modelling of wind-induced waves. Numerical calculations for studies of climate and designing hydrotechnical constructions. Dr.Sc. dissertation, avtoreferat, Moscow (In Russian)

Kosarev AN, Yablonskaya EA (1994) The Caspian Sea. SPB Academic Publishing, The Hague, 259pp

Kosarev AN (2005) Physico-geographical conditions of the Caspian Sea. In: Kostianoy AG, Kosarev AN (eds) The Caspian Sea Environment, Springer-Verlag, Berlin/Heidelberg/New York, pp 5–31. doi: 10.1007/698_5_002

Kosarev AN, Kostianoy AG (2005) Kara-Bogaz-Gol Bay. In: Kostianoy AG, Kosarev AN (eds) The Caspian Sea Environment. Springer-Verlag, Berlin/Heidelberg/New York, pp 211–221. doi: 10.1007/698_5_011

Kosarev AN, Kostianoy AG, Zonn IS (2009) Kara-Bogaz-Gol Bay: physical and chemical evolution. Aquat Geochem 15(1–2), Special Issue: Saline Lakes and Global Change. pp 223–236. doi:10.1007/s10498-008-9054-z

Kostianoy AG, Zavialov PO, Lebedev SA (2004) What do we know about dead, dying and endangered lakes and seas? In: Nihoul JCJ, Zavialov PO, Micklin PhP (eds) Dying and dead seas, NATO ARW/ASI Series. Kluwer, Dordrecht, pp 1–48

Kostianoy AG, Kosarev AN (eds) (2005) The Caspian Sea environment. The handbook of environmental chemistry. vol 5: water pollution, Part 5P. Springer-Verlag, Berlin/Heidelberg/New York, 271pp

Kostianoy AG, Terziev FS, Ginzburg AI, Zaklinsky GV, Filippov YuG, Lebedev SA, Nezlin NP, Sheremet NA (2008) The southern seas. In: Assessment report on climate change and its consequences in Russian Federation. vol 2: Consequences of the climate change. Research Center "Planeta," Moscow, pp 149–167 (in Russian)

Kouraev AV (1998) Hydrological conditions of the Northern Caspian during recent sea level rise. PhD dissertation, MGU, Moscow, 234pp (In Russian)

Kouraev AV, Papa F, Buharizin PI, Cazenave A, Crétaux J-F, Dozortseva J, Remy F (2003) Ice cover variability in the Caspian and Aral Seas from active and passive satellite microwave data. Polar Res 22(1):43–50

Kouraev AV, Papa F, Mognard NM, Buharizin PI, Cazenave A, Crétaux J-F, Dozortseva J, Remy F (2004a) Synergy of active and passive satellite microwave data for the study of first-year sea ice in the Caspian and Aral Seas. IEEE Trans Geosci Remote Sens (TGARS) 42(10):2170–2176

Kouraev AV, Papa F, Mognard NM, Buharizin PI, Cazenave A, Crétaux J-F, Dozortseva J, Remy F (2004b) Sea ice cover in the Caspian and Aral Seas from historical and satellite data. J Mar Syst 47:89–100

Kouraev AV, Zakharova EA, Samain O, Mognard-Campbell N, Cazenave A (2004) Ob' river discharge from TOPEX/Poseidon satellite altimetry data. Remote Sens Environ 93:238–245

Kouraev AV, Semovski SV, Shimaraev MN, Mognard NM, Legresy B, Remy F (2007a) Ice regime of lake Baikal from historical and satellite data: influence of thermal and dynamic factors. Limnol Oceanogr 52(3):1268–1286

Kouraev AV, Semovski SV, Shimaraev MN, Mognard NM, Legresy B, Remy F (2007b) Observations of lake Baikal ice from satellite altimetry and radiometry. Remote Sens Environ 108(3):240–253

Kouraev AV, Shimaraev MN, Buharizin PI, Naumenko MA, Crétaux J-F, Mognard NM, Legrésy B, Rémy F (2008) Ice and snow cover of continental water bodies from simultaneous radar altimetry and radiometry observations. Survey in geophysics – thematic issue "Hydrology from space." doi:10.1007/s10712-008-9042-2

Kouraev AV, Kostianoy AG, Lebedev SA (2009) Ice cover and sea level of the Aral Sea from satellite altimetry and radiometry (1992–2006). J Mar Syst 76(3):272–286. doi:10.1016/j.jmarsys.2008.03.016

Krapivin VF, Phillips GW (2001) A remote sensing-based expert system to study the Aral-Caspienne aquageosystem water regime. Remote Sens Environ 75(2):201–215

Krasnozhon GF, Lyubomirova KS (1987) Izucheniye ledovogo rezhima Severnogo Kaspiya po dannym meteorologicheskih sputnikov Zemli. (Study of ice cover in the northern Caspian from meteorological satellites.) Issledovaniye Zemli iz kosmosa (Study of Earth from Space) 5:27–32, Moscow (in Russian)

Lebedev SA, Kostianoy AG (2005) Satellite altimetry of the Caspian Sea. Sea, Moscow, 366pp (in Russian)
Lebedev SA, Kostianoy AG (2008a) Investigation of the Caspian Sea surface dynamics based on the satellite altimetry and drifter data. Abstracts, 37th COSPAR scientific assembly Montreal, Canada, 13–20 July 2008, Abstract CD, ISSN 1815-5619, A21-0033-08
Lebedev SA, Kostianoy AG (2008b) Integrated using of satellite altimetry in investigation of meteorological, hydrological and hydrodynamic regime of the Caspian Sea. J Terr Atmos Oceanic Sci 19(1–2):71–82
Maheu C, Cazenave A, Mechoso R (2003) Water level fluctuations in the La Plata basin (South America) from Topex/Poseidon altimetry. Geophys Res Lett 30(3):1143–1146
Medvedev PP, Lebedev SA, Tyupkin YS (1997) An integrated data base of altimetric satellite for fundamental geosciences research. In: Proceedings of the first East-European symposium on advances in data bases and information systems (ADBIS'97), St. Petersburg, Russia, 2–5 September 1997, St. Petersburg University, St. Petersburg, vol 2, pp 95–96
Mercier F, Zanife O-Z (2006) Improvement of the TOPEX/Poseidon altimetric data processing for hydrological purposes (CASH project). In: Proceedings of the symposium on 15 years of progress in radar altimetry, 13–18 March 2006, Venice, Italy
Mercier F, Cazenave A, Maheu C (2002) Interannual lake level fluctuations in Africa (1993–1999) from TOPEX-Poseidon: connections with ocean-atmosphere interactions over the Indian Ocean. Global Planet Change 32:141–163. doi:10.1016/S0921-8181(01)00139-4
Morris GS, Gill SK (1994a) Variation of Great Lakes water levels derived from GEOSAT altimetry. Water Resour Res 30:1009–1017. doi:10.1029/94WR00064
Morris GS, Gill SK (1994b) Evaluation of the TOPEX/POSEIDON altimeter system over the Great Lakes. J Geophys Res 99:24527–24540. doi:10.1029/94JC01642
Pobedonostsev SV, Abuzyarov ZK, Kopeikina TN (2005) On the quality of the Caspian Sea level observations (O kachestve nablyudeniy za urovnem Kaspiyskogo morya). In: proceedings of the State Hydrometeorological Center of the Russian Federation, issue 339, Sea and river hydrological calculations and foirecasts (Morskiye i rechniye gidrologicheskiye raschety i prognozy) (in Russian)
Proceedings of the 2nd space for hydrology workshop: surface water storage and runoff: modeling, in-situ data and remote sensing. 12–14 November 2007, Geneva, Switzerland Hydrospace '07 Workshop, European Space Agency, WPP-280 on CD-ROM
Roblou L, Lyard F, Le Henaff M, Maraldi C (2007) X-track, a new processing tool for altimetry in coastal oceans. In: ESA ENVISAT symposium, Montreux, Switzerland, 23–27 April 2007, ESA SP-636
Schrama E, Scharroo R, Naeije M (2000) Radar altimeter database system (RADS): towards a generic multi-satellite altimeter database system. Delft Institute for Earth-Oriented Space Research, Delft University of Technology, The Netherlands, 98
Shum C, Yi Y, Cheng K, Kuo C, Braun A, Calmant S, Chambers D (2003) Calibration of Jason-1 altimeter over lake Erie. Mar Geod 26(3–4):335–354
State water cadaster of the USSR (1930–1993). Issue 2, Marine yearbooks, Vol I, The Caspian Sea, mouths of Volga, Terek, Sulak and Kura rivers. Glavnoye Upravleniye gidrometeorologicheskoy sluzhby pri Sovete Ministrov SSSR. UGMS AzSSR (in Russian)
Terziev FS, Kosarev AN, Kerimov AA (eds) (1992) Gidrometeorologiya i gidrohimiya morey. (Hydrometeorology and hydrochemistry of seas.) vol VI: Caspian Sea, no. 1: Hydrometeorological conditions. Gidrometeoizdat, St. Petersburg (in Russian)
USDA Reservoir Database web site (accessed January 2009). http://www.pecad.fas.usda.gov/
Vasiliev AS, Lapshin VB, Lupachev YuV, Medvedev PP, Pobedonostsev SV (2002) Research of the Caspian Sea level by satellite altimetric measurements. In: Investigations of the oceans and seas, Proceedings of State Oceanographic Institute, Hydrometeoizdat, St. Petersburg, 208, pp 277–292 (in Russian)

Vignudelli S, Snaith HM, Cipollini P, Venuti F, Lyard F, Crétaux JF, Birol F, Bouffard J, Roblou L, Kostianoy A, Ginzburg A, Sheremet N, Kuzmina E, Lebedev S, Sirota A, Medvedev D, Khlebnikova S, Mamedov R, Ismatova K, Alyev A, Nabiyev T (2007) ALTICORE – a consortium serving European Seas with Coastal Altimetry. In: proceedings, XXIV International Union of Geodesy and Geophysics (IUGG) General Assembly, Perugia, Italy, 2–13 July 2007

Vignudelli S, Testut L, Calzas M, Mamedov R, Sultanov M, Alyev A, Crétaux J-F, Kostianoy A (2008) Installation of pilot sea level monitoring station for satellite altimetry calibration/validation in the Caspian Sea at Absheron port, Baku, Azerbaijan (8–13 June 2008). Vestnik Kaspiya (Caspian Sea Bulletin), N 4, pp 40–55

Zakharova EA, Kouraev AV, Cazenave A, Seyler F (2006) Amazon river discharge estimated from TOPEX/Poseidon satellite water level measurements. Comp Rend Geosci 338(3):188–196

Zakharova EA, Kouraev AV, Polikarpov I, Al-Yamani F, Crétaux J-F (2007) Radar altimetry for studies of large river basins: hydrological regime of the Euphrates-Tigris Rivers. In: Second Space for hydrology workshop "surface water storage and runoff: modelling, in-situ data and remote sensing," Geneva (Switzerland), 12–14 November 2007

Satellite Altimetry Applications in the Black Sea

14

A.I. Ginzburg, A.G. Kostianoy, N.A. Sheremet, and S.A. Lebedev

Contents

14.1 Introduction 368
14.2 Physical Processes in the Coastal Zone 369
14.3 Sea Level Measurements 371
    14.3.1 Long-Term Variability of Sea Level 371
    14.3.2 Eddy Dynamics 373
    14.3.3 Correspondence Between Altimeter-Derived and In situ Data on Sea Level 375
14.4 Wind Speed Measurements 378
14.5 Conclusions 384
References 386

Keywords Black Sea • Coastal anticyclonic eddies • Coastal upwelling • Rim Current • Sea level anomaly • Wind forcing

Abbreviations

ALTICORE	ALTImetry for COastal REgions
AVISO	Archiving, validation and interpretation of satellites oceanographic data
CCAR	Colorado Center for Astrodynamics Research
GDR B/C	Geophysical Data Record generation B or C
GFO	Geosat Follow-On

A.I. Ginzburg (✉), A.G. Kostianoy, and N.A. Sheremet
P.P. Shirshov Institute of Oceanology, Russian Academy of Sciences,
36 Nakhimovsky Pr., 117997, Moscow, Russia
e-mail: ginzburg@ocean.ru

S.A. Lebedev
Geophysical Center, Russian Academy of Sciences,
3 Molodezhnaya Str., 119296, Moscow, Russia

ISADB	Integrated Satellite Altimetry Data Base
J1	Jason-1
MGDR-B	Merged Geophysical Data Record Generation B
MS	Meteorological Station
NOAA	National Oceanic and Atmospheric Administration
PODAAC	Physical Oceanography Distributed Active Archive Center
PSMSL	Permanent Service for Mean Sea Level
RADS	Radar Altimeter Database System
RC	Rim Current
SLA	Sea Level Anomaly
T/P	TOPEX/Poseidon

14.1
Introduction

The Black Sea is a semi-enclosed basin at the southeast of Europe, keeping track of its ecological health and level is important for national-economic activity of six states on its shores. Because of limited water exchange with the open seas (with the Mediterranean Sea through the Turkish straits system Bosporus – Sea of Marmara – Dardanella Strait) as well as peculiarities of density stratification (Tuzhilkin 2008), the ecosystem of this basin is very sensitive to climate changes (in particular, to global warming) and anthropogenic forcing. An increase of anthropogenic stress determined by national-economic activity on the shores and in the coastal regions (river runoff, waste from cities and recreation zones, oil terminals, etc.) results in intensifying contamination and eutrophication of coastal waters. Therefore, the studies of the processes of horizontal water mixing in this zone and its water exchange with the deep-water basin represent an urgent task. On the other hand, national-economic activity on the shore necessitates monitoring the Black Sea level – which has a tendency, in common with the World Ocean level, to increase in connection with global warming – and retrieving information on wind speed. Altimeter-originated data on sea level have been already successfully used in studies of the Black Sea circulation and sea level changes (e.g., Korotaev et al. 2001; Sokolova et al. 2001; Ginzburg et al. 2003; Zatsepin et al. 2003; Goryachkin and Ivanov 2006; Kostianoy et al. 2008). It is hoped that radar altimetry, which provides along-track measurements of sea surface height and wind speed, can be an effective tool for handling the coastal zone problems.

For a proper interpretation of altimeter-derived data for the Black Sea, especially for its coastal zone, various physical processes and phenomena typical for this zone should be taken into account. So, first, we shall briefly review the physical processes in the coastal zone to investigate which satellite measurements of sea level changes and wind are important and which, on the other hand, can contribute to a correlation between altimeter-derived sea level anomaly (SLA) data and corresponding data from coastal meteorological stations (MS). Then, examples of using altimetry measurements of sea level will be considered. The comparison between observed (in situ) and altimeter-derived SLA and wind speed values for several coastal locations will be made as well and their correlation will be discussed.

14.2
Physical Processes in the Coastal Zone

The main elements of the Black Sea circulation are the basin-scale Rim Current (RC) that propagates over the continental slope in a cyclonic direction, one or two cyclonic gyres in the deep sea, and mesoscale (about 40–100 km in diameter) coastal anticyclonic eddies, which are formed between the RC and the coast and move generally in the same direction as the RC with typical velocity of 2–6.5 cm/s (Ovchinnikov and Titov 1990; Oguz et al. 1993; Ginzburg et al. 2008). These mesoscale eddies are observed along the entire sea periphery (Fig. 14.1). For example, two anticyclones over the northwestern continental slope and six eddies off capes of the Anatolian coast are seen in Fig. 14.1b. Coastal anticyclones serve as the main mechanism of coastal zone self-cleaning: as a result of involving the surrounding water in their orbital rotation, these eddies transport along their peripheries the polluted coastal water to the deep sea and, vice versa, cleaner open sea water to the coast. Therefore, tracing eddies' dynamics from the moment of their formation to dissipation is of particular value.

Mechanisms of generation of coastal anticyclonic eddies are not clearly understood (see discussion in Ginzburg et al. 2008). Observations of these features are more frequent in the summer season, when wind forcing is weak and the RC is the weakest. It implies an important role of wind in their formation (Zatsepin et al. 2002). So, regular information on wind speed and direction in the basin scale and near concrete coast locations is needed.

Another physical process, for which information on local wind speed and direction is required, is the uplifting of subsurface water (from depths to about 30 m) to the surface near the coast. Coastal upwelling is frequently observed in the summer months along the western, northern, northeastern, and Anatolian coasts of the Black Sea (Fig. 14.1b). It results in sharp temperature drop (by up to 10–15°C) near the coast. The transformed waters of upwelling driven from the shore by jets (about 10 km wide) and eddies over distances more than 100 km toward the deep-water part of the sea feature a temperature contrast of 1–2°C with respect to the adjacent waters (Ginzburg et al. 2000). It is believed that coastal upwelling is wind-driven in most regions of the Black Sea, although it is not evident for the Anatolian coast with marked topographic irregularities. Under upwelling, a negative surge is observed (sea level lowers) that can influence satellite-derived values of SLA.

Among the other processes which can also influence satellite measurements of sea level, the following can be named: river runoff (mouths of the largest rivers are located in the northwestern part of the sea (Mikhailov and Mikhailova 2008)); sea ice, which is formed in the northwestern part of the Black Sea and covers no more than 5% of the sea area even in severe winters (Kosarev et al. 2008); storm surges (sea level oscillations with values up to 30–40 cm in most cases and duration up to about 50 h) caused by coastal winds observed in the cold season (Kosarev et al. 2008); seiche oscillations up to 7 cm, with greatest period of 9.7 h (Kosarev et al. 2008); tidal semidiurnal oscillations that are practically absent in the Yalta and Sevastopol regions and have maximal values (up to 17 cm) in the Odessa Bay (Kosarev et al. 2008); and even tsunamis with observed wave height up to about 50 cm (Kosarev et al. 2008).

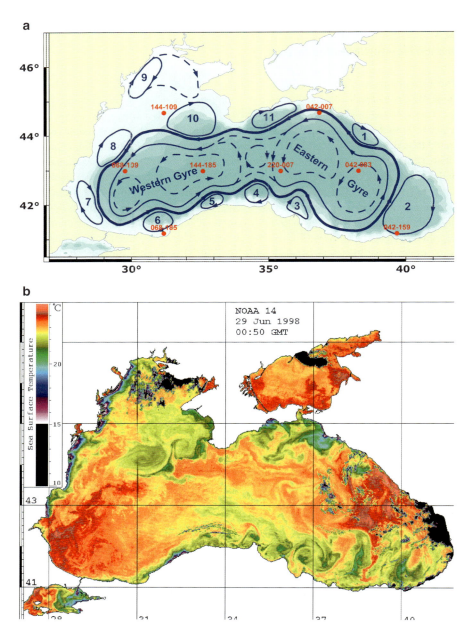

Fig. 14.1 Basin-scale and mesoscale circulation in the Black Sea: (**a**) scheme of circulation in accordance with (Oguz et al. 1993; Korotaev et al. 2001); Lighter shade of the sea surface corresponds to the depths of less than 200 m (**b**) National Oceanic and Atmospheric Administration (NOAA) infrared image for 29 June 1998. Thick line in (**a**) shows the mean position of the Rim Current (RC), circles illustrate places of frequent observations of coastal anticyclonic eddies; red points and numbers mark locations of crossover points of TOPEX/Poseidon (T/P) satellite tracks and corresponding numbers of these tracks

14.3
Sea Level Measurements

14.3.1
Long-Term Variability of Sea Level

The Black Sea level varies in response to climate changes. In accordance with in situ measurements, it increased by 20 cm in the last 100 years (Terziev 1991). Satellite altimetry enables tracing seasonal and interannual variability of sea level both of the sea as a whole and of individual regions from September 1992 to the present time (Vigo et al. 2005; Kostianoy et al. 2008). Examples of long-term (1992–2008) variability of SLA in the deep part of the sea and in its western and eastern regions obtained with processing of MGDR-B TOPEX/Poseidon (T/P) and GDR-B/C Jason-1 (J1) measurements are presented in Fig. 14.2a and b, respectively (Fig. 14.2a was obtained by averaging of SLA in four crossover points of satellite tracks 068-109, 144-109, 220-007, and 042-083; see Fig. 14.1a; for constructing Fig. 14.2b, corresponding pairs of these points of intersection of satellite tracks in the western and eastern parts of the sea were used). Information and software of the Integrated Satellite Altimetry Data Base (ISADB) developed in the Geophysical Center of Russian Academy of Sciences (Medvedev et al. 1997; Lebedev and Kostianoy 2005; Medvedev and Tyupkin 2005) were used for data processing and analysis. For satellite altimetry data, all necessary corrections are used without tidal and inverse barometer correction. Some details of the ISADB database are given in the chapter by Lebedev et al., this volume.

Calculations have shown that during the period from January 1993 to September 2008 the sea level in the sea as a whole increased with a velocity of 1.34 ± 0.11 cm/year (Fig. 14.2a); in the western and eastern regions it increased with a velocity of 1.42 ± 0.16 and 1.28 ± 0.12 cm/year, respectively (Fig. 14.2b). (Note that this value of velocity for the sea as a whole is slightly below the value 2.06 ± 0.3 cm/year reported in (Kostianoy et al. 2008) for the period from January 1993 to December 2006, which is conditioned by sea level decrease since 2005.) For comparison, from the middle 1920s to about 1985 the sea level increased with a velocity of 1.83 ± 0.7 mm/year (Goryachkin and Ivanov 2006). Thus, velocity of the sea level rise in the last 16 years was about seven times greater than in the preceding half a century. However, the character of SLA changes during 1993–2008 was not uniform: the sea level increased in general during 1993–1999 and from 2001 to 2005; marked decreases of the sea level followed its maxima in 1999 and 2005.

The sea level rise during 1993–1999 in Fig. 14.2a correlated well with the increase of the Danube River discharge and annual fresh water flux in the Black Sea in this period (Stanev and Peneva 2002; Mikhailov and Mikhailova 2008). The decrease of the sea level during 1999–2001 is also in agreement with data on the Danube River discharge in (Kostianoy et al. 2008) (We do not have similar data for the period after 2002). Interestingly, the character of SLA variability in crossover points of tracks 068-185 (Fig. 14.2c) and 042-159 (not shown here) located close to the coast differs from that in deep-sea areas (Fig. 14.2a and b): a steady decrease of the sea level is observed in these points since 1999 and 2001, respectively.

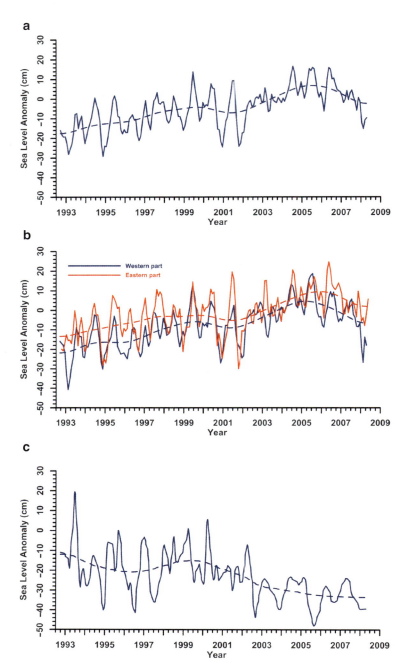

Fig. 14.2 Long-term variability of sea level: (**a**) in the Black Sea as a whole; (**b**) in its western and eastern parts; (**c**) in the crossover point of tracks 068–185 of the TOPEX/Poseidon (T/P) satellite (see Fig. 14.1a). Dashed lines show interannual variability, solid lines – seasonal variability

Combined altimeter data from satellites T/P and J1 (October 1992–August 2007) were also used for investigation of interannual variability of sea level in the interconnected Black Sea, Sea of Marmara, and Aegean Sea. In this case, data on SLA were obtained from the Radar Altimeter Database System (RADS) database (Schrama et al. 2000). Measurements along segments of track 109, which transects the southwestern Black Sea, Sea of Marmara, and eastern part of the Central Aegean practically simultaneously, were used (Fig. 14.3a) (time interval between initial and final points of measurements along track 109 was within 2 min; repeat cycle was 9.916 days). The character of SLA interannual variability in the southwestern Black Sea and Sea of Marmara was found to be in general similar (Fig. 14.3b): a rise during 1992–1999 and 2003–2005, a drop during 1999–2003 and 2005–2007; in the eastern part of the Central Aegean, a slower level rise (in comparison with the Sea of Marmara and Black Sea) during 1992–1999 and a drop after 2002 were observed, with small change in sea level during 1999–2002. Ranges of interannual variability of mean annual sea level within the period of 1992–2007 were 12 cm in the Black Sea, 11 cm in the Sea of Marmara, and 7 cm in the Aegean Sea. Note that SLA values in the Sea of Marmara exceeded those in the Black Sea during practically all the periods under consideration. In this case, the sea level difference between the Black Sea and Sea of Marmara was less than the average accepted value (about 30 cm, see Goryachkin and Ivanov 2006). It may be suggested, then, that flow of the Black Sea surface water in the Sea of Marmara during 1992–2007 was less than normal.

14.3.2
Eddy Dynamics

SLA maps like those produced by the Colorado Center for Astrodynamics Research (CCAR), USA, on the basis of the merged altimeter data from the T/P and ERS-2 satellites (repeat period of the maps is 3 days) can be used for tracing eddy evolution, from their formation to dissipation. Our experience using the CCAR SLA maps has shown (Ginzburg et al. 2003; Zatsepin et al. 2003) that they are very effective for investigation of dynamics of large and relatively long-lived (from 1 to several months) eddies such as the deep-sea anticyclone in the eastern part of the Black Sea in Fig. 14.1b (the corresponding SLA map is given in Fig. 14.4b). However, spatial resolution of the maps determined by a distance between satellite tracks (Fig. 14.4a) is inadequate to reveal coastal anticyclonic eddies or differentiate closely spaced eddy features. For example, two anticyclonic eddies over the northwestern continental slope in Fig. 14.1b are seen in Fig. 14.4b as a unified feature, while coastal eddies along the Anatolian and northeastern coasts are not manifested in this SLA map at all. Along-track altimeter measurements with high spatial resolution could be more appropriate for revealing such eddies. Note, however, that difference of dynamic height between an anticyclonic eddy center and its periphery is in most cases small (several centimeters), which is comparable with an error of altimeter measurements. On the other hand, coastal anticyclones can result in poor correlation between satellite-derived SLA values and in situ measurements in the regions of Tuapse, Sochi, Kaciveli, etc.

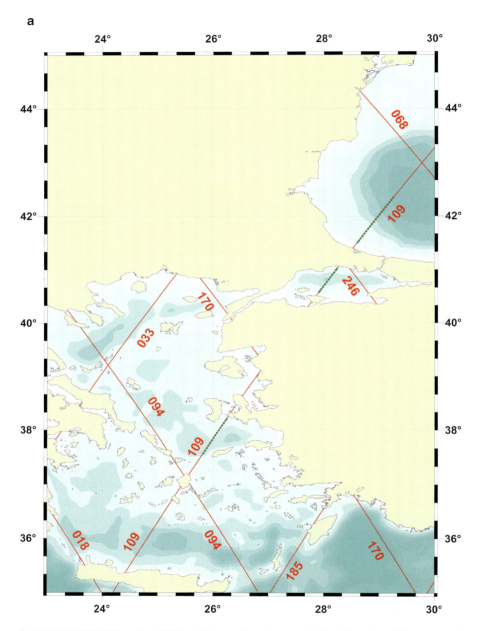

Fig. 14.3 (**a**) TOPEX/Poseidon (T/P) and J1 ground tracks over the Black Sea, Sea of Marmara, and Aegean Sea (points on the tracks mark segments to be processed for obtaining the sea level anomaly [SLA] variability); Lighter shade of the sea surface corresponds to the depths of less than 200 m (**b**) interannual variability of SLA in the Black Sea (*green line*), Sea of Marmara (*blue line*), and Aegean Sea (*red line*)

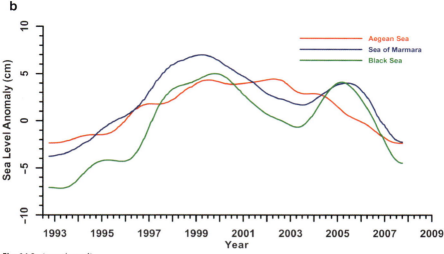

Fig. 14.3 (continued)

14.3.3
Correspondence Between Altimeter-Derived and In situ Data on Sea Level

Results of the comparison between the altimeter-derived Black Sea levels and corresponding in situ measurements were presented in a few papers. For example, a comparison between T/P data from the Archiving, Validation and Interpretation of Satellites Oceanographic data (AVISO) database and sea level gauge data provided by the Permanent Service for Mean Sea Level (PSMSL) (monthly mean data) was made for four coastal locations – three on the Bulgarian coast (Varna, 43.2°N, 27.9°E; Nesebar, 42.6°N, 27.8°E; and Burgas, 42.5°N, 27.5°E) and one on the Russian coast (Tuapse, 44.1°N, 39.1°E) for the period 1992–1996 (Stanev et al. 2000). The comparison has shown that relatively good correlation of two kinds of data (0.76) occurs only in the Tuapse case where the T/P track passes practically over meteorological station (Fig. 14.5a). For Varna, Nesebar, and Burgas, correlation coefficients were 0.65, 0.51, and 0.68, respectively.

The comparison between monthly mean SLA values from T/P altimeter and MS data for some coastal places (1992–1998) was made in (Goryachkin and Ivanov 2006). Altimeter-derived data were chosen in the region, which corresponded to 100–200 m isobath, and averaged over four along-track points (28 km) to filter small-scale disturbances of sea level. As a result of the comparison, the following correlation coefficients were obtained: 0.93 for Sevastopol, 0.92 for Yalta, and 0.77 for Tuapse.

In the framework of the ALTImetry for COastal Regions (ALTICORE) project, the consistency between altimeter-derived SLA values (from T/P and J1, the AVISO, and PODAAC databases) and measurements on coastal stations was also checked. The comparison of the two kinds of data made for several sea level gauges in the Black Sea (Fig. 14.5a) has shown

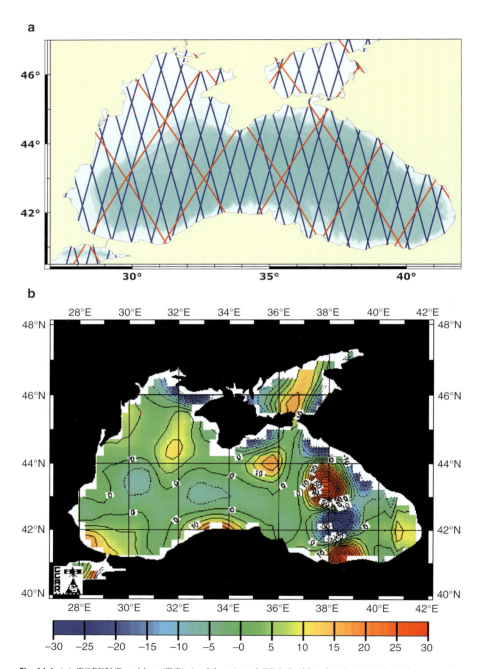

Fig. 14.4 (**a**) TOPEX/Poseidon (T/P) (*red lines*) and ERS-2 (*blue lines*) ground tracks over the Black Sea; (**b**) CCAR SLA (cm) map for 28 July 1998 Lighter shade of the sea surface corresponds to the depths of less than 200 m

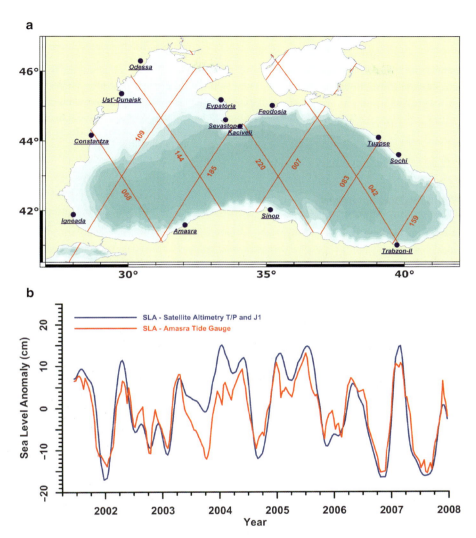

Fig. 14.5 (**a**) Satellite ground tracks of TOPEX/Poseidon (T/P) and J1 and locations of coastal places; (**b**) comparison of observed (*red line*) and altimeter-derived (*blue line*) sea level anomaly (SLA) variations from T/P and J1 at Amasra (current data) Lighter shade of the sea surface corresponds to the depths of less than 200 m

their poor correlation (less than 0.6) (Lebedev et al. 2008). An example of such comparison for almost 13 years (1993–2005) of coincident sea level gauge in Sevastopol and corresponding altimeter-derived sea level anomalies from T/P and J1 presented in (Vignudelli et al. 2008) has shown that the sea seasonal signal is well reproduced in the altimetry record; however, the correlation between the two records is poor (0.41). Another example of the comparison between altimeter-derived (segment of satellite track 185 closest to Amasra) SLA and MS data on sea level interannual variability for sea level gauge Amasra (Anatolian coast, Fig. 14.5a) is given in Fig. 14.5b. It may be seen that two curves in Fig. 14.5b are qualitatively similar; however, the correlation coefficient is not high: 0.67 and 0.57 when comparing current (hourly measurements) and monthly mean data, respectively.

Similar comparisons made for a number of sea level gauges gave correlation coefficients from 0.4 to 0.7, minimum values being obtained for sea level gauges at the western and eastern coasts, whereas maximum ones at the northern and southern coasts. It is unknown which was the reason for such poor correlation – sea level gauge data quality, altimetry data processing, distance between the sea level gauge and satellite track, and, at least in part, mesoscale water dynamics typical for the coastal regions of the Black Sea.

14.4
Wind Speed Measurements

In the framework of the ALTICORE project, the consistency between altimeter-derived wind speed and measurements on coastal stations was checked. Satellite data were retrieved from the RADS database. For every coastal place, satellites and their tracks were chosen so that the tracks were located close to the MS. Then, for every cycle of the selected satellite passes, spatial coordinates were determined, nearest to the concrete MS coordinates (within 15 miles), and corresponding values of satellite-originated wind speed together with corresponding time (UT) were selected. In situ wind speed data for the time nearest to the time moment determined in such a way were picked from the databases for every coastal MS. On the basis of the derived data on date, time, coordinates, and corresponding wind speed (satellite and in situ) files were formed for the following data processing with the aim to evaluate capacity of satellite altimetry to give reliable information on wind speed in various regional coastal areas.

A very significant volume of data necessitated computer processing of the formed data files. To this end, special programs (MatLab procedures) were developed. These programs retrieve the records from the RADS track data nearest to the MS points, choose the corresponding values of the observed wind speed (in situ database), form arrays for a comparison, calculate correlation coefficients, and draw the figures of data distributions and scatter plots. Note that the time interval between successive values of wind speed in the observed databases is rather wide – from 3 to 6 and even 12 h, which also involves difficulties when comparing satellite and observed data on wind speed. A linear interpolation was used to calculate the value of the observed wind speed at the time of satellite passing through the MS point.

For the comparison, six places on the Black Sea coast (MS locations) were selected (Tuapse, Sochi, Gelendzhik, Kaciveli, Constantza, and Istanbul). Satellites and their tracks chosen for these coastal places were as follows:

- For Tuapse, ERS/Envisat (tracks 872, 999), and T/P&J1 (track 083).
- For Sochi, ERS/Envisat (track 455).
- For Gelendzhik, ERS/Envisat (tracks 541, 958) and GFO (track 025).
- For Kaciveli, GFO (tracks 158, 255), T/P&J1 (tracks 185, 220), and ERS/Envisat (track 672).
- For Constantza, T/P&J1 (track 068) and ERS/Envisat (tracks 472, 599).
- For Istanbul, GFO (track 474), ERS/Envisat (track 141), and T/P&J1 (track 109).

The total number of satellite tracks processed for chosen places on the coast was about 30.

The ERS/Envisat and TOPEX/J1 satellite tracks, which were used when obtaining satellite-derived wind speed data for Sochi, Tuapse, Gelendzhik, and Kaciveli, are shown in Fig. 14.6; those of the GFO satellite used when processing data for Gelendzhik may be seen in Fig. 14.7. Examples of altimeter and observed data on wind speed and corresponding scatter plots for Sochi, Tuapse, and Gelendzhik are given in Fig. 14.8–14.10, respectively.

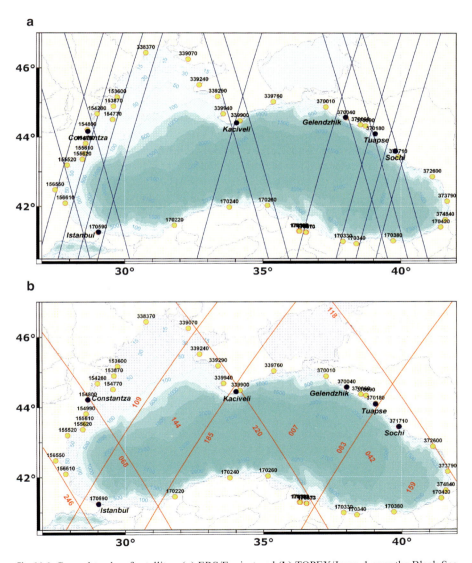

Fig. 14.6 Ground tracks of satellites: (**a**) ERS/Envisat and (**b**) TOPEX/Jason-1 over the Black Sea that we used for the analysis. Lighter shade of the sea surface corresponds to the depths of less than 200 m

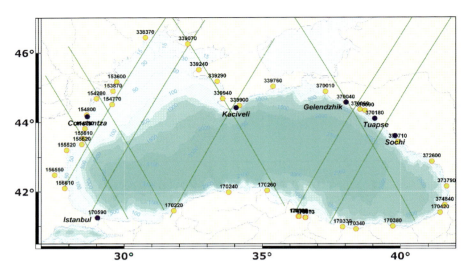

Fig. 14.7 Ground tracks of the Geosat Follow-On (GFO) satellite over the Black Sea that we used for the analysis. Lighter shade of the sea surface corresponds to the depths of less than 200 m

It may be seen from Fig. 14.8a that altimeter-derived wind speed values from Envisat for Sochi in most cases are considerably less than observed values. Correlation between altimeter and observed data (Fig. 14.8b) is rather poor. In the Tuapse case (also Envisat), however, correspondence between altimeter and observed data is satisfactory (Fig. 14.9). Note that Tuapse is the only place in the Black Sea where altimeter gives reliable information on wind speed (on the assumption of proper work of coastal MSs). For the Gelendzhik case, correlation is satisfactory for the GFO satellite (Fig. 14.10b, the correlation coefficient is 0.5); however, it is rather poor for the Envisat altimeter (pass 541) – satellite-derived wind speed values are considerably higher than in situ data. Poor correlations are also found for Kaciveli (altimeter-derived values are higher in comparison with observed ones), Constantza, and Istanbul (altimeter-derived values are less than observed ones).

Some increase of correlation between altimeter-derived and observed winds is obtained not at the point of satellite track nearest to the coast, but when several along-track seaward points (second, third, etc.) are chosen. However, further increase of a distance of along-track point from the coast results in a decrease of the correlation. For example, in the Gelendzhik case, the correlation coefficients at ten progressive along-track seaward points were as follows: 0.33, 0.47, 0.60, 0.64, 0.66, 0.63, 0.63, 0.60, 0.58, and 0.55.

For decreasing discrepancy between altimeter-derived and observed wind speed values, a new approach was proposed on the basis of the decomposition of all wind directions observed for a certain period in four quadrants in relation to the orientation of the coastline (Similar decomposition was also used for the Caspian Sea as well as for the Barents and White seas; see the corresponding chapters by Kouraev et al. and Lebedev et al., this volume). An example of such decomposition for the Gelendzhik case is shown in Fig. 14.11. The reference line of the composition is the line which is parallel to the coast in the region of Gelendzhik. This line and the line perpendicular to it become bisectors of four quadrants

14 Satellite Altimetry Applications in the Black Sea 381

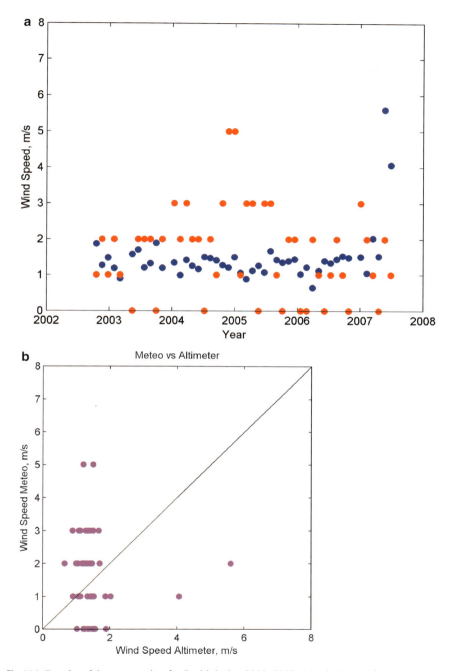

Fig. 14.8 Results of data processing for Sochi during 2002–2007: (**a**) wind speed from the Envisat altimeter (track 455, *blue circles*) and observations (*red*); (**b**) scatter plot of observed versus altimeter data

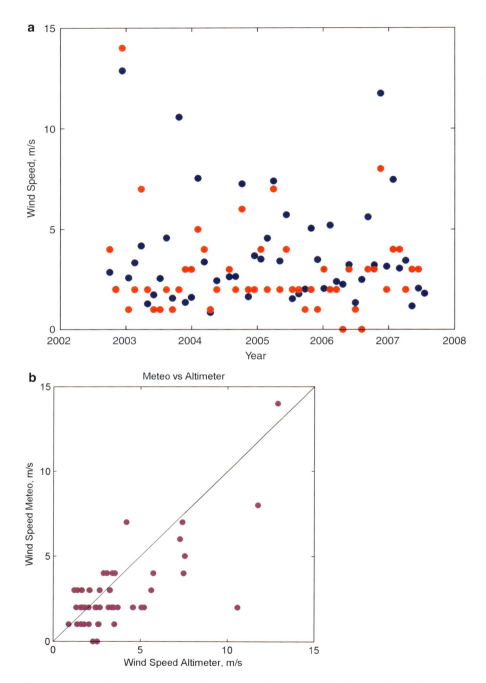

Fig. 14.9 Results of data processing for Tuapse during 2002–2007: (**a**) wind speed from the Envisat altimeter (track 999, *blue circles*) and observed (*red*) data; (**b**) scatter plot of observed versus altimeter data

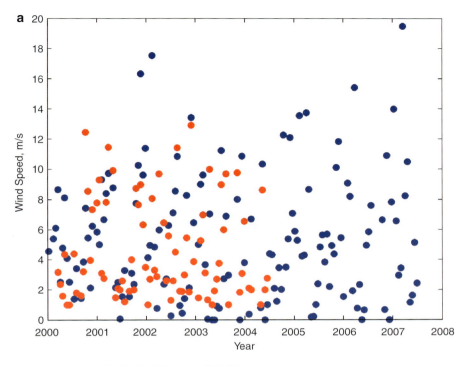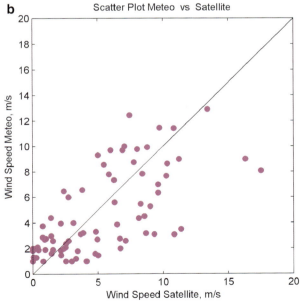

Fig. 14.10 Results of data processing for Gelendzhik during 2000–2004: (**a**) wind speed from the Geosat Follow-On (GFO) altimeter (track 025, *blue circles*) and observed (*red*) data; (**b**) scatter plot of observed versus altimeter data

Fig. 14.11 Scheme of decomposition of the winds in four quadrants (Q1–Q4) in relation to the coastline orientation for the Gelendzhik case. The appropriate correlation coefficients between altimetry-derived and in situ wind speed values are shown in the scheme. Arrows show prevailing wind direction for each quadrant

Q1–Q4, on which wind directions are divided (in this case, northerly to easterly winds fall into the Q1 quadrant, easterly to southerly into the Q2 quadrant, etc.).

It appeared that the best correlation ($r = 0.92$) was observed when wind blew from the sea to the coast (Q4 quadrant in this case) and the poorest one ($r = 0.44$) was for the winds in the opposite direction, i.e., from the coast. Correlations for two cases when wind blows along the coastline from left to right and vice versa were found to be 0.71 and 0.74, respectively. Scatter plots of observed and altimeter-derived winds for the directions corresponding to the Q1, Q2, Q3, and Q4 quadrants in Fig. 14.11 are presented in Fig. 14.12. Note that the correlation coefficient for all the observed data on wind directions without the decomposition was 0.5 (Fig. 14.10b).

14.5 Conclusions

One of the aims of this paper was to make an effort to apply available satellite altimetry data to the coastal zone of the Black Sea. Basically, the RADS database contains altimetry-derived information on sea level, wind speed, and wave height. Unfortunately, information on in situ measurements of wave heights for the Black Sea that could be compared with concurrent altimetry measurements is absent. (For the Caspian Sea, where such information is available, corresponding comparison was made; see the chapter by Kouraev et al., this volume.) So, only satellite data on sea level and wind speed are considered in this paper.

Altimeter-derived data on sea level are very effective for investigation of long-term variability of sea level in deep sea and tracing evolution of large open sea eddies in the Black Sea (Korotaev et al. 2001; Sokolova et al. 2001; Ginzburg et al. 2003; Zatsepin et al. 2003; Goryachkin and Ivanov 2006; Kostianoy et al. 2008). For the coastal regions, consistency between altimeter-derived sea level anomalies and wind speed and data of

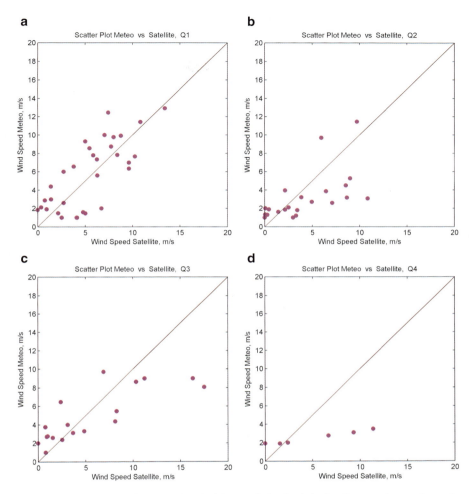

Fig. 14.12 Scatter plots of observed and altimeter-derived winds for four quadrants shown in Fig. 14.11: (**a**) Q1, (**b**) Q2, (**c**) Q3, and (**d**) Q4

meteorological stations appeared to be improper, at least when using satellite measurements averaging 1 cycle/s. In particular, the comparison carried out between satellite-derived and observed data on wind speed for several coastal places has shown their satisfactory correlation only for Tuapse. Possible reasons of such discrepancy, provided trustworthy measurements are made with altimeter, may be position of a satellite pass relative to a coastal meteorological station and influence of land on the results of satellite measurements, as well as lack of coincidence in time of these measurements.

To improve the correlation, some additional techniques must be used, e.g., calibration of measurement devices, a correct choice of the distance between coordinates of satellite track point of measurement and meteorological station location so as to exclude a possible influence of land, and the decomposition of the wind directions in four quadrants in relation to the orientation of the coastline. It is also possible that some increase of the correlation can be obtained by taking into account the angle, under which satellite track approaches the coast, as well as a direction of the flight – from the coast to the sea and vice versa.

Concerning the sea level analysis, we believe that further investigations should include an improvement of atmospheric corrections based on in situ meteo data, because real weather conditions in the Black Sea region may significantly differ from the model predictions. We expect that an increase in the frequency of averaging altimeter measurements will result in improvement of altimeter data for the coastal regions. Besides, better altimeter data are expected from future altimetric programs.

Acknowledgments The work was supported by International Project INTAS No 05-1000008-7927 (ALTICORE), by Russian Foundation for Basic Research (Grants NN 07-05-00141, 09-05-91221-CT_a, 10-05-00097-a), and by Russian Academy of Sciences Project "Small- and mesoscale dynamic processes in the ocean basing on satellite data and in situ measurements."

References

Ginzburg AI, Kostianoy AG, Soloviev DM, Stanichny SV (2000) Remotely sensed coastal/deep-basin water exchange processes in the Black Sea surface layer. In: Halpern D (ed) Satellites. oceanography and society. Elsevier, Amsterdam, the Netherlands

Ginzburg AI, Kostianoy AG, Sheremet NA (2003) Mesoscale variability of the Black Sea as revealed from TOPEX/Poseidon and ERS-2 altimeter data. Issledovanie Zemli iz kosmosa 3:34–46 (in Russian)

Ginzburg AI, Zatsepin AG, Kostianoy AG, Sheremet NA (2008) Mesoscale water dynamics. In: Kostianoy A, Kosarev A (eds) The Black Sea environment. The handbook of environmental chemistry, V5Q. Springer, Berlin/Heidelberg/New York

Goryachkin YuN, Ivanov VA (2006) The Black Sea level: the past, present and future. MHI NANU, Sevastopol (in Russian)

Korotaev GK, Saenko OA, Koblinsky CJ (2001) Satellite altimetry observations of the Black Sea level. J Geophys Res 106:917–933

Kosarev AN, Arkhipkin VS, Surkova GV (2008) Hydrometeorological conditions. In: Kostianoy A, Kosarev A (eds) The Black Sea environment. The handbook of environmental chemistry, V5Q. Springer, Berlin/Heidelberg/New York

Kostianoy AG, Terziev FS, Ginzburg AI et al (2008) Southern seas. In: Izrael YuA (ed) Estimating report about climate changes and their consequences on the Russian Federation territory, V. II. Consequences of climate changes. GU "NITS Planeta", Moscow (in Russian)

Lebedev SA, Kostianoy AG (2005) Satellite altimetry of the Caspian Sea. ""Sea", Moscow" (in Russian)

Lebedev SA, Sirota AM, Medvedev DP et al (2008) Verification of data of satellite altimetry in the coastal zones of the European seas. In: Modern problems of remote sensing of the Earth from space, OOO "Azbuka 2000", Moscow, Issue 5, vol II, pp 137–140 (in Russian)

Medvedev PP, Tyupkin YuS (2005) Satellite altimetry data system for geodynamics and oceanography studies: telematics aspects and applications. IST4Balt News J 1:39–41

Medvedev PP, Lebedev SA, Tyupkin YS (1997) An integrated data base of altimetric satellite for fundamental geosciences research. In: Proceedings of the first East-European symposium on advances in data bases and information systems (ADBIS'97), St. Petersburg, Russia, 2–5 September, St. Petersburg University, St. Petersburg, vol 2, pp 95–96

Mikhailov VN, Mikhailova MV (2008) River mouths. In: Kostianoy A, Kosarev A (eds) The Black Sea environment. The handbook of environmental chemistry, V5Q. Springer, Berlin/Heidelberg/New York

Oguz T, Latun VS, Latif MA et al (1993) Circulation in the surface and intermediate layers of the Black Sea. Deep-Sea Res 40(8):1597–1612

Ovchinnikov IM, Titov VB (1990) Anticyclonic vorticity of the currents in the coastal zone of the Black Sea. Doklady AN SSSR 314:1236–1239 (in Russian)

Schrama E, Scharroo R, Naeije M (2000) Radar altimeter database system (RADS): towards a generic multi-satellite altimeter database system. Delft Institute for Earth-Oriented Space Research. Delft University of Technology, The Netherlands

Sokolova E, Stanev EV, Yakubenko V et al (2001) Synoptic variability in the Black Sea. J Mar Syst 31:45–63

Stanev EV, Peneva EL (2002) Regional sea level response to global climatic change: Black Sea examples. Global Planet Change 32:33–47

Stanev EV, Le Traon PY, Peneva EL (2000) Sea level variations and their dependency on meteorological and hydrological forcing: analysis of altimeter and surface data for the Black Sea. J Geophys Res 105:17203–17216

Terziev FS (Ed) (1991) Hydrometeorology and hydrochemistry of the USSR seas. Project: Seas of the USSR. Vol IV. The Black Sea, 1, Hydrometeorological conditions. Gidrometeoizdat, St. Petersburg (in Russian)

Tuzhilkin VS (2008) Thermohaline structure of the sea. In: Kostianoy A, Kosarev A (eds) The Black Sea environment. The handbook of environmental chemistry, V5Q. Springer, Berlin/Heidelberg/New York

Vignudelli S, Kostianoy A, Ginzburg A et al (2008) Reprocessing altimeter data records along European coasts: lessons learned from the ALTICORE project. In: International geosciences and remote sensing symposium (IGARSS), Boston, MA, USA, 6–11 July, V. 3

Vigo I, Garcia D, Chao BF (2005) Changes of sea level trend in the Mediterranean and Black seas. J Mar Res 63(6):1085–1100

Zatsepin AG, Ginzburg AI, Kostianoy AG et al (2002) Variability of water dynamics in the north-eastern Black Sea and its effect on water exchange between near-shore and off-shore parts of the basin. Oceanology 42(Suppl 1):S1–S15

Zatsepin AG, Ginzburg AI, Kostianoy AG et al (2003) Observations of Black Sea mesoscale eddies and associated horizontal mixing. J Geophys Res 108(C8):2-1–2-27

15 Satellite Altimetry Applications in the Barents and White Seas

S.A. Lebedev, A.G. Kostianoy, A.I. Ginzburg, D.P. Medvedev, N.A. Sheremet, and S.N. Shauro

Contents

15.1	Introduction	390
15.2	Calibration/Validation of Satellite Altimetry Measurements	395
	15.2.1 Wind Speed	396
	15.2.2 Sea Surface Height	400
	15.2.3 Tidal Models	404
15.3	Long-Term Variability of Sea Level	406
15.4	Interannual Variability of Sea Ice Edge Position	411
15.5	Conclusions	413
References		413

Keywords Barents Sea • Ice cover • Regional tidal modelling • Sea level anomaly • White Sea • Wind speed

Abbreviations

ALTICORE	ALTImetry for COastal REgions
AVISO	Archiving, Validation and Interpretation of Satellites Oceanographic data
DBMS	Database Management System
GC RAS	Geophysical Center of Russian Academy of Sciences
GDR	Geophysical Data Record
GFO	Geosat Follow-On

S.A. Lebedev (✉), D.P. Medvedev, and S.N. Shauro
Geophysical Center, Russian Academy of Sciences, 3 Molodezhnaya Str.,
119296, Moscow, Russia
e-mail: lebedev@wdcb.ru

A.G. Kostianoy, A.I. Ginzburg, and N.A. Sheremet
P.P. Shirshov Institute of Oceanology, Russian Academy of Sciences,
36 Nakhimovsky Pr., 117997, Moscow, Russia

S. Vignudelli et al. (eds.), *Coastal Altimetry*,
DOI: 10.1007/978-3-642-12796-0_15, © Springer-Verlag Berlin Heidelberg 2011

GSFC	Goddard Space Flight Center NASA
HMRC	Hydrometeorological Research Center of Russian Federation
IGDR	Interim Geophysical Data Record
ISADB	Integrated Satellite Altimetry Data Base
J1	Jason-1
JPL	Jet Propulsion Laboratory
LSA	Laboratory for Satellite Altimetry
MGDR-B	Merged Geophysical Data Record Generation B
MS	Meteorological Station
NASA	National Aeronautics and Space Administration
NCAR	National Center for Atmospheric Research
NCEP	National Center for Environmental Prediction
NESDIS	NOAA's National Environmental Satellite, Data and Information Service
NOAA	National Oceanic and Atmospheric Administration
PODAAC	Physical Oceanography Distributed Active Archive Center
RADS	Radar Altimeter Database System
SLA	Sea Level Anomaly
SSH	Sea Surface Height
SSHA	Sea Surface Height Anomaly
T/P	TOPEX/Poseidon
TG	Tide Gauge
WMO	World Meteorological Organization

15.1
Introduction

The Barents Sea is one of the shelf Arctic seas. It is located on the continental European shelf between the northern coast of Europe and three archipelagoes – Novaya Zemlya, Franz Josef Land, and Spitsbergen (Fig. 15.1a). Its open water area is approximately 1,424,000 km^2 and total volume is 322,000 km^3. The Barents Sea shelf is rather deep. More than 50% of the area has a depth of 200–500 m. The average depth is 222 m; the maximum depth in the Norwegian trench reaches 513 m and in the Franz Josef Land straits it exceeds 600 m (Atlas of Arctic 1985). The Barents Sea watershed area is 668,000 km^2. The total river runoff is 163 km^3/year; 80–90% of it falls on the southeastern region of the sea. The largest river flowing into the Barents Sea is the Pechora River which has about a half of the watershed area, i.e., 322,000 km^2; its runoff is 130 km^3/year. The river runoff is essentially reflected in hydrological conditions only in a southeastern part of the sea. Therefore, this area is sometimes referred to as the Pechora Sea (Barents Sea 1985; Terziev et al. 1990; Kostianoy et al. 2004).

The White Sea is a semi-enclosed inland sea (Fig. 15.1b). The sea border with the Barents Sea is a line joining Cape Svyatoy Nos (northeastern coast of Kola Peninsula) with Cape Kanin Nos (northwestern extremity of Kanin Peninsula). The northern part of the sea is called the Voronka (funnel). The southern and central parts of the White Sea called the Basin are the largest and deepest parts of the sea. There are also several large and shallow bays in the area, namely the Dvinsky, Onega, Mezen, and Kandalaksha bays. The Gorlo (neck) is a narrow strait connecting the Basin and Voronka. The total water surface area is 90,873 km² including islands, and the total volume is 6,000 km³ including also the Voronka area opening to the Barents Sea. Thus, the White Sea covers approximately 6% of the total open water area of both seas and comprises only 2% of the total volume of marine water, but it assumes more than half of the river runoff in the region. The White Sea watershed area is 729,000 km² (Atlas of Arctic 1985). The total river runoff is 259 km³/year, which is about 4% of the total amount of the White Sea water volume. The main rivers are the Severnaya Dvina, Onega, and Mezen having runoff of 111, 18, and 26 km³/year correspondingly (White Sea 1989; Glukhovsky 1991).

Meteorological conditions in the Barents Sea are determined by atmospheric cyclones that are formed in the North Atlantic and move to the Barents Sea. In winter, southwesterly and southerly winds prevail in the southern part of the sea, whereas southeasterly and easterly winds are most frequent in the northern regions (Terziev et al. 1990). Monthly mean wind speed in the southern regions (in particular, nearby Kanin Nos) reaches about 10 m/s and in the north it decreases to 6 m/s. In summer, the pressure gradients are weakened. Homogeneous

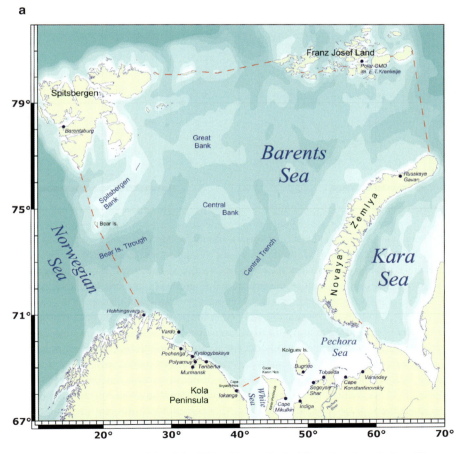

Fig. 15.1 Maps of the Barents (**a**) and the White (**b**) seas. Dashed lines show boundaries of the seas and their internal parts. Circles mark tide gauges location

b

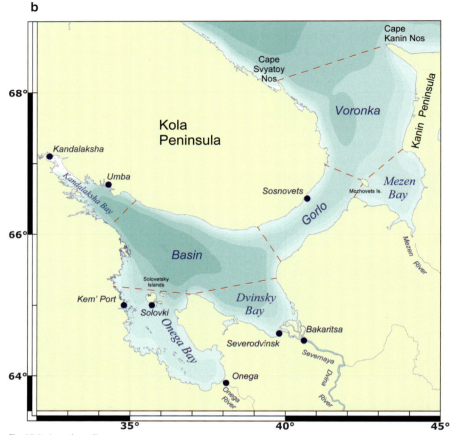

Fig. 15.1 (continued)

wind conditions, with predominance of winds of northern direction, dominate over almost the whole sea. Monthly mean wind speed in this season is about 6 m/s and on the north – 5 m/s.

In the White Sea, winds from the south, southwest, and west prevail from October to March, whereas in May–August winds from north, northeast, and east are most frequent (White Sea 1989). Southeasterly winds are frequently observed at the top of the bays (in ports Mezen, Kandalaksha, Onega). Monthly mean wind speed in the open sea and on islands is 7–10 m/s from September to April and 5–7 m/s from May to August. In the bays running deeply on land side, mean wind speed does not exceed 3–5 m/s during the whole year.

The general circulation and sea level variations in the Barents and White seas are formed under cumulative effect of wind forcing, water interaction, and exchanges among surrounding seas, strong tides, peculiarities of bottom topography, seasonal variability of river runoff, precipitation and ice cover, and other factors. Thus, sea level variations in the Barents and White seas have a complex nature and are characterized by a significant spatial and temporal variability.

The seasonal variations of the sea level in the Barents and White seas are caused by an impact of atmospheric pressure and wind, temperature and salinity, river runoff, precipitation, and ice cover (Glukhovsky 1991). For the Barents Sea, the maximum seasonal sea level variation is observed in November–December and its minimum is registered in May–June, that is in accordance with the atmospheric pressure effect upon the sea surface. For example, the difference between maximum and minimum mean sea level in Murmansk can amount to 40–50 cm (Terziev et al. 1990; Popov et al. 2000). A range of seasonal variations in the White Sea level are observed in the estuary zones of the Onega and Severnaya Dvina rivers and may reach 30–60 cm (Glukhovsky 1991; Filatov et al. 2005).

Sea level variations related to storm surge are formed by strong winds associated with passage of intense atmospheric cyclones. The western and northwestern cyclones dominate and amount to 88% of their total number. During extreme situations, the storm surge height can be comparable with the tidal height. The storm surge values are greatest in the southeastern part of the Barents Sea (up to 2 m). For the Kola coast and Spitsbergen area, storm surge heights are about 1 m; the smaller values of 0.5 m are observed near the Novaya Zemlya coast (Terziev et al. 1990). As concerns the White Sea, on the average the storm surge is of 30–70 cm, but in the Onega Bay it can reach about 80 cm (Glukhovsky 1991; Filatov et al. 2005).

Tides play the key role in the coastal water dynamics in the Barents and White seas, and dominate in total sea level variations (Terziev et al. 1990; Glukhovsky 1991). In the southern part of the Barents Sea, the semidiurnal tide reaches amplitudes up to 2 m, which is the largest in the Arctic seas. In the northern and northeastern Barents Sea, the tidal height decreases and for the Spitsbergen coast equals to 1–2 m. For the southern coast of Franz Joseph Land, it is only 40–50 cm. It is explained by the peculiarities of the bottom relief, the coastline configuration, and the interference of tidal waves, which come from the North Atlantic and Arctic oceans. The tidal heights in the White Sea vary from 8 m in the Mezen Bay to 1 m in the Dvinsky Bay (Terziev et al. 1990; Glukhovsky 1991; Semenov and Luneva 1996; Popov et al. 2000).

As a result of nonlinear tidal phenomena and nonlinear interaction of the main tidal components (M_2, S_2, N_2, K_2, K_1, O_1, P_1, Q_1), there is a set of additional tidal harmonics (Kagan and Romanenkov 2007). According to the results of numerical simulation (May 2005), amplitude of nonlinear harmonics on the average is about 5% of the total tide height for the Barents Sea and about 50% for the White Sea with the exception of the Voronka (10–25%). The maximum value (more than 25%) is observed between the Spitsbergen Bank and Central Bank in the Barents Sea, and in the Pechora Sea. In the White Sea, the maximum value is about 100% for the Severnaya Dvina and Kandalaksha bays (Fig. 15.2).

Ice cover is also a cause of seasonal variability of the White Sea level and the southeastern Barents Sea level (or the Pechora Sea level) when most of the marine areas are frozen in winter (Terziev et al. 1990; Glukhovsky 1991).

Because of the incoming warm Atlantic waters, the Barents Sea is never (even in the most severe winters) covered completely with ice. This is its basic difference from other Arctic seas. Significant interannual and seasonal variability of its ice cover extent is observed. The greatest ice cover extent is usually seen in the middle of April, and the least one is at the end of August and the first half of September. In the most severe winters, more than 90% of the sea surface is covered with ice, and in especially warm winters the greatest ice cover extent even in April does not exceed 55–60% (Barents Sea 1985; Terziev et al. 1990).

In August–September of anomalously warm years, the sea is completely free from ice, and in anomalously cold years the ice cover in these months is stored on 40–50% of its area, lying mainly in the northern regions. The location of the ice edge during summer can vary by hundreds of kilometers from year to year, and there is also variability on longer time scales, up to the century scale, correlated with North Atlantic Oscillation (Vinje 2001). These variations reflect the interannual dynamics of inflowing Atlantic water and atmospheric forcing.

Recent studies indicate that the sea ice cover is undergoing significant climate-induced changes, affecting both its extent and thickness. The extent of the sea ice cover is effectively monitored from satellite platforms using passive microwave imagery. Satellite-derived estimate of maximum ice cover (seasonal and multiyear ice) trend was about −2.5% per decade (or −3.4 ± 0.3*10^3 km^2/year) during 1979–1996 (Parkinson et al. 1999). During 1979–2007, it was about −11% per decade for the multiyear ice extent and area (Comiso et al. 2008). Decreases occur in all seasons and on a yearly average basis, although they are largest in spring and smallest in autumn.

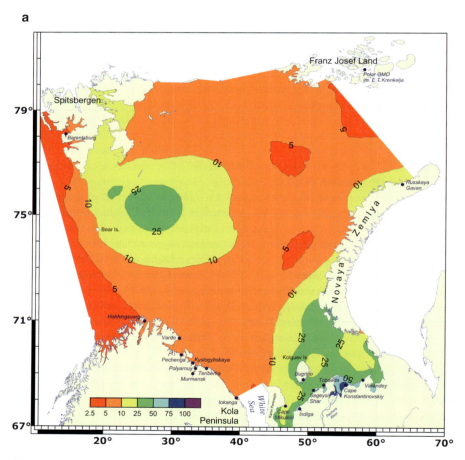

Fig. 15.2 The ratio (%) of total nonlinear harmonic amplitudes to total main tidal component amplitudes for the Barents (**a**) and White (**b**) seas (May 2005)

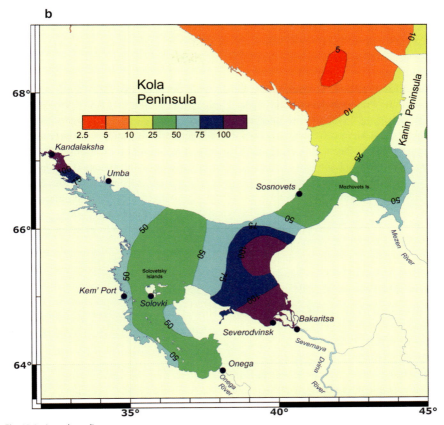

Fig. 15.2 (continued)

Historically, sea level variability in the Barents and White seas (tidal, storm surge, seasonal and interannual variations) has been investigated on the basis of the tide gauge data and so, such studies were limited within a coastal zone. Analysis of sea level variations offshore has become possible with a development of mathematical and numerical hydrodynamic simulation. Traditional research of Arctic ice cover was on the basis of infrared and microwave radiometry data. Since 1992, a new reliable source of information about variability of the sea level and ice cover extent, as well as of wind speed and wave height – satellite altimetry data has become available.

15.2
Calibration/Validation of Satellite Altimetry Measurements

In the framework of the ALTICORE Project (ALTImetry for COastal REgions, http://www.alticore.eu), we have analyzed altimetric measurements (1 Hz data) from the satellites TOPEX/Poseidon (T/P), ERS-1/2, Envisat Geosat Follow-On (GFO), and Jason (J1) from

September 1992 to December 2008 in order to compare satellite-derived data on sea level and wind speed in the coastal zone with in situ measurements at coastal meteorological station (MSs) of the Barents and White seas (Lebedev et al. 2008). Satellite altimetry data were received from the following data bases:

- T/P, ERS-1, ERS-2, Envisat, GFO, and J1 were obtained from the Ocean Altimeter Pathfinder Project at the Goddard Space Flight Center (GSFC) NASA (Koblinsky et al. 1999a, b).
- Besides, T/P Merged Geophysical Data Records generation B (MGDR-B) were obtained from the NASA Physical Oceanography Distributed Active Archive Center (PODAAC) at the Jet Propulsion Laboratory (JPL) of California Institute of Technology (Benada 1997).
- The J1 Interim Geophysical Data Records (IGDR) and Geophysical Data Records (GDR), generation A, B, C, were obtained from Archiving, Validation, and Interpretation of Satellite Oceanographic data (AVISO) and PODAAC (Picot et al. 2006).
- T/P, ERS-1, ERS-2, Envisat, GFO, and J1 data were obtained also from Radar Altimeter Database System (RADS) (Schrama et al. 2000).
- GFO GDR data were received from Laboratory for Satellite Altimetry (LSA) at NOAA's National Environmental Satellite, Data and Information Service (NESDIS) (GEOSAT Follow-On GDR User's Handbook 2002).

Information and software of the Integrated Satellite Altimetry Data Base (ISADB) developed in the Geophysical Center of Russian Academy of Sciences (GC RAS) (Medvedev et al. 1997, 2002; Lebedev and Kostianoy 2005; Medvedev and Tyupkin 2005) have been used for data processing and analysis.

ISADB has three levels of information: input altimeter data, the supplementary geophysical and geodetic information, and the results of the problem-oriented preliminary processing. The database management system (DBMS) has the problem-oriented modes of a complex analysis of data and provides the space-time graphic presentation of results of processing in addition to the routine DBMS functions. It also allows recalculating the corresponding corrections, when new or updated geophysical models are linked to the ISADB.

In situ tide gauge (TG) data were obtained from Hydrometeorological Research Center (HMRC) of Russian Federation.

In situ meteorological data on 6-hourly (0, 6, 12, 18 GMT) wind speed at coastal MSs (World Meteorological Organization [WMO] weather stations) were obtained from Russian Weather Server (http://meteo.infospace.ru/main.htm).

15.2.1
Wind Speed

In the framework of the ALTICORE project, the consistency between altimeter-derived wind speed and measurements on coastal meteorological stations was checked. Satellite-derived wind data used were obtained from the RADS data base. Methods of choosing satellites and their tracks, spatial coordinates, and time of satellite measurements of wind speed for every coastal place under consideration as well as peculiarities of computer processing of the formed data files are described in the chapter by Ginzburg et al., this volume.

For the comparison, 30 MSs on the Barents Sea coast and 21 MSs on the White Sea coast were selected. As an example, we show results of the comparison for two MSs

– Kem' Port (White Sea) and Kanin Nos (Barents Sea). Satellites and their tracks chosen for these coastal places were as follows:

- For Kem' Port, ERS & Envisat (tracks 025, 816), T/P & J1 (track 061), and GFO (tracks 074, 425).
- For Kanin Nos, ERS & Envisat (tracks 558, 797) and GFO (track 165).

The ERS & Envisat, GFO, and T/P & J1 satellite tracks, which were used when obtaining satellite-derived wind speed data for Kem' Port and Kanin Nos are shown in Fig. 15.3. Examples of altimeter and observed (in situ) data on wind speed and corresponding scatter plots for these MSs are given in Fig. 15.4 and 15.5, respectively.

In the cases of both Kem' Port and Kanin Nos (Fig. 15.3), MS databases for 2000–2007 were used (Fig. 15.4a and 15.5a, respectively). Altimeter wind speed values for Kem' Port were in most cases noticeably less than meteo ones (Fig. 15.4a and b) and correlation

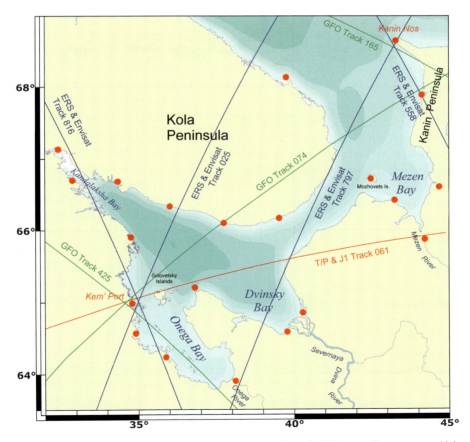

Fig. 15.3 ERS & Envisat GFO, and T/P & J1 satellites tracks in the White and Barents seas which were used for the comparative analysis between altimetric and MS in situ data. Red circles mark locations of MSs

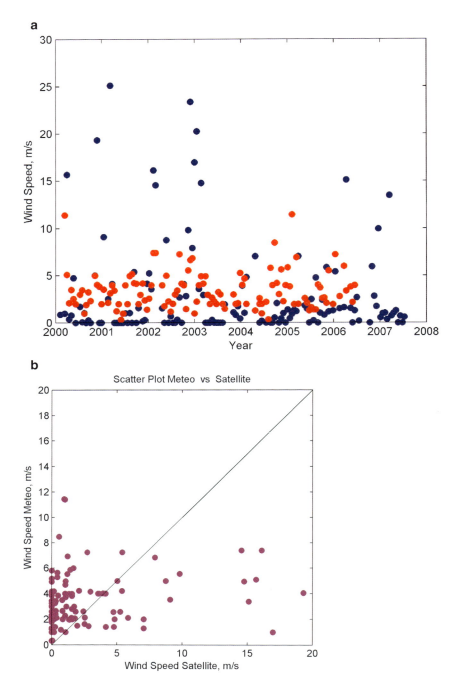

Fig. 15.4 Temporal variability of in situ MS (red circles) and altimetry (blue circles) wind speed (**a**) and scatter plot of observed versus GFO altimeter data (track 425) (**b**) for Kem' Port

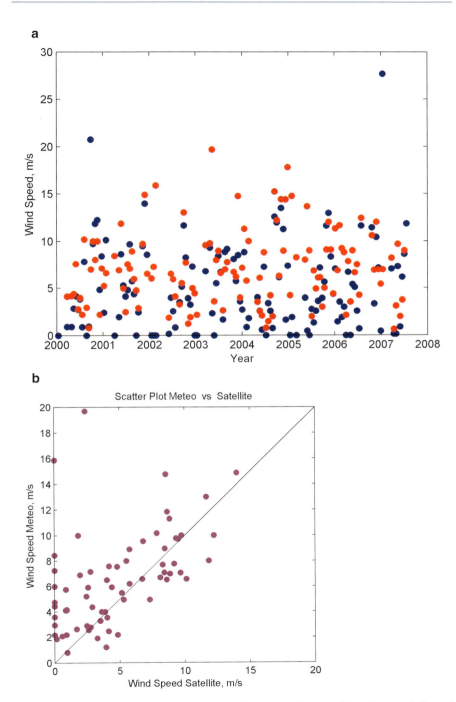

Fig. 15.5 Temporal variability of in situ MS (red circles) and altimetry (blue circles) wind speed (**a**) and scatter plot of observed versus GFO altimeter data (track 165) (**b**) for Kanin Nos

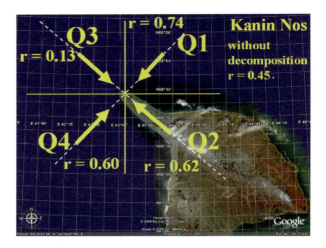

Fig. 15.6 Scheme of decomposition of wind directions in the four quadrants (Q1–Q4) in relation to the orientation of the coastline for the Kanin Nos case. The appropriate correlation coefficients between wind speed from altimetry and in situ MS records are shown in the scheme. Arrows show prevailing wind direction for each quadrant

between them was practically absent ($r = 0.16$). Influence of land may be supposed as a possible reason of such poor correlation. Better correlation ($r = 0.45$) was observed in the Kanin Nos case (Fig. 15.5b).

Improvement in correlation between meteo and altimeter data on wind speed can be obtained when using decomposition of the winds in four quadrants according to wind direction relative to coastline direction. The principle of this decomposition is described in the chapter by Ginzburg et al., this volume. An example of its application for the Kanin Nos case is shown in Fig. 15.6 and 15.7. The reference line was chosen to be parallel to the axis of this Peninsula. This line and perpendicular to it became bisectors of the four quadrants, on which wind directions were divided.

Such a "wind decomposition procedure", which was successfully tested in the Black and Caspian seas also (see corresponding chapters by Ginzburg et al. and Kouraev et al., this volume), seems to be very promising in the calibration and validation of altimeter-derived data in coastal regions of the World Ocean.

15.2.2
Sea Surface Height

Also in the framework of the ALTICORE project, the consistency between altimeter sea surface height (SSH) data acquired by the satellites T/P, ERS-1/2, Envisat GFO, and J1, and measurements of 18 TGs located in the Barents Sea and 8 TGs in the White Sea was investigated (Fig. 15.1). Processing of satellite altimetry data included calculation of sea surface height anomaly (SSHA) by using all corrections without tidal and inverse barometer correction. Then, for every cycle of the selected satellite passes, spatial coordinates of the point nearest to the concrete TG location (within 15 miles) were determined.

For the Barents Sea, correlation between SSHA derived from altimetry and tide gauge data can be fairly high (>0.85) when using ERS-1/2 and Envisat data. For GFO data the correlations are noticeably lower, probably because of orbit errors and inaccurate

Fig. 15.7 Scatter plots of observed and altimeter-derived winds for the directions Q1 ($r = 0.74$) (**a**), Q2 ($r = 0.62$) (**b**), Q3 ($r = 0.13$) (**c**), and Q4 ($r = 0.60$) (**d**)

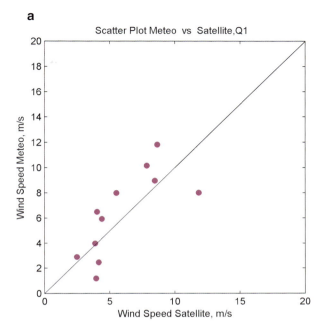

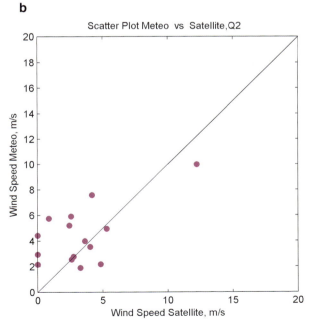

Fig. 15.7 (continued)

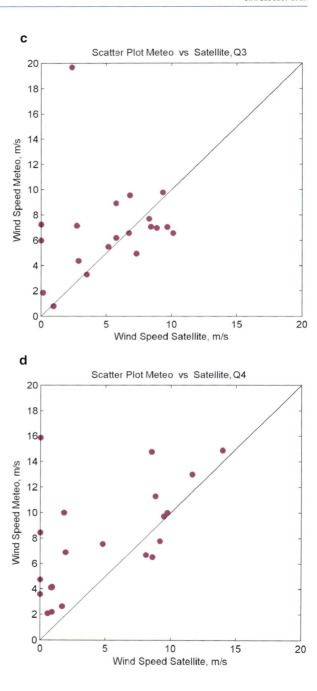

corrections. Correlations are extremely high (0.99) when ERS-1/2 and Envisat data are used for the comparison with Vardo and Hohhingsvarg TGs for which there are long and high quality time series (Lebedev et al. 2008). These high correlations might be explained by the tide effect, which is very significant in the Barents Sea (see above). Moreover, these TGs are located in the area where nonlinear and residual tidal phenomena are not important. For TGs in Murmansk, Polyarny, Indiga and others which are located in fjords or estuaries, the correlations are small (<0.4). Absence of 1 Hz altimetric measurements in these areas or strong effect of nonlinear and residual tidal phenomena may be a cause of it. Details of SSHA comparison are shown in Fig. 15.8 and Table 15.1. This result is in good agreement with our previous investigations (Lebedev and Tikhonova 2000; Lebedev et al. 2003).

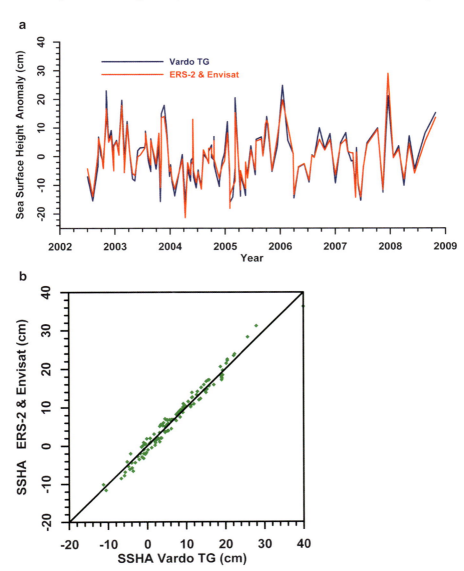

Fig. 15.8 Time variability of Vargo TG and ERS-2 & Envisat SSHA (**a**) and scatter plot of TG data versus altimetry data (**b**)

Table 15.1 Correlation between satellite and TG SSHA data for the Barents Sea

Tide Gauge	GFO		ERS-1/2 & Envisat	
	Track number	Correlation	Track number	Correlation
Hohhingsvarg	048, 079, 134, 165, 192, 251, 278, 309, 364, 395, 450, 481	0.69	025, 111, 302, 388, 569, 655, 760, 846	0.99
Vardo	076, 079, 162, 165, 248, 392, 395, 478, 481	0.71	311, 416, 502, 855, 874, 960	0.99
Teriberka	104, 190, 395, 481	0.43	072, 158, 225, 530, 616, 683	0.97
Iokanga	132, 218, 304, 395	0.37	053, 186, 272, 511, 730, 816	0.89
Pechenga	162, 248, 309, 334, 395	0.45	330, 874,	0.93
Tobseda	193, 279, 332	0.44	042, 195, 281, 500, 586, 739, 825	0.87
Varandey	049, 102, 135, 221	0.39	242, 395, 853, 939	0.89

For the White Sea, a comparison between satellite and TG SSHA data also shows significant correlations (>0.6) for all the satellites except GFO. Greater values of correlation for TG Onega (0.76 for T/P and J1 data; 0.96 for ERS-1/2 and Envisat data) and TG Severodvinsk (0.97 and 0.98, respectively) are conditioned by their location. Both TGs are situated in river estuaries; therefore, river runoff has a strong influence on the hydrological regime in these parts of the White Sea. Correlation minimum (0.66 and 0.61, respectively) is observed at TG Kem' Port where nonlinear and residual tidal phenomena are important on shallow water. Some details of sea level comparison are shown in Table 15.2. This result is also in accordance with our previous investigations (Lebedev and Tikhonova 2000; Lebedev et al. 2003).

According to the results obtained, satellite altimetry SSHA data are in good consistency with TG SSHA.

15.2.3
Tidal Models

Peculiarities of the tidal regime in the Barents and White seas cause inaccurate use of global tidal models for a processing (tidal correction) of satellite altimetry data. Therefore, to improve altimetry data processing in these regions, it is very important to use regional tidal model.

Global tidal models: At present there are more than 11 global tidal ocean models (Le Provost 2001). They differ by spatial resolution and procedures of assimilation of TG and satellite altimetry data. A variety of tests indicate that all these tidal models agree within 2–3 cm in the deep ocean, and they represent a significant improvement over the classical Schwiderski

Table 15.2 Correlation between satellite and TG SSHA data for the White Sea

Tide Gauge	T/P & J1 Track number	Correlation	GFO Track number	Correlation	ERS-1/2 & Envisat Track number	Correlation
Kem' Port	090, 239, 242	0.66	023, 074, 160, 425	0.35	272, 483, 730, 816, 941	0.61
Onega	090, 239, 242	0.76	023, 074, 160, 425	0.41	272, 483, 730, 816, 941	0.96
Severo-dvinsk	137, 213, 242	0.97	195, 281, 332, 418	0.56	100, 311, 769, 558	0.88
Solovki	061, 064, 239, 242	0.76	023, 074, 109, 160, 246, 425	0.59	272, 397, 483, 730, 816, 941,	0.65
Kanda-laksha	–	–	195, 218, 304, 390	0.45	358, 941	0.58
Sos-novets	–	–	051, 137, 160, 223, 246, 453	0.44	014, 100, 472, 683	0.77
Umba	–	–	109, 195, 218, 304, 390	0.67	358, 397, 816, 941	0.95

1980 model (Shum et al. 1997). But global tidal models ignore nonlinear, residual, and other tidal phenomena which are inherent to coastal and shallow water areas of marginal and semi-enclosed seas (e.g., see Hwang and Chen 1997; He et al. 2001). In our investigation global tidal models CSR4.0 (Eanes and Bettadpur 1995) and GOT00.2 (Ray 1999) were used.

Regional tidal model: A three-dimensional baroclinic model with free surface was developed in the Laboratory of Sea Applied Research at HMRC of the Russian Federation (Popov et al. 2000). This model reproduces sufficiently accurate storm surges and tides in the Barents and White seas. It uses the time step which is equal to 120 s for the coarse grid while the mesh size is 6 nautical miles. For the fine grid, the time and spatial steps are 60 s and 3 nautical miles, accordingly.

For the Barents and White seas only M_2, S_2, and K_2 main tidal components were used in calculation. Mean temperature and salinity fields, as well as atmospheric fields from the NCEP/NCAR Reanalysis project (Kalnay et al. 1996; Kistler et al. 2001) were used in calculation of the Barents Sea level and currents for 1948–2008. In contrast to global models, results of the HMRC model represent fields of total tide height for each time moment with a given time step.

For comparison of global and regional tidal models, spectral density of the sea level was calculated for all TGs and for different main tidal constituents – M_2, S_2, and K_2 with a time step of 1 h. Analysis of SSHA was on the basis of the results of harmonic analysis of temporal series. Unbalances of main tidal constituents amplitudes and phases were calculated as $\Delta H = |H_1 - H_2|/H_{max}$ and $\Delta g = |g_1 - g_2|/q$, where H_{max} is maximum value of main tidal constituents amplitude (H_1) of the model or TG data; g_1 is a phase of simulation result or

Table 15.3 Comparison of main tidal components (amplitude and phase) between TG records and the results of tide hydrodynamic simulation

Tide gauge	Tidal component	Global tidal model				Regional tidal model	
		Model CSR4.0		Model GOT00.2		Model HMRC	
		ΔH (%)	Δg (h)	ΔH (%)	Δg (h)	ΔH (%)	Δg (h)
Solovki	M_2	53.1	0.49	30.5	0.21	27.3	0.15
	S_2	31.8	0.23	77.3	0.89	29.0	0.19
	K_1	5.9	0.02	58.5	0.54	26.1	0.29
Pechenga	M_2	76.2	0.78	87.4	1.25	7.5	0.05
	S_2	73.9	0.84	58.6	0.47	26.3	0.17
	K_1	28.3	0.15	14.1	0.16	17.1	0.09
Vardo	M_2	76.8	0.45	45.8	0.69	22.2	0.65
	S_2	25.7	0.29	45.1	0.84	19.8	0.28
	K_1	43.9	0.93	27.2	0.45	4.2	0.09
Varandey	M_2	68.3	2.01	34.6	0.78	16.6	1.39
	S_2	73.7	1.34	57.3	1.17	26.4	0.79
	K_1	24.6	0.98	39.8	1.04	15.0	1.10

TG data, and q is an angular velocity. Results of the comparison of global and regional tidal models are shown in Table 15.3.

Spectral density of TG Solovki (Fig. 15.9a) includes main tidal constituents (M_2, S_2, and K_2), their nonlinear interference (see peaks with periods less than 8.5 h), and periods related to synoptic dynamics. According to initial conditions, the results of calculation based on global tidal models CSR4.0 (Fig. 15.9b) and GOT00.2 (Fig. 15.9c) contain only three main tidal constituents: M_2, S_2, and K_2. Appearance of complementary tidal constituents (period of 4.3 h) in spectral density resulting from the HMRC regional model (Fig. 15.9d) proves that this model correctly describes the Barents and White seas tides. Thus, a regional tidal model (e.g., HMRC) must be used for tidal correction of altimetry data.

15.3
Long-Term Variability of Sea Level

Sea level reflects changes in practically all dynamic and thermodynamic processes of terrestrial, oceanic, atmospheric, and cryospheric origin. During the period 1954–1989 the observed sea level at TGs over the Russian sector of the Arctic Ocean rose at a rate of approximately 0.12 ± 0.06 cm/year; taking into consideration corrections for the process of Earth's crust uplift and glacial isostatic adjustment, this rate was approximately +0.18 ± 0.08 cm/year (Proshutinsky et al. 2004). According to other investigations, this value was +0.12 ± 0.08 cm/year (Vorobyov et al. 2000).

However, the trend of the Barents Sea level estimated from TG observational data is −0.05 ± 0.05 cm/year (Proshutinsky et al. 2004) after all corrections. It may be supposed that negative value of the Barents Sea level trend is a result of a slow rise of Fennoscandia and Novaya Zemlya. The coast of Kola Peninsula on the Barents and White seas rises with a rate of 4–8 mm/year (Scherneck et al. 2003). The rate of rise of the Earth's crust in the

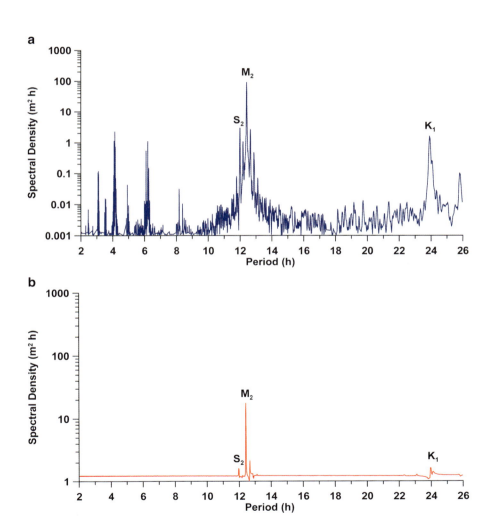

Fig. 15.9 Spectral density of the SSHA time series at the TG Solovki (**a**) and the superposition of main tidal constituents M_2, S_2, and K_2 calculated with tidal models: (**b**) CSR4.0, (**c**) GOT00.2, and (**d**) HMRC

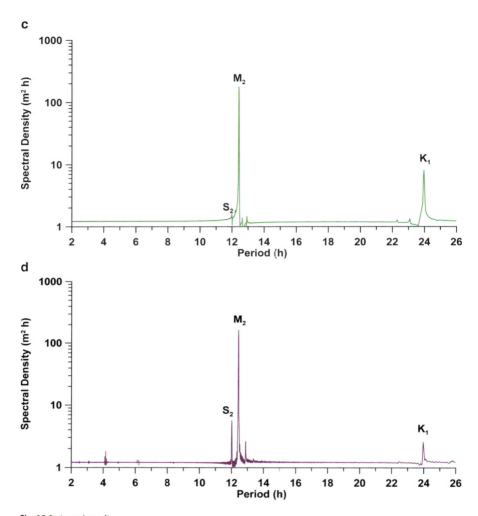

Fig. 15.9 (continued)

region of Novaya Zemlya Island is 1–2 mm/year (Pobedonostsev and Rozanov 1971; Nikonov 1980). For interpretation of climatic sea level change, it is desirable to keep in mind that the Barents Sea shelf is sagging (Krapivner 2006).

At the end of investigation of long-term variability of sea level, SLA (sea level anomaly) was calculated with the HMRC regional tide model at three points (Fig. 15.10). Points A and B are located near Scandinavia and Kola Peninsula. Their positions coincide with the isoline 2 cm of the climatic dynamic topography field obtained with the HMRC hydrodynamic model without accounting for tide-generating force (Popov et al. 2000). Third point C is located on the boundary between the Barents and White seas. Altimetric data from the satellites ERS/Envisat and GFO were used for processing.

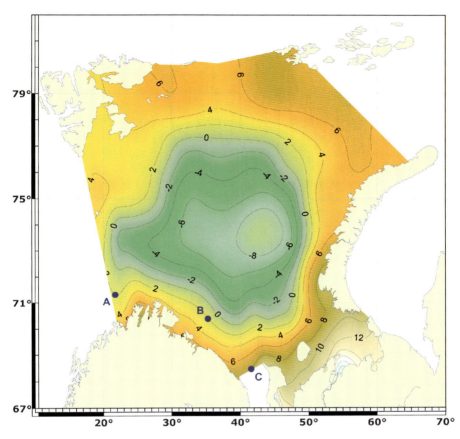

Fig. 15.10 Location of points A, B, and C superimposed on a climatic dynamic topography field obtained with the HMRC hydrodynamic model for time period 1958–2008 (Popov et al. 2000)

Analysis of temporal SLA variations at point A shows that SLA increases during March 1998–May 1999, May 2001–August 2001, and June 2005–September 2007 with a rate of 4.58, 9.62, and 2.00 cm/year, respectively (Fig. 15.11a). Ice cover in these time periods is less than in the following years. So it is possible that this phenomenon was associated with intensification of the Norwegian Coastal Current.

SLA at point B also increased with a rate of 5.43 cm/year (January 1999–September 1999), 5.80 cm/year (June 2001–December 2001), and 5.43 cm/year (August 2004–May 2005) (Fig. 15.11b). Interesting feature of the temporal variability in this figure is zero trend from June 2002 to March 2003. This point is located in the stream of the Murmansk Coastal Current. So it may be supposed that (in general) similar character behaviour of SLA temporal variation at both points is conditioned by the influence of the Norwegian Coastal Current on the Murmansk Coastal Current.

Zero trend is obtained on SLA temporal variability at point C from September 2005 to the present time (Fig. 15.11c). Values of SLA positive trend were as follows:

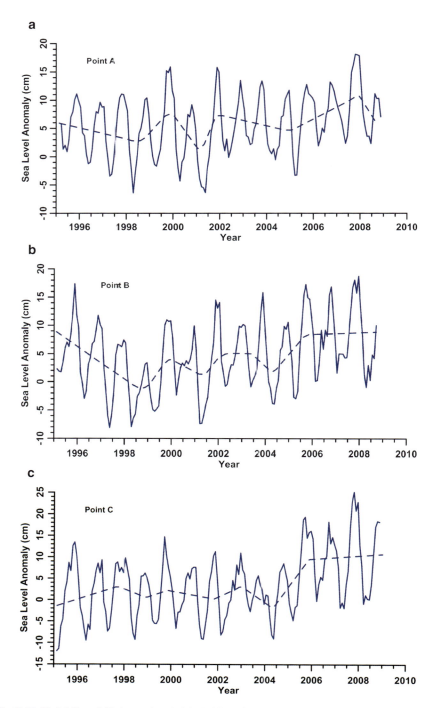

Fig. 15.11 Variability of SLA at points A (a), B (b), and C (c) based on ERS and Envisat data

2.22 cm/year during January 1999–September 1999, 2.50 cm/year during November 2001–September 2002, and 7.0 cm/year during May 2004–September 2005. However, periods of SLA increase differed from those for points A and B. Averaged values of long-term SLA trend for the period from January 1995 to the present time are 0.35, 0.48, and 0.74 cm/year for points A, B, and C, respectively. Thus, tendency of an increasing SLA is observed in the Barents Sea. Values of these trends obtained call for further refinements.

15.4
Interannual Variability of Sea Ice Edge Position

Traditional research of Arctic ice cover and ice edge location are on the basis of infrared and microwave radiometry data. But satellite altimetry data may be also used to estimate ice cover extent or position of an ice edge.

For this purpose, 34 descending tracks of the satellites ERS-2 and Envisat have been chosen (Fig. 15.12). These tracks are located under the optimal angle to mean climatic sea ice edge position. Significant SSH records along the satellite tracks over the sea surface

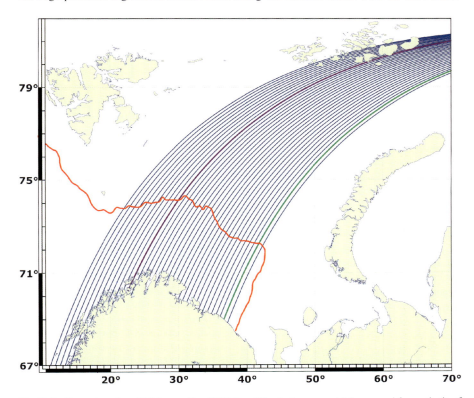

Fig. 15.12 Ground tracks of 34 descending ERS-2 and Envisat passes which are used for analysis of the along-track sea ice edge position. Red line is the mean climatic sea ice edge position (Lebedev and Kostianoy 2008). Green line shows the position of track 988 and purple line of track 932

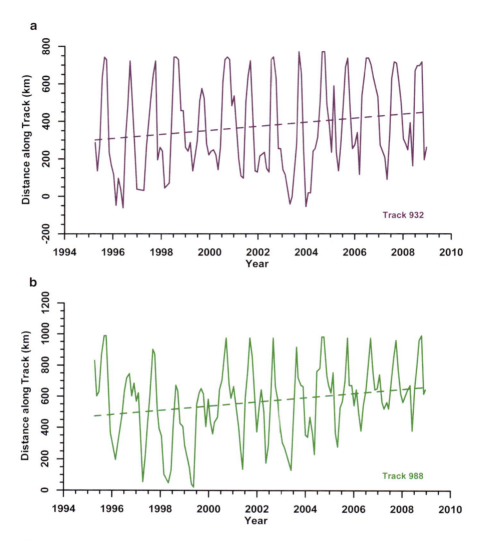

Fig. 15.13 Temporal variability of sea ice edge position along 932 (**a**) and 988 (**b**) descending ERS-2 and Envisat tracks. Dashed line shows interannual trend. Zero corresponds to the mean climatic sea ice edge position

and over the ice fields strongly differ. This point of the signal change located on the track was identified as an ice edge. Some temporal variations of ice edge position along tracks are shown in Fig. 15.13.

From 1995 to 2008, the sea ice edge didn't reach its mean climatic position along track 988. For track 932, the sea ice edge observed in the winters of 1996, 2003, and 2004 was displaced southwestward by 60, 40, and 53 km, respectively, from its mean climatic position. For both tracks, the 2000 and 2007 winters were very warm.

According to our calculations, interannual trends of sea ice position northeastward along tracks 988 and 932 were 13.2 ± 0.6 and 10.7 ± 0.4 km/year, respectively. For summer months, similar trends were 24.8 ± 1.2 and 8.2 ± 0.5 km/year, respectively. Averaged velocity of the ice edge along-track displacement for all tracks was 11.8 ± 0.5 km/year.

15.5 Conclusions

Spatial and temporal coverage of the Barents and White seas by satellite altimetry missions offers the possibility of continuous high-precision monitoring of sea level, ice cover, and sea surface wind speed. Influence of tides must be successfully estimated using numerical simulation based on the regional tidal models rather than global ones. The results obtained seem to be promising in application of satellite altimetry data on SLA and wind speed for monitoring a number of environmental parameters of the Barents and White seas.

Acknowledgements The work was supported by International Project INTAS No 05-1000008-7927 (ALTICORE) and by Russian Foundation for Basic Research (Grants No 06-05-64871, No 07-05-00415, and No 08-05-97016).

References

Atlas of Arctic (1985) Publisher of the Main Office of Geodesy and Chartography, Moscow (in Russian)
Barents Sea (1985) Hydrometeorological conditions of a shelf zone of the USSR seas Hydrometeoizdat, Leningrad (in Russian)
Benada RJ (1997) Merged GDR (TOPEX/Poseidon), Generation B users handbook, Version 2.0. Physical Oceanography Distributed Active Archive Center (PODAAC), Jet Propulsion Laboratory, Pasadena
Comiso JC, Parkinson CL, Gersten R, Stock L (2008) Accelerated decline in the Arctic sea ice cover. Geophys Res Lett 35:L01703. doi:10.1029/2007GL031972
Eanes RJ, Bettadpur SV (1995) The CSR 3.0 Global Ocean Tide Model. Center for Space Research, Technical Memorandum CSR-TM-95-06, Austin, University of Texas
Filatov N, Pozdnyakov D, Ingebeikin Ju, Zdorovenov R, Melentyev V, Tolstikov A, Pettersson L (2005) Oceanographic regime. In: Filatov N, Johannessen O, Pozdnyakov D, Bobylev L, Pettersson L. (eds) White Sea. Its marine environment and ecosystem dynamics influenced by global change. Springer, Berlin. doi:10.1007/3-540-27695-5_4
GEOSAT Follow-On GDR User's Handbook (2002) http://ibis.grdl.noaa.gov/SAT/gfo/
Glukhovsky BKh (ed.) (1991) Hydrometeorology and hydrochemistry of the seas of the USSR, Vol. 2, White Sea, Issue 1, Hydrometeorological conditions, Leningrad, Hydrometeoizdat (in Russian)
He Y, Zhao J, Lu X (2001) Shallow sea tides estimated using T/P data and adjoint assimilation model. In: Geoscience and Remote Sensing Symposium, 2001, IGARSS apos; 01. IEEE 2001 International 3:1073-1075. doi:10.1109/IGARSS.2001.976750
Hwang C, Chen CZ (1997) Improving the M2 tide model over shallow waters using TOPEX/Poseidon and ERS-1 altimetry. Mar Geodesy 20:291–305

Kagan BA, Romanenkov DA (2007) Effect of the nonlinear interaction of tidal harmonics on their spatial structure as applied to the system of the White and Barents seas. Izvestiya Atmos Ocean Phys 43(5):655–662. doi:10.1134/S0001433807050131

Kalnay E, Kanamitsu M, Kistler R, Collins W, Deaven D, Gandin L, Iredell M, Saha S, White G, Woollen J, Zhu Y, Chelliah M, Ebisuzaki W, Higgins W, Janowiak J, Mo KC, Ropelewski C, Wang J, Leetmaa A, Reynolds R, Jenne R, Joseph D (1996) The NCEP/NCAR 40-year reanalysis project. Bull Am Meteor Soc 77(3):437–471

Kistler R, Kalnay E, Collins W, Saha S, White G, Woollen J, Chelliah M, Ebisuzaki W, Kanamitsu M, Kousky V, van den Dool H, Jenne R, Fiorino M (2001) The NCEP–NCAR 50-year reanalysis: monthly means CD–ROM and documentation. Bull Am Meteor Soc 82(2): 247–267

Koblinsky CJ, Ray R, Becley BD, Wang YM, Tsaoussi L, Brenner A, Williamson R (1999a) NASA ocean altimeter pathfinder project. Report 1: data processing handbook. NASA Goddard Space Flight Center, NASA/TM-1998-208605

Koblinsky CJ, Ray R, Becley BD, Wang YM, Tsaoussi L, Brenner A, Williamson R (1999b) NASA ocean altimeter pathfinder project. Report 2: data set validation. NASA Goddard Space Flight Center, NASA/TM-1999-209230

Kostianoy AG, Nihoul JCJ, Rodionov VB (2004) Physical oceanography of frontal zones in the Subarctic seas. Elsevier Oceanography Series, V. 71, Elsevier, Amsterdam

Krapivner RB (2006) Rapid sagging of the Barents shelf over the last 15–16 ka. Geotectonics 40(3):197–207. doi:10.1134/S0016852106030046

Le Provost JM (2001) Ocean tides. In: Fu LL, Cazenave A (eds) Satellite altimetry and earth sciences. A handbook of techniques and applications. Academic, San Diego, CA

Lebedev SA, Kostianoy AG (2005) Satellite altimetry of the Caspian Sea. "Sea", Moscow (in Russian)

Lebedev SA, Kostianoy AG (2008) Seasonal and interannual sea level and ice fraction variability in the Barents and White seas based on remote sensing data. Processing of IX Biennal Pan ocean remote sensing conference (PORSEC), Guangzhou, China, 2–6 December 2008, pp 34–35

Lebedev SA, Tikhonova OV (2000) Application of a satellite altimetry at research of a sea level of a southeastern part of the Barents Sea. In: Proceeding of the VI international scientific and technical conference "Modern methods and means of oceanological researches", Moscow, Russia, 10–14 November 2000, vol 2, pp 58–64 (in Russian)

Lebedev SA, Zilberstein OI, Popov SK, Tikhonova OV (2003) Analysis of temporal sea level variation in the Barents and the White seas from altimetry, tide gauges and hydrodynamic simulation. In: Hwang C, Shum CK, Li JC (eds) International workshop on satellite altimetry. IAG Symposia, V. 126. Springer, Berlin/Heidelberg

Lebedev SA, Sirota A, Medvedev D, Khlebnikova S, Vignudelli S, Snaith HM, Cipollini P, Venuti F, Lyard F, Bouffard J, Crétaux JF, Birol F, Roblou L, Kostianoy A, Ginzburg A, Sheremet N, Kuzmina E, Mamedov R, Ismatova K, Alyev A, Mustafayev B (2008) Exploiting satellite altimetry in coastal ocean through the ALTICORE project. Russ J Earth Sci 10:ES1002. doi: 10.2205/2007ES000262

May RI (2005) Simulation of climate significant nonlinear tidal phenomena in the Euro Arctic seas. In: IEEE OCEANS'05 EUROPE conference proceedings. Oceanography: modeling & data processing, Brest, France, 2005, vol I, pp 401–406. doi:10.1109/OCEANSE.2005.1511748

Medvedev P, Tyupkin Yu (2005) Satellite altimetry data system for geodynamics and Oceanography studies: telematics aspects and applications. IST4Balt News J 1:39–41

Medvedev PP, Lebedev SA, Tyupkin YuS (1997) An integrated data base of altimetric satellite for fundamental geosciences research. In: Proc. first East-European Symp. Adv. in data bases and information systems (ADBIS'97), St. Petersburg, Russia, 2–5 September 1997, St. Petersburg University, St. Petersburg, vol 2, pp 95–96

Medvedev PP, Nepoklonov VB, Medvedev DP (2002) The application of satellite altimetry data for the studies of the level variations of the Sea of Okhotsk, Baikal Lake and of the geodynamics of the Western Pacific Belt. Information system for satellite geodesy data interpretation in geodynamics problems. In: Proceedings of the international seminar "On the use of space technologies for Asia-Pacific regional Crustal movements studies", APSG-IRKUTSK 2002, Irkutsk, Russia, 5–10 August 2002, Moscow, GEOS, pp 234–248

Nikonov AA (1980) Modern vertical movements of the coast of northern and far eastern seas of the USSR. Geologia i geofizika 12:71–77 (in Russian)

Parkinson CL, Cavalieri DJ, Gloersen P, Zwally HJ, Comiso JC (1999) Arctic sea ice extents, areas, and trends, 1978–1996. J Geophys Res 104(C9):20837–20856. doi:10.1029/1999JC900082

Picot N, Case K, Desai S, Vincent P (2006) AVISO and PODAAC user handbook, IGDR and GDR Jason Products, SMM-MU-M5-OP-13184-CN (AVISO), JPL D-21352 (PODAAC), Edition 3

Pobedonostsev SV, Rozanov LL (1971) Modern vertical movements of the White and Barents seas coast. Geomorfologia 3:57–61 (in Russian)

Popov SK, Safronov GF, Zilberstein OI, Tikhonova OV, Verbitskaya OA (2000) Density and residual tidal circulation and related mean sea level of the Barents Sea. In: Mitchum GG (ed) Ocean circulation science derived from the Atlantic, Indian and Arctic sea level networks, Toulouse, France, 10–11 May 1999, Intergovernmental Oceanographic Commission Workshop Report, 171, UNESCO

Proshutinsky A, Ashik IM, Dvorkin EN, Hakkinen S, Krishfield RA, Peltier WR (2004) Secular sea level change in the Russian sector of the Arctic Ocean. J Geophys Res 109(C3):C03042. doi:10.1029/2003JC002007

Ray RD (1999) A global ocean tide model from TOPEX/Poseidon altimetry: GOT99.2, NASA Tech Memo 209478, Goddard Space Flight Center

Scherneck HG, Johansson JM, Koivula H, van Dam T, Davis JL (2003) Vertical crustal motion observed in the BIFROST project. J Geodyn 35(4):425–441. doi:10.1016/S0264-3707(03)00005-X

Schrama E, Scharroo R, Naeije M (2000) Radar altimeter database system (RADS): towards a generic multi-satellite altimeter database system. Delft Institute for Earth-Oriented Space Research, Delft University of Technology, The Netherlands

Semenov EV, Luneva MV (1996) Numerical model for tidal and density driven circulation of the White Sea. Izv Akad Navk USSR, Phys Atmos Ocean 32(5):704–713 (in Russian)

Shum CK, Woodworth PL, Andersen OB, Egbert G, Francis O, King C, Klosko S, Le Provost C, Li X, Molines JM, Parke M, Ray R, Schlax M, Stammer D, Temey C, Vincent P, Wunsch C (1997) Accuracy assessment of recent ocean tide models. J Geophys Res 102(C11):25173–25194. doi:10.1029/97JC00445

Terziev FS, Girduk GV, Zykova GG, Dzhenyuk SL (eds) (1990) Hydrometeorology and hydrochemistry of the seas of the USSR, Vol 1, Barents Sea, Issue 1, Hydrometeorological conditions, Hydrometeoizdat, Leningrad (in Russian)

Vinje T (2001) Anomalies and trends of sea-ice extent and atmospheric circulation in the Nordic seas during the period 1864–1998. J Clim 14(3):255–267. doi:10.1175/1520-0442(2001) 014<0255:AATOSI>2.0.CO;2

Vorobyov VN, Kochanov SY, Smirnov NP (2000) Seasonal and multiyear sea-level fluctuations in the Arctic Ocean. Ministry of Education of Russian Federation, St. Petersburg (in Russian)

White Sea (1989) Hydrometeorological conditions of a shelf zone of the USSR seas, vol 5. Hydrometeoizdat, Leningrad (in Russian)

Satellite Altimetry Applications off the Coasts of North America

16

W.J. Emery, T. Strub, R. Leben, M. Foreman, J.C. McWilliams, G. Han, C. Ladd, and H. Ueno

Contents

16.1	Background	418
16.2	East Coast of North America	419
	16.2.1 The Labrador Sea and Coastal Eastern Canada	419
	16.2.2 The Scotian Shelf and the North Atlantic Bight	421
16.3	The Gulf of Mexico and Caribbean Sea	425
16.4	West Coast of North America	429
	16.4.1 California Current System	431
16.5	Alaskan Stream	443
16.6	Tidal Energy in the Bering Sea	447
16.7	Future Work	448
References		448

W.J. Emery (✉) and R. Leben
CCAR, University of Colorado, Boulder, CO 80309-0431, USA
e-mail: william.emery@colorado.edu

T. Strub
COAS, School of Oceanography, Oregon State University, Corvallis, OR, USA

M. Foreman
Institute of Ocean Sciences, Sidney, B.C., Canada

J.C. McWilliams
Institute of Geophysics and Planetary Physics, UCLA, Los Angeles, CA, USA

G. Han
Fisheries & Oceans Canada, Northwest Atlantic Fisheries Center, St. John's, NL, Canada

C. Ladd
Pacific Marine Environmental Lab, NOAA, Seattle, WA, USA

H. Ueno
Institute for Observational Research into Global Change, Japan Agency Marine-Earth Science and Technology, Yokosuka, Kanagawa, Japan

Keywords Satellite altimetry • U.S. coasts • California Current • Gulf of Mexico • Gulf Stream • Labrador Current • Loop Current

Abbreviations

ADCP	Acoustic Doppler Current Profiler
AVHRR	Advanced Very High Resolution Radiometer
AVISO	Archiving, Validation and Interpretation of Satellite Oceanographic data
CHL	CHLorophyll
CTD	Conductivity-Temperature-Depth
DEM	Digital Elevation Model
DOT	Dynamic Ocean Topography
DUACS	Data Unification and Altimeter Combination System
ECMWF	European Centre for Medium-Range Weather Forecasts
EKE	Eddy Kinetic Energy
ENSO	El Niño Southern Oscillation
EOF	Empirical Orthogonal Function
GOES	Geostationary Operational Environmental Satellite
JPL	Jet Propulsion Laboratory
MCC	Maximum Cross Correlation
NCEP	National Center for Environmental Prediction
RMS	Root Mean Square
ROMS	Regional Oceanic Modeling System
SCB	Southern California Bight
SSALTO	Segment Sol multimissions d'ALTimétrie, d'Orbitographie et de localisation précise
SSH	Sea Surface Height
SSHA	Sea Surface Height Anomaly
SST	Sea Surface Temperature
TG	Tide Gauge
T/P	TOPEX/Poseidon
VACM	Vector-Averaging Current Meter
WOCE	World Ocean Circulation Experiment
XBT	eXpendable BathyThermograph

16.1 Background

A number of studies have used satellite altimeter data, alone or in combination with other satellite, in situ or model fields, to investigate the circulation of the boundary currents surrounding North America. In nearly all cases, these investigations have concentrated on the variability in the seasonal jets and mesoscale circulation features found in the several hundred kilometers next to the coast. Only recently have there been attempts to retrieve and use altimeter sea surface height (SSH) data in the traditional domain of "coastal oceanography," over the continental shelf and/or within several tens of kilometers of land. The problems encountered off North America in retrieving altimeter data near land are the same as those described in other regions – primarily distortions of the altimetric waveforms, non-linear tidal currents over continental shelves, and a lack of atmospheric water vapor correction due to interaction of the land with the passive microwave radiometer used for this correction (Vignudelli et al. 2005). The methods used to investigate the mesoscale circulation fields in the larger-scale boundary currents have been instructive, however, in establishing the nature and limitations of traditional altimetric approaches and suggesting improvements and modifications that allow altimeter data to be extended closer to land. Here we describe those earlier altimetric studies and note how they evolved into the more coastal investigations that are now under way.

Another complexity of these coastal zones is the contribution of the ageostrophic components of the coastal currents, which cannot be mapped using satellite altimetry. The simplest of these is the surface Ekman component, which applies to both the Deep Ocean and coastal currents and may be estimated from satellite or in situ wind measurements. Unlike the open ocean, however, other ageostrophic currents induced by the wind and tides in coastal regions are often non-linear and frequently overwhelm the geostophic currents that can be mapped using altimeter data. Other non-altimetric measurements, such as drifting buoy trajectories, current meter moorings, Acoustic Doppler Current Profiler (ADCP), currents mapped by coastal radars and currents inferred from satellite imagery, are needed to resolve the often-significant ageostrophic currents in the coastal region. These then need to be combined with the altimetric estimates of the geostrophic currents (Matthews and Emery 2009).

In the following sections we review some important results of coastal altimetry studies off the coasts of North America. We begin with a brief summary of some results of studies of the East Coast of North America starting in the Labrador Sea and then moving south to Nova Scotia and then finally to the east coast of the U.S. Studies of the Gulf of Mexico are then reviewed followed by some example studies from the west coast of North America where quite a few studies of coastal altimetry have been carried out. In between there is a short review of a paper on tides in the Bering Sea. We finish up with a short section on future plans for coastal altimetry studies in this region.

16.2
East Coast of North America

16.2.1
The Labrador Sea and Coastal Eastern Canada

For reviews of the shelf, slope, and offshore boundary circulation characteristics of the NW Atlantic, see Loder et al. (1998), Boicourt et al. (1998), Townsend et al. (2006), and Lohrenz and Verity (2006). An overly simplified depiction of the circulation along the U.S. east coast north of Cape Hatteras consists of an equatorward flow of cooler and fresher water over the wide shelf and slope, bounded on the offshore edge by a warm and salty poleward current over the continental slope. The fresh and cool water originates to the north in the Labrador Sea, augmented by coastal inputs and diminishing in strength toward the south (Loder et al. 1998). It's southward flow is driven, at least in part, by buoyancy forcing. The Gulf Stream is relatively far offshore in this region but may contribute to the poleward slope current through its influence on offshore SSH. South of Cape Hatteras, the flow over the shelf appears less organized and the Gulf Stream stays closer to the slope.

Traditional altimeter analyzes are not usually attempted over the wide shelf, where tidal motions are strong and errors in tidal models used to remove the tidal signals from altimeter data contaminates the signal. Dong and Kelly (2003) tested four tidal models over the mid-Atlantic Bight shelf and found that none succeeded in removing the tidal signal from the altimeter data. They also found no significant correlations between the altimeter-derived velocities and current meter records over the shelf (<200 m water depth), but good correlations with measured currents over the slope (water depths of 500 m and deeper). They attributed the lack of correlations over the shelf to unresolved tidal motion and restricted their altimeter analyzes to examinations of the boundary currents over the slope and farther offshore. This has been the case for all altimeter studies along the U.S. east coast, as characterized by the selection of papers summarized below.

The early study of Kelly and Gille (1990) developed a method to estimate the Gulf Stream's position, width, and strength as it crossed repeated altimeter tracks. By assuming a Gaussian shape to the velocities and an SSH shape consistent with the geostrophic relation, they were able to estimate both the temporal mean and instantaneous absolute velocity and SSH profiles from the time series of altimeter sea surface height anomalies (SSHA, with the temporal mean removed from the altimeter record at each alongtrack location). The method relies on the fact that the Gulf Stream meanders and changes position along the track over a range of distances as large as its width. This allows them to use an iterative process and least squares fit to estimate the three parameters that define the Gaussian – its center position, width, and height. Although this may seem far removed from the coast, it comes into play in examining causes for changes in the slope currents (below).

Moving upstream to the source of the cool, fresh water flowing equatorward over the shelf, Han and Tang (2001) use 6 years of TOPEX/Poseidon (T/P) altimeter data along with a WOCE hydrographic section to compute volume transports in the western Labrador Sea (Han and Tang 2001). Two ascending altimeter tracks (Fig. 16.1) were selected that overlap the WOCE hydrographic section, which was sampled repeatedly in each of the years between 1993 and 1998. The density sections for 1994 and 1996 clearly show large horizontal density gradients over the shelf break and upper slope where the traditional Labrador Current is located. These gradients decrease sharply towards the shore and deep water. To compute seasonal transports, the bottom boundary current was assumed zero by neglecting the bottom boundary layer. Over seasonal time scales the wind-driven near surface Ekman currents are small and can also be neglected. Thus, the total cumulative seasonal volume transport can be calculated by concurrent altimetric and hydrographic data. Due to the paucity of hydrographic data the instantaneous CTD sections were used instead of seasonal averages with the seasonally averaged altimeter data. Thus, spring and summer transports were then averaged to obtain a mean April to Sept. transport.

The resulting cumulative southward transports for the time series from 1993 to 1998 are presented here in Fig. 16.2 where the total transport is computed from the barotropic and

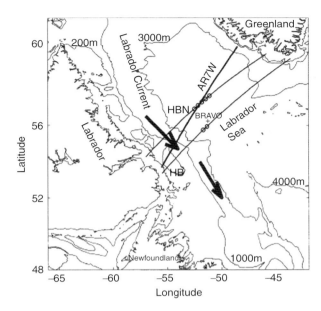

Fig. 16.1 Map showing the study area: HB and HBN are two TOPEX/Poseidon ascending ground tracks that straddled the AR7W hydrographic section in the western Labrador Sea. The 200, 1,000, 3,000, and 4,000 m isobaths (*thin lines*) and the location of the Labrador Current (*thick arrows*) are also shown. The five open circles on HBN indicate the outer limits of integration in the calculation of the volume transport, from the deepest sea westward at an interval of 25 km. The corresponding outer limits for HB (for clarity only the easternmost and westernmost locations are shown) are determined from the values of H/f (H is the water depth and f is the Coriolis parameter; Han and Tang 2001)

baroclinic components. Errors in the barotropic component were computed from errors in both the altimeter surface height anomaly and the WOCE hydrographic section dynamic height. Baroclinic errors are computed from the hydrographic data. Here a positive anomaly corresponds to a larger than average southward transport. The interannual range of the total transport is about 6.2 Sv, which is comparable to the seasonal change of about 10 Sv. The years 1993, 94, 95, and 1997 have larger than average transports while the years 1996 and 1998 have smaller than average transports. The barotropic and baroclinic transports behave oppositely, compensating for each other and resulting in a total transport anomaly with interannual variations smaller than either of the two components. The total transport in the Labrador Sea circulation is positively correlated with the winter North Atlantic Oscillation index.

16.2.2
The Scotian Shelf and the North Atlantic Bight

Dong and Kelly (2003) made use of the above results in their analysis of seasonal and interannual variability in SSHA and corresponding cross-track geostrophic velocity anomalies across six T/P tracks in the mid-Atlantic Bight (~37°–43°N) during 6 years (~1993–1998).

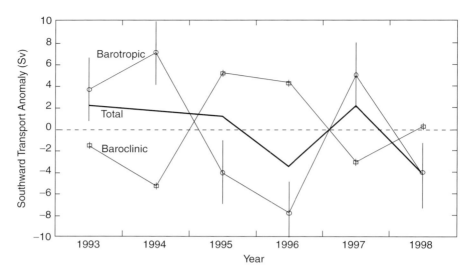

Fig. 16.2 Cumulative southward transport anomaly in the western Labrador Sea (west of the open circles in Fig. 16.1) as a function of year. The total transport (*thick lines*) is the sum of the barotropic (*circles*) and baroclinic (*squares*) components. The standard errors associated with the barotropic and baroclinic components are shown as vertical lines (Han and Tang 2001)

Their results show a seasonal reversal of the coherent southwestward surface currents along the continental slope, flowing to the northeast during the summer. A shorter record of current meters verified this reversal. During the summer of 1996, however, flow to the southwest strengthened rather than reversed. Comparisons to the strength of the outflow from the Labrador Sea (Han and Tang 2001) do not explain this anomalous behavior. Using the method of Kelly and Gille (1990), however, Dong and Kelly (2003) demonstrate that the Gulf Stream moved to the south in 1996 and they attribute the lack of northeastward flow to this shift in Gulf Stream location and lower offshore SSHA values during summer 2006, reducing the cross-slope pressure gradient and the associated along-slope geostrophic current.

Slightly farther upstream, Han (2004) analyzed a longer time series (through 2000) of T/P altimeter sea levels in conjunction with frontal analysis from infrared satellite imagery to study the temporal and spatial variability of sea-surface currents over the Scotian Slope. Geostrophic surface current anomalies normal to altimeter ground tracks are derived from sea level anomalies relative to local temporal means for the study period between 1992 and 2000. The variability of these geostrophic currents is plotted along the altimeter tracks in Fig. 16.3 as thick grey lines. Also plotted are the positions of the shelf slope front (thick dashed line) and the northern edge of the Gulf Stream (thick dash dot line). The geostrophic current anomaly including the Han et al. (1997) model mean is shown as darker, thinner lines inside the lines of geostrophic current anomaly variability.

In general, the RMS values increase westward and offshore. On individual tracks, values at the shelf break are large relative to the ones just farther offshore. The westward intensification is more evident over the western Scotian Slope and the offshore enhancement occurs mainly between the shelf-slope front and the Gulf Stream northern boundary. Typical

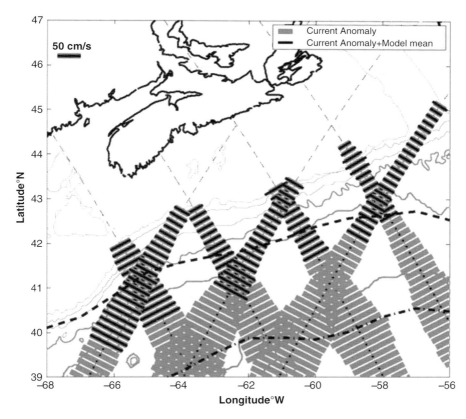

Fig. 16.3 Twice RMS values of the altimetric geostrophic current anomalies (subsampled spatially) from 1992 to 2000 (thick gray lines). The thin darker lines represent twice the RMS values when Han et al.'s (1997) model means are included. Also depicted are the positions of the shelf-slope front (*thick dashed line*) and the northern boundary of the Gulf Stream (*thick dashed dot line*; Han et al. 1997)

RMS values of the cross-track current anomalies are 20–30 cm/s over the upper and lower slope and 20–50 cm/s over the continental rise. The maximum cross-track current variability of 70–75 cm/s is located close to the Gulf Stream northern boundary. The offshore intensification is associated with the proximity of the Gulf Stream. Also, Gulf Stream rings are known to be important to the slope water circulation variability. Part of the westward intensification may also be attributed to increased warm core ring activity to the west. Crossover analyzes of altimetric current anomalies indicate that the total current variability can be estimated by a factor of 1.5 from the cross-track current variability and the variance is isotropic to within 14 %. The rotational speed of the Gulf Stream rings can reach 1–2 m/s over the Scotian Slope, with the RMS variability of ~60 cm/s.

Han (2007) extended his earlier study and that of Dong and Kelly (2003) to cover the period from 1992–2002, combining T/P geostrophic surface current anomalies with a climatological mean circulation field of a finite element model to construct nominal absolute currents over the Scotian Slope. In Fig. 16.4 we present the annual mean climatological currents interpolated onto the satellite ground tracks. These show an equatorward flow along the shelf edge and the upper continental slope. The crossover point of the ascending and descending tracks is the only location where we can estimate the total currents from altimetry alone. These points are also shown on Fig. 16.5 as yellow spots.

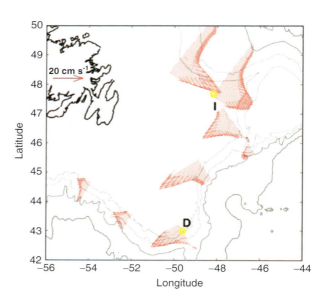

Fig. 16.4 Climatological annual mean surface currents over the Newfoundland Slope derived from Han's (2004) model results. On cross-track components are shown. The yellow point I (47.67°N, 48.19°W) and D (42.99°N, 49.60°W) represent particular crossover points of the satellite tracks (Han 2007)

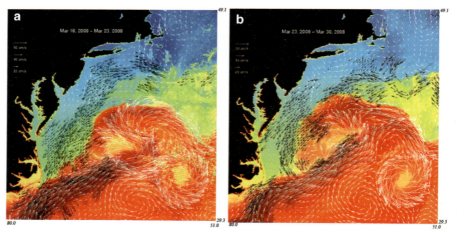

Fig. 16.5 (**a**) Week-long (Mar. 16–Mar. 23, 2008) composites of MCC currents (*black*) using both infrared and ocean color imagery and average satellite altimeter geostrophic currents (*white*) from all available altimeter satellites. (**b**) Week-long (Mar. 23–Mar. 30, 2008) composites of MCC currents (*black*) using both infrared and ocean color imagery and average satellite altimeter geostrophic currents (*white*) from all available altimeter satellites

The authors of the papers cited above use along-track altimeter data and are careful not to extend that data over the shelf. Groups such as the AVISO (Archiving, Validation, and Interpretation of Satellite Oceanographic data) project now make globally gridded SSH and geostrophic velocity fields available, combining data from multiple altimeters, as described by Le Traon et al. (2003). These fields often extend into coastal regions where the alongtrack SSH data have been flagged as unreliable, for several reasons (unresolved tides, poor tracking, and wet troposphere corrections). As such, they represent extrapolated values of SSH and velocity, which have been shown off the U.S. west coast to have no significant correlation to currents measured by current meters (Saraceno et al. 2008). On a wide continental shelf such as found in West Florida and the Gulf of Mexico the gridded AVISO SSH has been shown to be a good estimate of SSH seaward of the 50 m isobath (approximately 100 km from shore; Liu and Weisberg 2007).

To examine coastal currents over shallow shelves or where land interferes with the SSH signals and their corrections, altimeter surface velocity fields must be combined with current information derived from other sources. The primary method for the calculation of surface velocities from satellite data other than altimeters is to use sequences of surface temperature or color images, either tracking small scale features using the "maximum cross correlation" (MCC) technique (Emery et al. 1986) or inverting equations for the conservation of heat to solve for the advective term (Kelly 1989) in SST image sequences. Both these methods produce similar fields and can be formally combined with altimeter cross-track velocities (Kelly and Strub 1992).

To demonstrate the information contributed by each technique, detailed maps of the mean-weekly upper ocean circulation are computed from the altimeter heights and from composites of MCC currents (black vectors) for weekly time periods. In Fig. 16.5a and b two week-long surface vector fields from MCC and altimetry depicting the circulation just east of Chesapeake Bay are represented by white vectors, corresponding to the geostrophic altimeter currents, and black vectors, representing the total current estimated from sequential infrared and ocean color imagery. The black and white vectors agree well in the core regions of the northeastward flowing Gulf Stream, which splits into counter-rotating vortices. In the northern section of the cyclonic eddy that is being created to the north of the Gulf Stream, currents in both fields agree in flowing toward the southwest. Currents in the MCC field (black arrows) extend coherently over the wide shelf to the coast. The altimeter velocities, however, are more confused in the region next to the coast. Much work remains to be done in developing methods that make use of information from multiple sources in deriving optimal velocity fields over the wide shelves off the U. S. east coast.

16.3
The Gulf of Mexico and Caribbean Sea

The general circulation in the Gulf of Mexico and Caribbean Sea (sometimes called the Intra-Americas Sea) is described in reviews by Boicourt et al. (1998), Mooers and Maul (1998), and Lohrenz and Verity (2006). The topography includes a number of basins, separated by sills and shelves, and numerous islands. The primary circulation consists of water entering in the southeast from South America to Puerto Rico, continuing west through the Caribbean Sea and then north through the Strait of Yucatan into the Gulf of Mexico, where it forms the anticyclonic Loop Current and exits the Gulf through the Strait of Florida. East of the Mississippi outflow the currents along the wide northern Gulf and Florida Shelves are generally to the east and south around Florida; west of the Mississippi outfall they are toward the west. This is again, a greatly oversimplified picture.

Altimetry in the Gulf of Mexico, as off the U.S. east coast, has not been used to map the currents over the wide shelf. In the open ocean, however, altimetric measurements offer a unique satellite capability of monitoring the strong mesoscale circulations during periods of time when infrared and ocean color surface signatures are masked by the shallow layer that responds more directly to solar insolation and thus masks the circulation driven patterns just beneath it. The microwave radar character of the satellite altimetry makes it ideal for mapping strong surface meanders such as the Loop Current, which dominates the Gulf of Mexico upper layer circulation. The Loop Current pinches off eddies that impact the shelf currents in the northwest corner of the Gulf of Mexico; the Loop Current itself can sometimes (though rarely) impact the northern Gulf shelf.

Leben (2005) developed metrics to describe the Loop Current intrusions into the Gulf of Mexico and to quantify the eddy separation for the period from Jan. 1, 1993 through July 1, 2004. Time series of Loop Current statistics such as extent, boundary length, area, circulation, and volume are found by tracking the 17-cm contour in the daily surface height maps. This contour closely tracks the edge of the high-velocity core of the Loop Current (Fig. 16.6).

This altimetric Loop Current tracking is both qualitatively and quantitatively evaluated by direct comparisons with Loop Current thermal fronts observed in sea surface temperature imagery during times of good thermal contrast and cloud-free conditions. Eddy separation events (Fig. 16.7) are identified by the changes in the Loop Current length associated with the breaking of the tracking contour.

A total of 16 Loop Current eddies generated by the Loop Current over this 11.5-year period led to significant changes in the Loop Current extent during this same period of time. Statistical and spectral analyzes indicate that eddy separation occurs most frequently at 6, 9, and 11.5 months, with little or no power at the annual frequency. Significant low frequency spectral power is found at a period of 17 to 22 months, which is associated with a far southern retreat of the Loop Current and separation periods near 18 months.

A bimodal distribution of the retreat after eddy separation is identified that influences the duration period. When the Loop Current retreats below 25°N the average separation period following retreat is 16.2 months, much longer than the 5.5 month average for the cases where part of the Loop Current remains north of 25°N. Of greater interest is the near linear relationship between the Loop Current retreat and subsequent separation periods. Thus, the irregular retreat of the Loop Current after separation may be linked to why the Loop Current exhibits such an irregular eddy separation period. In a similar study of eddy variability farther south, Carton and Chao (1999) analyzed three years of T/P altimeter SSH anomalies in the Caribbean Sea. Large cyclonic and anticyclonic eddies were found and shown to be quite regular appearing at nearly 3-month intervals west of the southern Lesser Antilles. An example shown here in Fig. 16.8 compares the mid-January 1993 SSH anomalies from T/P with coincident SSH observed by the ERS-1 satellite altimeter.

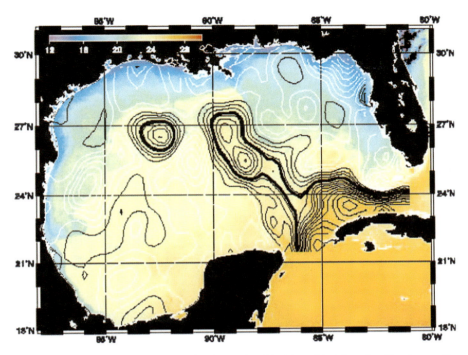

Fig. 16.6 Contoured sea surface height (*white* is negative, *black* is positive) from the altimetry map for Mar. 13, 2002 overlaid on the night-time composite SST image from the GOES 8 satellite for the same day (courtesy of Nan Walker, Louisiana State University). The 17-cm contour used to track the Loop Current boundary and eddies is shown by the bold line. The westernmost eddy separated from the Loop Current on Feb. 28, followed closely by a second eddy on Mar. 15 (Leben 2005)

Here the eddy sizes and locations are similar in both altimeters but the ERS-1 with its increased spatial resolution has eddy amplitudes that are increased by a factor of 2. The distribution of sea levels simulated by a 1/6° general circulation numerical model also bears some similarities to the T/P observations in structure but with increased amplitudes.

Many of these eddies dissipate in the coastal waters off Nicaragua half a year after they first appear. This life cycle is well reproduced in the general circulation model. Analysis of the model output suggests that these eddies are mainly limited to the thermocline and above, with little phase lag in the vertical. The westward propagation of these eddies is clearly shown in the Hovemueller diagram of Fig. 16.9. East of the Caribbean Sea level, anomaly variations are primarily annual; but within the Caribbean the sea level field is even more strongly dominated by westward propagating eddies than observed. The average propagation speed is 12 cm/s with typical spatial scales of 200–300 km.

Off the West Florida coast Liu and Weisberg (2007) proposed a method to estimate absolute SSH near the coast by integrating in situ coastal ocean observations (currents, hydrography, bottom pressure, coastal sea level, and winds) along an across-shelf transect. This method provides an independent coast SSH estimate to directly compare with calibrated or corrected coastal altimeter data. Based on a 3-year (1998–2001) time series

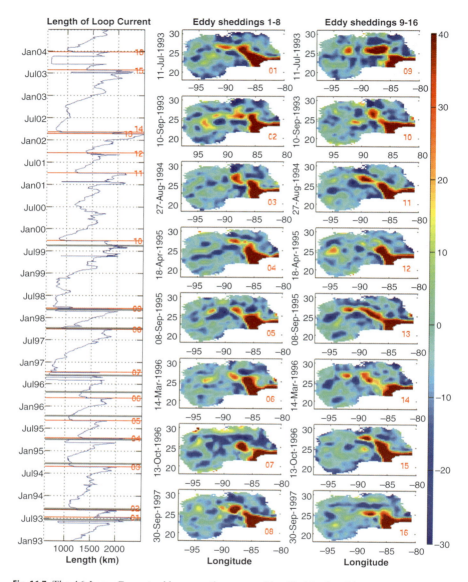

Fig. 16.7 The 16 Loop Current eddy separation events identified in the altimeter record are presented here. Sea surface height maps on the separation dates are shown in the panels to the right (note that the values above 40 cm and below −30 cm have been clipped). Eddy separation times were objectively determined by breaking of the 17-cm tracking contour, which causes a discrete change in the Loop Current length (*left panel*). The length time series is overlaid with red lines corresponding to the 16 events (Leben 2005). Gray lines show the ten separation times determined using the subjective method described in Sturges and Leben (2000)

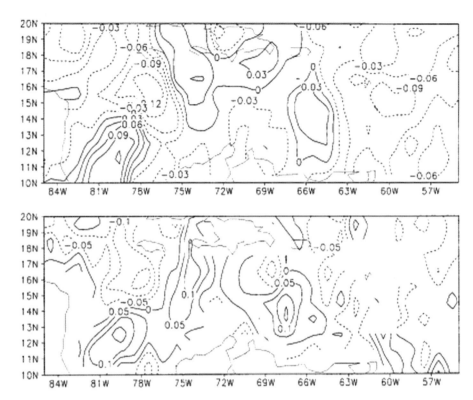

Fig. 16.8 Sea level anomaly in mid-January, 1993 as observed by T/P (*top*) and ERS-1 (*bottom*, Carton and Chao 1999)

comparison between satellite measured and dynamically estimated SST, this method appears to work well.

16.4
West Coast of North America

Circulation along the west coast of North America is summarized in reviews by Hickey (1998), Mackas et al. (2006), Royer (1998), Stabeno et al. (1998), and Badan-Dangon (1998). At the largest scales, the eastward North Pacific Current bifurcates offshore of the coast near 50°N, flowing into the poleward Alaska Current to the north and the equatorward California Current to the south. The California Current eventually turns westward off Mexico. Seasonal reversals also occur in the California Current, with poleward currents next to the coast in fall and winter. A cyclonic gyre often occupies the Southern California Bight, while the core of the California Current flows south along the western edge of the Bight. To the south of the California Current, a boundary current flows along Central

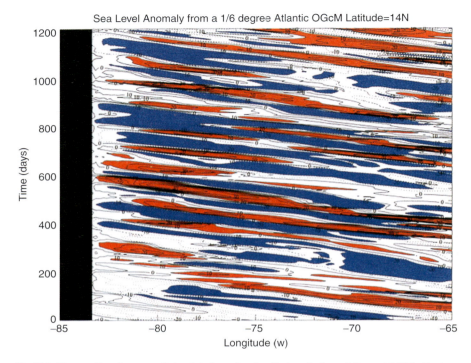

Fig. 16.9 Hovmueller diagram of simulated sea level with longitude and time at 14°N. Contour interval is 5 cm. Anomalies exceeding 10 cm are shaded (Carton and Chao 1999)

America and mainland Mexico, sometimes called the Costa Rica Current, with seasonal reversals that have not been systematically studied.

As for the U.S. east and Gulf coasts, altimetric studies have primarily looked at the mesoscale structure of currents and eddies in the core of the California Current, west of the continental shelf. Unlike the east and Gulf coasts, however, this shelf is narrow (20–50 km), allowing altimetric measurements to approach the traditional "coastal" zone. Recent investigations have attempted to push even closer to the coast, as described below.

The earliest altimetric studies used Geosat altimeter data to describe westward propagation of SSH signals in terms of annual Rossby Waves (White et al. 1990). Although these results were somewhat uncertain, due to the aliasing properties of the Geosat sampling, westward propagation of signals in the California Current has been demonstrated by numerous studies using T/P altimeter and in situ data, as well as by numerical models (Kelly et al. 1998; Strub and James 2000; Haney et al. 2001; Marchesiello et al. 2003). These signals originate next to the coast due to the annual change in the alongshore winds. Thus along the U.S. west coast, the annual and interannual variability in the coastal currents and water mass properties are transferred westward, into the deep ocean (Chereskin et al. 2000). This is unlike the east and Gulf coasts of North America, where open ocean currents and eddies may move onshore and affect the coastal circulation. Altimeter data played a key role in clarifying this aspect of coastal circulation along the west coast, as described below.

16.4.1
California Current System

The ability of along-track data from precision altimeters to resolve mesoscale SSH and surface velocity features in the California Current was evaluated by Strub et al. (1997) using a cluster of current meters, moored under an offshore crossover of the T/P satellite tracks. An ADCP was located directly under the crossover, while four vector-averaging current meters (VACM's) were located 2 km east and 15 km north, west, and south of the crossover. In addition, infrared satellite imagery from the Advanced Very High Resolution Radiometer (AVHRR) provided surface temperature fields that helped to interpret cross-track altimeter velocities and quantified the resolution of along-track SSH gradients.

Fig. 16.10 (left) shows geostrophic cross-track velocities along T/P tracks from one 10-day cycle, overlaid on an SST image from the same period. Extensive field surveys demonstrate that a coherent equatorward jet is found along the outer boundary of the cold water filaments and this jet is captured by the altimeter velocities, wherever the tracks cross that boundary. Because the SST features are composed of recently upwelled water, the colder temperatures indicate denser water, with lower dynamic heights. Using this relation, Strub et al. (1997) compare along-track SSH data to nearly coincident AVHRR SST data and conclude that the altimeter is capable of resolving SSH features with horizontal scales of 50–80 km in the along-track direction.

Comparing the geostrophic vector velocities from the altimeter at the cross-over at approximately 37.1°N, 127.6°W with the coincident ADCP and surrounding VACM currents leads to the following conclusions: (1) Ageostrophic Ekman currents (discussed further below) contribute to the measured currents above 48 m; (2) RMS differences of about 7–8 cm/s are found between the altimeter and ADCP measurements at 48 m; (3) Much of the RMS velocity difference comes from unresolved small-scale (15–30 km) variability in the oceanic currents – the altimeter velocities are more similar to spatially averaged (over the cluster) currents, with an RMS magnitude of 3–5 cm/s or less between the altimeter and spatially averaged current meter velocities; and (4) variance ellipses calculated from the altimeter velocities at the cross-over are in good agreement with those calculated from the subsampled current meter. The third point is consistent with the fact that a 62 km length scale is used to calculate the SSH gradient in the geostrophic altimeter velocity calculation.

These analyzes established confidence in the ability of the T/P altimeter to sample the SSH and velocity field in the California Current system. In addition, variance ellipses calculated over the entire California Current demonstrate that the velocity components are not isotropic but are polarized and that the energetic region of the California Current is surrounded and isolated by a region of lower energy (an "eddy desert") starting 500–700 km offshore. This provides evidence that the dominant energy within about 500 km of the coast is not brought into the region from a deep-ocean eddy field to the west, but must be generated within the California Current.

Adding fields of gridded SSH to the along-track data analyzes, Strub and James (2000) describe the seasonal cycle of SSH, currents and eddy statistics off the west coast of the U.S. and Baja California, extending the region considered to include the eastern North Pacific between the Equator and the Gulf of Alaska (Fig. 16.11). These studies build on an earlier look at gridded SSH fields using Geosat data from 1987 and 1988 (Strub and James 1995).

Fig. 16.10 (a) A field of T/P cross-track velocities from cycle 2 (October 5–11, 1992), characteristic of late summer and fall, overlain on an SST image from October 9, 1992. The current meter moorings are located under the crossover of tracks 54 and 69. (b) SSH contours from a combination of altimeter (T/P and ERS-1) and tide gauge data, centered on August 27, 1993 and overlaid on an AVHRR SST image from the same day. Four tide gauges are used, located at approximately 40.8°, 39.0°, 38.0°, and 36.6°N (Strub et al. 1997)

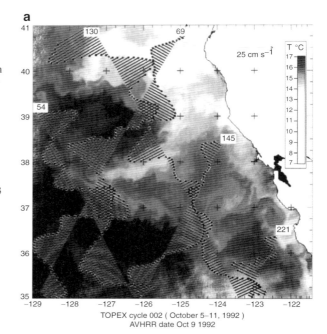

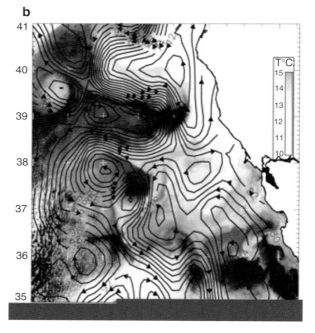

16 Satellite Altimetry Applications off the Coasts of North America

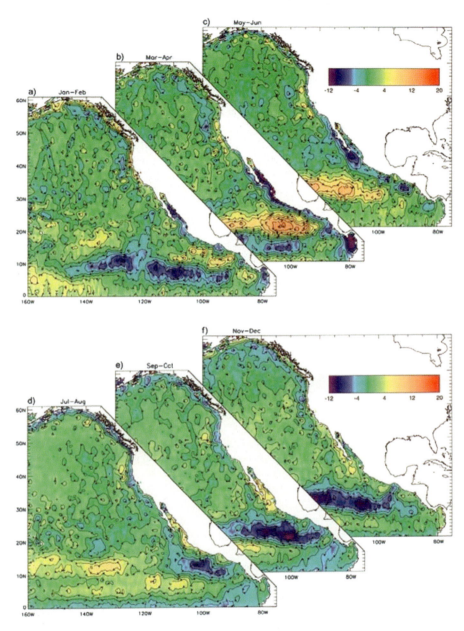

Fig. 16.11 Two-month seasonal SSH residuals based on Geosat, T/P, ERS-1, and ERS-2. Arrows indicate the direction of geostrophic flow and geostrophic surface transports are inversely proportional to the contour spacing. The contour interval is 2 cm (Strub and James 2002a)

Those fields demonstrated that the summer filaments and jets sampled off central California by numerous studies are part of a continuous jet extending from Washington to Southern California; they did not, however, establish the origin of that jet.

The later studies (Strub and James 2000, 2002a) use SSH data from a combination of multiple altimeters (Geosat, T/P and ERS-1) and tide gauges (TG), which allow the gridded SSH fields and alongshore transports to extend to the coast, bridging the traditional coastal gap in altimeter data. In these fields, the dominant variability in SSH and transport north of 20°N occurs along the boundaries, within several hundred kilometers of the coast. The eastward North Pacific Current is found to vary only weakly on seasonal time scales. Bograd et al. (1999) found a weak increase of flow in the North Pacific Current in winter on the basis of drifting buoy velocities, but this seasonal change was much weaker than the variations found in the boundary currents of the California Current System, consistent with the analysis of Strub and James (2002a).

The large-scale SSH fields along the U.S. West Coast clarify that the alongshore bands of high and low SSH form next to the coast, driven by seasonal downwelling- and upwelling-favorable winds, then move westward. Wave number spectra are useful in tracking the larger scales of the meanders formed near the coast as they move westward into the deep ocean, where the ambient spatial scales are smaller. Previous studies using drifter and altimeter data (Kelley et al. 1998) also document the offshore seasonal movement of eddy kinetic energy (EKE) generated in the vicinity of the seasonal jet. Interannual variability of the EKE field has been studied using wavelet analysis by Keister and Strub (2008). Although they could analyze details of the EKE field in the middle of California Current, 100–400 km from the coast, they could not look within the 100 km closest to the coast to verify the origin of the EKE in the very nearshore jets that form each year. However, combined with the appearance of the "eddy desert" farther offshore, the westward propagation of signals points to the seasonally developing coastal jet as the source of eddy energy in the mesoscale field of the California Current. A similar development, with different timing, occurs in the Gulf of Alaska, where the nearshore EKE is strongest in winter, moving offshore in summer. This implies that the EKE is generated in the poleward Alaska Current and Alaska Stream, which are strongest in winter next to the northern and northeastern boundary of the Gulf of Alaska.

The larger-scale fields of SSH in Strub and James (2002a) also demonstrate a coherent seasonal evolution of low and high coastal SSH signals that move from low to high latitudes around the basin. Low SSH values occur next to the coast off Central America during boreal winter, while coastal SSH values are high north of 30°N, from southern California to the Gulf of Alaska. These signals move counter-clockwise around the basin, reversing the fields by boreal summer, roughly following the expanding and contracting high and low pressure systems in the North Pacific atmospheric patterns.

The innovation in the above investigations is the use of tide gauge data to span the coastal gap in altimeter data. The same procedure was used in a subsequent study of the residual transports between the deep ocean and the coast of Oregon, Washington, and British Columbia during the 2000–2002 period, using T/P data (Strub and James 2003). Anomalous eastward transports into the boundary region along British Columbia between 52° and 54°N occurred first, between mid-2000 to mid-2001, followed by eastward transport into the boundary between 50° and 52°N in 2001. At the same time, anomalous transports were southward in the 300 km closest to the coast from 2000–2002, strongest in 2001 and the first half of 2002. These altimeter-derived transports suggest that both onshore and southward displacement anomalies off British Columbia and the Pacific Northwest contributed to the subarctic characteristics of the water observed off Oregon in this event. This supports the conclusions of Freeland et al. (2003) regarding the equatorward and/or onshore advection of subarctic water that created the cold, fresh, and rich anomalies found in the pycnocline off the Pacific Northwest in summer of 2002. Bograd and Lynn (2003) found a similar (fresh, cold and nutrient-rich) subsurface anomaly in the core of the California Current (150–350 km offshore) off southern California (33°–35°N) in summer 2002. If the anomaly reported by Freeland et al. (2003) at 50°N in mid-2001 moved 1,500 km south in 1 year, it would account for the Bograd and Lynn observations.

In another large-scale study with implications for the coastal currents in the boundaries of the North Pacific, Foreman et al. (2008) calculated dynamic ocean topography (DOT) fields using a high-resolution numerical model driven by seasonal climatologies of temperature, salinity, and wind stress. The in situ 3-D temperature and salinity climatologies were calculated from the available archive of CTD, XBT, and ARGO hydrographic data. The mean wind fields are from the NCEP reanalysis. Their average summer mean DOT field is shown in Fig. 16.12. No altimeter data are used in their model calculation.

Comparisons between their dynamic topography fields and those from JPL and AVISO (incorporating altimeter and satellite gravity measurements) showed reasonable agreement

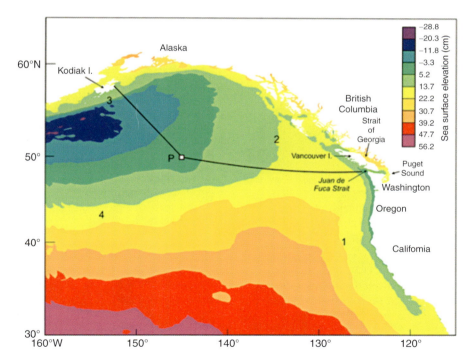

Fig. 16.12 Average summer mean dynamic topography (cm) computed from temperature and salinity climatology and NCEP wind stress. Consistent with T/P and Jason altimeter data the white square at the location of former weather station PAPA; the elevation is adjusted to have a value of 3.15 cm (Foreman et al. 2008)

in the deep ocean and poor agreement over the continental margins, within approximately 500 km of land. Their model compared well with alongtrack altimeter SSH differences near the margins, especially in the 200 km next to land where model and altimeter data show strong features (10–15 cm changes). Comparisons between the model and coastal tide gauge measurements were also good. The importance of this study for coastal analysis lies in the fact that models of this type could be run with very high resolution and extended into the 50 km next to the coast. They would then provide an absolute mean SSH field to replace the missing temporal mean in altimeter time series, resulting in time-variable fields of absolute SSH and currents in coastal domains.

Considering interannual time scales, Strub and James (2002b, c) used gridded and alongtrack altimeter data to describe the paths by which anomalous ocean conditions appeared along the mid-latitude west coasts of both North and South America (Strub and James 2002b, c) during the 1997–1998 El Niño. SSH data from the T/P tracks within ~100 km of the coast were used as proxy tide gauge data over the path between the equator and the mid-latitudes in both hemispheres, allowing lagged correlations between SSH at different locations, along with correlations between SSH and local winds to separate distant and local forcing effects. The results show that SSH signals traveled quickly down the coast of South America in May 1997. Off North America, large-scale signals traveled between the equator and the mouth of the Gulf of California (22°N) in May, and then slowly filled the Gulf until July–August, when they continued up the coast of North America (Fig. 16.13).

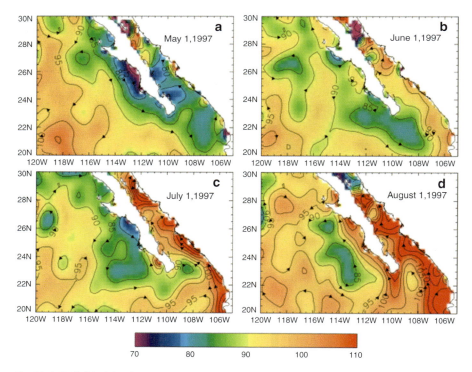

Fig. 16.13 Individual 35-day snapshots of SSH around the mouth of the Gulf of California and southern Baja California during May-August, 1997 (Strub and James 2002b)

This was not accomplished by classical "Kelvin wave" dynamics, but by a more advective process, by which they filled the mouth of the Gulf and created an artificial "coast" of high SSH, which allowed the northward movement to continue. North of 30°N, local fall and winter winds became more important in forcing the SSH anomalies, becoming the dominant factor off British Columbia and Alaska.

With the availability of gridded, multi-satellite altimeter SSH and velocity fields from the AVISO project, produced by SSALTO/DUACS, it has become less common for individual investigators to grid their own alongtrack data. The AVISO data do not incorporate tide gauge data, since they are global in scope, and do not accurately represent the region within 50 km of the coast as mentioned earlier. Even when investigators regrid the along-track data for a specific study, tide gauge data are often not available. For example, when Espinosa-Carréon et al. (2004) show seasonal and non-seasonal EOF's of SSH, SST, and satellite surface chlorophyll (CHL) off Baja California, most of the information in the 50–100 km next to the coast is provided by the SST and CHL. The SSH fields are more useful in describing the mesoscale circulation in the 500 km next to the coast. Off Oregon and Washington, Venegas et al. (2008) present similar seasonal and non-seasonal fields of SSH, SST, CHL, and surface winds, which extend over the narrow (20–50 km) shelf. One must realize, however, that both altimeter and scatterometer are extrapolated into the 30–50 km region next to the coast. The interpretation of the nearshore SSH fields is aided, in this study, by the SST data, with much higher spatial resolution and a close relation to the SSH data. This brings out both the limitations of the altimeter and the utility of multi-sensor analyzes in coastal ocean studies.

Even with the inclusion of TG SSH, the gridded SSH fields cannot resolve the smaller eddies (tens of kilometers) found next to the coast. This can be seen in Fig. 16.10 (right), where contours of gridded SSH are overlain on an SST field (Strub and James, 1997, unpublished). The SSH field provides reasonable representation of the meandering jet and associated eddies at distances of 100 km from the coast and with scales of approximately 100 km or more. It misplaces some of the eddies, as Pascual et al. (2006) have shown is the case when less than four altimeters are used, but the combination of SST and SSH provides a realistic view of the mesoscale circulation field. However there is a smaller (10–30 km) cyclonic eddy, next to the coast in the SE part of the field that cannot be resolved by the gridded field. This is typical of most comparisons between SST and SSH fields examined in this region and raises the question of whether there is additional information from the alongtrack altimeter data that could be used on smaller scales, next to the coast. It also reinforces the benefit of combining SSH data with SST fields, especially in upwelling regions where the SST fields serve as a proxy (albeit imperfect) for SSH.

One of the few attempts to look in detail at the SSH data along tracks approaching the coast of North America is reported by Saraceno et al. (2008), who investigated tracks off Northern California, Oregon, and Washington. Fig. 16.14 shows SSH data along two tracks approaching the Oregon coast (both ascending and descending T/P and Jason tracks approach the coast off the U.S. west coast), overlaid from a coincident SST image.

Standard along-track altimeter data are flagged as unreliable east of the cross-over, 50–60 km from the coast. In this case, we have simply replaced the wet tropospheric correction from the onboard radiometer with an estimate of the correction from the ECMWF atmospheric model water vapor field. Although the model correction is overly smooth, it allows us to see the

Fig. 16.14 SST (colors) and alongtrack SSH from the Jason-1 altimeter during June 27–29, 2005. Clouds are indicated by gray or speckled patterns in SST. Altimeter tracks are the "straight" lines and SSH is shown as a line plot "above" (east) or "below" (west) the tracks. Sharp drops in SSH (13 cm and 25 cm) are seen along the track moving from NW to SE, as it approaches the coast and passes over recently upwelled, colder, and denser water (Saraceno et al. 2008)

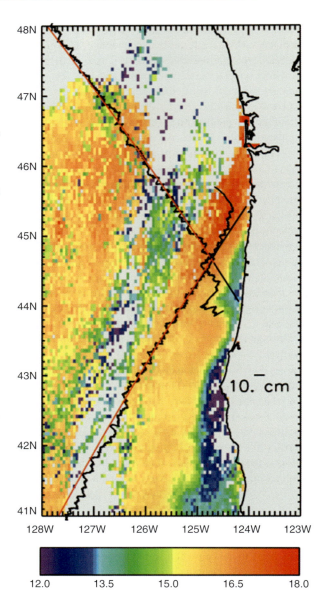

sharp drops in SSH along the track the moves from the NW to the SE, as the track moves over two fronts, one hidden by an upper layer of warm water and the second clearly visible in the SST field. Closer to the coast, other corrections (tides, tracking) eliminate the data at distances of approximately 20 km from the coast. The challenge is to first retrieve the along-track SSH data and then to use it in a realistic fashion to improve SSH fields next to the coast. Tracks are separated by distances and times too great to resolve the rapidly changing small spatial scales of coastal features. Use of the alongtrack data where and when it is sampled is one option, combining with SST and surface color satellite fields or in situ data. Assimilation into coastal circulation models is the other approach under investigation.

16.4.1.1
Ageostrophic Currents in the California Current System

One of the more difficult problems in an eastern boundary current region like the California Current is estimating the ageostrophic component of the current. Since the altimeter cannot measure this component of the current the approach here was to have an independent measure of the total current and subtract off the geostrophic current as estimated from the coincident altimeter data. Total currents were computed both from drifting buoys and infrared and ocean color satellite imagery (Matthews and Emery 2009). Data from Jan. 1994 to December, 2005 were used in this study and Fig. 16.7 shows a comparison between these three different types of current estimators for four fall examples in the central California Current region (between San Francisco and Los Angeles). During this time three satellite altimeters were operating (ERS-1, ERS-2, and T/P). The altimetry derived geostrophic currents are mapped as green, the image-derived currents by red, and the drifting buoy currents are in blue.

Throughout the sequence, the meander offshore of Point Reyes closes off to form a zonally elongated cyclonic eddy, while the alongshore jet is displaced westward. Closer examination of the coincident MCC (called MCC currents for the maximum cross-correlation method used to compute them) and altimetry observations shows a steady westward displacement of the MCC velocities relative to the geostrophic flow. The strong offshore current, relative to the geostrophic flow, suggests that a shallow locally wind driven current is possibly being measured by the MCC method.

This example time series demonstrates the close correlation between the image derived currents and those from the trajectories of the drifting buoys while the geostrophic currents computed from the altimeter data often agree but also exhibit places when they do not agree at all. These are likely instances of strong ageostrophic flow in this strongly wind-driven coastal region.

Time averaged MCC currents exhibit strong offshore jet-like currents extending offshore and coinciding with major topographic and coastline features such as promontories. These jets are not generally in the geostrophic altimeter currents. This suggests that these oceanic features are likely ageostrophically driven and are tied to local wind and bottom topography. In addition eddy statistics computed from the MCC velocities indicate dramatically different dynamics in the near-shore and offshore regions similar to those results discussed earlier for the California Current System. Near-shore regions show a dramatic decrease in Eddy Kinetic Energy (EKE), with major axes of variance that are modulated by the shape of the coastline. Offshore, EKE levels increase and major axes of variance are aligned in the meridional direction.

It should be observed that the mesoscale geostrophic currents depicted in particular by the satellite altimetry, is composed largely of eddies and meanders that dominate the California Current System during all seasons (Fig. 16.15).

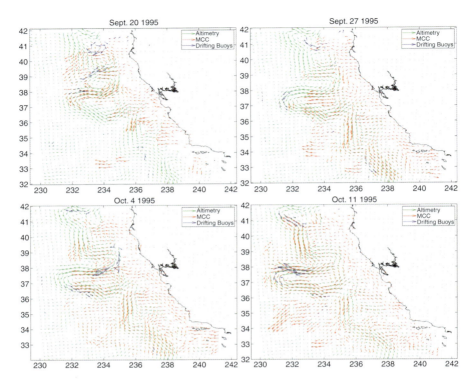

Fig. 16.15 A sample time series of current fields from satellite altimetry (*green*), drifting buoys (*blue*) and infrared + ocean color sequential images (red) using data from Sept. 20, 1995 to Oct. 11, 1995 (Matthews and Emery 2009)

All the different current observation methods exhibit clearly this mesoscale circulation structure. So these mesoscale features are characteristic both of a western boundary current region such as the Gulf Stream and an eastern boundary current region such as the California Current System. They are rather ubiquitous features whenever there is a moderately strong ocean surface current.

16.4.1.2
Numerical Modeling of the Structure and Dynamics of the California Current System

The mesoscale variability in the California Current System was modeled by Marchesiello et al. (2003) using the Regional Oceanic Modeling System (ROMS) model. The model spanned the entire California Current System with a 3.5 km horizontal resolution and was forced by mean-seasonal momentum, heat and water fluxes at the surface, and adaptive nudging to gyre-scale fields at the open water boundaries. The model equilibrium solutions show realistic mean and seasonal conditions of vigorous mesoscale eddies, fronts, and filaments.

Fig. 16.16 Longitude-time (Hovmueller) diagram of the sea level anomaly (SLA) along the northern boundary of the Pacific. The SLA was averaged within 2° south of the 1,000 m depth contour (Ueno et al. 2009)

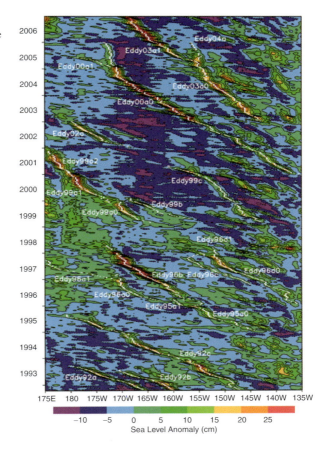

Seasonal variations of model sea surface height (SSH) from both northern and southern latitudes are plotted along with that from the T/P altimeter and the alongshore wind stress in Fig. 16.20. The maximum in wind-driven upwelling season is in summer in the northern region but earlier in spring off California (consistent with observations discussed by Strub and James 2000). Model and altimeter data both show nearly coincident wind and oceanic response suggesting that the coastal seasonal cycle is locally forced with no evident oceanic teleconnections between southern and northern regions.

The seasonal westward propagation in the upwelling region of the California Current System is demonstrated in Fig. 16.21 for both the SSH (left panel) and for surface and depth-integrated Eddy Kinetic Energy (EKE). Extrema in all these three quantities move offshore at a speed of about 2 cm/s in quasi-recurrent seasonal patterns. This is the speed that would be expected from baroclinic Rossby waves as pointed out for this region in observational studies (Kelly et al. 1998; Chelton and Schlax 2007).

The primary eddy generation mechanism is the baroclinic instability of upwelling, alongshore currents. There is a progressive movement of mean-seasonal currents and eddy energy offshore and downward into the oceanic interior in an annually recurrent cycle.

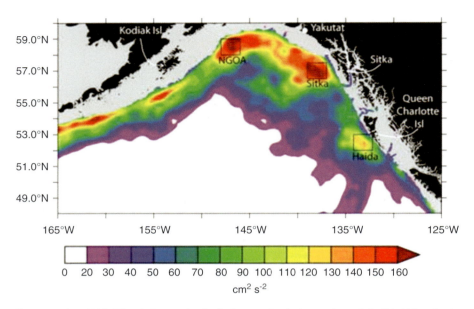

Fig. 16.17 Mean Eddy Kinetic Energy (cm^2 s^{-2}). Gray region is the continental shelf (<300 m depth. Ladd 2007)

16.4.1.3
Numerical Modeling of the Southern California Bight and Comparisons with Satellite Altimetry Data

The circulation in the Southern California Bight (SCB) is influenced by the background California Current farther offshore, by tropical remote forcing through the coastal waveguide alongshore, and local atmospheric forcing. The region is characterized by local complexity in its topography and the coastline. ROMS is applied to the SCB circulation, its multiple-scale variability, and interannual, seasonal, and intraseasonal variations. Three nested horizontal grid resolutions of 20, 6.7, and 1 km are forced with momentum, heat, and freshwater fluxes at the surface and nudging to gyre-scale reanalysis fields at the boundaries. This oceanic model starts in an equilibrium state from a multiple-year cyclical climatology run and is integrated from 1996 to 2003. The 1 km resolution results are compared against HR radar data, current meters, ADCP) data, hydrographic measurements, tide gauges, drifting buoy trajectories, satellite altimeter currents, and satellite radiometer data. T/P SSHA data are used to assess the low-frequency variability in the ROMS model simulation of the SCB circulation. One altimeter pass through the 1 km SCB model domain (Fig. 16.22) which are labeled as Lines 43, 199, and 206; this altimeter data paucity is why these authors decided not to compare their model outputs with any of the gridded altimeter products. SSH anomalies from these three ground tracks are plotted in Fig. 16.23 for both the T/P observations and for the ROMS model simulations. The T/P data are 10-day SSH anomalies relative to a 9 year (1993–2001) mean sea surface. Fortunately the 1997–1998 ENSO event is contained within this short time series. ROMS SSH anomalies are also referenced to the same 9-year mean sea surface. There are clear interannual events that are

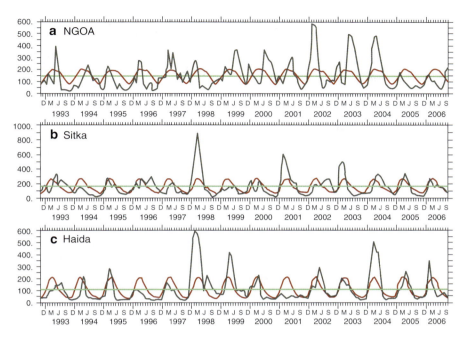

Fig. 16.18 Time series of monthly averaged (*black*), monthly climatological mean (*red*), and mean (*green*) EKE (cm^2/s^2) averaged over the areas around Kodiak (**a**), Sitka (**b**), and Haida (**c**) (Ladd 2007)

coherent between the tracks and often between T/P and ROMS. One that is not even coherent is in late 2000 when a low-SSH in T/P is not matched in ROMS. The largest positive SSH anomalies are during 1997–1998 in both T/P and ROMS. The ROMS anomalies are only slightly larger than the T/P data.

16.5
Alaskan Stream

Eddies also dominate the Alaskan Stream region (Ueno et al. 2009) again reflected in satellite altimetry data between 1992 and 2006 and profiling float data. Fifteen long-lived eddies were identified and tracked during their westward migration through the North Pacific (Fig. 16.16).

Of these, three eddies were present at the start of the satellite altimetry observations, three formed in the eastern Gulf of Alaska off Sitka, Alaska, and four were first detected at the head of the Gulf of Alaska near Yakutat, Alaska. The other five eddies formed along a line between 157° and 169°W. Most of the eddies formed in the Gulf of Alaska decayed before exiting the Gulf while those eddies formed farther south managed to cross 180° reaching the western subarctic gyre. Four of the five-southern eddies formed under negative or weakly positive wind stress curls. Comparison of the southern eddy propagation with the local bottom topography suggests that eddies propagated faster over steeper

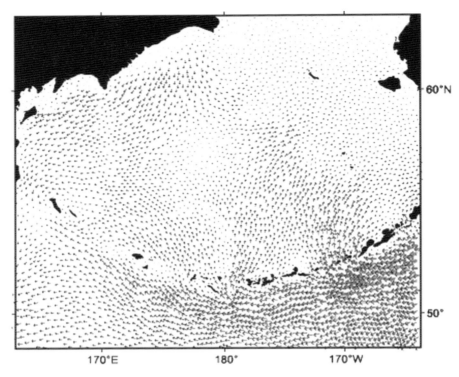

Fig. 16.19 Vertically integrated energy flux for M_2 (post assimilation). Each full shaft in a multi-shafted vector represents 100 KW/m. The latitudinal scale has been selected and vectors have been sub-sampled so that the minimum separation is 20 km (Foreman et al. 2006)

bottom slopes although eddy speeds observed were slower than those predicted by the topographic planetary wave dispersion relation.

In a related study Ladd (2007) used satellite altimetry data to map eddy kinetic energy in the Gulf of Alaska (Fig. 16.17).

High EKE is associated with the Alaskan Stream which is seen as a narrow structure of high EKE that reflects and strongest and most unstable component of the circulation in the Gulf of Alaska. Note the pockets of high EKE in the waters surrounding Kodiak, Sitka, and Haida (Queen Charlotte Islands). Of these the Sitka and Haida are eddy formation regions while the area around Kodiak is an area through which all eddies formed in the eastern Gulf of Alaska are forced to pass and interact with bottom and coastal topography. Thus, the variability in the Sitka and Haida regions is due to the formation of mesoscale eddies and the variability at Kodiak is due to the passage of mesocale eddies formed farther to the east. The SLA pattern at Sitka suggests that eddies form here, flow seaward then turn northward in the northwestward flowing Alaska Current. Eddies formed in the Haida region do not move as much offshore and appear to simply advect to the north.

Time series of monthly averaged (black) EKE's at these three locations exhibit a very clear climatological mean signal that has a maximum in March and minimum in November (Fig. 16.18).

Fig. 16.20 Near-shore seasonal cycle of simulated and T/P and ERS SSH anomalies and alongshore wind stress. The variables are averaged over 100 km wide alongshore strips (*top*) between 40° and 46°N and (*bottom*) between 34.5° and 40°N (Marchesiello et al. 2003)

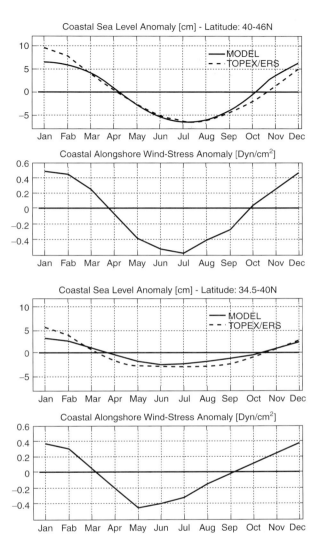

Maximum EKE occurs as the downwelling winds start to weaken. The strong downwelling favorable winds in December/January drive a coastal current flowing northwest with associated cross-shelf density gradients and storage of potential energy. As these winds weaken periods of upwelling favorable winds become more likely. These wind driven events convert this potential energy to kinetic energy through baroclinic instability (Ikeda et al. 1986).

These time series also clearly indicate the effects of El Nino on the altimeter derived EKE. There are however, some weaker El Nino events that do not show up in these time series. The black monthly averaged line in Fig. 16.19 exhibits a peak at the 1997–1998 El Nino event. It is interesting that this peak is present at Haida and Sitka but by the time the SLA had reached the Kodiak region the El Nino signal had dissipated.

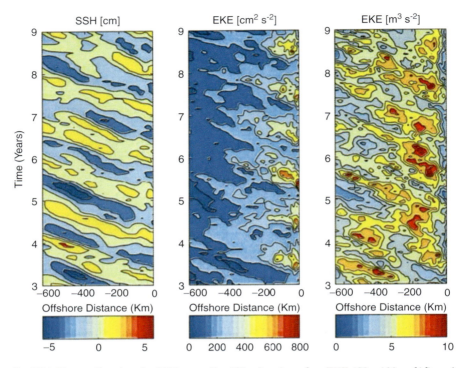

Fig. 16.21 Hovmoeller plots for SSH anomalies (CI = 2 cm), surface EKE (CI = 100 cm²/s²), and depth-integrated EKE (CI=1 m³s⁻²) from the model with 5 km resolution and averaged alongshore in the upwelling region (34.5°–43°N; Marchesiello et al. 2003)

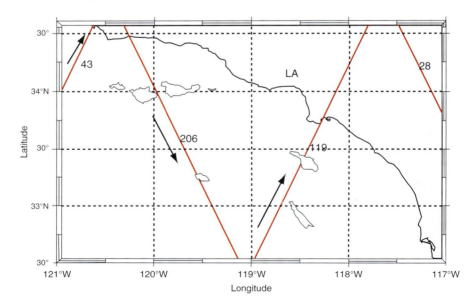

Fig. 16.22 T/P satellite tracks over the SCB. The arrows are the direction the satellite moves along the track (Marchesiello et al. 2003)

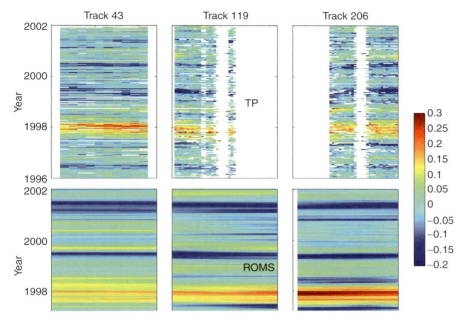

Fig. 16.23 Along-track SSH anomalies (m) from T/P (*top row*) and ROMS (*bottom row*), each relative to its own time and track mean. The distance on the abscissa starts from the beginning point of a track in the domain (Fig. 16.22; Marchesiello et al. 2003)

16.6
Tidal Energy in the Bering Sea

Although unresolved tidal motions limit the use of altimetry in coastal regions over shallow continental shelves, tides and their energy also represent important oceanographic processes for mixing and transport. Thus, the generation of strong tidal motions by topographic features in shallow regions and the propagations of that energy along the coast or into the deep ocean is a topic of interest in coastal oceanography. Combinations of tidal models and altimeter measurements have proved useful in studying the flux of tidal energy, as demonstrated by Foreman et al. (2006). In this case, tidal harmonics are computed from T/P altimetry and assimilated into a barotropic, finite element model of the Bering Sea and its boundaries. After evaluating its accuracy through comparisons with independent bottom pressure gauges, the model is used to estimate energy fluxes through each of the Aleutian Passes and the Bering Strait and to construct an energy budget for the major tidal constituents. The finite element model does not conserve mass locally and this gives rise to an additional error term in the energy budget, whose contribution is significant for the model prior to assimilation but is reduced substantially with the assimilation technique.

The M_2 tidal constituent is estimated to have the largest net energy flux into the Bering Sea but the sum of the first three largest diurnal constituents is found to be greater than the sum of the largest three semi-diurnals. The vertically integrated energy flux for the M_2 tide is presented here in Fig. 16.19 as a case after altimeter data assimilation. Samalga and Amutka Passes in the Aleutians are found to be the primary conduits for the influx of semi-diurnal energy while Amchitka Pass is the primary conduit for diurnal energy. A significant portion of the diurnal energy is seen to exist in the form of continental shelf waves trapped along Bering Sea slopes.

16.7
Future Work

Most of the problems facing coastal oceanography off North America are common to other regions in the world. First the nature of the altimeter waveforms in the coastal ocean need to be investigated and a retracking method selected and implemented. It is widely accepted that the last 10 km close to the shore requires retracking of the altimeter waveform. Retracked altimetry needs to be compared with higher resolution coastlines and land digital elevation models (DEM's). It is not clear that there is one single retracking method that will be useful in all regions. Instead the retracking method may vary with geographic location as the fairly narrow shelf of the west coast may not impact the altimeter waveform too greatly while the much wider shelf off the east coast may contribute to changes in the altimeter waveforms that have to be corrected. It is important that these corrections be developed to the best of our ability because the coastal zone is where the bottom topography changes very rapidly.

The atmospheric moisture correction for the altimeter path-length is also a problem in the coastal ocean where the large spot sizes of the passive microwave radiometers are easily contaminated by land and restrict their utility to farther offshore than many coastal oceans. Many land decontamination methods have been recently developed and used to reprocess earlier altimeter data. It is difficult to evaluate these reprocessed water vapor data as there are few independent direct measurements of atmospheric water vapor in the coastal atmosphere. Improvements have been and are being made to the passive microwave radiometers to increase spatial resolution making it possible to bring this measurement closer to the coast. Still there is need to continually improve the water vapor corrections in the coastal zone. Finally there is real need to get the tidal corrections very precise in the coastal zone as here the tidally forced currents can become non-linear and difficult to forecast. Thus, we need the best possible calibration/validation systems to provide corrections to the altimeter data in the dramatic space/time variability in the coastal zone such as that carried out by Liu and Weisberg (2007) for the West Florida coast.

References

Badan-Dangon A (1998) Coastal circulation from the Gal´apagos to the Gulf of California. In: Robinson A, Brink KH (eds) The sea, vol 11, pp 315–343

Bograd SJ, Lynn RJ (2003) Long-term variability is the southern California current system. Deep-Sea Res II 50:2355–2370

Bograd SJ, Thomson RE, Rabinovich AB, LeBlond PH (1999) Near-surface circulation of the northeast Pacific Ocean derived from WOCE-SVP satellite-tracked drifters. Deep-Sea Res II 46:2371–2403

Boicourt WC, Wiseman WJ, Valle-Levinson A, Atkinson LP (1998) Continental shelf of the southeastern United States and the Gulf of Mexico: in the shadow of the western boundary current. In: Robinson AR, Brink KH (eds) The sea, vol II: the global coastal ocean-regional studies and syntheses. Wiley, New York, pp 135–182

Carton JA, Chao Y (1999) Caribbean Sea eddies inferred from TOPEX/Poseidon altimetry and a 1/6° Atlantic Ocean model simulation. J Geophys Res 104:7743–7752

Chelton DB, Schlax MG, Samelson RM, de Szoeke RA (2007) Global observations of large oceanic eddies. Geophys Res Lett 34:L15606. doi:10.1029/2007GL030812

Chereskin TK, Morris MY, Niiler PP, Kosro PM, Smith RL, Ramp SR, Collins CA, Musgrave DL (2000) Spatial and temporal characteristics of the mesoscale circulation of the California current from eddy-resolving moored and shipboard measurements. J Geophys Res 105:1245–1269

Dong S, Kelly KA (2003) Seasonal and interannual variations in geostrophic velocity in the Middle Atlantic Bight. J Geophys Res 108(C6):3172. doi:10.1029/2002JC001357

Emery WJ, Thomas AC, Collins MJ, Crawford WR, Mackas DL (1986) An objective procedure to compute advection from sequential infrared satellite images. J Geophys Res 91(color issue): 865–879

Espinosa-Carréon TL, Strub PT, Beier E, Ocampo-Torres F, Gaxiola-Castro G (2004) Seasonal and interannual variability of satellite-derived chlorophyll pigment, surface height and temperature off Baja California. J Geophys Res 109:C03039

Foreman MGG, Cummins PF, Cherniawsky JY, Stabeno P (2006) Tidal energy in the Bering Sea. J Mar Res 6:797–818

Foreman MGG, Crawford WR, Cherniawsky JY, Galbraith J (2008) Dynamic ocean topography for the northeast Pacific and its continental margins. Geophys Res Lett 35:L22606. doi: 10.1029/2008GL035152

Freeland HJ, Gatien G, Huyer A, Smith RL (2003) Cold halocline in the northern California current: an invasion of subarctic water. Geophys Res Lett 30(3):114L

Han G (2004) Scotian Slope circulation and eddy variability from TOPEX/Poseidon and frontal analysis data. J Geophys Res 109:C03028. doi:10.1029/2003JC002046

Han G (2007) Satellite observations of seasonal and interannual changes of sea level and currents over the Scotian Slope. J Phys Oceanogr 37:1051–1065. doi:10.1175/JPO3036.1

Han G, Tang CL (2001) Interannual variations of volume transport in the western Labrador Sea based on TOPEX/Poseidon and WOCE data. J Phys Oceanogr 1:199–211

Han G, Hannah CG, Smith PC, Loder JW (1997) Seasonal variation of the three-dimensional circulation over the Scotian Shelf. J Geophys Res 102:1011–1025

Haney RL, Hale RA, Dietrich DE (2001) Offshore propagation of eddy kinetic energy in the California current. J Geophys Res 106:11,709–11,717

Hickey BM (1998) Coastal oceanography of western North America from the tip of Baja California to Vancouver Island. In: Robinson AR, Brink KH (eds) The sea, Vol 11. Wiley, New York

Ikeda M, Emery WJ, Mysak LA (1986) Seasonal variability of the meanders in the California current system off Vancouver Island. J Geophys Res 89:3487–3505

Keister JE, Strub PT (2008) Spatial and interannual variability in mesoscale circulation in the northern California current system. J Geophys Res 113:C04015. doi:10.1029/2007JC004256

Kelly KA (1989) An inverse model for near-surface velocity from infrared images. J Phys Oceanogr 19:1845–1864

Kelly KA, Gille ST (1990) Gulf Stream surface transport and statistics at 69°W from the Geosat altimeter. J Geophys Res 95:3149–3161

Kelly KA, Strub PT (1992) Comparison of velocity estimates from advanced very high resolution radiometer in the coastal transition zone. J Geophys Res 97(C6):9653–9668

Kelley KA, Beardsley RC, Limeburner R, Brink KH, Paduan JD, Chereskin TK (1998) Variability of the near-surface eddy kinetic energy in the California Current based on altimeter, drifter and moored current meter data. J Geophys Res 103:13067–13083

Ladd C (2007) Interannual variability of the Gulf of Alaska eddy field. Geophys Res Lett 34:L11605. doi:10.1029/2007GL029478

Leben RR (2005) Altimeter-derived loop current metrics. In: Sturges W, Lugo-Fernandez A (eds) Circulation in the Gulf of Mexico: observations and models. AGU Geophys Monograph, vol 161, pp 181–202

Le Traon PY, Faugere Y, Hernandez F, Dorandeu J, Merttz F, Ablain M (2003) Can we merge GEOSAT-follow-on with TOPEX/Poseidon and ERS-2 for an improved description of the océan circulation? J Atmos Oceanic Technol 20:889–895

Liu Y, Weisberg RH (2007) Ocean currents and sea surface heights estimated across the West Florida Shelf. J Phys Oceanogr 37:1697–1713

Loder JW, Boicourt WC, Simpson JH (1998) Overview of western océan boundary shelves. The Sea, vol 11. Wiley, New York

Lohrenz SE, Verity PG (2006) Regional oceanography: Southeastern United States and Gulf of Mexico. In: Robinson AR, Brink KH (eds), The Sea, Vol 13, Wiley, New York

Mackas DL, Peterson WT, Ohman MD, Lavanlegos BE (2006) Zooplankton anomalies in the California current system before and during the warm ocean conditions of 2005. Geophys Res Lett 33:L22S07. doi:10.1029/2006GL027930

Marchesiello P, McWilliams JC, Shchepetkin AF (2003) Equilibrium structure and dynamics of the California current system. J Phys Oceanogr 33:753–783

Matthews DK, Emery WJ (2009) Velocity observations of the California current derived from satellite imagery. J Geophys Res 114:C08001. doi:10.1029/2008JC005029

Mooers CNK, Maul GA (1998) Intra-Americas Sea coastal ocean circulation. In: Robinson AR, Brink KH (eds) Global coastal ocean. The sea, vol 11, Wiley, New York, pp 183–208

Pascual A, Gaugere Y, Larnicol G, Le Traon PY (2006) Improved description of the ocean mesoscale variability by combining four satellite altimeters. Geophys Res Lett 33:L02611. doi:10.1029/2005GL024633

Royer TC (1998) Coastal processes in the northern North Pacific. In: Robinson AR, Brink KH (eds) The sea, vol 11, chap 13, pp 395–414. Wiley, New York, 1062 p

Saraceno M, Strub PT, Kosro PM (2008) Estimates of sea surface height and near-surface alongshore coastal currents from combinations of altimeters and tide gauges. J Geophys Res 113: C11013. doi:10.1029/2008JC004756

Stabeno PJ, Bond NA, Kachel NB, Salo SA, Schumacher JD (1998) On the temporal variability of the physical environment over the south-eastern Bering Sea. Fisheries-Oceanogr 10:81–98

Strub PT, James C (1995) The large-scale summer circulation of the California current. Geophys Res Lett 22:207–210

Strub PT, James C (2000) Altimeter-derived variability of surface velocities in the California currne system: 2. Seasonal circulation and eddy statistics. Deep-Sea Res II 47:831–870

Strub PT, James C (2002a) Altimeter-derived surface circulation in the large-scale NE Pacific Gyres. Part 1: seasonal variability. Prog Oceanogr 53:163–183

Strub PT, James C (2002b) Altimeter-derived surface circulation in the large-scale NE Pacific Gyres. Part 2: 1997–1998 El Niño anomalies. Prog Oceanogr 53:185–214

Strub PT, James C (2002c) The 1997–1998 oceanic El Niño signal along the southeast and northeast Pacific boundaries – an altimetric view. Prog Oceanogr 54:439–458

Strub PT, James C (2003) Altimeter estimates of anomalous transports into the northern California current during 2000–2002. Geophys Res Lett 30(15):8025. doi:10.1029/2003GL017513

Strub PT, Chereskin TK, Niiler PP, James C, Levine MD (1997) Altimeter-derived variability of surface velocities in the California current system: 1. Evaluation of TOPEX altimeter velocity resolution. J Geophys Res 102:12727–12748

Sturges W, Leben RR (2000) Frequency of ring separations from the loop current in the Gulf of Mexico: a revised estimate. J Phys Oceanogr 30:1814–1819

Townsend D, Thomas A, Mayer L, Thomas M, Winlan J (2006) Oceanography of the northwest Atlantic Shelf, The sea: the global coastal ocean: interdisciplinary. Harvard University Press, Harvard

Ueno H, Freeland HJ, Crawford WR, Onishi H, Oka E, Sato K, Suga T (2009) Anticyclonic eddies in the Alaskan Stream. J Phys Oceanogr 00:1–18

Venegas RM, Ted Strub P, Beier E, Letelier R, Thomas AC, Cowles T, James C, Soto-Mardones L, Cabrera C (2008) Satellite-derived variability in chlorophyll, wind stress, sea surface height, and temperature in the northern California Current System. J Geophys Res 113:C03015. doi:10.1029/2007JC004481

Vignudelli S, Cipollini P, Roblou L, Lyard F, Gasparini GP, Manzella G, Astraldi M (2005) Improved satellite altimetry in coastal systems: case study of the Corsica Channel (Mediterranean Sea). Geophys Res Lett 32:L07608. doi:10.1029/2005GL022602

White WB, Tai CK, DiMento J (1990) Annual Rossby wave characteristics in the California current region from the Geosat exact repeat mission. J Phys Oceanogr 20:1297–1310

Evaluation of Retracking Algorithms over China and Adjacent Coastal Seas

L. Yang, M. Lin, Y. Zhang, L. Bao, and D. Pan

Contents

17.1	Introduction	454
17.2	Data Set	457
17.3	Waveform Analysis	458
	17.3.1 Ocean to Land	458
	17.3.2 Land to Ocean	460
17.4	Retracking Algorithms	461
	17.4.1 Inter-comparison of Jason-1 SLA at Crossover Points	461
	17.4.2 Jason-1 Ground Comparison of SSH	463
	17.4.3 Jason-1 Ground Comparison of SWH	466
17.5	Conclusions and Perspectives	469
References		470

Keywords Altimeter • China coastal seas • Waveform retracking

---

L. Yang (✉) and D. Pan
State Key Laboratory of Satellite Ocean Environment Dynamics, Second Institute of Oceanography, State Oceanic Administration, Hangzhou 310012, China

and

Computer Science Department, Nanjing University of Science and Technology, Nanjing 210094, China
email: yle1024@gmail.com

M. Lin and Y. Zhang
National Satellite Ocean Application Service, State Oceanic Administration, Beijing 10008, China

L. Bao
Institute of Geodesy and Geophysics, China Academy of Sciences, Wuhan 430077, China

Abbreviations

AVISO   Archiving, Validation and Interpretation of Satellite Oceanographic data
DORIS   Dual-frequency Doppler Orbitography and Radio positioning by Satellite
ECMWF   European Centre for Medium-Range Weather Forecasts
GDR   Geophysical Data Record
GM   Geodetic Mission
J1   Jason-1
MSS   Mean Sea Surface
OCOG   Offset Center Of Gravity
Rms   Root Mean Square
SGDR   Sensor Geophysical Data Record
SLA   Sea Level Anomaly
SSH   Sea Surface Height
SSN, SYA, HLD, CST, XMD, RZH, WZH, NZU, FCN, YGH, DFG
      names of tide gauge stations
Std   Standard Deviation
SWH   Significant Wave Height
TG   Tide Gauge

17.1 Introduction

Satellite altimetry products are used in research and routine operations to monitor ocean circulation and to improve our understanding of the role of the ocean on climate and weather. However, most coastal altimetry data are unusable because the measurements are affected by the noisier radar echoes from the land. In addition, the presence of land makes it impossible to perform some geophysical corrections, such as wet troposphere range correction, ocean tide correction, and atmospheric high-frequency forcing correction, as used for the open ocean. Dealing with the problem of radar altimeter measurements in coastal areas is our primary concern in this study.

Hancock and Lockwood (1990) discussed the influence of an island on the Geosat waveform retracking technique and on the retrieved geophysical parameters. They found that islands can locally affect sea surface height (SSH) by more than 4 m. Deng et al. (2002) investigated the behavior of ERS-2 and Poseidon altimeter waveform data in coastal regions and estimated that in a region up to 22 km from the coast of Australia the altimeter range may be poorly estimated by the on-satellite tracking software.

Different retracking techniques (see the chapter by Gommenginger et al., this volume) have been developed, such as the Ocean retracking algorithm (Callahan and Rodriguez 2004) for the open ocean, which is used to obtain better range estimates in ground processing than those obtained with the onboard tracking algorithm; the Beta5/9 algorithm (Martin et al. 1983), the OCOG (offset center of gravity) technique (Wingham et al. 1986), the Ice-2 algorithm (Legresy and Remy 1997), S/V algorithm (Davis 1993) for continental ice sheet; and the threshold algorithm (Bamber 1994; Laxon 1994; Davis 1996) for sea ice. Although those retracking algorithms are proposed for specific surfaces, many researchers employ these algorithms for other surfaces, such as land (Berry et al. 1997) and inland lakes and rivers (Berry 2006; Frappart et al. 2006), to derive proper altimetry ranges. For the coastal ocean, Brooks et al. (1998) analyzed the TOPEX/Poseidon waveforms when the altimeter transited from water to land and from land to water. They showed that waveform retracking can be used to extend the altimeter-derived sea surface topography several kilometers shoreward. However, the threshold waveform retracking method employed by them is not automated for selecting a threshold level. Berry et al (1997) and Freeman and Berry (2006) developed an expert system to classify the coastal waveforms into ten categories, and then four retracking methods were used correspondingly. By combining the Beta5 and OCOG retracking algorithms to process TOPEX/Poseidon and ERS altimeter waveforms in Chinese coastal seas, Bao et al. (2004, 2007) showed that retracking can improve the accuracy of estimated SSH and resolution in the coastal region. Guo et al. (2006) and Hwang et al. (2006) proposed a retracker, named the improved threshold retracker, based on the leading edge detection and sub-waveform extraction. Results using the altimeter Geosat/GM data showed that the improved threshold retracker is better than the Beta5 and threshold methods in the coastal sea around Taiwan Island in comparison with geoid model results.

China and adjacent seas (14–45°N, 105–130°E) are typical marginal seas located on the western Pacific Ocean. China Sea includes the South China Sea, the East China Sea, the Yellow Sea, and the Bohai Sea. The South China Sea encompasses an area from Singapore to the Strait of Taiwan of around 3,500,000 km^2, which includes the shallow shelf, the shelf break, and the deep basin. The East China Sea and the Yellow Sea connect to the Japan/East Sea through the Korea/ Tsushima Strait to the northeast and to the South China Sea through the Taiwan Strait to the southwest. A line running northeastward from the mouth of the Yangzi River to the southwestern tip of Korea is defined as the boundary between the Yellow Sea to the north and the East China Sea to the south. The Bohai Sea is a semi-enclosed coastal ocean on the northeastern China Sea with 78,000 km^2 in area. Fig. 17.1 shows the coastal line and bathymetry of this region. The China Sea with the adjacent seas is chosen for this coastal study with altimeter waveforms for the following reasons. First, the coastal line is extremely irregular with multiple islands and inlets, especially in the southeastern part of the Chinese mainland. There are various kinds of coastal terrain types: hill, mountain, coral reef, mangrove, plain, delta, beach, gritty, muddy, cape, and fault (Chen 2007). Second, most areas are over the continental shelf with shallow water depth of 50–200 m, except the southern part of the South China Sea. The Chinese Continental Shelf occupies the area running southwestward from Taiwan Strait to the Gulf of Tonkin and connects to the Sunda Shelf and Gulf of Thailand in the south through a narrow steep slope around the Vietnamese coast. Third, the flow in the China Sea is dominated by strong semidiurnal M2 tidal currents with superimposition of semidiurnal S2 and diurnal O1 and K1 currents. The amplitude of the M2 is as high as 1.8 m in the Taiwan Strait and the coast of the East China Sea, while in the coasts of the Bohai Sea and South China Sea, the amplitude of M2 tide is relatively weak (Fang et al. 2004).

Coastal dynamics of the China Sea is complex because of influences from the freshwater input, the depth contours, the shoreline, Kuroshio, the monsoon, and so on. According to historical knowledge and traditional in situ measurements, multiple coastal currents along the Chinese Coast exist (Liu et al. 2005), for example, the Minze Coastal Current, the Yellow Sea Coastal Current, and the South China Sea Coastal Current. The Minze Coastal Current, along the coast of the East China Sea, is strong and southward during winter with the speed of around 20 cm/s, and weak and northward in the summer. From 1405 to 1433,

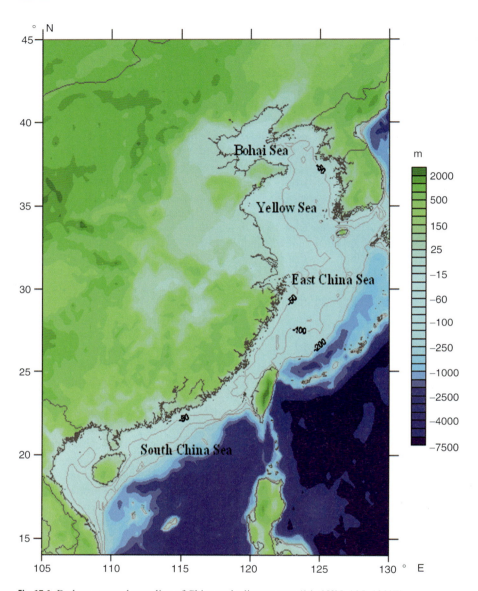

Fig. 17.1 Bathymetry and coastline of China and adjacent seas (14–45°N, 105–130°E)

the Chinese explorer He Zheng, who sailed to Arabia, East Africa, India, Indonesia, and Thailand seven times with a huge fleet of ships, chose to begin his voyages in winter possibly to use the "highway" of the Minze Coastal Current. Moreover, the water exchange between the Minze Coastal Current with cold low-salinity water and the Taiwan Warm Current with high-salinity water, a branch of Kuroshio, forms Zhoushan Fishery, which is the biggest coastal fishery in China. Pu et al. (2004) investigated the alongshore current near the Qingdao and Jiaozhou Bay (Yellow Sea) based on many years of monthly mean sea level measurements from tide gauges. Yang et al. (2003) showed the characteristics and mechanisms of coastal currents near the Qiongzhou Strait (South China Sea) using the in situ drifting of bottle track data, and measurements during cruise investigations. However, in order to study the inter-seasonal and inter-annual variability of currents and flows in large regions, in situ measurements are far from sufficient, and remote sensing is a practical solution. By dealing with problems of coastal altimetry, dynamical processes in the coastal China Sea might be revealed in new detail, which will be helpful to improve our understanding of the mechanism of coastal circulation, shipping safety, fishery, and so on.

17.2 Data Set

The Jason-1 SGDR (Sensor Geophysical Date Record) product was provided by AVISO (Archiving, Validation and Interpretation of Satellite Oceanographic data), which includes not only the geophysical parameters, but also additional information of tracker data, instrumental corrections and 20-Hz 104 sampled waveform data (Zanife et al. 2001). J1 is the follow-on to TOPEX/Poseidon, whose main features J1 has inherited. The ground track velocity of J1 is 5.8 km/s. The ground track pattern repeats, within +1 km, every 9.92 days (Picot et al. 2003). Thirty-one cycles (March 2006 to February 2007, cycle155 to cycle188) of J1 measurements over the ocean off China are used. The data of cycles 177, 178, 179 are missing because the altimeter was in safe mode. J1 ground tracks cross our study region (14–45°N, 105–130°E) with 21 passes (9 ascending passes and 12 descending passes) every cycle (Fig. 17.2).

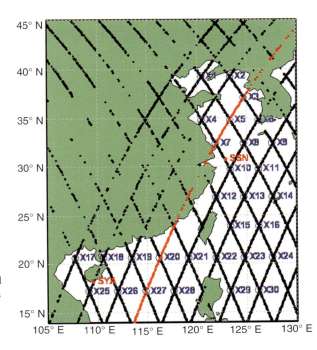

Fig. 17.2 Twenty-one ground tracks and 30 crossovers of J1 in region of (14–45°N, 105–130°E). The red ground track is pass 229. Red points indicate the locations of tide gauge stations (SYA and SSN)

The shoreline (1:250,000 map scale) and ETOP02 (2' resolution) bathymetry data were provided by the NOAA (National Oceanic and Atmospheric Administration). In situ tide gauge measurements were provided by the Chinese National Marine Data and Information Service of the State Oceanic Administration.

17.3 Waveform Analysis

The typical ocean return echo corresponds to the Brown model (Brown 1977) waveform. The altimeter determines the Range and SWH (significant wave height) by tracking the midpoint and estimating the slope of the leading edge of the return waveform, respectively.

In a coastal region, because the reflection from land or islands is much stronger than that from the ocean, when an altimeter ground track approaches, recedes, or runs parallel to the coastline, even though the altimeter's nadir point is over the sea, the altimeter tends to track the off-nadir return from the land. The affected distance is related to the altimeter footprint size. J1, at Ku band (13.575 GHz), has an antenna beamwidth of 1.28°, with an average altitude of about 1,336 km. The beam-limited footprint radius can be calculated; it is about 12.3 km. The radius of the pulse-limited footprint r is derived from (Chelton et al. 2001),

$$r = \left[\frac{(c\tau + 2H_w)h}{1 + h/R} \right]^{1/2} \quad (17.1)$$

where h is the satellite altitude, 1,336 km; R is the radius of the earth, 6371 km; c is light speed; τ is pulse duration, 3.125 ns; H_w is the sea surface wave height. For example, if the wave height is 2 m, the pulse-limited footprint radius is 2.33 km. Because the beam-limited footprint radius is much larger than the pulse-limited footprint radius, land returns might be recorded in the waveform when the distance of J1 ground track to land is about 10 km, and the land return might contaminate the entire waveform when the distance is less than the radius of the pulse-limited footprint.

The waveforms of J1 cycle164 pass 229 (pass with red colour in Fig. 17.2) moving from southwest to northeast across the eastern mainland of China were examined. Fig. 17.3 shows the ground track and three series of successive 20-Hz waveforms. Each waveform was separated by 0.05 s, corresponding to about 300 m in the ground, given the ground speed of the satellite is 5.8 km/h. On the basis of visual inspection, waveforms in Fig. 17.3c are the typical land returns, which are sharp and strong, while waveforms in Fig. 17.3b and d consist of returns from both ocean and land.

17.3.1 Ocean to Land

A series of 20-Hz waveforms of J1 approaching the Fujian province of China from the ocean is shown in Fig. 17.3b. The beginning waveforms around latitude 23.68°E are

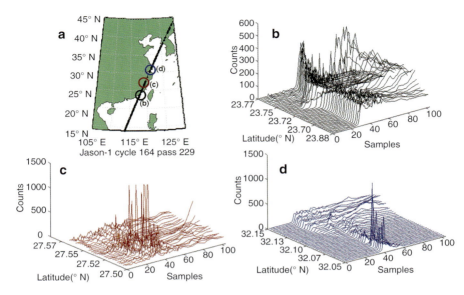

Fig. 17.3 Series of 20-Hz return waveforms of J1 cycle 164 pass 229 across east mainland of China. (**a**) Black line indicates the ground track of 229. Three circles (*black, red, blue*) indicate the locations where waveforms in (**b**), (**c**), and (**d**) are recorded, respectively; (**b**) A series of waveforms when J1 moved from ocean to land; (**c**) A series of waveforms when J1 is over land; (**d**) A series of waveforms when J1 moved from land to ocean

typical ocean return and the distance from land is 10.63 km. As the J1 ground track gets much closer to the land, the stronger land return appeared on the right of the waveform and gradually moved forward. At latitude 23.74°E (distance to land, 5.03 km), the leading edge of the ocean return is completely contaminated by the land return. Fig. 17.3b is a typical series of waveforms showing the process of land return contamination as the ground track approaches land from the ocean. Brooks et al. (1998) described this kind of waveform series in detail.

However, this phenomenon was not observed for all J1 ground tracks transitioning from ocean to land. Land effects on the altimetry waveforms also depend on the type of terrain along the shoreline. The terrain of the area of 23.68°N 117.57°E where pass 229 arrived is muddy coast, with relatively weak reflection. The spherically expanding pulse emitted from the altimeter antenna reached the ocean surface first and the leading edge of the waveform is recorded. The expanding circle wave then reached the brighter land surface, and another leading edge with higher energy appeared in the late samples of the waveform. As the ground track approached closer to land, the stronger land return gradually moved forward to the leading edge of the waveform. Therefore, as long as the leading edge contributed by the ocean return is still visible, the SSH of this ground track can be extended several kilometers shoreward by retracking the waveforms. If the slope of the coastal terrain is steep and land reflection is strong, the pulse reaches the land surface before the ocean, forming the leading edge of the waveform. The trailing edge of waveform is noisy with many spikes, which might be contributed by the ocean return or by the undulation of the land terrain. Because the land reflection is much stronger than the ocean, the SSH is difficult to recover in such case.

17.3.2 Land to Ocean

A series of waveforms was recorded when J1 cycle 164 pass 229 left the mainland of China at the Yangtze River Delta to the Ocean, as shown in Fig. 17.3d. As the ground track moved northeastward, the spike waveforms were transformed to typical Brown model ocean waveforms. However, track loss occurred on other ground tracks, especially those from land to ocean, because of the undulation of terrain: no waveforms can be used in these coastal areas. It is found that J1 loses track more often from land to ocean than from ocean to land, which can be explained by the fact that after the J1 lost track over land (Fig.17.3c) it did not switch back to the track return signal immediately, even though the altimeter nadir point was already over the ocean. Table 17.1 shows the flagged distances of J1 ground tracks in our study region for cycle 164 according to the alt_echo_type when the altimeter approached and left land, respectively. The mean flagged distances of J1 are 15.5 km from land to ocean, and 12.3 km from ocean to land.

To summarize, because the microwave return signal from the land is much stronger than that from the ocean, the altimeter waveforms in the coastal sea are different from those in the open ocean, and more complex. The effects vary with the type of coastal terrain, shoreline shape, and the moving direction of the altimeter in relation to the shoreline.

Table 17.1 The flagged distances (in kilometers) of tracks in cycle 164 of J1 according to the alt_echo_type when the altimeter approached and left the land respectively

Pass No.	Flagged distance (km)		Pass No.	Flagged distance (km)	
	Land to ocean	Ocean to land		Land to ocean	Ocean to land
001	5.9	3	190	17.7	
077	5.9	5.9	012	5.9	
	35.4	5.9	088	17.7	
153	23.6	11.8	164	47.2	17.7
	17.7	17.7		5.9	
229		11.8	240	11.8	
	11.8	23.6	062	11.8	11.8
051		35.4		11.8	
	17.7	53.1	138	11.8	29.9
038	5.9			5.9	
114	29.9	29.9	214	11.8	
	11.8		Mean	15.5	12.3

17.4
Retracking Algorithms

Accurate range estimates are obtained using refined procedures known as altimeter waveform retracking. As long as the ocean signal is still distinguishable in the waveform, although it might be flagged as unusable, these waveforms can be used to derive the coastal SSH. Many retracking algorithms have been developed for specific surfaces (see the chapter by Gommenginger et al., this volume). The Envisat altimeter (ESA 2005), employing four retrackers (Ocean, Ice-2, OCOG, and Threshold), is aimed at operationally providing accurate height estimates over various surfaces. However, the four results are provided without statement of their accuracy.

We performed Ocean (Callahan and Rodriguez 2004), Ice-2 (Legresy and Remy 1997), OCOG (Wingham et al. 1986), Threshold (Bamber 1994), and Beta5 (Martin et al. 1983) retracking algorithms using 1-year J1 waveforms (20 Hz) in China and adjacent seas. The Ocean, Ice-2, and Beta5 retracking algorithms are implemented as statements of references. The difference between the OCOG and Threshold retracking algorithms is the way to determine the amplitude. There are different threshold values (10, 25, 50, or 75%) for Threshold and OCOG algorithms. Davis (1996) stated that the threshold of 10% is appropriate for ice surfaces dominated by volume scattering, while the threshold of 50% is appropriate when surface scattering dominates the return waveform for the Threshold algorithm. Bamber (1994) suggested that the threshold of 25% is significantly less noisy than the 50% when there is a significant surface slope. The thresholds of 50% for the Threshold algorithm and 25% for the OCOG algorithm had been adopted in our study.

17.4.1
Inter-comparison of Jason-1 SLA at Crossover Points

Twenty-one passes forming 30 crossover points were selected (Fig. 17.2). According to the distance from the nearest land or island, crossovers are categorized into four classes, two crossovers whose distances from land are less than 10 km, seven crossovers between 10 and 50 km, eight crossovers between 50 and 100 km, and 13 crossovers larger than 100 km. The sea level anomaly (SLA) at each ascending and descending pass at the 30 crossovers was calculated as

$$SLA = H_{alt} - H_{range} - \Delta H_{atm} - \Delta H_{SS} - \Delta H_{tide} - \Delta H_{gp} - MSS$$
$$\Delta H_{atm} = h_{wet\_tropo} + h_{dry\_tropo} + h_{iono}$$
$$\Delta H_{tide} = h_{ocean\_tide} + h_{solid\_earth\_tide} + h_{loading\_tide} + h_{pole\_tide}$$
$$\Delta H_{gp} = h_{inv\_bar} + h_{hf\_fluc} \tag{17.2}$$

where:

H_{alt} is the satellite altitude;

H_{range} is the range from altimeter to the ocean surface, which is derived from the waveform retracking algorithms;

ΔH_{atm} is the atmosphere correction, which consists of three parts, wet troposphere correction h_{wet_tropo}, dry troposphere correction h_{dry_tropo}, and ionosphere correction h_{iono}. The ECMWF (European Centre for Medium-Range Weather Forecasts) wet troposphere corrections and DORIS (Dual-frequency Doppler Orbitography and Radio positioning by Satellite) ionosphere corrections were used instead of wet tropospheric corrections measured by Jason Microwave Radiometer measurements, and ionospheric corrections measured by altimeter; ΔH_{ss} is the sea state correction;

ΔH_{tide} is the tidal correction, which includes four parts, geocentric ocean tide correction h_{ocean_tide}, loading tide correction $h_{loading\_tide}$ from GOT00.2 model, pole tide correction h_{pole_tide}, solid earth tide correction $h_{solid\_earth\_tide}$

ΔH_{gp} represents the other geophysical corrections, which include inverse barometer correction h_{inv_bar}, and high-frequency wind respond correction h_{hf_fluc}; and MSS is the mean sea surface from the CLS01 model. Except for H_{range}, values of other parameters are the same for different retracking algorithms, and can be found in SGDR product.

To compare the four retracking algorithms quantitatively, the bias, root mean square (rms) and standard deviation (std) of the SLA difference of ascending and descending passes at the crossovers were calculated (Table 17.2). The numbers in brackets in column one are numbers of compared pairs in that category. N represents the percentages of valid result (−2.5 m<SLA<2.5 m).

When the distances are larger than 50 km, the most accurate retracking algorithm is Ocean, then Threshold, OCOG, and Ice-2. The rms and std of the Ocean retracking algorithm are about 0.23 m and 0.23 m, respectively. When the altimeter ground tracks get close to the land, the accuracy and the valid result percentage of the Ocean algorithm decreased. When the distance is between 10 and 50 km, the accuracy of the Threshold algorithm is better than that of the Ocean algorithm. The valid-result percentages of the Ocean algorithm decrease as the coast becomes closer, from 92.6% when the distance is larger than 100 km to 90.3% when the distance is between 10 and 50 km. Furthermore, when the distance is less than 10 km, the Ocean retracking algorithm rejects most of the data. This is because a significant number of ocean return waveforms do not conform to the Brown model as the coast is approached. Thus, the standard processing algorithm cannot be used in the coastal altimeter waveform process.

Among the other three retracking algorithms, the Ice-2 retracking algorithm performs less well than the other two. The valid result percentage is small and error is large. Because the Ice-2 algorithm fits the leading edge and trailing edge of the waveform separately, empirical identification of the leading and trailing edge and setting of a fitting step and range of parameters are crucial to the retracking results. The valid result percentages of OCOG and Threshold algorithms are both 19.4% when the distance is less than 10 km, considerably larger than the 1.6% of the Ocean algorithm. When the distance is between 10 and 50 km, the rms and std of the Threshold algorithm are 0.395 and 0.353 m, respectively, which are much better than 0.95 and 0.951 m of OCOG. However, when the distance is less than 10 km, the accuracy of the Threshold algorithm decreased and the OCOG algorithm is better than the Threshold algorithm since both the rms and std of the OCOG are smaller, namely 1.179 and 0.845 m, respectively.

To summarize, comparisons of the retracked SLA of four retracking algorithms at the crossover points between the ascending and descending passes show that (1) when

Table 17.2 Number of valid results of four retracking algorithms: Ocean, Ice-2, OCOG, and Threshold, as well as bias, root-mean square-deviation and relative root-mean-square deviation of the difference between ascending and descending passes for four retracking algorithms. The number in the brackets in column one is the number of compared pairs in that category

Distance			Ocean	Ice2	OCOG	Threshold
d(62) <10 km		N	1.60%	0.00%	19.40%	19.40%
		bias(m)	0.05	NaN	0.823	0.814
		rms(m)	0.05	NaN	1.179	1.286
		std(m)	0	NaN	0.845	0.995
10 km <d(217)< 50 km		N	90.30%	6.90%	74.20%	90.30%
		bias(m)	0.184	0.875	0.194	0.178
		rms(m)	0.417	1.328	0.951	0.395
		std(m)	0.375	0.998	0.931	0.353
50 km<d(248)< 100 km		N	92.30%	9.30%	81.50%	93.10%
		bias(m)	0.037	0.892	0.158	0.037
		rms(m)	0.232	1.086	0.895	0.326
		std(m)	0.229	0.62	0.88	0.324
d(403)>100 km		N	92.60%	8.20%	79.90%	92.30%
		bias(m)	0.084	0.764	0.148	0.11
		rms(m)	0.238	1.106	0.73	0.286
		std(m)	0.223	0.794	0.714	0.263

the distances of crossovers from the land are larger than 50 km, the Ocean algorithm is the best, because the ocean return waveforms are not yet affected by land; (2) When distances are between 10 and 50 km, both the Ocean algorithm and Threshold retracking algorithm are appropriate; and (3) When the distances are less than 10 km, the OCOG retracking algorithm is appropriate considering both the accuracy and valid result percentage.

17.4.2
Jason-1 Ground Comparison of SSH

The hourly-averaged SSH observations of two tide gauge stations have been compared with the retracking results of 1-year J1 measurements. The locations of tide gauge stations and the corresponding Jason-1 ground tracks are shown in Fig. 17.2. Station SYA and SSN are located at the South China Sea and East China Sea, respectively. The sea surface variation of SSN is much bigger than that of SYA, because of the tidal effect. The detailed information is given in Table 17.3.

Table 17.3 Information of tide gauge stations and the corresponding J1 ground tracks

Name	Latitude (°N)	Longitude (°E)	Distance to Land (km)	J1 Pass no.	Distance to J1 track (km)
SSN	30.75	122.8	1.897	062	4.08
SYA	18.23	109.53	0.053	138	3.73

The sea surface height is calculated as

$$SSH = H_{alt} - H_{range} - \Delta H_{atm} - \Delta H_{ss} - h_{solid\_earth\_tide} - h_{loading\_tide} - h_{pole\_tide}$$
$$\Delta H_{atm} = h_{wet\_tropo} + h_{dry\_tropo} + h_{iono} \qquad (17.3)$$

where the terms are the same as in Eq. 17.2. The solid earth tide correction $h_{solid\_earth\_tide}$, tide loading correction $h_{loading\_tide}$, and pole tide correction h_{pole_tide} are movements of the earth's crust that should be subtracted before comparing the SSH measured by J1 and tide gauge stations. Furthermore, the SSH calculated by Eq. 17.3 is relative to the reference ellipsoid, whereas the SSH measured by tide gauge station is relative to the local mean sea level. Because buoys at the tide gauge stations are not equipped with GPS antenna, the difference of absolute sea surface variation cannot be derived. Alternatively, we examined the relative temporal variation of the altimetry and in situ SSH.

Fig. 17.4 shows the SSH derived from J1 measurements of pass 062 and 077 near tide gauge station SSN and SYA using five retracking algorithms. For all the algorithms, when the nadir point of altimeter moved close to land (the water depth decreases), the altimetry SSH decreased significantly. Fig. 17.5 is 1 year SSH of five retrackers and in situ measurements at location of 30.86°N 122.72°E with distance to land of 1.15 km near station SSN. The correlation between in situ SSH and Ocean, Ice2, OCOG, Threshold, Beta5 retrackers are 0.946, 0.802, 0.952, 0.941, 0.935, respectively. The std are 0.267, 1.283, 0.259, 0.279, and 0.294, respectively. The correlation and std between the in situ and altimetry SSH are similarly calculated at nadir points along the ground tracks of Jason-1 near the tide gauge stations. Fig. 17.6 and 17.7 present the results for stations SSN and SYA without the Ice-2 algorithm, respectively. It can be seen that as the ground track approaches land, the correlations increase, and std decreases. However, when the nadir points get too close to land, or even over the land, the N decrease significantly and values of correlation and std are out of the reasonable range. We averaged the correlation and std of three records (60 waveforms) within reasonable range around the stations. The results are given in Table 17.4. Because tide gauge station SSN is located on an island, results approaching and leaving the station are given separately in Table 17.4 under the columns for SSN (north) and SSN (south). It can be seen that the correlation coefficient of OCOG is the highest, whereas standard deviation of OCOG is the lowest, for both stations. Furthermore, the std of Station SSN are much higher than those of SYA for all retracking algorithms, because the influence of tides at SSN is much stronger than that at SYA when the locations of the tide gauge station and J1 ground track do not coincide.

Thus, SSH comparisons between J1 and in situ measurements show that the OCOG algorithm is the best-performing among the five algorithms, and the absolute value of std depends locally on sea surface variation extent.

17 Evaluation of Retracking Algorithms over China and Adjacent Coastal Seas

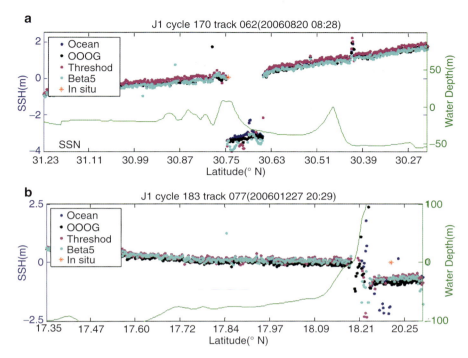

Fig. 17.4 The along-track SSH derived from J1 measurements using five retracking algorithms of cycle 164 pass 062 around tide gauge station SSN (**a**) and of cycle 164 pass 077 around station SYA (**b**). The green lines are water depth

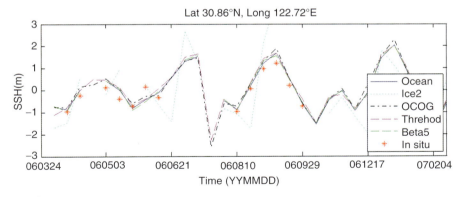

Fig. 17.5 The time series of SSH derived from J1 measurements using five retracking algorithms, and in situ measurements at 30.86°N 122.72°E near station SSN

However, the smallest std of the OCOG retracking algorithm in Table 17.4 is 9.57 cm. To achieve more accurate coastal altimetry SSH, proper coastal wet troposheric and tidal corrections are needed.

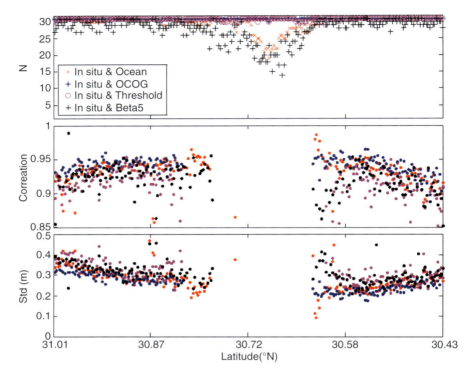

Fig. 17.6 The number of valid results, correlation coefficient, and standard deviation between tide gauge station measurements and results of Ocean, OCOG, Threshold, and Beta5 retracking algorithms for station SSN

17.4.3 Jason-1 Ground Comparison of SWH

SWH is an important property measured by satellite altimetry, besides SSH and ocean surface backscatter coefficient, which is used to validate wave models or to be assimilated into the global wave forecasting systems (Lefevre and Skandrani 2003). The altimetry SWH is determined from the slope of the waveform leading edge without correcting for the effects of atmosphere and sea state, which is a different procedure to that for altimetry SSH. Therefore, it is possible to assess retracking algorithms by comparing the SWH without considering problems of correction terms in coastal seas, such as wet troposphere range correction, ocean tide correction, and atmospheric high-frequency forcing correction. However, except for the Ocean retracking algorithm, other algorithms generate only SSH without SWH by determining the midpoint position of the waveform leading edge. Because the midpoint position is one of three unknown parameters (leading edge midpoint position, leading edge slope, and amplitude) for implementing the Ocean retracking algorithm, SWH can be generated for every algorithm by combining that algorithm with the Ocean retracker. For example, we first performed the OCOG retracking algorithm to derive the midpoint positon, which was then used as a known parameter to initiate the Ocean retracker with only two unknown parameters. Therefore, the additional SWH of other retrakers could be derived.

17 Evaluation of Retracking Algorithms over China and Adjacent Coastal Seas   467

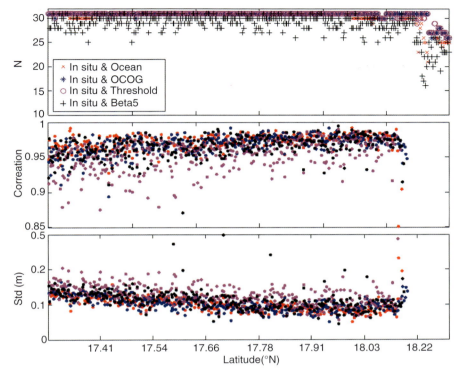

Fig. 17.7 The number of valid results, correlation coefficient, and standard deviation between tide gauge station measurements and results of Ocean, OCOG, Threshold, and Beta5 retracking algorithms for station SYA

Table 17.4 The averaged correlation coefficient and standard deviation between tide gauge station measurements and results of Ocean, Ice-2, OCOG, Threshold, and Beta5 retracking algorithms of 60 waveforms for station SSN (north), SSN (south), and SYA

	Retracker	SSN (north)	SSN (south)	SYA
Correlation	Ocean	0.929	0.935	0.943
	Ice-2	0.397	0.287	0.421
	OCOG	0.943	0.946	0.974
	Threshold	0.916	0.850	0.941
	Beta5	0.866	0.836	0.875
Standard deviation (cm)	Ocean	32.1	23.9	17.3
	Ice-2	197.0	228.0	105.9
	OCOG	28.8	22.3	9.57
	Threshold	34.9	45.4	17.6
	Beta5	45.9	44.9	18.1

Table 17.5 Information about tide gauge stations whose SWH measurements were used for comparison with altimetry SWH

Name	Latitude (°N)	Longitude (°E)	Jason-1 Pass No.	Distance to Jason-1 track (km)
HLD	40.70	120.98	077	41.75
CST	37.38	122.68	153	47.40
XMD	36.05	120.42	062	55.48
RZH	35.40	119.55	062	47.26
WZH	28.00	120.88	240	47.96
NZU	20.95	110.58	077	11.92
FCN	21.62	108.33	001	17.40
YGH	18.52	108.68	038	8.608
DFG	19.10	108.63	038	27.23

For comparisons, in situ SWH measurements of nine tide gauge stations were selected. The information of locations of tide gauge stations and corresponding Jason-1 ground tracks are given in Table 17.5. Collocated pairs of in situ and altimetry SWH measurement are determined when the time difference is less than 1 h. The altimetry SWH are averaged over 50 km along track.

Because of the limited in situ SWH measurements, the number of colocated pairs is only 73. Fig. 17.8 shows the results. The SWH of OCOG, Threshold, and Beta5 retracking algorithms are systematically higher than others. The correlations are 0.3269, −0.0406, −0.1323, −0.0689, −0.0957, and 0.2702 for Ocean, Ice-2, OCOG, Threshold, Beta5 retracking algorithms and

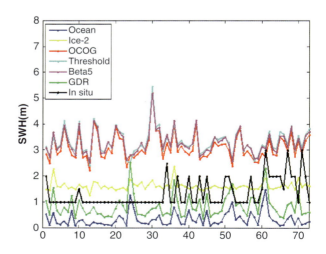

Fig. 17.8 The SWH derived from five retracking algorithms and in situ tide gauge station measurements

GDR product, respectively. It can be seen that all the correlations are unexpectedly low, and results of the Ocean retracking algorithm and GDR product are much better than others. There are two reasons to account for the result. First, the in situ SWH from the tide gauge stations are inappropriate to be used to compare the retracking algorithms, because the tide gauge stations are located on the land or islands, where the altimetry waveforms are already badly contaminated. Second, the time and space window(1h, 50 km) of comparisons is too large, whereas the effected distance of the altimetry waveform by the land signal is only around 10km. Therefore, to validate the retracking of the SWH on coastal seas, SWH measurements from near-shore buoy whose distance to land is around 10 km should be used, as well as smaller time and space comparison windows.

17.5
Conclusions and Perspectives

To explore the capability of coastal altimetry, the Ocean, Ice-2, OCOG, Threshold, and Beta5 waveform retracking algorithms were implemented using the 1-year (March, 2006 to February, 2007; cycle 155 to cycle 188) J1 data of ocean water off China (14°N–45°N,105°E–130°E). We compared the retracked SLA differences of J1 ascending and descending passes at crossovers, as well as the SSH variation and SWH between J1 measurements and ground tide gauge station measurements.

A significant number of ocean return waveforms do not conform to the Brown model near coasts. On the basis of the waveform analysis, the effect of land on the coastal altimetry waveform varies with the type of coastal terrain, shoreline shape, and whether the altimeter is approaching or going away from the shoreline. When the slope of the coastal terrain is small and the reflection is not very strong, the process of land contamination on ocean surface return can be found in a series of waveforms. When the slope of the coastal terrain is steep and reflection is strong, J1 altimeter loses track: this happens much more often when J1 is transiting from land to ocean than from ocean to land.

As comparisons demonstrated, the Ocean algorithm is most accurate when the distance of the altimeter nadir measured point is more than 50 km from the nearest land. But the accuracy and the valid result percentage of the Ocean algorithm decrease rapidly and the Ocean algorithm cannot provide valid results when the altimeter ground track gets close to the land, especially when the distance is less than 10 km. Thus, the standard processing algorithm cannot be used in the coastal altimeter waveform processing. Among the other four retracking algorithms, the Ice-2 retracking algorithm performs least well because its retracking results depend on empirical knowledge of the return waveform. The Beta5 retracker is based on a parametric waveform model, which is not appropriate to the complex coastal waveforms. Both Threshold and OCOG retracker are easy to implement, but the OCOG is better than Threshold retracker, considering both the accuracy and valid result percentage of the Jason-1 SLA crossover inter-comparisons and Jason-1 in situ SSH comparisons.

Calculating the SWH by combining two retrackers is a way to evaluate different retracking algorithms without taking into account the error of the atomosphere and sea state corrections. Small time and space comparison windows, as well as proper in situ SWH measurements, should be used considering that the distance of land effects on altimetry waveform is only around 10 km.

Together with further validation for longer term in situ SSH and SWH data, future work hopes to include waveform recognition and extraction, which might be helpful to retrack coastal altimetry waveforms.

However, to provide accurate coastal altimetry products, the wet troposphere correction and ocean tide model are also big contributors to the error in retracked altimetry, even if the waveforms are properly retracked.

Acknowledgements The authors would like to thank the AVISO data team for providing the Jason-1 altimeter waveform data, and the National Marine Data and Information Service of State Oceanic Administration P.R. China for providing the in situ tide gauge station measurements. Thanks also go to the anonymous reviewers for their constructive comments on an earlier version of this article. We also thank T. Platt for improving the English of this paper.

References

Bamber JL (1994) Ice sheet altimeter processing scheme. Int J Remote Sens 15(4):925–938

Bao LF, Lu Y, Xu HZ (2004) Waveform retracking of TOPEX/Poseidon altimeter in Chinese offshore. Chinese J Geophys 47(2):216–221 (in Chinese)

Bao LF, Lu Y, Wang Y, Xu HZ (2007) Coastal mean sea surface height by retracking ERS-1 altimeter waveform data. Prog Geophys 22(2):427–431 (in Chinese)

Berry PA (2006) Two decades of inland water monitoring using satellite radar altimetry. In: Proc. of 15 years of progress in radar altimetry Venice, Italy

Berry PAM, Bracke H, Jasper A (1997) Retracking ERS-1 altimeter waveforms over land for topographic height determination: an expert systems approach. In: Proceedings of the 3rd ERS symposium on space at the service of our enviroment, ESA SP-414, vol 3, pp 17–21, Florence, Italy

Brooks RL, Lockwood DW, Lee JE et al (1998) Land effects on TOPEX radar altimeter measurements in Pacific Rim Coastal Zones. Remote Sensing of the Pacific Ocean by satellites. NSW, Australia, pp 175–198

Brown GS (1977) The average impulse response of a rough surface and its applications. IEEE Trans Antennas Propag 25(1):67–74

Callahan PS, Rodriguez E (2004) Retracking of Jason-1 data. Mar Geod 27(3):391–407

Chelton DB, Ries JC, Haines BJ et al (2001) Satellite altimetry. In: Fu LL, Cazenave A (eds) Satellite Altimetry and Earth Sciences. A handbook of techniques and applications. Academic, San Diego, CA, pp 1–131

Chen JY (2007) Investigation and research of China estuary coasts, Higher Education, Beijing (in Chinese).

Davis GH (1993) A surface and volume scattering retracking algorithm for ice sheet satellite alitimetry. IEEE Trans Geosci Remote Sens 31(4):811–818

Davis CH (1996) A robust threshold retracking algorithm for extracting ice-sheet surface elevation from satellite radar altimeters. IGARSS 3:1783–1787

Deng X, Featherstone W, Hwang C et al (2002) Estimation of contamination of ERS-2 and Poseidon satellite radar altimetry close to the coasts of Australia. Mar Geod 25(1):249–271

ESA (2005) ENVISAT RA2/MWR Product handbook, Issue 2, http://envisat.esa.int/dataproducts/. Accessed 19 Jan. 2009

Fang GH, Wang Y, Wei Z et al (2004) Empirical cotidal charts of the Bohai, Yellow, and East China Seas from 10 years of TOPEX/Poseidon altimetry. J Geophys Res 109:C11006

Frappart F, Calmant S, Cauhope M et al (2006) Preliminary results of ENVISAT RA-2-derived water levels validation over the Amazon basin. Remote Sens Environ 100:252–264

Freeman J, Berry PAM (2006) A new approach to retracking ocean and coastal zone multimission altimetry. In: Proceedings of 15 years of progress in radar altimetry, Venice, Italy

Guo JY, Hwang CW, Chang XT, Liu YT (2006) Improved threshold retracker for satellite altimeter waveform retracking over coastal sea. Prog Natural Sci 16(7):732–738

Hancock DW III, Lockwood DW (1990) Effects of island in the Geosat footprint. J Geophys Res 95(C3):2849–2855

Hwang CW, Guo JY, Deng XL, Hsu HY, Liu YT (2006) Coastal gravity anomalies from retracked Geosat/GM altimetry: improvement, limitation and the role of airborne gravity data. J Geod 80:204–216

Laxon S (1994) Sea ice altimeter processing scheme at the EODC. Int J Remote Sens 15(4):915–924

Lefevre JM, Skandrani C (2003) Impact of using several altimeters for improving numerical wave analyses and forecasts. In Proc IGARSS'03 Symp, pp 827–829

Legresy B, Remy F (1997) Surface characteristics of the Antarctic ice sheet and altimetric observations. J Claciol 43(114):197–206

Liu ZC, Liu BH, Huang ZZ et al (2005) China coastal and adjacent sea submarine topography and Geomorphy. Ocean, Beijing (in Chinese)

Martin TV, Zwally HJ, Brenner AC, Bindschadler RA (1983) Analysis and retracking of continental ice sheet radar altimeter waveform. J Geophys Res 8:1608–1616

Picot N, Case K, Desai S, Vincent P (2003) AVISO and PODAAC User Handbook. IGDR and GDR Jason Products, SMM-MU-M5-OP-13184-CN (AVISO), JPL D-21352 (PODAAC)

Pu SZ, Cheng J, Zhang YJ et al (2004) Monsoon and annual variability of sea surface slope and their effects on alongshore current. Marine Sci Bull 23(2):1–7 (in Chinese)

Wingham DJ, Rapley CG, Griffiths H (1986) New technique in satellite altimeter tracking systems. In: IGASS'86 symposium digest, vol 1, pp 185–190

Yang SY, Bao XW, Chen CS, Chen F (2003) Analysis on characteristics and mechanism of current system in west coast of Guangdong Province in the summer. Acta Oceanol sin 25(6):1–8

Zanife OZ, Dumont JP, Stum J, Guinle T (2001) Jason-1 user products. Ssalto products specifications, vol. 1, SMM-ST-M-EA-10879-CN

18 Satellite Altimetry for Geodetic, Oceanographic, and Climate Studies in the Australian Region

X. Deng, D.A. Griffin, K. Ridgway, J.A. Church, W.E. Featherstone, N.J. White, and M. Cahill

Contents

18.1	Introduction	475
18.2	The Coastal Currents Around Australia	475
	18.2.1 Surface Geostrophic Currents	476
	18.2.2 Annual Cycle of Currents Along Southern Australia	479
	18.2.3 Transport of the EAC from Altimetry	481
	18.2.4 Net Transport	482
	18.2.5 Transport Along the Section	482
18.3	Mapping Coastal Ocean Currents Using Altimetry, Tide Gauges, Drifters, and Thermal Imagery	484
	18.3.1 Coastal Sea Level Mapping	484
	18.3.2 Coastal Dynamics	485
18.4	Sea level Rise	488
	18.4.1 Global and Coastal Sea level Rise	488
	18.4.2 Implications for Understanding Coastal Sea level Rise	492
18.5	The Role of Satellite Altimetry in Determining the Australian Marine Gravity Field	493
	18.5.1 How Accurate are Altimeter Data near the Australian Coast?	494
	18.5.2 Using Altimetry to Detect Ship-track Gravity Data Errors	498
	18.5.3 Assessment of AUSGeoid98 offshore of Western Australia	501
18.6	Conclusions and Perspectives	503
References		504

X. Deng (✉)
Centre for Climate Impact Management and School of Engineering,
The University of Newcastle, Australia
e-mail: xiaoli.deng@newcastle.edu.au

D.A. Griffin, K. Ridgway, J.A. Church, N.J. White, and M. Cahill
CSIRO Wealth from Ocean Flagship, and the Centre for Australian Weather and Climate Research, Australia

W.E. Featherstone
Western Australian Centre for Geodesy and The Institute for Geoscience Research,
Curtin University of Technology, Australia

Keywords Australian coastal regions · Gravity field · Ocean currents · Satellite altimetry · Sea level rise

Abbreviations

AVHRR HRPT	Advanced Very High Resolution Radiometer High Resolution Picture Transmission
AVISO	Archiving, Validation and Interpretation of Satellite Oceanographic data
CARS	CSIRO Atlas of Regional Seas
CHAMP	Challenging Mini-Satellite Payload
CSIRO	Commonwealth Scientific and Industrial Research Organisation
DOT	Dynamic Ocean Topography
DUACS	Data Unification and Altimeter Combination System
EAC	East Australian Current
EGM2008	Earth Gravitational Model 2008
EGM96	Earth Gravitational Model 1996
ENSO	El Niño Southern Oscillation
Envisat	Environmental Satellite
ERS	European Remote Sensing Satellite
GA	Geoscience Australia
GDR	Geophysical Data Record
Geosat	Geodetic Satellite
GFO	Geosat Follow-On
GGM	Global Geopotential Model
GIA	Glacial Isostatic Adjustment
GOCE	Gravity Field and Steady-State Ocean Circulation Explorer
GPS	Global Positioning System
GRACE	Gravity Recovery and Climate Experiment
GRS80	Geodetic Reference System 1980
IGDR	Interim Geophysical Data Record
IPCC AR4	Intergovernmental Panel on Climate Change Fourth Assessment Report
IPCC	Intergovernmental Panel on Climate Change
LC	Leeuwin Current
MSS	Mean Sea Surface
NSW	New South Wales
OFAM	Ocean Forecasting Australia Model
PISTACH	Prototype Innovant de Système de Traitement pour l'Altimétrie Côtière et l'Hydrologie (i.e., Coastal and Inland Water Innovative Altimetry Processing Prototype)
RHS	Right Hand Side
SLA	Sea Level Anomaly
SSH	Sea Surface Height
SST	Sea Surface Temperature
STD	Standard Deviation
Sv	Sverdrup

SynTS	Synthetic Temperature and Salinity
T/P	TOPEX/Poseidon
XBT	eXpendable BathyThermograph

18.1
Introduction

Australia is surrounded by the Pacific, Indian, and Southern Oceans. Its coastal zone exhibits a broad and complicated range of topographical features. Its ocean dynamics is dominated by several major ocean currents including the East Australian Current (EAC), Leeuwin Current (LC), Indonesian Throughflow, and to the south the South Australian Current and Antarctic Circumpolar Current (e.g., Church and Craig 1998). Understanding the physical processes and dynamic features of oceans boarding the Australian continent using satellite altimetry is thus of great importance for assessing how the ocean's variation responds to the regional and global climate change, which has occupied both geodetic and oceanographic communities since the 1970s.

The relatively homogeneous and continuous global coverage of sea surface measurements from satellite altimeter missions have been used in Australian studies of the marine gravity field, ocean current systems, and sea level change. These missions are the Geodetic Satellite (Geosat), European Remote Sensing satellites (ERS-1 and ERS-2), TOPEX/Poseidon (T/P), Geosat Follow-On (GFO), Environmental Satellite (Envisat), Jason-1, and Jason-2. The applications are advanced by the new developments in modelling and assimilation algorithms, improvement of knowledge of the Earth's gravity, and combination of altimetry with other data sources from satellite remote sensing techniques, drifting buoys, tide gauges, and hydrographic observations.

This chapter overviews the recent results from a number of altimetric applications in Australia. The ocean current systems and their seasonal and interannual variations around Australia are first addressed in Sect. 18.2. This is followed by results on real-time coastal ocean currents (Sect. 18.3) using a sea level mapping system, which has provided a web-based service for public access to maps of sea level, geostrophic velocity, and sea surface temperature in the Australasian region. Section 18.4 estimates recent trends of sea level rise in the Australian region. The assessment of the quality of the coastal altimetric data and marine gravity field, and development of a coastal waveform retracking system to improve altimetric data in Australian coastal zones are provided at the end of the text (Sect. 18.5).

18.2
The Coastal Currents Around Australia

The Australian region is dominated by several major ocean current systems (Church and Craig 1998). The EAC, one of the world's major western boundary currents, flows southward adjacent to the east coast transporting warm, high-salinity water (Ridgway and Dunn 2003). Along its path, large eddies (300 km in diameter) are generated and separated off into the Tasman Sea (Bowen et al. 2002; Mata et al. 2006a). The LC originates off the North West Shelf and flows down the Western Australian coast in winter, bringing warm, relatively fresh water (Cresswell and Golding 1980). The current turns around the south coast and penetrates eastward as far as Western Tasmania (Ridgway and Condie 2004). As well as influencing the regional climate along its path, the LC transports many tropical organisms into Southern Australia (Maxwell and Cresswell 1981). The Indonesian Throughflow provides a connection between the Pacific and Indian Oceans through the Indonesian archipelago. This flow has a major influence on both the climate of the Indian Ocean and the global oceans (Godfrey et al. 1995).

In this section, the coastal currents around Australia are examined using results from altimetric observations and other supporting datasets. Some simple methods are applied to the standard along-track and gridded altimetry data to examine the coastal circulation around Australia. Firstly, the full coastal current regime is determined for typical summer and winter conditions. The seasonal anomalies are derived from the altimeter dataset. The mean ocean state is obtained from an ocean model mean field that has been strongly nudged towards ocean climatology. A case study is presented around Southern Australia and Tasmania to demonstrate how the seasonal behavior may be explained using monthly patterns of the annual cycle of altimetric sea level. Finally a time series of the volume transport of the EAC is obtained from a method which merges in situ and altimeter observations.

18.2.1
Surface Geostrophic Currents

18.2.1.1
Seasonal Sea Level Anomalies

The standard along-track altimeter data used are from a subset of the constellation of satellites flying in the period October 1992 to December 2008. These include T/P, Jason, ERS-1/2, and Envisat missions. Data from the T/P interleaved and the GFO missions were not used. Standard environmental corrections have been applied to the data (Benada 1997).

A seasonal climatology was produced from the altimetry dataset. Annual and semiannual components were estimated from the sea level anomaly (SLA) time series using least-squares harmonic analysis. The components were determined at each point along the tracks of all of the satellites. The spatial resolution is higher than that obtained for an individual 10-day field as all the tracks are used in the mapping rather than the more limited data available for the 10-day window. The downside is that any differences in the periods covered by each satellite may introduce interannual biases into the results. For this reason, we have used only data from satellite missions that cover the most study period. The along-track height estimates of each component were then interpolated onto a regular grid (0.25° × 0.25°) using a locally-weighted least squares method or "loess" filter (Cleveland and Devlin 1988) with a minimum radius of 350 km. The half amplitude cut-off wavenumber is thus 210 km as it approximates a simple block average of the same width. The uniform fields of amplitude (Fig. 18.1) and phase of the two seasonal components are obtained from this procedure.

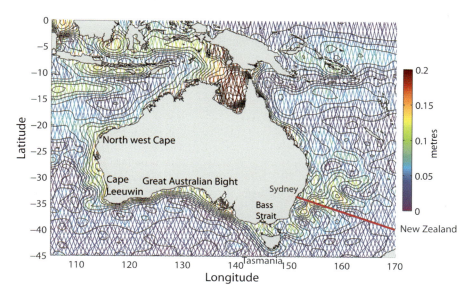

Fig. 18.1 The raw along-track altimeter data for the amplitude of the annual cycle compared to the corresponding gridded field (contours, units in metres). The Sydney to Wellington, New Zealand, eXpendable BathyThermograph (XBT) transect is shown as a red line. The XBT and altimetric data along this transect will be used to estimate the transport of the EAC in Sect. 18.2.3

18.2.1.2
Absolute Mean Sea Level Topography

Mean temperature, T, and salinity, S, fields from a high-resolution ocean climatology of CSIRO (Commonwealth Scientific and Industrial Research Organisation) Atlas of Regional Seas (CARS, Ridgway et al. 2002) are assimilated into a high resolution ocean model, Ocean Forecasting Australia Model, (OFAM, Schiller et al. 2008). In a 2-year run, OFAM is forced by seasonal winds and strongly nudged to the climatological fields. The resulting absolute mean topography (without altimeter data) agrees closely with existing observationally based estimates at large spatial scales (Vossepoel 2007) while retaining the realistic mesoscale structure derived from the assimilated T and S.

18.2.1.3
Surface Current Fields

Summer and winter surface current fields are presented in Fig. 18.2. These fields are constructed by adding the seasonal anomalies generated by the altimeter-derived seasonal components (cf. Sect. 18.2.1.1) to the absolute mean topography described in Sect. 18.2.12. The geostrophic current is then determined from the resulting seasonal surface height fields. The results demonstrate that the major coastal currents experience significant seasonal changes that are almost 180° out of phase from east to west coasts.

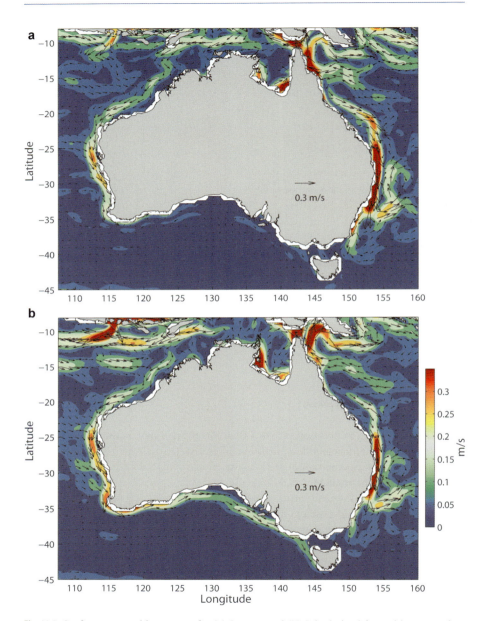

Fig. 18.2 Surface geostrophic currents for (**a**) January and (**b**) July derived from altimeter surface SLAs (1992–2008) combined with an absolute mean surface topography. The vectors show the current direction and the filled contours indicate the current magnitude

In summer (January, Fig. 18.2a), the EAC shows its peak flow. It has a strong current core from its formation region at ~18°S, just south of the South Equatorial Current bifurcation (Church 1987; Webb 2000), down to at least 38°S to the east of Bass Strait (Ridgway and Godfrey 1997). In fact, there is a weaker flow east of Tasmania which represents the surface manifestation of the summer EAC flow (Ridgway 2007). Off the western coast, there is evidence of south-eastward, shelf-edge flow from ~17°S around Cape Leeuwin (115°E; 34°S) to 125°E. This represents the LC in its weaker summer state. At this time, there are no indications of any coherent alongshore flow over the southern shelf region enclosing the Great Australian Bight and Western Tasmania.

The pattern is transformed in winter (Fig. 18.2b) with the Leeuwin system now in its strongest mode (Cresswell and Golding 1980; Smith et al. 1991; Feng et al. 2003) and the EAC reverting to a relatively weaker pattern (Ridgway and Godfrey 1997). A continuous shelf-edge flow is evident from at least 12°S around the western and southern coasts to Southern Tasmania. The flow along the northern portion of Western Australia which feeds the LC has been named the Holloway Current (D'Adamo et al. 2009). Only limited observations have been made in this region but current meter moorings have shown a winter enhanced flow along this portion of the shelf-edge (D'Adamo et al. 2009). In this winter period, the EAC is weaker and has a smaller meridional extent. The northern component of the South Equatorial Current bifurcation is presented throughout the year (Fig. 18.2b) but is intensified during winter (Kessler and Gourdeau 2007). The pattern in Fig. 18.2b suggests that there may be increased flow into the eastward meander of the Tasman Front with a consequent reduction in the southward transport of the EAC Extension. There is a similar transition between the Tasman Front and EAC Extension at decadal timescales (Hill et al. 2009).

18.2.2
Annual Cycle of Currents Along Southern Australia

The seasonal changes to the coastal currents described here are complex and arise from a variety of forcing mechanisms. To identify some of these processes we focus on a specific region, the southern coast and consider the annual cycle of coastal sea level generated from the altimeter-derived gridded components (Fig. 18.3).

Along the southern coast, the annual signal is dominant with the January and July fields representing the seasonal extremities. The coastal sea level is depressed by some 15 cm along the entire southern coast in summer and raised by the same amount in winter. Pariwono et al. (1986) extracted a similar pattern from coastal tide gauges and observed that the transition at the coastal boundary from summer to winter states occurred almost simultaneously (within several days) over the entire coast. The results here show that there is a phase delay over the coastal waveguide from west to east indicating that there is a propagating sea level pulse. We observe that in January a coastal drop in sea level drives a meandering westward flow located some 150–200 km from the southern coastal boundary. This may represent a surface expression of the Flinders Current, a year-round upwelling favorable subsurface boundary current, flowing from east to west along Australia's southern shelves (Middleton and Cirano 2002).

As the LC off the west coast strengthens in autumn, the southern height regime begins a transition to its winter state. By May a positive sea level anomaly has been established along the entire southern shelf and an associated eastward current is located at the

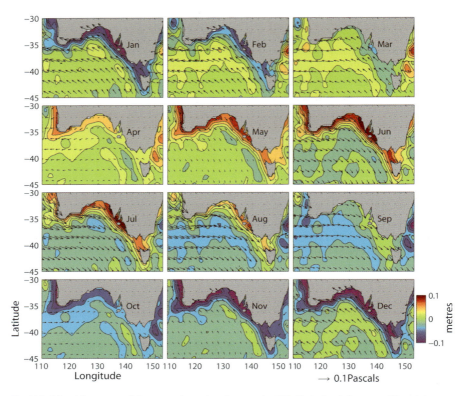

Fig. 18.3 Monthly maps of the annual sea level anomaly (SLA) derived from satellite altimetry (units in meters). The vectors show the direction and magnitude of the annual cycle of wind stress from the Quikscat satellite (units in Pascals)

shelf-edge. For the following four months there is a continuous warm current on the continental slope that flows from Northwest Cape (not shown), turns to the southeast around Cape Leeuwin (115°E; 34°S), continues across the Great Australian Bight, and finally flows southward down the west coast of Tasmania. Output from a model study (Cirano and Middleton 2004) forced by typical winter winds showed similar eastward flow. The results presented here provide the linking element between observations in the western (Godfrey et al. 1986), central (Hahn 1986), and eastern sectors (Baines et al. 1983) of this region.

A previous study strongly suggested that the changes in coastal sea level and hence coastal currents were forced by the variations in alongshore winds (Ridgway and Condie 2004). In summer, a high-pressure ridge is maintained over the South Australian Basin which induces a consistent pattern of southeasterly wind. In contrast, in winter the anticyclone moves to the north over Central Australia resulting in a predominantly westerly wind regime. The transition from an easterly to a westerly alongshore pattern occurs from March to May and this is also the period that coastal sea level changes from a negative to a positive anomaly. The westerly wind stress at the coast drives an onshore Ekman mass transport which leads to coastal downwelling and an eastward shelf-edge flow (Middleton and Cirano 2002; Cirano and Middleton 2004).

However, from Fig. 18.3, it is observed that in April the first indication of an increased component of coastal sea level appears to enter the region from the west. A plume of high sea level extends around Cape Leeuwin and eastward to 135°E. At this time, the easterly winds have relaxed but still have not switched into their downwelling favorable mode. In the following months, the winds do develop a strong alongshore westerly component which would promote onshore Ekman flux and reinforce the already established high coastal sea level. The results indicate that while the winds appear to provide an important contribution to the enhanced coastal sea level and consequent eastward shelf-edge flow, the main forcing mechanism is a propagating pulse of sea level which is generated farther to the northwest.

In addition to the growth and decay cycle of the coastal anomalies, there is a coherent westward propagation of the positive and negative structures. For example, the positive anomaly which is fully developed in June, has moved some 300 km to the west by November. As the positive anomaly strengthens from May to July, the laminar nature of the flow begins to break down. Despite the smoothing inherent in the interpolation procedures, two centers of instability (41°S, 44°S) in July can be identified which develop into distinct anticyclonic eddies. The appearance of the southern eddy is in phase with a further anti-cyclonic eddy that originates to the southeast of Tasmania. The eddies remain embedded in the larger ridge as it propagates westward. Similarly, Cirano and Middleton (2004) reported the growth of high and low features off Western Tasmania in their seasonal model. The evolution of a high off Northwest Tasmania over the first 56 days of their model run mirrors the growth of the eddy from July to September. There is analogous behavior observed in the negative anomalies from November to July.

18.2.3
Transport of the EAC from Altimetry

The EAC can be inferred directly from satellite altimetry observations (Ridgway et al. 2008). $T(z)$ and $S(z)$ are determined from satellite surface properties and used to calculate the baroclinic transport. A multiple linear regression technique, "Synthetic Temperature and Salinity" (SynTS), is used which is on the basis of vertical statistics generated from the historical ocean data archive. At each grid-point,

$$T(z) = <T(z)> + \alpha(z)\delta h + \beta(z)\delta T_0 \qquad (18.1)$$

The first term on the RHS is a climatological mean (1992–2006), $\delta h = \delta h_{0/2000}$ is the steric height anomaly, δT_0 is the surface temperature anomaly, and z is a set of standard depths, 0–2,000 m. Similar expressions to Eq. (18.1) are obtained for salinity. The coefficients α and β are continuous spatially dependent functions, determined by fitting weighted least-squares regression to the historical in situ data. The weights are a function of data-grid separation, data-density, land barriers, and ocean depth (Ridgway et al. 2002). Within the EAC region, altimetric sea level, $\delta h_a \approx \delta h_{0/2000}$ (correlation = 0.95, regression coefficient = 0.96), and so the surface inputs to Eq. (18.1) come from satellite quantities (Ducet et al. 2000; Ridgway et al. 2008).

18.2.4
Net Transport

A 15-year time series of the net transport across the Tasman Sea, between Sydney and Wellington along the XBT section (cf. Fig. 18.1) estimated from altimetric observations is shown in Fig. 18.4a. The time series shows a rich spectrum of frequencies from eddy-scale, seasonal, interannual to decadal. This is the first time that such an extended record of the EAC flow has been determined.

The mean southward transport is 10.10 Sv (Sverdrup, Sv = 10^6 m^3/s) with a standard deviation (STD) of some 7.14 Sv. Despite the substantial meridional cancellation across the section arising from the EAC retroflection, Fig. 18.4a indicates that the net transport retains a high degree of variability. Peak to trough changes in transport are typically 20–25 Sv with the largest individual fluctuation of 38 Sv occurring in December 1994. There is a peak southward flow of 28.09 Sv (8% of the time series is above 20 Sv) but a maximum of 16.64 Sv to the north (the December 1994 event). The transport is overwhelmingly biased in the poleward direction – only for some 8% of the 1992–2006 period are northward flow reversals observed. Of these, most occur early in the record (<1998) where the fraction is 16%. After this date, it falls to just 3%. This apparent change of state is related to a quasi-decadal signal. While a 15-year record is insufficient to truly resolve a decadal signal, the net southward flow through the region weakened in the middle of the 1990s and then strengthened again in 2001. The net transport ranged from 5 Sv in 1995 to 16 Sv in 2001. This decadal-like change is in phase with variations observed in the strength of the South Pacific Gyre (Roemmich et al. 2007).

There are prominent higher frequency components evident in the series including a substantial seasonal signal. It has an amplitude of 4 Sv which has a peak expression in mid February. This agrees in phase with the results of Roemmich et al. (2005) who found a similar phase but with a much reduced amplitude (1.4 Sv) for transport above 12°C and referenced to 800 dB. Further to the north at 28°S, Ridgway and Godfrey (1997) also found an annual amplitude of 5 Sv in net meridional transport (2000-dB reference). Even larger seasonal southward flows are directly associated with the EAC jet adjacent to the boundary (154.6°E) and the secondary meander at 157.7°E, with amplitudes of more than 7 Sv (not shown).

18.2.5
Transport Along the Section

The evolution in time of what may be described as the EAC coastal southward jet (west of 157°E) is examined. Fig. 18.4b shows time-longitude plots of the cumulative transport along the XBT section (cf. Fig. 18.1) within 250 km of Sydney, at full 10-day resolution and filtered to isolate longer term temporal signals (1, 2 and 5-year loess filter). The 10-day raw data field (panel 1) is seen to be dominated by eddies which occur about three to four times per year with clear evidence of interannual variations. This corresponds to the peak of 100–115 days observed in previous analyses of the EAC eddy field (Mata et al. 2000; Bowen et al. 2005; Mata et al. 2006a, b). Panel 2 (1-year low-pass) shows that while the

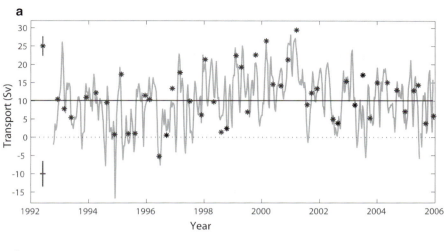

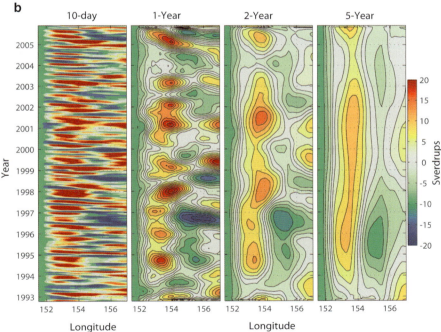

Fig. 18.4 (**a**) Time series of EAC between Sydney and Wellington, from SynTS (grey line) and in situ XBT measurements (asterisks). The error bars on the LHS show the standard error of each estimate. (**b**) Time-longitude plot of EAC cumulative transport (units in Sv) from Sydney to 157E. The four panels (*left* to *right*) show the transport at full 10-day resolution and filtered using a loess smoother with 1-year, 2-year, and 5-year time windows

seasonal cycle is the dominant temporal signal, there is also significant interannual variability. In this frequency band, the EAC flow occurs as seasonal eddies peaking in summer; in fact, the temporal evolution of the seasonal eddies shown previously in Ridgway and Godfrey (1997) is observed. In 1993 and 2004 the flow is anomalously weak with a greatly suppressed summer maximum flow. The final two panels (2-year and 5-year low-passed series) reveal a more temporally continuous poleward structure west of 154.5°E with the EAC maximum varying both in magnitude and location.

The coastal ocean currents and their flow in seasonal and interannual scales have been estimated in this section. Their real-time flow is to be investigated in next section.

18.3 Mapping Coastal Ocean Currents Using Altimetry, Tide Gauges, Drifters, and Thermal Imagery

Since 2004, a publicly-accessible, daily-updated website showing maps of sea level, geostrophic velocity, and sea surface temperature for the Australasian region (http://www.cmar.csiro.au/remotesensing/oceancurrents/) has been maintained by CSIRO. A key feature of the sea level mapping system is that the maps extend right to the coast because of the inclusion of coastal tide-gauge data interpolated continuously around the Australian coastline. This distinguishes it from the Data Unification and Altimeter Combination System (DUACS) run by Archiving, Validation, and Interpretation of Satellite Oceanographic team (AVISO) (http://www.aviso.oceanobs.com/en/data/product-information/duacs/). Specially-processed ("coastal") altimetry data (e.g., see Sect. 18.5.1) have not, to date, been used routinely in the system but the potential value of doing this can be illustrated by the effect of including the tide gauges. Trials of incorporating the project PISTACH (Prototype Innovant de Système de Traitement pour l'Altimétrie Côtière et l'Hydrologie, i.e., Coastal and Inland Water Innovative Altimetry Processing Prototype) in the Centre National d'Etudes Spatiales (CNES), France, version of the Jason-2 data are underway at the time of writing.

18.3.1 Coastal Sea Level Mapping

Australia has some 64 tide gauges located at irregular intervals around the coastline. Some form the national Australian Baseline Array, while others are operated by various state agencies. For the purposes discussed here, the recorded sea levels are iteratively leveled so that all are referenced to an estimate of their mean for the 1992–1999 epoch. Diurnal and semi-diurnal tides and inverse barometer signals are removed, as they are for the altimeter data. Almost all pairs of neighboring gauges, from Darwin (130.8°E; 12.5°S) around the south to Cape York (142.3°E; 10.7°S), record highly coherent signals, so simple interpolation can be used to fill the gaps with some confidence. Indeed, at any chosen gauge, interpolation between the closest gauges either side explains more of the variance than the altimeter data from the nearest point beyond the 200 m isobath.

For maps made in near-real time, the standard altimetry Interim Geophysical Data Records (IGDRs) are used, with an asymmetric data window admitting data three days after the analysis time, and up to 15 days before it. For maps made in delayed-mode, the analysis is centered, and uses the Geophysical Data Records (GDRs). In both cases, data from as many altimeters and tide gauges as possible are used: T/P interleaved with Jason-1, Envisat, and GFO in 2004 and 2005, but only Jason-2 and Envisat by the end of 2008. The 1 Hz (7-km interval) SLA data are edited for quality control (outliers or tracks with large orbit errors), smoothed along track using a Hanning filter of half-amplitude width 63 km, then subsampled at 0.25° latitude intervals.

The along-coast interpolated tide-gauge SLA data are merged with the along-track smoothed altimeter SLA data using a mapping scheme similar to that of Ducet et al. (2000). Our temporal and spatial covariance functions go to zero at 20 days and 350 km at the equator and 20 days and 200 km at 60°S. The white noise error of the SLA is assumed to be 3 cm. The (Bowen et al. 2002) method of using velocities from pairs of sea surface temperature (SST) images as a gradient constraint in the SLA mapping has also been implemented but we have only used it for specific case studies at present.

18.3.2
Coastal Dynamics

Australia has several distinct regimes, as far as the variability of coastal and oceanic sea level is concerned. Here we discuss just two, briefly, that differ greatly in terms of the energy levels in the coastal and nearby deep ocean: Australia's southern and eastern coastlines.

Across the south coast, there is as much or more variability of sea level at the coast as there is in the deep ocean. For example, Fig. 18.5 shows the SLA and SST maps for a day after a strong south-easterly wind event caused upwelling to occur off the Eyre Peninsula (135°E) and the Bonney Coast (138°E–142°E), as it often does there during summer. The SST map clearly shows the reduced temperatures while the tide-gauges and SLA map show the reduced coastal sea level. The geostrophic current is in the direction of the wind, even though the wind was not used for the mapping. The movement of the plume of cold water over time (not shown) confirms this estimate to be qualitatively correct.

During winter, in contrast, the winds are predominately westerly and the LC comes around Cape Leeuwin at the south-west corner of Australia. This flow is clearly seen in Fig. 18.6 as a plume of warm water. The eastward flow can be inferred from raised sea level along the coast. What the SLA map does not show, because it includes no data between the deep sea and the coast, is that the LC is a narrow jet at the edge of the shelf, as is sometimes clear from the SST imagery and the tracks of drifting buoys. It is expected that including "coastally-processed" altimetry across the shelf will concentrate the sea level gradient in the maps at the outer shelf.

Off the south-east coast of Australia, sea level variations at the coast are much weaker than those offshore, where steep gradients occur in association with the EAC and its eddies (Fig. 18.7 and 18.8). The continental slope is a very effective barrier to the propagation of the oceanic sea level signal, so it is only on rare occasions that coastal tide gauges record

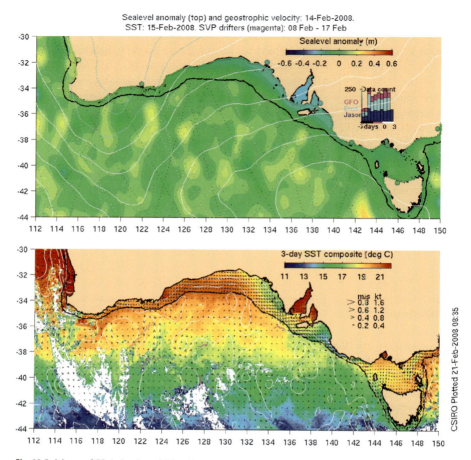

Fig. 18.5 Maps of SLA (*top*) and SST (*bottom*) for 14 February 2008, made on 21 February 2008. In the top panel, filled circles show the location of tide gauges contributing data to the map, while lines of dots show the ground tracks of the altimeters. The filled colour of the tide gauges shows the SLA averaged for the day shown. The colours of the dots along the altimeter tracks denote the age of the data with respect to the analysis time, white for the three days more recent than the analysis date, magenta for the three days after the analysis date, and black for the oldest (least-weighted) data. Atmospheric pressure contours are shown in white for high pressure and blue for low. The sea level map is repeated in the lower panel as contours, and the geostrophic velocity derived from it is shown as black arrow heads. Magenta arrow heads show the daily positions of Global Drifters. The SST is the 75th percentile of all NOAA (National Ocean and Atmospheric Administration) environmental satellite AVHRR HRPT (Advanced Very High Resolution Radiometer High Resolution Picture Transmission) data received in a 3-day data window. Cloud-clearing can be seen to be incomplete

transient signals that are clearly associated with offshore features. More frequently, the sea level variations are associated with the much more short-lived forcing provided by the passage of atmospheric systems. Fig. 18.7 shows the situation just before the passage of a storm, while Fig. 18.8 shows how the westerly winds in Bass Strait, and south-westerly

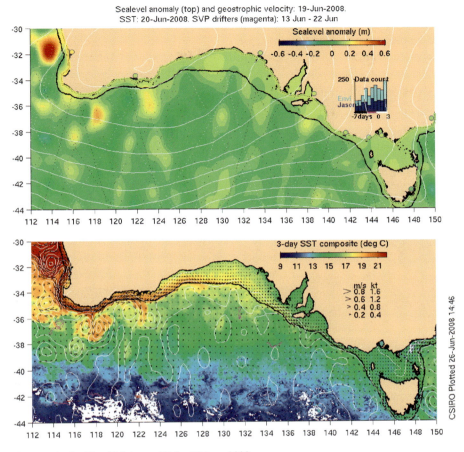

Fig. 18.6 As for Fig. 18.5, but valid for 19 June 2008

winds along the coast drove a positive coastal trapped wave of 0.2 m amplitude along the coast, raising coastal sea level coherently for 1,000 km between Bass Strait (39°S) and Newcastle (33°S). In the two days between these two maps, only two new altimeter tracks intersect the coast along the state of New South Wales (NSW). These confirm that the coastal signal is either restricted to the continental shelf as expected (Church and Freeland 1987; Griffin and Middleton 1991), or insignificant compared to the amplitudes associated with the ocean eddies.

Our approach of interpolating the tide gauges first onto a line along the coast which is then mapped along with the altimeter data is a simple but effective expedient. We plan to improve this by using two- (x, y) or three- (x, y, t) dimensional covariance functions determined point-wise around the country from a high-resolution ocean model, to more accurately represent the highly non-isotropic nature of the variability of sea level.

To conclude, the potential value of using specially processed, on-shelf retrievals of SLAs from altimeters has been demonstrated simply by including data from coastal tide gauges in a merged product. In future, the continuous sampling of the tide

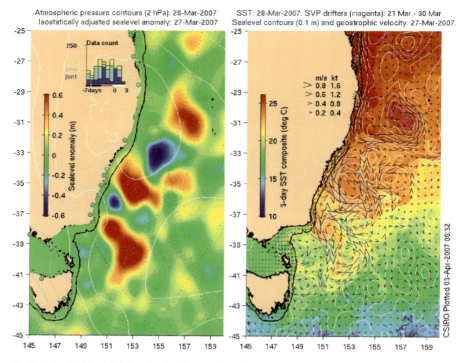

Fig. 18.7 As for Fig. 18.6, but for the Tasman Sea off the state of NSW, valid for 27 March 2007

gauges, combined with the across-shelf sampling by the altimeters, should yield a product that is superior to one using a single data source.

18.4
Sea level Rise

In this section, the rate of sea level rise in the Australian region is estimated using data from satellite altimetry, tide gauges, and the Bluelink Ocean Reanalysis (Schiller et al. 2008; Oke et al. 2008).

18.4.1
Global and Coastal Sea level Rise

18.4.1.1
Global Averaged Sea level Rise

Sea level rise is an important consequence of climate change but regional and local sea level is also affected by climate variability and the variability of coastal currents. Over the twentieth century, global averaged sea level rose by 1.7 ± 0.3 mm/year, increasing to

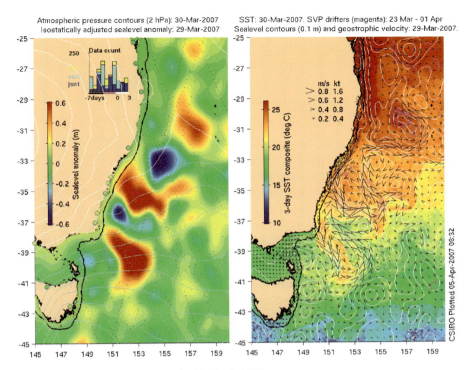

Fig. 18.8 As for Fig. 18.7, but valid for 29 March 2007

3.3 ± 0.4 mm/year from the early 1990s (Leuliette et al. 2004; Church and White 2006). The Intergovernmental Panel on Climate Change Fourth Assessment Report (IPCC AR4) projections of sea level rise for the last decade of the twenty-first century (approximately equivalent to projections for 2095) compared to the 1980–1999 period are for a global-averaged sea level rise of 18–59 cm if potential rapid dynamical changes in ice sheet flow are excluded (IPCC 2007; Meehl et al. 2007). If the observed ice sheet flow from 1993 to 2003 is scaled with global average temperatures, the projected sea level rise could increase by an additional 10–20 cm (or more; IPCC 2007) effectively giving the total range 18–79 cm or more (76 cm in Meehl et al. 2007). However, a lack of models adequately simulating the poorly understood dynamical ice-sheet processes means that this additional contribution could be an under or over estimate (IPCC 2007).

Satellite and in situ observations (Rahmstorf et al. 2007) indicate that since the start of the projections in 1990, sea level has been rising near the upper end of the IPCC Third Assessment Report projections (Church et al. 2001) and thus effectively at the upper end of the IPCC AR4 projections with the dynamic ice sheet component included. In addition, there have been suggestions that the IPCC AR4 projections underestimated the amount of sea level rise and that somewhat larger rises should be expected (Rahmstorf 2007; Horton et al. 2008).

While models do project different regional and coastal rates of sea level rise, there are significant differences between models and inadequate understanding of these regional distributions. Also, it is important to recognize that the impacts of sea level rise depend on coastal rather than deep ocean sea level rise and it is clearly important to be able to relate changes in offshore sea levels to coastal conditions.

18.4.1.2
Coastal and Global Averaged Sea level Rise

Examination of satellite-altimeter data has significantly improved our understanding of regional variations in the rate of sea level rise, mostly related to interannual climate variability. There has been some controversy over how well coastal sea levels agree with global-mean (i.e., deep water) sea levels (see, e.g., Holgate and Woodworth 2004; White et al. 2005). However, White et al. (2005) demonstrated that, *averaged over time*, trends in mean coastal and mean global sea level tend to agree, although there is higher variability in coastal sea level, largely induced by inter-annual variability, than in global averaged sea level. An example of this is Fremantle (115.7°E; 32.1°S), Western Australia (near Perth) that has a relatively small tidal range, but large interannual and decadal variations in sea level, a clear El Niño Southern Oscillation (ENSO) signal (Church et al. 2004). Other locations (e.g., on the east coast of Australia) have smaller interannual and decadal signals, but a higher tidal range. These local variations in sea level trends can last for decades (e.g., the relatively low sea level rise on the north-west Australian coast for much of the latter half of the twentieth century) but still tend to average out over longer periods.

18.4.1.3
Recent Sea level Rise in the Australian Region

The rate of sea level rise in the Australian region from January 1993 to December 2007, as estimated from satellite altimeter data demonstrates several features that are local manifestations of the global distribution of sea level rise (Fig. 18.9). There is a maximum of sea level rise to the north of Australia in the western equatorial Pacific. This feature is dominated by the change from more El Niño-like conditions at the start of the record to more La Niña-like conditions near the end of the record. These high rates of sea level rise are transmitted through the Indonesian Archipelago to the eastern Indian Ocean and the north and north-west of Australia (Wijffels and Meyers 2004) and decay counter-clockwise around the Australian coastline. Off Southeast Australia, there is a maximum at latitudes of about 35°S in the Tasman Sea. This is consistent with the spin up of the South Pacific sub-tropical gyre by increased wind stress curl (Roemmich et al. 2007). The minimum in the rate of sea level rise off the east coast at about 25°S is likely to be a dynamic response associated with the southward movement of water and a shallowing of the subtropical thermocline at these latitudes.

Sea level rise around the Australian coastline has been measured since the early 1990s by an array of carefully quality-controlled acoustic tide gauges, the Australian Baseline Array, the locations of which are shown in Fig. 18.9 (http://www.bom.gov.au/oceanography/projects/abslmp/abslmp.shtml). The tide-gauge records used here have been corrected

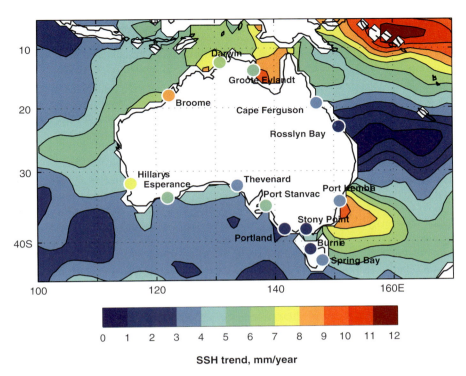

Fig. 18.9 Sea level rise in the Australian region. The offshore linear trends of sea level rise (units in mm/year) are determined from T/P and Jason-1 altimeter data over the period January 1993 to December 2007. The altimeter data have all standard corrections applied except that there is no adjustment for atmospheric pressure changes. The linear trends in coastal sea level data are shown by the coloured circles. A GIA correction has been applied to both data sets

for vertical land motion associated with changes in surface loading of the Earth using results from the Glacial Isostatic Adjustment (GIA) model of Milne et al. (1999). No additional correction has been made at this stage for local tectonic vertical movements as the time series of appropriate data (e.g., from Global Positioning System (GPS) receivers) are only now becoming long enough and high enough quality to contribute to useful estimates of sea level trends. The satellite altimeter data have been corrected for GIA using a model developed by M.E. Tamisiea (personal communication 2005). These adjustments are typically less than 0.4 mm/year in the Australian region and are not significant in the results shown here. The coastal sea level data (and the satellite data) do not include any adjustment for changes in atmospheric pressure (i.e., the "inverse barometer correction") as it is actually sea level change that is important for coastal impacts.

The coastal sea level data have both significant similarities and some striking differences to the offshore satellite record. On the north and north-west coasts of Australia, sea level is rising at rates well above the global average (rates of up to 10 mm/year), similar to the offshore satellite data. These rates decrease clockwise around Australia, similar to the satellite data. Typically the offshore and coastal rates of sea level rise are within about 2 mm/year of each other in this region. One anomaly is the gauge at Hillarys (115.7°E; 31.8°S), which shows relative sea level rising at over 7 mm/year compared with the offshore rate of about 4 mm/year. This 3 mm/year difference is the result of sinking of the tide gauge (P. Tregoning, personal communication 2008) as a result of compaction of the coastal sandy location associated with significant ground-water extraction in the region. Less than 30 km to the south at the Fremantle tide gauge (situated on a raised limestone reef), the rate of coastal relative sea level rise (not shown) is about 3.7 mm/year, in approximate agreement with the offshore altimeter data.

On the south eastern and eastern Australian coastline, the rates of sea level rise are typically 2–4 mm/year, close to or less than the global averaged rate of rise. Strikingly the coastal record at Port Kembla (150.9°E; 34.5°S) shows little indication of the offshore peak in sea level rise that is so prominent in the satellite data at about 35°S in the Tasman Sea. Recent analysis (Hill et al. 2008, 2009) shows that the strength of the EAC, particularly its southward extension, varies on interannual time scales and has strengthened over the 1993–2003 period. Thus, the strong sea level zonal gradient across the southward flowing EAC has increased over this period, consistent with the different offshore and coastal rates of sea level rise at this location.

The rate of sea level rise in the Australasian region, as estimated using the Bluelink Ocean Reanalysis (Oke et al. 2008; Schiller et al. 2008), is shown in Fig. 18.10. The Bluelink Reanalysis assimilated coastal tide-gauge data (without GIA correction) and offshore altimeter data, both adjusted for the inverse barometer effect. The data assimilation methodology was designed for transient features but the model seems to accurately capture the higher rate of offshore sea level rise compared to the coastal rate and as a result the increase in strength of the southward extension of the EAC over this period (Fig. 18.9 and 18.10). Along the south coast of Australia at Esperance (121.9°E; 33.9°S), Thevenard (133.7°E; 32.1°S), and Port Stanvac (138.5°E; 35.1°S), the greater rise of coastal than offshore sea level rise again seems to be related to the increase in strength of eastward shelf and slope currents. However here, the differences are not as large away from the strong dynamical features associated with the western boundary current.

18.4.2
Implications for Understanding Coastal Sea level Rise

The impacts of sea level rise depend on coastal rather than deep ocean sea level rise and it is clearly important to be able to relate changes in offshore sea levels to coastal conditions. As demonstrated here, changes in coastal currents can be an important part of determining coastal sea level change. This is likely to be especially important in western boundary current regions where there are energetic boundary currents. There are a number of approaches that can be used but observations of both off-shore and coastal sea level would seem to be crucial elements of any study. Of course, it is also critically important to consider the stability of coastal tide gauges. Their vertical motion needs to be carefully assessed using direct estimates of vertical velocity as estimated for example from GPS data. It will also be valuable to be able to use satellite-altimeter data as close to shore as possible. Other

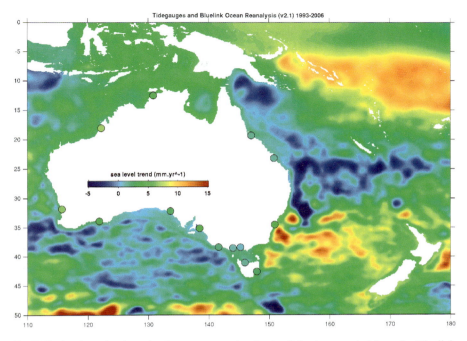

Fig. 18.10 As above but here the deep ocean sea level rate of rise is computed from the Bluelink Ocean Reanalysis that assimilated all available altimeter and tide-gauge data

effects, such as storm surges, are also of importance, and can greatly exacerbate the impact of changes in sea level.

18.5
The Role of Satellite Altimetry in Determining the Australian Marine Gravity Field

Marine gravity anomalies are a primary data source for modelling the geoid (i.e., loosely the mean sea level) and geological structure of continental shelves. The short wavelengths (<400 km) of the marine gravity field can be obtained from ship-track gravity measurements near the coast and from satellite radar altimeter measurements of sea surface height (SSH) for virtually any marine area in the world (e.g., Hwang and Parsons 1996; Sandwell and Smith 1997; Andersen and Knudsen 1998; Hwang et al. 1998, Andersen and Knudsen 2000; Hwang et al. 2002). The longer wavelength gravity field (>400 km) is more accurately measured at orbital altitude using spacecraft such as the Challenging Mini-Satellite Payload (CHAMP, Reigber et al. 2002), Gravity Recovery and Climate Experiment (GRACE, Tapley et al. 2005), and Gravity Field and Steady-State Ocean Circulation Explorer (GOCE, Klees et al. 2000).

For recovery of the static marine gravity field, measurements of the slope of the ocean surface are usually used, which are computed by differencing altimeter-derived SSH along satellite altimeter ground tracks. However, it is well known that the current standard altimeter data products (e.g., GDR) cannot provide precise SSHs in coastal areas, where altimeter SSH measurements are contaminated significantly by reflections from land and inaccurate geophysical corrections (e.g., ocean tides). As such, the accuracy of altimeter-derived gravity anomalies in coasts is usually lower than those over open oceans (Fairhead et al. 2001; Sandwell and Smith 2005; Hwang et al. 2006).

The altimeter data in coastal areas can be improved by retracking (or reprocessing) the raw altimeter measurement of waveforms (e.g., Deng and Featherstone 2006; Hwang et al. 2006). The waveform is the profile of backscattered power recorded by a satellite-borne radar altimeter (e.g., Brown 1977). In this section, the contaminated distances from the Australian coastline, within which the altimeter SSH data are inaccurate, are first quantified from analyzing altimeter waveforms. A coastal waveform retracking system is then developed to improve the altimeter data near the coast. The system reprocesses altimeter waveforms including those across-shelf and nearshore data, and is different from the sea-surface mapping system developed in Sect. 18.3. Here for a comprehensive review of altimeter waveforms and retracking methods, the reader is referred to the chapter by Gommenginger et al., this volume.

The precision of the marine gravity field is also investigated in the Australian coastal areas in this section. The quality of Australian ship-track gravity data is examined using altimeter-derived gravity anomalies. The precision of the existing AUSGeoid98 (Featherstone et al. 2001) and some recent GRACE-derived gravimetric geoid models are assessed using an altimeter-derived mean sea surface (MSS) and the ocean climatology of CARS (Ridgway et al. 2002).

18.5.1
How Accurate are Altimeter Data near the Australian Coast?

18.5.1.1
Altimeter Coastal-contaminated Distance Around Australia

The altimeter SSHs are often in error near the coast, principally because the raw altimeter measurements of waveforms are contaminated by the complex nature of echoes returned from rapidly varying coastal topographic surfaces (both land and sea). To quantify the affected distance from the Australian shoreline, five cycles of Poseidon (January 1998 to January 1999) and one cycle of ERS-2 (March to April 1999) 20 Hz waveforms over the Australian coast are used. The distance is estimated by threshold retracking the waveforms and computing descriptive statistics of mean waveforms across a 350-km zone offshore from the Australian coastline (Deng et al. 2002).

The threshold retracking algorithm is based on a rectangular box (cf. Fig. 4.7) that has the same centroid position as the waveform samples. The dimensions of the rectangle (i.e., amplitude, width, and centre of gravity) are computed using the Offset Centre of Gravity method (Wingham et al. 1986). The amplitude estimate is then used to threshold retrack the waveform's leading edge at 25%, 50%, or 75% of the rectangle's amplitude (Davis 1995). Following the assumptions of statistical homogeneity of open ocean surfaces defined by Brown (1977), the 50% threshold-tracking gate from ocean waveforms should correspond to the position of the tracking reference gate in the range window. Here, the tracking reference gate is predefined in the range window, which is related to the midpoint of waveform's leading edge, such as gate 32.5 for ERS-2 and gate 31.5 for Poseidon. However, this is not always the case for waveforms sampled over near-coastal surfaces.

To investigate the contaminated zones, the Australian coast was subdivided into ten geographical areas of 15° × 15°. Within these areas, ERS-2 20 Hz waveforms were threshold-retracked at 50%. If the estimated tracking gate was located between gates 31–33, the waveform was considered not to be affected by land or coastal sea surface states; otherwise the data were considered to be contaminated. The criterion of gates 31–33 is referenced to the statistics of 50% retracking gates over ocean areas 50–350 km offshore Australia, where most retracking gates (96%) lie within 31–33.

From the threshold-retracked results around Australia, it was found that the ERS-2 contaminated waveforms comprise about 0.2–2.9% of all data depending on different sub-areas, but they are all in close proximity to the coast (cf. Deng et al. 2002). The longer contaminated distances occur off the north Australian coast (~19 km) and off the south Australian coast (~22 km) near Tasmania, where there are complex shoreline and coastal topographies. The eastern and western Australian coasts show less contaminated data and shorter contaminated distances (~8 km).

In order to provide a statistically meaningful assessment of contaminated areas, mean waveforms and standard deviations were also computed over three distance bands (0–350 km, 0–50 km, and 0–10 km). The contaminated distance was determined by analyzing shape variations between the mean coastal waveform and theoretical waveform (Brown 1977). As an example, mean waveforms and standard deviations in five 2-km-wide bands are shown in Fig. 18.11. Waveforms within 10 km of the coastline are distorted, especially in the 0–4 km distance band. The closer to the coastline, the larger the standard deviations. Beyond 8–10 km (to 350 km), the mean waveform shapes for both altimeters tend towards the ocean waveform.

Overall, the maximum contaminated distances vary from ~7 km to ~22 km, depending upon locations and features of the shoreline, as well as the altimeter mission used. A mean contaminated distance of ~8 km for Poseidon and ~10 km for ERS-2 is typical for most of the Australian coastal regions.

It is therefore recommended that such contaminated measurements must first be detected and then improved (or removed) prior to their being included in geodetic and oceanographic solutions in near-coastal areas. Until then, altimeter SSH or range data should be interpreted with some caution for distances less than, say, ~22 km from a coastline, and discarded altogether for distances less than 4 km.

To improve the altimeter range precision in coastal regions, retracking the raw altimeter waveform is an attractive option. This will be discussed next.

18.5.1.2
Coastal Retracking System for Altimeter Waveforms

Retracking the altimeter waveforms can refine the parameter estimates routinely given as part of GDR products. Of these parameters, we are primarily interested in the arrival time

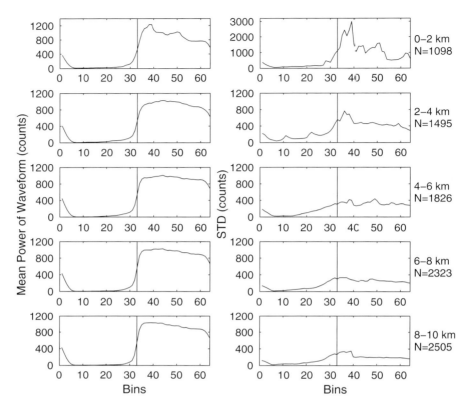

Fig. 18.11 Mean waveform (*left*) and standard deviation (*right*) for ERS-2 in five 2-km-wide bands from the Australian coastline

that is defined as the position of the actual tracking gate in the waveform sample (or range) window. The arrival-time estimate can be used to infer the range correction by computing the offset with respect to the tracking reference gate. This correction is then applied to the range calculated by the on-board algorithm to improve its precision.

An analysis of coastal waveform shapes (Deng 2004) showed that raw waveforms present diverse shapes in the contaminated zone around Australia, suggesting that they cannot be retracked by using only a single retracking algorithm. Deng (2004) and Deng and Featherstone (2006) have therefore developed a coastal waveform retracking system. The system consists of two retracking techniques and seven retrackers.

These two techniques are the iteratively reweighted least-square fitting and threshold retracking. The retrackers include the Brown (1977) model, five- and nine-parameter models (both with linearly and exponentially decaying trailing edges), and threshold retracking algorithm with 30% and 50% threshold levels. The iterative least-squares fitting procedure estimates parameters by making the altimeter waveform coincide with an analytical return power model (cf. Sect. 4.7.3.1). The use of different fitting models and threshold levels ensures that the system can deal with various waveforms in coastal areas.

Two issues are critical for designing a retracking system: waveform classification and bias determination. Correctly classifying waveforms enables their shape to be retracked by an appropriate model (or algorithm), while accurately detecting the bias between fitting and threshold retrackers will produce consistent range estimates. The biases are caused by the different algorithms (cf. Fairhead et al. 2001). Threshold retracking takes no account of noise in the waveform (especially in the trailing edge), and how this noise affects the leading edge from one waveform sample to the next. The fitting algorithm is affected by noise in the trailing edge. Thus, the bias results from the noise in the fitting algorithm, as well as too simplistic assumptions in the threshold retracking.

As a case study, this system was applied to the Australian coasts using two cycles (42 and 43) of ERS-2 20 Hz waveform data (March to May 1999). The contaminated waveforms in ten 15° × 15° Australian coastal regions were categorized (cf. Deng 2004; see also Fig. 4.4 of Sect. 4.3.5 in Chapter by Gommenginger et al., this volume). Most contaminated waveforms (~90%) have ocean-like shapes and can be retracked using the least-squares fitting procedure. The remaining waveforms (~10%) have non-ocean-like shapes, such as single pre-peaked, high-peaked, single mid-peaked, post-peaked, double pre-peaked, double post-peaked, and multi-peaked waveforms (Fig. 4.4). Although these non-ocean-like waveforms only comprise a small percentage of total contaminated waveforms, they are found much closer to the coastline (1.73–3.12 km on average). Some of them cannot be fitted to a model and should be retracked using other retrackers that are the threshold retrackers with 30% or 50% levels in this particular system.

The biases among different retrackers were investigated using ERS-2 waveforms over an open ocean area (50–350 km from the coastline, 30°S−15°S, 105°E−120°E). The waveforms were retracked using the 50% threshold level, the ocean model, and a five-parameter model. Retracked SSHs were used to compute the SSH differences between different retrackers at each location. A cubic polynomial function was fitted to all SSH differences. The linear term was used as the bias estimate. The bias (< 1 cm) between the five-parameter and ocean fitting models is small and almost statistically insignificant. However, significant bias of ~56.7 cm was found between the 50% threshold algorithm and both the five-parameter and ocean models. These relative biases among retracking algorithms were then added to or subtracted from the threshold-retracked altimeter ranges.

To assess the precision of the retracked data and to determine how the retracked altimeter SSH profiles represent the true sea surface, the SSH was compared before and after retracking with an external quasi-independent reference from the AUSGeoid98 gravimetric geoid model (Featherstone et al. 2001). The term quasi-independent reflects the fact that altimeter-derived gravity anomalies were one of the many data sources for AUSGeoid98. The comparison assumed that AUSGeoid98 was correct in the coastal zone, which was the case at the time of this research in 2004 though more accurate geoid models have been recently released (see Sect. 18.5.3).

Strictly, the instantaneous SSH is not equivalent to the geoid. The geoid is the equipotential surface of the Earth's gravity field. The difference between the altimeter SSH and geoid includes the time-variant component (e.g., tides) and the time-invariant component (e.g., the mean dynamic ocean topography, DOT). Assuming that the time-variant component can be removed through averaging and the mean DOT is a bias in the coastal zone, the verification of any improvement in the SSH will be seen in the standard deviations (STDs), with a smaller value indicating an improvement in the SSH.

Descriptive statistics of the differences between SSH data (20 Hz) before and after retracking with AUSGeoid98 were computed (Fig. 18.12). For ERS-2 cycles 42 and 43, Fig. 18.12 indicates that the values of STDs decrease after retracking. There is a significant improvement over distances 0–15 km from the coastline, with the STDs dropping to ~1 m. Beyond 15 km, there is a slight decrease in the STDs. These results suggest that a more precise SSH profile can be obtained from retracking altimeter waveforms, especially 0–15 km from the coastline.

Using this particular coastal waveform retracking system, the contaminated altimeter SSHs have been improved and extended ~10–15 km shoreward in Australian coastal regions compared with unretracked data.

18.5.2
Using Altimetry to Detect Ship-track Gravity Data Errors

The principal problem with marine gravity measurements made onboard ships (ship-track gravimetry) is that they are subject to biases because of drift in relative gravimeters, as well as incorrect positioning and velocity, resulting in incorrect Eötvös corrections (e.g., Wessel and

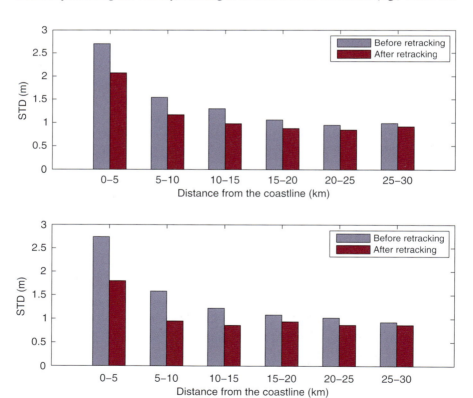

Fig. 18.12 STD of the mean difference between the SSHs and AUSGeoid98 heights before and after retracking in six 5-km-wide distance bands (*top*: cycle 42, and *bottom*: cycle 43)

Watts 1988). In practice, this bias problem is reduced by observing ship-track gravimetry in a hashed pattern (cf. Fig. 18.13) that allows for a crossover adjustment to enforce consistency at the cross points. Without this, ship-track gravimetry should not be relied upon. Ship-track gravity data used to be included in the Australian national gravity database from Geoscience Australia (GA), the acquisition of most of which is described in Symonds and Willcox (1976), Mather et al. (1976), and Murray (1997). It will be shown here that these data were unadjusted and have since been withdrawn from the GA database (Featherstone 2009).

Using unadjusted, and thus biased, ship-track gravity data will invalidate any geological, geophysical, or geodetic interpretation or application. For instance, a biased ship-track could indicate spurious features that may lead to incorrect follow-up surveys, needlessly taking additional resources. Another example is the [then-incorrect] assumption of crossover-adjusted GA ship-track data in the computation of AUSGeoid98 (Featherstone et al. 2001), which has corrupted this geoid model in some coastal regions (e.g., Kirby 2003; Featherstone 2008; Claessens et al. 2009). The unadjusted state of Australian ship-track data also raises questions as to the validity of other studies that have utilized them (e.g., Mather et al. 1976; Zhang 1998).

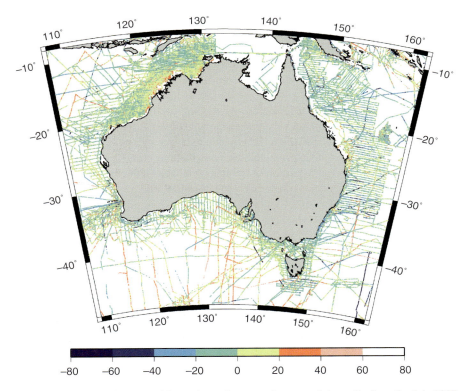

Fig. 18.13 Differences between ship-track gravity anomalies around Australia from the July 2007 data release from GA and marine gravity anomalies from version 16.1 of the Sandwell and Smith (2005) 1 arc-minute grid (units in mGal; Lambert projection)

Petkovic et al. (2001) described a "re-leveling" of the GA ship-track gravity data.[1] GA ship-track data were adjusted to fit onto Sandwell's v7.2 grid of the then-available multi-mission satellite altimeter-derived marine gravity anomalies from one group (Sandwell and Smith 1997). This seems to have introduced errors in the coastal zone, as demonstrated indirectly by Featherstone (2003) and confirmed by P. Petkovic (2002, personal communication) for the Bass Strait region. The problem with adjusting ship-track data in this way is the implicit assumption that the satellite altimetry is correct, which is especially not the case in the coastal zone as shown in Sect. 18.5.1 (cf. Deng and Featherstone 2006; Hwang et al. 2006) or in continental shelf and shallow-sea areas where tides and tropospheric corrections to the altimeter ranges are poorly modeled (cf. Andersen and Knudsen 2000). When compared with independent external data, they become even more pronounced.

Marine gravity anomalies can be deduced from satellite radar altimetry, where the measured and time-averaged SSH can be converted to gravity using a variety of inverse methods (Haxby et al. 1983; Olgiati et al. 1995; Sandwell and Smith 1997; Andersen and Knudsen 1998; Hwang et al. 1998; Wang 2001; Hwang et al. 2002; Sandwell and Smith 2005). The benefit of altimeter-derived marine gravity anomalies is that they are derived from relatively homogeneous data coverage and several different satellite missions can be merged (cf. Hwang and Parsons 1996). They are also not subject to drift- and navigation-based errors. Importantly in the context of this study, these altimeter-derived gravity anomalies are totally independent of the ship-track data.

The altimeter-derived marine gravity anomalies used here around Australia come from the v16.1 grid of multi-mission satellite altimetry produced by Sandwell and Smith (2005; http://topex.ucsd.edu/WWW_html/mar_grav.html). The key improvements over the original treatise (Sandwell and Smith 1997) are the use of more and recent altimeter data, a different gridding algorithm, and the use of re-tracked altimeter waveforms that can improve the gravity anomalies in the coastal zone (cf. Hwang et al. 2006).

Fig. 18.13 shows the deficiencies in the GA ship-track gravity data, where biases of over ±60 mGal (most < −60 mGal) in magnitude are evident among crossing tracks, showing that no crossover adjustment has been applied. Computing the descriptive statistics of the differences for all 149,961 observations gives max = 972.201 mGal, min = −181.905 mGal, mean = −1.383 mGal, and STD = 13.492 mGal. This very large range necessitated the use of a truncated z-scale in Fig. 18.13. In some areas, however (e.g., over parts of the North West Shelf), the ship-track data do appear to be more homogeneous, suggesting that some crossover adjustment may have been applied to these data, but this cannot be confirmed at present.

Therefore, the retracked satellite altimeter data are recommended as a superior alternative source of marine gravity anomalies around Australia, which might also be applied to other coastal regions in the world. At the very least, users of ship-track gravity data should carefully screen ship-track gravity data to ensure that they have been crossover-adjusted before they are relied upon in any Earth-science study.

---

[1] Note that an attempt to crossover-adjust the GA ship-track data in the classical way failed, because the scarcity of the tracks in most regions rendered the least-squares adjustment ill-conditioned and thus unreliable. This may explain the approach taken by Petkovic et al. (2001) in the study contracted by GA to Intrepid Geophysics.

18.5.3
Assessment of AUSGeoid98 offshore of Western Australia

This section assesses the quality of AUSGeoid98 and recently GRACE-derived geoid models in the LC region offshore of Western Australia (20°S–45°S, 108°E–130°E).

AUSGeoid98 (Featherstone et al. 2001) is a regional gravimetric geoid model, which consists of a 2′ × 2′ grid (approximately 3.6 km) of geoid-ellipsoid separations relative to the geodetic reference system 1980 (GRS80) ellipsoid. In the LC region, the long-wavelength components of AUSGeoid98 are computed using the Earth gravitational model 1996 (EGM96) complete up to degree and order 360. The short-wavelength components were computed mainly from two sources. These were the 1996 release of the Australian ship-track gravity measurements (mostly without cross-adjustment) around Australia within the continental shelf region, and satellite altimeter-derived marine gravity anomalies (without retracking procedure) for open ocean regions and within some 15–30 km of the coastline. As discussed in Sect. 18.5.1 and 18.5.2, these unadjusted ship-track gravity data and unretracked altimeter SSHs introduce errors into AUSGeoid98. The assessment is therefore to quantify how AUSGeoid98 is affected by the errors.

The recently released global geopotential models (GGMs) from GRACE have been significantly improved in terms of accuracy and resolution (e.g., Tapley et al. 2005; Förste et al. 2008). These GRACE-derived GGMs provide now a precise determination of the long-wavelength components of the geoid on a global scale. However, the error estimates for these GGMs are global values and thus not necessarily representative of the GGM in a particular region. As such, it is informative to assess the precision of GRACE-derived GGMs to understand their error characteristics in the region of interest.

The assessment of precision of the geoid models is accomplished by computing the mean DOT. The mean DOT is the time-invariant part of the SSH that arises from the ocean's circulation. Its gradient is related to the ocean surface velocity field through geostrophy. The mean DOT relative to a reference ellipsoid can be determined from subtracting an estimate of the Earth's geoid from an altimeter-observed MSS (e.g., Tapley et al. 2003), provided that adequate estimates of the geoid are known. The mean DOT (or steric height) referenced to an ocean surface of "no motion" can also be derived from a hydrographical climatology (e.g., Jayne 2006). Therefore, the accuracy of the geoid model can be assessed by statistical comparison between computed mean DOTs when considering their spatial and regional difference.

Four GGMs were assessed over a range of wavelengths (Deng et al. 2009). These geoid models include AUSGeoid98, EGM96, GRACE-derived GGM02C (Tapley et al. 2005), and EIGEN-GL04C (Förste et al. 2008). In addition, the high-resolution ocean climatology of CARS (Ridgway et al. 2002) and an altimeter-derived MSS grid (KMS04, http://www.spacecenter.dk) were used. CARS was chosen because it is a representation of the mean field of the ocean as indicated by the historical data from 1900 to 2006 around Australia. The mean DOT grids were computed from the geoid and KMS04 MSS at various spatial scales. The mean DOT computed from CARS is used as an independent in situ "truth."

Because the various grids and models are of different spatial resolutions, it is important to smooth them to a comparable level before computing the mean DOT (cf. Bingham et al. 2008). All data sets of GGMs, MSS, and CARS were smoothed in the spatial domain using the Hamming filter (cf. Jayne 2006) for different spherical harmonic degrees or spatial scales corresponding to degrees of 60, 120, 150, 200, and 360. The fields of mean DOTs and CARS steric heights with the same spatial scale were then compared to each other.

As the CARS steric height is referenced to the 2000-m depth of "no motion," while the GGMs are referred to the GRS80 ellipsoid, there is an offset between mean DOTs from CARS and a GGM. The offset was estimated by averaging the difference between the CARS's steric height and mean DOT from a GGM related to the same filter radius. The offset is different for each GGM and the chosen spatial resolution. As an example, the mean of offsets of GGMs is 2.240 ± 0.017 m for degree $N = 60$ (~333 km) in the LC region. The offset was removed before statistical comparison.

The standard deviations of the differences between the mean DOTs from CARS and KMS04 MSS referred to the geoid at various spatial scales are shown in Fig. 18.14. Models of GGM02C+EGM96, obtained by using EGM96 coefficients to fill in above the degree-200 GGM02C geoid grid, and EIGEN-GL04C show higher precision than EGM96 and AUSGeoid98 around all degrees. They have the best accuracy (~6.6 cm) around degree-60, decreasing down to ~9.5 cm at degree 360. AUSGeoid98 is the worst, with the STDs 11.3 cm at degree-60 decreasing to ~13.1 cm for most other degrees. EGM96 is slightly better than AUSGeoid98 by ~1 cm for all degrees tested.

The degradation of AUSGeoid98 observed in this section (compared to EGM96) confirms that the unadjusted ship-track gravity data and unretracked altimeter-derived gravity

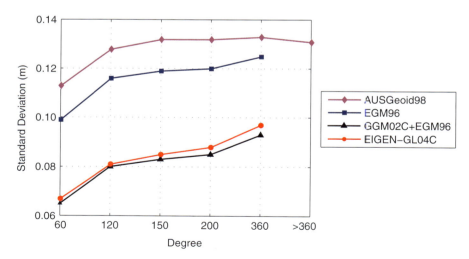

Fig. 18.14 Standard deviation of the CARS steric height minuses mean DOT from a geoid model at various spatial scales. GGM02C+EGM96 complete to degree/order 360 is obtained using EGM96 coefficients to fill in above the degree-200 GGM02C geoid grid

anomalies are inaccurate. This agrees with results obtained in previous Sect. 18.5.1 and 18.5.2. Results also suggest that AUSGeoid98 needs to be re-estimated using a latest GRACE-derived GGM, such as the recently available EGM2008 (Pavlis et al. 2008). This work is underway at present.

18.6
Conclusions and Perspectives

As illustrated in this chapter, together with other data sources (e.g., from remote sensing techniques and in situ observations) and modelling techniques, satellite altimetry has made a unique contribution to a number of key research areas in the Australian region, including observing and understanding sea level rise, the structure and variability of the major boundary current systems, and determining and verifying the marine gravity field.

The Australian sea level mapping system does not yet routinely use the new 'coastally processed' altimeter data products discussed in this book, and in our Sect. 18.5. Instead, it merges the off-shelf satellite altimeter SLA data with along-coast interpolated tide-gauge SLA data to construct a SLA field that extends to the coast, enabling estimation of average geostrophic velocities over the shelf and upper slope. Incorporation of PISTACH data in this system has recently commenced and will be discussed elsewhere.

A waveform retracking system, based upon a systematic analysis of satellite radar altimeter waveforms around Australia, has been developed. It uses both the iteratively weighted least squares fitting and threshold retracking algorithms. It improves the altimeter SSH through reprocessing altimeter waveforms. The improved altimeter SSH fields are extended from beyond ~22 km to beyond ~2.5 km from the Australian coastline. These results demonstrated that both systems have potential to be applied to other coastal regions.

The ocean surface currents around Australia have been characterized on seasonal-to-interannual scales by combining a seasonal climatology from altimeter SLA with an absolute mean surface topography from assimilating the CARS temperature and salinity climatology into the OFAM model. This provides a new look into the dynamics of the major boundary current systems. The results show that the major coastal currents experience significant seasonal and annual variations. Specifically, the baroclinic transport of the EAC has, for the first time, been systematically examined using a 15-year (1992–2006) time series from satellite altimetry along a XBT sections. The net EAC flow across the Tasman Sea shows a range of temporal signals from eddy-scale, seasonal, interannual to decadal. The net EAC transport ranges from 5 Sv in 1995 to a maximum of 16 Sv in 2001.

Satellite altimeter data for the period 1993–2007 clearly indicate a rise in sea level in the Australian region. The north coasts of Australia have the maximum of sea level rise up to ~10 mm/year. The south-eastern and eastern Australian coasts have shown the minimum of sea level rise, at 2–4 mm/year. A significant finding in this region is the higher rate of offshore sea level compared with coastal sea level rise as a result of variations in the coastal ocean currents.

It has been shown that marine gravity anomalies from unretracked altimeter SSH data and unadjusted ship-track gravity measurements are inaccurate around Australia. Using these data has resulted in the degradation of AUSGeoid98 in the LC region. It is recommended that the unadjusted marine gravity data should be neglected, used with extreme caution, or replaced by retracked satellite altimeter data for geological, geophysical, and geodetic studies.

This chapter also presents important issues that are to be addressed in the future. For example, the coastal altimeter SSH field is expected to be further improved by adding retracked altimeter data to the data sources of the sea level mapping system. The improved coastal altimeter SSH data are also expected to be used in the examination of sea level rise and coastal ocean currents. The correction of tectonic vertical deformation at tide gauges should be studied and applied to the tide-gauge sea level data. The improved knowledge of the Earth's time varying gravity and satellite altimetry is providing an opportunity of interdiscipline research for geodetic and oceanographic communities. The centimeter precision of geoid at a spatial resolution of 100 km is expected in the near future from satellite gravity solutions from CHAMP, GRACE, and GOCE. For the oceans, this means that absolute sea level fields and associated ocean currents can be independently estimated from satellite altimetry with increased accuracy.

With continuous altimeter missions (e.g., Jason-2, Altika and CryoSat-2), other data sets (e.g., in situ temperature and salinity profiles and other remote sensing data), and recent (and near future) substantial improvement of the geoid model, new efforts will improve our description and understanding of the variability in sea level and ocean boundary currents in coastal oceans, and identifying the propagation of oceanic signals on climate timescales.

Acknowledgements We thank Prof Richard Coleman for valuable discussions and comments on various portions of this chapter. We would also like to thank the anonymous reviewers, CSIRO reviewers (Ms Anne-Elise Nieblas and Dr Sophie Bestley), and editors (Dr Stefano Vignudelli, Dr Andrey Kostianoy, Dr Paolo Cipollini, and Dr Jérôme Benveniste) for their helpful comments on this chapter.

References

Andersen OB, Knudsen P (1998) Global marine gravity field from the ERS-1 and Geosat geodetic mission altimetry. J Geophys Res Oceans 103(C4):8129–8137

Andersen OB, Knudsen P (2000) The role of satellite altimetry in gravity field modelling in coastal areas. Phys Chem Earth 25(1):17–24

Baines PG, Edwards RJ, Fandry CB (1983) Observations of a new baroclinic current along the western continental slope of Bass Strait. Aust J Fresh Res 34:155–157

Benada JR (1997) PO.DAAC merged GDR (TOPEX/Poseidon) generation B user's handbook, Version 2.0. JPL PO.DAAC 068.D002

Bingham RJ, Haines K, Hughes CW (2008) Calculating the ocean's mean dynamic topography from a mean sea surface and a geoid. J Atmos and Ocean Technol. doi:10.1175/2008JTECHO568.1

Bowen M, Emery WJ, Wilkin J et al (2002) Extracting multi-year surface currents from sequential thermal imagery using the Maximum Cross Correlation technique. J Atm Ocean Technol 19: 1665–1676

Bowen M, Wilkin JL, Emery WJ (2005) Variability and forcing of the East Australian Current. J Geophys Res. doi:10.1029/2004JC002533

Brown GS (1977) The average impulse response of a rough surface and its applications. IEEE Trans Antenns Propag AP-25(1):67–74

Church JA (1987) The East Australian Current adjacent to the Great Barrier Reef. Aust J Mar Fresh Res 38:671–683

Church JA, Craig PD (1998) Australia's shelf seas – diversity and domplexity. In: Robinson AR (ed) The sea, vol 11. Wiley, New York, pp 933–964

Church JA, Freeland HJ (1987) The energy source for the coastal-trapped waves in the Australian coastal experiment region. J Phys Oceanogr 17:289–300

Church JA, White NJ (2006) A 20th century acceleration in global sea level rise. Geophys Res Lett. doi:10.1029/2005GL024826

Church JA, Gregory JM, Huybrechts P et al (2001) Changes in sea level. In Climate change 2001: the scientific basis. In: Houghton JT, Ding Y, Griggs DJ, Noguer M, van der Linden P, Dai X, Maskell K, Johnson CI (eds) Contribution of working group 1 to the third assessment report of the Intergovernmental Panel on Climate Change, Cambridge University Press, pp 639–694

Church JA, White NJ, Coleman R et al (2004) Estimates of the regional distribution of sea level rise over the 1950 to 2000 period. J Climate 17(13):2609–2625

Cirano M, Middleton JF (2004) Aspects of the mean wintertime circulation along Australia's southern shelves: numerical studies. J Phys Oceanogr 34:668–684

Claessens SJ, Featherstone WE, Mujitsarama IM et al (2009) Is Australian data really validating EGM2008, or is EGM2008 just in/validating Australian data? Newton's Bull 4:207–251

Cleveland WS, Devlin SJ (1988) Locally weighted regression: an approach to regression analysis by local fitting. J Am Stat Ass 83:596–610

Cresswell GR, Golding TJ (1980) Observations of a south-flowing current in the southeastern Indian Ocean. Deep-Sea Res 27A:449–466

D'Adamo N, Fandry C, Buchan S et al (2009) Northern sources of the Leeuwin Current and the "Holloway Current" on the North West Shelf. J Roy Soc West Austr 92(2): 53–66

Davis CH (1995) Growth of the Greenland ice sheet: a performance assessment of altimeter retracking algorithms. IEEE Trans Geosci Remote Sens 33(5):1108–1116

Deng X (2004) Improvement of geodetic parameter estimation in coastal regions from satellite radar altimetry. PhD thesis, Curtin University of Technology, Perth, Australia, p. 248

Deng X, Featherstone WE, Hwang C et al (2002) Estimation of contamination of ERS-2 and POSEIDON satellite radar altimetry close to the coasts of Australia. Mar Geod 25(4):249–271

Deng X, Featherstone WE (2006) A coastal retracking system for satellite radar altimeter waveforms: application to ERS-2 around Australia. J Geophys Res – Oceans. doi:10.1029/2005JC003039

Deng X, Coleman R, Featherstone WE et al (2009) Assessment of geoid models offshore Western Australia using in-situ measurements. J Coastal Res. doi:10.2112/07-0972.1

Ducet N, Le Traon PY, Reverdin G (2000) Global high-resolution mapping of ocean circulation from TOPEX/Poseidon and ERS-1 and -2. J Geophys Res 105:19477–19498

Fairhead JD, Green CM, Odegard ME (2001) Satellite-derived gravity having an impact on marine exploration. Lead Edge 20:873–876

Featherstone WE (2003) Comparison of different satellite altimeter-derived gravity anomaly grids with ship-borne gravity data around Australia. In: Tziavos IN (ed) Gravity and Geoid. Ziti Editions, Thessaloniki, pp 326–331

Featherstone WE (2008) GNSS-based heighting in Australia: current, emerging and future issues. J Spatial Sci 53(2):115–133

Featherstone WE (2009) Only use ship-track gravity data with caution: a case-study around Australia. Aust J Earth Sci 56(2). doi:10.1080/08120090802547025

Featherstone WE, Kirby JF, Kearsley AHW et al (2001) The AUSGeoid98 geoid model of Australia: data treatment, computations and comparisons with GPS-levelling data. J Geod. doi:10.1007/s001900100177

Feng M, Meyers G, Pearce A et al (2003) Annual and semi-annual variations of the Leeuwin Current at 32°S. J Geophys Res. doi:10.1029/2002JC001763

Förste C, Schmidt R, Stubenvoll R et al (2008) The GeoForschungsZentrum Potsdam/Groupe de Recherche de Gèodésie Spatiale satellite-only and combined gravity field models: EIGEN-GL04S1 and EIGEN-GL04C. J Geod. doi:10.1007/s00190-007-0183-8

Godfrey JS, Vaudrey DJ, Hahn SD (1986) Observations of the shelf-edge current south of Australia, Winter 1982. J Phys Oceanogr 16:668–679

Godfrey JS, Alexiou A, Ilahude AG et al (1995) The role of the Indian Ocean in the global climate system: recommendations regarding the global ocean observing system, Report of the Ocean Observing System Development Panel. Texas A&M University, College Station, TX, USA, 89 pp

Griffin DA, Middleton JH (1991) Local and remote wind forcing of New South Wales inner shelf currents and sea level. J Phys Oceanogr 21:304–322

Hahn SD (1986) Physical structure of the waters of the South Australian continental shelf. Res Rep 45, FIAMS, Flinders University, South Australia

Haxby WF, Karner GD, Labrecque JL et al (1983) Digital images of combined oceanic and continental data sets and their use in tectonic studies. EOS Trans Am: Geophys Union 64(52): 995–1004

Hill KL, Rintoul SR, Coleman R et al (2008) Wind forced low frequency variability of the East Australia Current. Geophys Res Lett. doi:10.1029/2007GL032912

Hill KL, Rintoul SR, Ridgway KR et al (2009) Decadal changes in the South Pacific western boundary current system revealed in observations and reanalysis state estimates. J Geophys Res (in press).

Holgate SJ, Woodworth PL (2004) Evidence for enhanced coastal sea level rise during the 1990s. Geophys Res Lett. doi:10.1029/2004GL019626

Horton R, Herweijer C, Rosenzweig C et al (2008) Sea level rise projections for current generation CGCMs based on the semi-empirical method. Geophys Res Lett. doi:10.1029/2007GL032486

Hwang C, Parsons B (1996) A optimal procedure for deriving marine gravity from multi-satellite altimetry. J Geophys Int 125:705–719

Hwang C, Kao EC, Parsons BE (1998) Global derivation of marine gravity anomalies from Seasat, Geosat, ERS-1 and TOPEX/Poseidon altimeter data. Geophys J Int 134:449–459

Hwang C, Hsu HY, Jang R-J (2002) Global mean sea surface and marine gravity anomalies from multi-satellite altimetry: applications of deflection-geoid and inverse Vening Meinesz formulae. J Geod. doi:10.1007/s00190-002-0265-6

Hwang C, Guo J, Deng X et al (2006) Coastal gravity anomaly from retracked Geosat/GM altimetry: improvement, limitation and the role of airborne gravity data. J Geod. doi:10.1007/s00190-006-0052-x

IPCC (2007) Climate change 2007: the physical science basis. Contribution of working group 1 to the fourth assessment report of the Intergovernmental Panel on Climate Change. In: Solomon S, Qin D, Manning M et al (eds) Intergovernmental Panel on Climate Change, Cambridge/New York

Jayne SR (2006) Circulation of the North Atlantic Ocean from altimetry and the gravity recovery and climate experiment geoid. J Geophys Res. doi:10.1029/2005JC003128

Kessler W, Gourdeau L (2007) The annual cycle of circulation of the southwest Subtropical Pacific, analyzed in an ocean GCM. J Phys Oceanogr 37:1610–1627

Kirby JF (2003) On the combination of gravity anomalies and gravity disturbances for geoid determination in Western Australia. J Geod. doi:10.1007/s00190-003-0334-5

Klees R, Koop R, Visser P et al (2000) Efficient gravity field recovery from GOCE gradient observations. J Geod 74(7–8):561–571

Leuliette E, Nerem SR, Mitchum T (2004) Calibration of TOPEX/Poseidon and Jason altimeter data to construct a continuous record of mean sea level change. Mar Geod 27:79–94

Mata MM, Tomczak M, Wijffels S et al (2000) East Australian Current volume transports at 30°S: Estimates from the WOCE hydrographic sections PR11/P6 and the PCM3 current meter array. J Geophys Res 105:28,509–28,526

Mata MM, Wijffels S, Church JA et al (2006a) Eddy shedding and energy conversions in the East Australian Current. J Geophys Res. doi:10.1029/2006JC003592

Mata MM, Wijffels S, Church JA et al (2006b) Statistical description of the East Australian Current low-frequency variability from the World Ocean Circulation Experiment PCM3 current meter array. J Marine Fresh Res 57:273–290

Mather RS, Rizos C, Hirsch B et al (1976) An Australian gravity data bank for sea surface topography determinations (AUSGAD76). Unisurv G25, School of Surveying, University of New South Wales, Sydney, pp 54–84

Maxwell JGH, Cresswell GR (1981) Dispersal of tropical marine fauna to the Great Australian Bight by the Leeuwin Current. Aust J Fresh Res 32:493–500

Meehl GA, Stocker TF, Collins W et al (2007) Global climate projections. Climate change 2007: the physical science basis. In: Solomon S, Qin D, Manning M (eds) Contribution of working group 1 to the fourth assessment report of the Intergovernmental Panel on Climate Change. Cambridge University Press, United Kingdom/New York

Middleton JF, Cirano M (2002) A northern boundary current along Australia's southern shelves: the Flinders Current. J Geophys Res. doi:10.1029/2000JC000701

Milne GA, Mitrovica JX, Davis JL (1999) Near-field hydro-isostasy: the implementation of a revised sea level equation. Geophys J Int 139:464–482

Murray AS (1997) The Australian national gravity database. AGSO J Aust Geol Geophys 17(1):145–155

Oke PR, Brassington GB, Griffin DA et al (2008) The Bluelink ocean data assimilation system (BODAS). Ocean Model. doi:10.1016/j.ocemod.2007.11.002

Olgiati A, Balmino G, Sarrailh M et al (1995) Gravity anomalies from satellite altimetry: comparison between computation via geoid heights and via deflections of the vertical. Bull Géodésique 69(3):252–260

Pavlis NK, Holmes SA, Kenyon SC et al (2008) Earth gravitational model to degree 2160: EGM2008, paper presented to the European Geosciences Union General Assembly, Vienna, Austria, April 2008

Pariwono JI, Byre JAT, Lennon GW (1986) Long-period variations in sea level in Australasia, Geophys. J Royal Astron Soc 87:43–54

Petkovic P, Fitzgerald D, Brett J et al (2001) Potential field and bathymetry grids of Australia's margins. Proceedings of the ASEG 15th geophysical conference and exhibition, Brisbane, August [CD-ROM]

Rahmstorf S (2007) A semi-empirical approach to projecting future sea level rise. Science. doi:10.1126/science.1135456

Rahmstorf S, Cazenave DA, Church JA et al (2007) Recent climate observations compared to projections. Science. doi:10.1126/science.1136843

Reigber C, Balmino G, Schwintzer P et al (2002) A high quality global gravity field model from CHAMP GPS tracking data and accelerometer (EIGEN-1S). Geophys Res Lett. doi: 10.1029/2002GL015064

Ridgway KR (2007) Seasonal circulation around Tasmania – an interface between eastern and western boundary dynamics. J Geophys Res. doi:10.1029/2006JC003898

Ridgway KR, Condie SA (2004) The 5500-km long boundary flow off western and southern Australia. J Geophys Res. doi:10.1029/2003JC001921

Ridgway KR, Dunn JR (2003) Mesoscale structure of the mean East Australian Current system and its relationship with topography. Prog Oceanogr 56:189–222

Ridgway KR, Godfrey JS (1997) Seasonal cycle of the East Australian Current. J Geophys Res 102:22,921–22,936

Ridgway KR, Dunn JR, Wilkin JL (2002) Ocean interpolation by four-dimensional least squares application to the waters around Australia. J Atmos Ocean Tech 19:1357–1375

Ridgway KR, Coleman RC, Bailey RJ et al (2008) Decadal variability of East Australian Current transport inferred from repeated hi-density XBT transects, a CTD survey and satellite altimetry. J Geophys Res. doi:10.1029/2007JC004664

Roemmich D, Gilson J, Willis J et al (2005) Closing the time-varying mass and heat budgets for large ocean areas: the Tasman Box. J Clim 18:2330–2343

Roemmich D, Gilson J, Davis R et al (2007) Decadal spin-up of the South Pacific subtropical gyre. J Phys Oceanogr 37(2):162–173

Sandwell DT, Smith WHF (1997) Marine gravity anomaly from Geosat and ERS 1 satellite altimetry. J Geophys Res Solid Earth 102(B5):10039–10054

Sandwell DT, Smith WHF (2005) Retracking ERS-1 altimeter waveforms for optimal gravity field recovery. Geophys J Int 163(1):79–89

Schiller A, Oke PR, Brassington GB et al (2008) Eddy-resolving ocean circulation in the Asian-Australian region inferred from an ocean reanalysis effort. Prog Oceanogr. doi:10.1016/j.pocean.2008.01.003

Smith RL, Huyer A, Godfrey JS et al (1991) The Leeuwin Current off Western Australia, 1986–1987. J Phys Oceanogr 21:323–345

Symonds PA, Willcox JB (1976) The gravity field offshore Australia. BMR J Aust Geol Geophys 1:303–314

Tapley BD, Chambers DP, Bettadpur S et al (2003) Large scale ocean circulation from the GRACE GGM01 geoid. Geophys Res Lett. doi:10.1029/2003GL018622

Tapley BD, Ries J, Bettadpur S et al (2005) GGM02 – an improved Earth gravity field model from GRACE. J Geod. doi:10.1007/s00190-005-0480-z

Vossepoel FC (2007) Uncertainties in the mean ocean dynamic topography before the launch of the gravity field and steady-state ocean circulation explorer (GOCE). J Geophys Res. doi:10.1029/2006JC003891

Wang YM (2001) GSFC00 mean sea surface, gravity anomaly, and vertical gravity gradient from satellite altimeter data. J Geophys Res Oceans 106(C12):31167–31174

Webb D (2000) Evidence for shallow zonal jets in the south equatorial current region of the southwest Pacific. J Phys Oceanogr 30:706–720

Wessel P, Watts AB (1988) On the accuracy of marine gravity measurements. J Geophys Res Solid Earth 94(B4):7685–7729

Wingham DJ, Rapley CG, Griffiths H (1986) New techniques in satellite tracking systems. In: Proceedings of IGARSS'88 symposium, Zurich, Switzerland, September, 1339–1344

White NJ, Church JA, Gregory JM (2005) Coastal and global averaged sea level rise for 1950 to 2000. Geophys Res Lett. doi:10.1029/2004GL021391

Wijffels S, Meyers GA (2004) An intersection of oceanic wave guides: variability in the Indonesian throughflow region. J Phys Oceanogr 34:1232–1253

Zhang K (1998) Altimetric gravity anomalies, their assessment and combination with local gravity field. Finn Geod Inst Rep 98(4):137–144

Lakes Studies from Satellite Altimetry

19

J.-F. Crétaux, S. Calmant, R. Abarca del Rio,
A. Kouraev, M. Bergé-Nguyen, and P. Maisongrande

Contents

19.1	Altimetry Over Lakes	510
19.2	Lakes and Climate Change	513
19.3	Andean Lakes	515
19.4	East African Lakes	519
	19.4.1 Climate Change and the East African Lakes	519
	19.4.2 Lake Victoria	521
19.5	Central Asian Lakes	525
19.6	Conclusions	529
References		530

Keywords Climate change · Hydrology · Lakes · Satellite altimetry · Water management

Abbreviations

AO	Antarctic Oscillation
AVISO	Archiving, Validation and Interpretation of Satellite Oceanographic data
CNES	Centre National d'Études Spatiales

J.-F. Crétaux (✉), M. Bergé-Nguyen, and P. Maisongrande
CNES/LEGOS, 14 av. Edouard Belin, F-31400 Toulouse, France
e-mail: jean-francois.cretaux@cnes.fr

S. Calmant
IRD, LEGOS, CP 7091, Lago Sul, 71619 – 970, Brasilia (DF), Brasil

R.A. Rio
Departamento de Geofísica (DGEO), Facultad de Ciencias Físicas y Matemáticas,
Universidad de Concepción, Concepción, Chile

A. Kouraev
Université de Toulouse ; UPS (OMP-PCA), LEGOS, 14 av. Edouard Belin, F-31400
Toulouse, France and
State Oceanography Institute, St. Petersburg branch, Vasilyevskiy ostrov, 23 liniya, 2a,
St. Petersburg, Russia

CTOH	Centre de Topographie des Océans et de l'Hydrosphère
ECMWF	European Centre for Medium-Range Weather Forecasts
ENSO	El Niño/Southern Oscillation
ESA	European Space Agency
GCM	Global Climate Model
GDR	Geophysical Data Record
GFO	Geosat Follow-On
IOD	Indian Ocean dipole
IPCC	Intergovernmental Panel on Climate Change
ITCZ	InTertropical Convergence Zone
NAO	Northern Atlantic Oscillation
NASA	National Aeronautics and Space Administration
NATO	North Atlantic Treaty Organization
NCEP	National Center for Environmental Prediction
PDO	Pacific Decadal Oscillation
PODAAC	Physical Oceanography Distributed Active Archive Center
RMS	Root Mean Square
SO	Southern Oscillation
SST	Sea Surface Temperature
T/P	TOPEX/Poseidon
UNESCO	United Nations Educational, Scientific and Cultural Organization
USDA	US Department of Agriculture

19.1
Altimetry Over Lakes

So far, the measuring of water levels using satellite altimetry has been developed and optimized for open oceans. Nevertheless, numerous studies have been already published in different fields of continental hydrology based on coupled satellite altimetry/*in-situ* gauge measurements, from a general point of view (Birkett 1995; Calmant and Seyler 2006; Crétaux and Birkett 2006) or to be more specific with respect to lakes (Cazenave et al. 1997; Birkett 2000; Mercier et al. 2002; Aladin et al. 2005; Coe and Birkett 2005; Crétaux et al. 2005; Medina et al. 2008; and Kouraev et al. 2009; Chapter by Kouraev et al., this volume). These studies show that radar altimetry is currently an essential technique for different applications: assessment of hydrological water balance (Cazenave et al. 1997; Crétaux et al. 2005), prediction of lake level variation (Coe and Birkett 2005), studies of anthropogenic impact on lake water storage (Aladin et al. 2005), and correlation of interannual fluctuations of lake levels on a regional scale with ocean-atmosphere interaction (Mercier et al. 2002).

Several satellite altimetry missions have been launched since the early 1990s: ERS-1 (1991–2000), T/P (1992–2005), ERS-2 (1995–present), GFO (2000–2008), Jason-1 (2001–present), Jason-2 (2008–present) and Envisat (2002–present). ERS-1, ERS-2, and Envisat have a 35-day temporal resolution and an 80 km inter-track spacing at the Equator. T/P, Jason-1 and Jason-2 have a 10-day temporal resolution and 350 km inter-track spacing at the equator. GFO has a 17-day temporal resolution and 170 km inter-track spacing at Equator. The combined global altimetry historical data set now span more than a decade and is intended to be continuously updated in the coming decade. Combining altimetry data from several in-orbit altimetry missions also increases the spatiotemporal resolution of the remotely-sensed hydrological variables. Radar altimetry has, however, a number of limitations over land because radar waveforms (e.g., raw radar altimetry echoes reflected from the land surface) are complex and multi-peaked due to interfering reflections from water, vegetation canopy, and rough topography. These effects result in data having decreased validity, compared to data collected over oceans. Moreover, the instruments can only operate in a *profiling mode* and do not have a true global view. Large numbers of lakes are thus not seen by current altimetry satellites. Stage accuracies are also dependent on target size, which will limit worldwide surveying and limnological applications. However, there is adequate scope for systematic monitoring at continental scales and the provision of new stage information where gauge data are absent. Typically, altimetry measurements can range in accuracy from a few centimeters (e.g., Great Lakes, USA) to tens of centimeters depending on size and wind conditions. Minimum target sizes depend on a number of factors including footprint size, topography complexity, and instrument-tracking logic. Table 19.1 summarizes the expected accuracy for each satellite mission depending on the target characteristics (taken from Crétaux and Birkett 2006). Summing the squares of the individual RMS values gives the total height error. The range precision will be improved, as mean lake heights (across the profile) are determined, effectively averaging all valid height measurements (N) from coastline to coastline.

The derivation of time series of surface height variations involves the use of repeat track methods. This methodology employs the use of a mean (Crétaux and Birkett 2006) lake height profile. This is derived from averaging all height profiles across the lake expanse within a given time interval, effectively smoothing out the varying effects of tide and wind setup. The height differences between the reference pass and each repeat pass enable the time series of lake height variation to be created.

The accuracy of a single lake height measurement will vary depending on the knowledge of the range, the orbit and the various corrections. Range precision will vary, according to surface roughness and the selected satellite mission, as more complex tracking logic and algorithms try and maximize the quantity of data collected over continental surfaces. The precision of the measurement will then strongly depend on the capability to retrieve the time that corresponds to the actual height at the nadir of the antenna. Smith (1997) stated that the major difficulty in retrieving ranges over continental waters results from the variability in the shape of the return waveforms when onboard trackers

Table 19.1 Summary of satellite altimetry general characteristics

Satellites	Operation period	Orbital cycle (days)	Accuracy (cm RMS)	Minimum target area and width
ERS-2	1995–2002	35	>9	>100 km^2 > 500 m
Envisat RA-2	> 2002	35	>9	>100 km^2 > 500 m
T/P	1992–2005	10	>3	>100 km^2 > 500 m
Jason-1	>2002	10	>3	>100 km^2 > 500 m

are designed for a typical ocean waveform. Regular "ocean-type" trackers expect long tail shapes of energy distribution while echoes bouncing off rivers are often specular (Guzkowska et al. 1990) or a combination of specular echoes. Thus, in the best case, energy is received but the range estimate may be erroneous; in the worst case, the altimeter loses tracking and subsequent echoes are lost. The antenna-reflector range is determined by fitting the waveform with a predefined analytical function (so-called waveform "tracking"). If the analytical function is not well suited to the waveform, shape tracking leads to wrong estimates of the height value (or even no estimate at all). Here, it is worth noting that the radar altimetry data collected by the ongoing Envisat mission are nominally retracked with four algorithms.

The most widely used for all missions since Geosat is the "OCEAN" tracker (Rodriguez and Chapman 1989), tuned for deep ocean surface. Because the shape of the echoes that is reflected by other surfaces such as ice caps and continental waters can be very different from the ones reflected by the oceanic surface, other trackers have been developed and applied to retrack the radar waveforms. The OCOD/ICE-1 (Wingham et al. 1986) and ICE-2 (Legresy and Remy 1997) trackers have been developed to recover heights over sea ice; SEA-ICE (Laxon 1994) to recover heights over the sea ices too. Envisat is the first mission for which ranges are provided nominally with more than one tracker.

Based on a study made over the Amazon River, Frappart et al. (2006) have established some quality assessment of each tracker available with Envisat data by comparison with *in-situ* data. They have demonstrated that ICE-1 is the most suitable one for lakes and rivers among all algorithms of retracking.

The air density, the amount of water vapor and the content of free electrons in the ionosphere also modify the travel time of radar waves throughout the atmosphere. The electronic content in the ionosphere and air pressure is given by independent data sets. However, the amount of water vapor is estimated using microwave radiometers embarked onboard together with the radar altimeters. The current microwave radiometers fail to estimate the atmospheric content of water vapor over continents because the signature of the atmospheric water vapor is mixed with that of ground wetness. Thus, this effect ranging from a few centimeters to tens of centimeters cannot be accurately corrected for over land waters and corrections can only be estimated from large-scale global data sets such as ECMWF or NCEP. Yet, these model-based corrections – although satisfactory in must cases – can suffer large errors, in particular in the case of mountain lakes. Crétaux et al. (2009a) have shown in the case of Lake Issykul that the dry troposphere corrections delivered in the T/P, Envisat, Jason and GFO GDRs were wrong either because the altitude of the lake was not accounted for to compute the reference air pressure or this altitude varies within the lake surface, producing an artificial variation of the correction along the tracks crossing the lake. In this case, the largest difference between corrections at a given day exceeds 60 cm. Such an error is particularly impacting when merging stages from different missions in order to increase the time sampling.

Finally, the quality of the orbit directly enters into the error budget of the water height. Yet, orbit modelling did progress during the last decade and the orbital errors can be considered of second order for continual waters.

The techniques are now so well validated that lake and reservoir altimetry measurements can be found in several web sites:

Hydroweb: Developed at the LEGOS laboratory in Toulouse, this web database contains time series over water levels of large rivers, lakes and wetlands at the global scale. The lake levels are based on merged T/P, Jason-1, Envisat, ERS-2 and GFO data freely available through the NASA and the CNES data services (PODAAC and AVISO). The monthly level variations deduced from multi-satellite altimetry measurements for almost 150 lakes and reservoirs are freely provided. Potentially, the number of lakes monitored could be significantly increased. These lake and reservoir level time series are available at: http://www.legos.obs-mip.fr/en/soa/hydrologie/hydroweb/

The USDA has developed a database of water level variations on a 10-day basis for almost 100 lakes monitored in a near real time from the satellite missions T/P and Jason-1. The data are freely available directly on the web site: http://www.pecad.fas.usda.gov/

River&Lakes database: The European Space Agency, in cooperation with the De Montfort University in UK has developed an algorithm for altimetry wave form retracking, which is now operating automatically. They have set up a database which provides instantaneous water level products on a near real time basis, over the whole Earth. This database does not provide historical altimetry data that can, however, be obtained on demand from ESA. The web site is at the following address: http://earth.esa.int/riverandlake/

19.2
Lakes and Climate Change

The volume of water stored within lakes and reservoirs is a very special and sensitive proxy primarily for precipitation; secondly, for other climatic parameters through evaporation, temperature, surface pressure, wind stress, radiation (both short and long waves); and thirdly, for hydrologic parameters such as groundwater and runoff. Lakes, thus, may be used to study the combined impact of climate change and water resources management. *Climate change* affects several aspects of the hydrological cycle. Water management generally directly affects the water balance parameters of a lake, through irrigation or hydropower industry, while climate change alters the water balance of a lake through long-term changes in temperature and precipitation.

A lake system represents complex interactions between the atmosphere, surface and underground water, a system that responds to climatic conditions, yet is tempered by upstream water uses for agriculture, industry or human consumption. Thus, the overall water volume (a function of surface area and lake level) depends on the balance between the water inputs and outputs. The inputs are the sum of direct atmospheric precipitation over the lake and surface runoff of the surrounding areas. Underground seepage can also provide inflow. Volume outputs are the sum of the direct evaporation from the lake's surface in addition to river outflow and groundwater seepage. Some of these water balance components can be, generally, easily measured (river runoff, precipitation, lake level), depending on the availability of *in-situ* measurements while others can only be deduced globally from hydrological or climatic models (evaporation, underground water) or partially from local field campaigns. To close the water budget of a lake, the measurement of level/volume variations with time is a key parameter, as this can provide estimates of volume variation (when the relation between level and volume is known).

The assessment of lake water balance could provide better knowledge of regional and global climate change, while a *quantification* of the human stress on water resources across all continents is also an important factor. Moreover, this assessment can be of particular interest for all these regions where human habitation is still highly dependent on water variability, whether as direct water consumption or as a source of hydropower. In many Sub-Saharan African countries, water resources are one of the major limiting constraints on development. The region's electricity supply is dependent largely on hydropower; inland fisheries and navigation are also often dependent on the hydrological regime in lakes and rivers; a reliable supply of good quality water is required for human consumption, industry, and irrigation. The hydrological regime of the major lake and river systems in these regions is, therefore, of particular interest.

Indeed, in this regard, the choice of lake is of particular importance. One has to distinguish between closed (endorheic) lakes with no outflow, and open lakes (exorheic) that have one or more outflowing rivers. The first class of lakes depends closely on the balance of inflows (precipitation and runoff of rivers) and evaporation. They, hence, could be considered as good proxies of climate change, since a small change in the temperature or precipitation in the lake basin can have a direct consequence on the lake water level.

Moreover, Hostetler (1995) noted that deep lakes with steep shore topography are good proxies for high amplitude-low frequency changes, while shallow water basins are better indicators for rapid low-amplitude changes. The latter are extremely sensitive for revealing decreased water input and rising evaporation.

Observations of many lakes in different climatic and regional zones can be used to discriminate between regional and global climates as well as help in assessing the impact of anthropogenic forcing occurring on different spatial and timescales:

— The Southern Andean lakes (Chilean and Argentinean lakes on both sides of the Andes): the Chilean lakes (Ranco, Villarica, Panguipugui, Rupanco, etc.); lakes on the limit between Chile and Argentina (Carrera-Buenos-Aires, Cochrane, San-Martin-Ohiggins, etc.); the Argentinean lakes (Argentino, Viedma, and some smaller lakes) and finally, the world's southernmost lake (excluding Antarctica), Lake Fagnano.
— The African lakes of the Eastern Rift (Tanganyka, Victoria, Malawi, Mweru, Kyoga, Albert, Tana, Kiwu, Bangweulu, and Edouard).
— The central Asian lakes (Aral Sea, Issykkul, Balkhash, and some large reservoirs along the Syr Darya and Amu Darya rivers).

In its great majority, lake oscillations from one year to the next are associated with the local climatic regimes that govern the precipitation variability over the watershed of a given lake. In many parts of the world, interannual variations in rainfall and runoff are strongly influenced by the ENSO (New et al. 2001), which are characterized by anomalous and large deviations from mean annual levels of precipitation. Lake level variability depends on the magnitude and frequency of ENSO events given that, in many parts of the world, peak discharges occur during ENSO years. However, other climatic oscillations do also play a role: PDO, NAO, and SO.

In this regard, the first two lake groups are interesting because their level variation is closely associated with climatic forcing. The El Niño-Southern Oscillation as the main driver worldwide of interannual climate variability (New et al. 2001) imprints lake level variability at these timescales. However, through ocean-atmosphere couplings and climate tele-connections, lake levels are subject to different climate oscillations: African lake levels are influenced by the African monsoons, the North Atlantic Oscillation and the Indian Oceanic Dipole. Lake level in South America is affected by the North Atlantic Oscillation, the southern annular mode (also called the Southern Antarctic Oscillation) as well as the southern American monsoon. Across Central Asia lakes are influenced by a mix of direct human factors (through irrigation) and climate change and variability.

In the following part of this chapter, we will show some recent results obtained over these groups of lakes in order to illustrate the potential uses of radar altimetry for the monitoring and study of these kinds of water bodies.

19.3
Andean Lakes

The large inland water bodies of South America are of crucial importance for the complex and diverse economical environments in several countries: tourism in Altiplano and Argentina, fish breeding in Chile (Fig. 19.1), water consumption and hydropower in most of these countries. Lakes and reservoirs also form high-altitude aquatic ecosystems, including ecological interaction with the surrounding terrestrial areas that are highly dependent on the water cycle and interannual variability of water storage of the lakes themselves.

Some lakes located in the Altiplano Andean regions (e.g., Titicaca, Poopo) are highly receptive to ENSO (Garreaud and Aceituno 2001; Zola and Bengtsson 2006). Associated with the PDO (Garreaud and Battisti 1999), ENSO drives interannual and decadal variations of precipitation, which has a direct impact on lakes of this region. Consequently, they are also indirectly influenced by global climate change, which strongly interacts with and modifies the climatic oscillations. For South America, the IPCC change scenarios predict an increase of temperature going up from 1°C to 6°C degrees, and an increase of precipitation anomalies by about 20% before the end of the twenty-first century (Bates et al. 2008). More precisely, the projected mean warming for Latin America for 2100, according to different climate models, ranges from 1°C to 4°C for the B2 emissions scenario and from 2°C to 6°C for the A2 scenario.[1] Most GCM (Global Climate Model) projections indicate larger (positive or negative) rainfall anomalies for the tropical region, and smaller-to-diminished ones for the extra-tropical part of South America. In the case of the tropical regions, the increase of anomalies will severely impact the runoff of the rivers that feed the lakes in the Altiplano, especially Lake Titicaca. Confidence in model estimates is higher for some climate variables (e.g., temperature) than for others (e.g., precipitation). Models still show significant errors although these are generally greater at smaller scales. Despite such uncertainties, there is considerable

---

[1] A2 and B2 are two different scenarios, though not extremes, of what Earth might be like by the end of the twenty-first century. B2 scenario emphasizes environmental preservation and social equity while A2 scenario anticipates higher CO_2 concentrations, larger human population, greater energy consumption, more changes in land use and scarcer resources, and less diverse applications of technology.

Fig. 19.1 Map of South America with the biggest lakes of the Andean Mountains

confidence that climate models do provide credible quantitative estimates of future climate change, particularly at continental scales and above (Randall et al. 2007).

However, despite these predicted high precipitation anomalies, no one can ascertain whether the lakes in the Altiplano will ultimately rise or shrink during the next decades, given that in addition to the precipitation uncertainties, other climate parameters must also be taken into account. The predicted higher temperature will enhance the role of evaporation in the water balance of those lakes. Nowadays, Lake Titicaca is well monitored by satellite altimetry (Fig. 19.2b), with the availability of very high precision data (Fig. 19.2a). Even in the case where *in-situ* gauges are operational, data are, in most of the cases, not freely available to the scientific community due to national data policies. Continuous monitoring of this lake from radar altimetry in the future (with the next mission: Jason-2, Jason-3, Sentinel-3, AltiKa, etc.) will contribute to better understanding of its water balance under the influence of diverse climatic forcing.

19 Lakes Studies from Satellite Altimetry 517

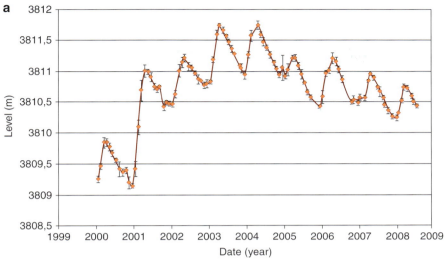

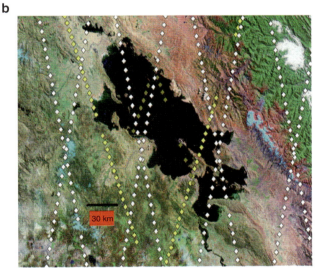

Fig. 19.2 (**a**) Lake Titicaca level variations from 1 Hz radar altimetry data. (**b**) Altimetry satellite tracks over the Lake Titicaca (Envisat: *white dots*, GFO: *green dots*, T/P interleaved orbit: *yellow dots*)

In the case of extratropical climate in South America, even though precipitation is thought to change little or even diminish in terms of anomalies, it is already the case with several lakes that are affected by global warming in the form of changing glacial feed. Examples include lakes Argentino and Viedma located on the border of Chile and Argentina (Depetris and Pasquini 2000; Coudrain et al. 2005). All of these lakes are also well monitored by altimetry (Fig. 19.3a and b).

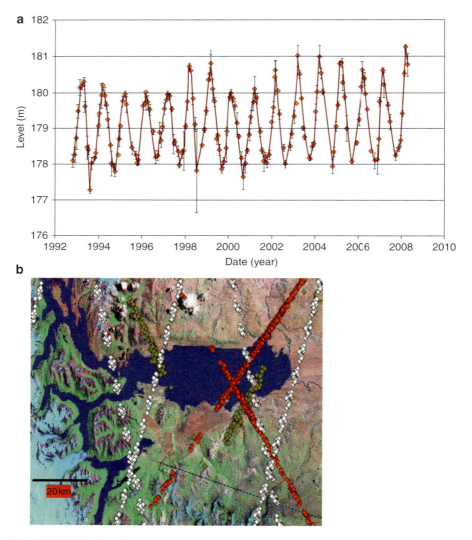

Fig. 19.3 (**a**) Lake Argentino level variations from 1 Hz radar altimetry data. (**b**) Altimetry satellite tracks over the Lake Argentino (Envisat: *white dots*, GFO: *green dots*, T/P initial orbit and Jason-1: *red dots*)

Ongoing research provides an example of the complexity of the diverse sources affecting lakes variability over southern America. The General Carrera/Buenos Aires Lake, a Chilean/Argentinean lake located in Patagonia (46°30′S 72°0′W), is recognized as representing the second greatest hydrologic lake basin for South America (after that of Lake Titicaca). Although the river inputs to the lake are driven by direct rainfall, large portions of river discharges come only from snow accumulation on different mountains of the Andean cordillera and its subsequent melting 3 months later. In addition, part of the water inflow also arises from some of the melting glaciers belonging to Northern Patagonian Ice fields. Its glaciers are largely in retreat and with only an area of 4,200 km², it is still the largest continuous mass of ice outside of the Polar regions. Also, it is suspected that direct input through groundwater plays an important role in the water balance. Further, the climatic regimes affecting the western and eastern sides of the lake (mainly represented by the precipitation amounts) are different. To the west, in the Andes portion of the watershed, precipitation is affected directly by ENSO regimes with moisture transported by the south westerly winds, while over its eastern part, in the Pampa catchment's region, the regimes present a complex signature where ENSO is not the leading mode. In addition to these complexities, *in-situ* measurements are necessary for completing the water balance. Unfortunately, the current in-situ monitoring of lake level variability involving river inflow and outflows and meteorological variables is intermittent at best, which precludes a complete and correct establishment of the water balance. Models and *in-situ* campaigns are also necessary.

Therefore, once again it is essential to have accurate water level measurements in order to understand the water balance of these lakes as well as others in the region, and thereby quantify the impact of climate change on the water resources represented by the major lakes of the Andean chain.

In particular, in the case of South Andean lakes, the combined effect of ENSO and global warming on the lake level has not been studied as yet. First of all, the cause of long-term variability of precipitation at the boundary between Chile and Argentina has just recently been understood to represent a combination of ENSO and AO on interannual timescales, as well as the direct effect of the PDO for decadal timescales (Fogt and Bromwich 2006; Gillett et al. 2006; and Fagel et al. 2008). Moreover, in this region, global warming is likely to cause decreasing precipitation and increasing temperature (Bates et al. 2008) with consequently increasing runoff from glacier thawing on the one hand, and decreasing direct precipitation over the lakes on the other. Lakes, thus, may be considered as very sensitive proxies for determining the balance between those two opposite impacts. The level variations of a lake like Argentino (Fig. 19.3a) thus reflect those different phenomena.

19.4 East African Lakes

19.4.1 Climate Change and the East African Lakes

Understanding the climate variability of Africa, and therefore tropical variability and its prominent role as the heat engine of the global climate system is one of the key goals in climate research. Also, African lakes themselves modify the local climate to a large extent: Nicholson and Yin (2002) showed that precipitation over Lake Victoria is more intense than over its catchment area by more than 30%. In addition, African lake managers and all those who are dependent economically and socially (fisheries, water availability, etc.) on lake resources, wish to understand the potential long-term controls that climate exerts on lake level and productivity (Stager et al. 2005). From short instrumental records it is evident that, as over the seas, climate variability can impact primary production through, for example, changes in wind-driven upwelling or seasonal intensification and weakening of stratification linked to temperature variability (Plisnier and Coenen 2001; Cohen et al. 2006).

In addition to understanding present climates with modern measurements (in situ, or satellite), one has to try to reconstruct the historical climatology (i.e., interannual–decadal–centennial timescale, beyond instrumental records). However, scarcity of instrumental and documentary records prior to the colonial period (Nicholson 2001) coupled with the limited potential of traditional high-resolution climate-proxy archives such as tree rings or ice cores in Africa, impedes a correct reconstruction of past climates (Verschuren 2003). Fortunately, sedimentary climate records accumulating at the bottom of Africa's climate-sensitive lakes demonstrated a great potential to document the continent's patterns of past climate change.

But for understanding these sedimentary records, on seasonal timescales at least, a correct understanding of the complexities of the water balance of a given lake is required, as well as validation of the proxy indicators. However, although a dense network of natural lakes covers the East African Rift region, only the greatest lakes of the region have been monitored by ground gauges (Lake Victoria, Lake Tanganyika) up to the present moment. This suggests that some of the water balance mechanism may still be poorly understood.

Thanks to satellite altimetry, most of the lakes of Africa are now being monitored. Study of 7-year satellite radar altimetry data from the T/P mission and the utilization of precipitation data over East Africa and the Indian Ocean, (Mercier et al. 2002) show evidence of hydro-meteorological linkages between the Indian Ocean SST and precipitation over the East African continent, and changes in lake levels. However, the main components of the water balance (precipitation and evaporation) over the East African lakes are still very uncertain and its connection with regional climate variability is an ongoing debate (Nicholson and Yin 2002).

For example, analyses of level records of the Great East African Lakes such as Victoria, Tanganyika, and Nyassa-Malawi, and regional climatic variability (precipitation amount) show synchronicity, which can only be explained by large-scale mechanisms (Ogallo 1988; Nicholson 1996; Ropelewski and Halpert 1996), associated initially with ENSO cycles. Sharp increase in the water levels of Lake Victoria (Fig. 19.5a), Lakes Tanganika, Kariba, and Turkana in 1998 may reflect this ENSO impact on lakes in the African Rift (Mercier et al. 2002). However, some studies (Beltrando and Cadet 1990; Hastenrath et al. 1993; Richard 1994; and Becker et al. 2010) have shown already that these could also be related to the Indian oceanic variability. Only recently the stronger relationship with the IOD, a particular oceanic circulation system taking place in the Indian Ocean (Saji et al. 1999) has been explained thanks to equatorial zonal atmospheric circulation (Bergonzini et al. 2004). The IOD is partly responsible for driving climate variability of the surrounding landmasses, and therefore, over eastern Africa (Marchant et al. 2006). Also, episodically or non-linearly, ENSO modulates the climate in eastern Africa and therefore, lake level variability (Mistry and Conway 2003). However, whether (and how) the IOD is an artifact of the El Niño–Southern Oscillation (ENSO) system, or it is a separate and distinct phenomenon is not the subject of this chapter.

For decadal to centennial timescales a strong and significant link between lake levels and solar activity (possibly amplified by sea surface temperature, ENSO and PDO) has been found for most of the great lakes of eastern Africa over much of the last millennia (Stager et al. 2005; Garcin et al. 2007; and Stager et al. 2007). In addition, it has also been shown that the Indian monsoon and the migration of the ITCZ[2] do play a role (Tierney and Russell 2007).

---

[2]The location of the inter-tropical convergence zone (ITCZ) varies over time. Over land, it moves back and forth across the equator following the sun's zenith point. Over the oceans, where the convergence zone is better defined, the seasonal cycle is more subtle, as the convection is constrained by the distribution of ocean temperatures.

Fig. 19.4 Lakes of East African Rift

Considering now the anthropogenic role, General Circulation Models (GCM) have shown that one of the consequences of global warming could be significant changes in precipitation and surface air temperature (Xu et al. 2005; De Wit and Stankiewicz 2006). This would have an impact on regional hydrology and ultimately affect lake and reservoir levels, especially in a region like the East African Rift, which is covered by a variety of lakes of different sizes and morphologies.

Current satellite radar altimetry missions (Jason-1/2, GFO, and Envisat) have increased the number of lakes observable over East Africa (around 15 big lakes, in these regions (Fig. 19.4 and 19.5f) are monitored currently from satellite radar altimetry with an accuracy of a few centimeters in the best cases, up to 10–20 cm in the worst ones). This would be a valuable tool to look carefully at climate variability impact on those water bodies.

19.4.2 Lake Victoria

Lake Victoria is the world's second largest lake in terms of surface area (68,000 km^2), and is very shallow (Mean depth of 40 m with maximum of 84 m), which makes it sensitive to sudden and even relatively small changes in the water balance parameters, such as precipitation, evaporation, or surface runoff. Lake Victoria can hence be considered as a good indicator of climate change in this region. It is located in an arid to semiarid region which is, however, subjected to a high a mean annual precipitation of 1,200–1,600 mm, while evaporation is estimated at around the same order of magnitude (Nicholson and Yin 2002). Lake Victoria has exhibited very high level fluctuations over the last 15 years (Fig. 19.5a). Since

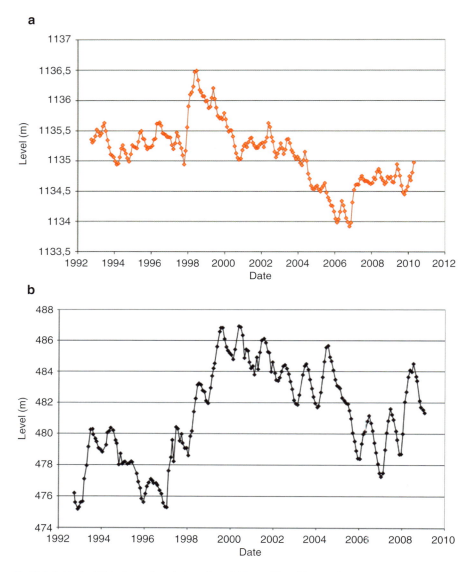

Fig. 19.5 (a) Lake Victoria level variations from 1 Hz satellite altimetry data. (b) Lake Kariba level variations from 1 Hz satellite altimetry data used in a multi-satellite processing mode. (c) Lake Turkana level variations from 1 Hz satellite altimetry data used in a multi-satellite processing mode. (d) Lake Tanganyika level variations from 1 Hz satellite altimetry data used in a multi-satellite processing mode. (e) Lake Malawi level variations from 1 Hz satellite altimetry data used in a multi-satellite processing mode. (f) Altimetry satellite tracks over the East African Lakes (Envisat: *white dots*, GFO: *green dots*, T/P initial orbit and Jason-1: *red dots*, T/P interleaved orbit: *yellow dots*)

19 Lakes Studies from Satellite Altimetry

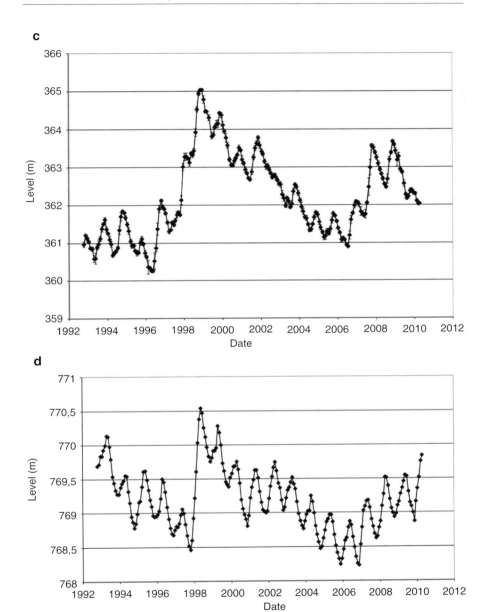

Fig. 19.5 (continued)

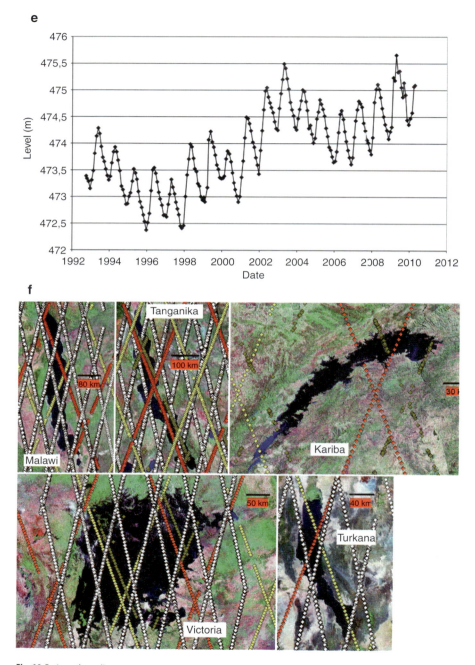

Fig. 19.5 (continued)

water from Lake Victoria flows into the Nile River, this high water level variability can also seriously alter the water uses for downstream countries. From 1999 to 2007, the level of Lake Victoria has declined around 2.5 m (Fig. 19.5a). This represents a loss of water of 20 km^3 per annum. During the same period, for the other large lakes of the region, water level variations deduced from altimetry measurements (Fig. 19.5b–e) can provide interesting information on the regional hydrology: Lake Tanganyika level time series is very similar to that of Lake Victoria, which reveals regional climate forcing on water storage. For lakes Kariba and Turkana, the patterns are also similar to that of Lake Victoria but an increase of the level in 1998 is less abrupt and longer in duration, while for Lake Malawi, water level variations followed a totally different pattern. The combinations of different factors (natural and anthropogenic) that could explain these observations would be a valuable source for understanding the lake hydrological regime within the Eastern African Rift (Swenson and Wahr 2009). Moreover, they are excellent proxies of tropical climate variability.

19.5
Central Asian Lakes

The major central Asian lakes are in the drainage basins of two large rivers, the Amu Darya and the Syr Darya, shared by five different countries: Uzbekistan, Kazakhstan, Tajikistan, Kyrgyzstan, and Turkmenistan. Irrigation water is extracted on a large scale from the rivers that drain into the remnants of the Aral Sea. Consequently, water management in this region has a transboundary dimension which has political and societal ramifications. A political agreement was signed in the mid-1990s for a rational use of the water resources across the region. Through the use of large reservoirs, both Kyrgyzstan and Tajikistan agreed to retain upstream water over the winter period for later use by Kazakhstan and Uzbekistan which require a large amount of water downstream for irrigation in the dry season. As compensation, Kazakhstan, and Uzbekistan agreed to provide gas and electricity to Kyrgyzstan and Tajikistan. The agreement is a source of contention, which highlights the need to reinforce the collaboration between the water management agencies, as well as the maintenance of the water resources monitoring and sharing of the river runoff, reservoir storage, and hydro climatologic data. In fact, UNESCO, NATO, and other international organizations aim to strengthen this agreement by providing the framework for enhancing the collaboration among political managers, hydrologists and scientists in general.

In this context, the information given by altimetry data can be very useful for water management monitoring in a large area like Central Asia, since all large reservoirs (Chardarya, Karakul, Sarykamish,and Toktogul) are well monitored by this technique (Fig. 19.5a–d). This is particularly interesting when *in-situ* data are also available. In fact, several of these lakes are monitored by ground-based gauges, though access to current information (post 2001) is still restricted. When there is an overlap period between gauge and altimeter observation, validation exercises reveal the typical accuracy of the satellite measurements. For example, an accuracy of 5 cm RMS (Fig. 19.6d, Lake Issykkul), to 10 cm RMS (Fig. 19.6b, Lake Chardarya), is obtained. In some other cases, one does not have a common period of

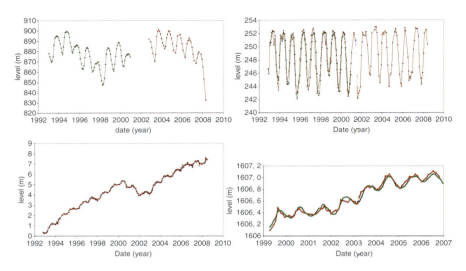

Fig. 19.6 (**a**) Toktogul reservoir (Kyrgyzstan) level variations from altimetry (*red*) and in-situ data (*green*). (**b**) Chardarya reservoir (Kazakhstan) level variations from altimetry (*red*) and in-situ data (*green*). (**c**) Sarykamish reservoir (Turkmenistan) level variations from altimetry. (**d**) Lake Issykkul (Kyrgyzstan) level variations from altimetry (*red*) and in-situ data (*green*)

observation as in the case for the Toktogul reservoir (Fig. 19.6a). Due to political restriction, *in-situ* data are not available after 2001, while altimetry provides continuous monitoring of this reservoir from the launch of Envisat in 2002. One can see that for this reservoir the level variation is very high (it has lost 50 m in less than 1 year). Due to political issues (transboundary water management of the Syr Darya River) information on the water level monitoring of reservoirs is quasi nonexistent, but one can suspect that Kyrgyz Republic have provoked water release from the Toktogul reservoir for hydropower generation during winter 2007–2008.

For another reservoir of the region, the Sarykamish, there are no gauge data available at all. Lake Sarykamish is located in Central Asia, south of the Aral Sea, in the territory of Turkmenistan. It is filled by diverted water from the Amu Darya River, through ancient natural channels, with a continuous increase since the beginning of the 1990s. Every year the lake receives approximately 5–6 km³ of drainage water (Aladin et al. 2005). Thanks to altimetry measurement one can easily see the consequences of diverting the flow of the Amu Darya River (Fig. 19.6c). This political decision, made by Turkmenistan Republic in the early 1990s has contributed, with the enhancement of irrigation in the Amu Darya river basin, to the shrinking of the Big Aral Sea.

It is therefore evident that altimetry provides a source of important independent information complementary to that produced by the ground-based networks and perhaps more critically, can provide hydrological information where gauges are lacking. In this respect it provides an additional tool for decision-makers in the field of water management.

In parallel to the issues of transboundary water management, the Central Asian lakes (Aral Sea, Balkhash, Issykkul, and many artificial reservoirs) are extremely sensitive to global change and are exposed to growing anthropogenic pressure; some of them have already suffered severe ecological modifications. Monitoring their level variations thus also helps to infer the impact of climate change on the lakes and reservoirs, though a complete understanding would enable an interpretation of the many other complexities involved. Lake levels may drop due to potential increases in evaporation, but levels may increase due to the increase in thaw rates of the surrounding glacier and mountain snow. In the case of Issykkul Lake, located in Kyrgyzstan, the water level has continuously decreased in the twentieth century up to 1998 at a rate of around 4 cm/year. During the past 8 years, the level has considerably increased at a rate of 10 cm/year (Fig. 19.6d). Perhaps this increase can be attributed to the more rapid rate of recent high glacier melt linked to regional warming (Romanovski 2002). The Aral Sea which mainly loses water through evaporation is also affected to a lesser extent by global warming. Studies made by (Small et al. 2001) showed that global warming has accelerated the shrinking of the Aral Sea by around 10 cm/year from 1960 to 1990. With altimetry monitoring it is possible to quantify the impact of global warming on the Aral Sea during the last 15 years.

Several articles have already been published about the Aral Sea crisis, and a few of them specifically about water level monitoring from satellite altimetry (Aladin et al. 2005; Crétaux et al. 2005; Crétaux et al. 2009b; and Kouraev et al. 2009). Since the separation of Aral Sea in two independent basins in 1989, the Big Aral in the south and the Small Aral in the north, the Big Aral has continuously shrunk, varying from 20 cm/year to 1 m/year (Fig. 19.7b). The Small Aral has evolved in a different way during the same period. As the runoff of the Syr Darya River has more or less compensated the evaporation on this small water basin, its level has oscillated around a mean level value (Fig. 19.7a). The high water stage has even been enhanced through the construction of a dam in Berg's Strait (Fig. 19.8a), in particular in 1997 and in 2005.

Using a combination of a precise digital bathymetry map (DBM) of the basin with level variation deduced from altimetry, Crétaux et al. (2005) computed the resulting volume variation of the Big Aral Sea for the period 1993–2004. They showed that the reduction of lake volume as measured by T/P, GFO, and Jason is smaller than that deduced from examination of the hydrological budget; they assume that possible significant underground water inflow

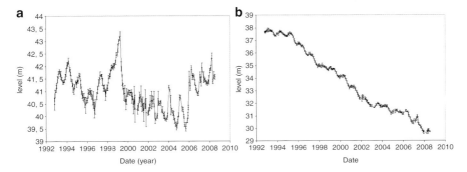

Fig. 19.7 (**a**) Small Aral Sea level variations from satellite altimetry. (**b**) Large Aral Sea level variations from satellite altimetry

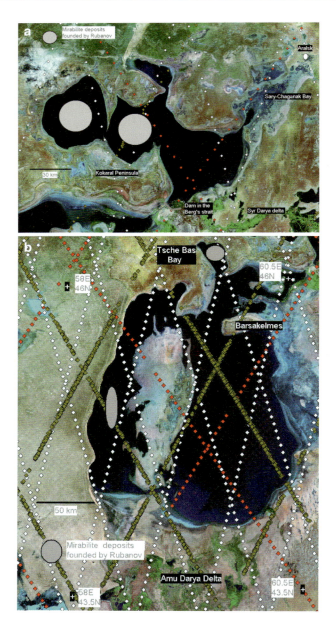

Fig. 19.8 (**a**) Altimetry satellite tracks over the Small Aral (Envisat: *white dots*, GFO: *green dots*, T/P initial orbit and Jason-1: *red dots*). The mirabilite deposits location found by I.V. Rubanov is also marked on the figure (*gray circle*). The satellite image is from Landsat image taken in 2000. (**b**) Altimetry satellite tracks over the Large Aral (Envisat: white dots, GFO: *green dots*, T/P initial orbit and Jason-1: *red dots*). The mirabilite deposits location found by I.V. Rubanov is also marked on the figure (*gray circle*). The satellite image is from Landsat image taken in 2000

cannot be ignored. This conjecture needs to be further assessed by hydrogeological modelling and more accurate data collected on the evaporation and precipitation rates in the region.

Even if this can be accurately established, the groundwater flow will likely only slow the desiccation of the Big Aral. To reverse the process or even to stabilize the sea level to that of the mid-1990s would require more underground flow than could be supplied in even the most optimistic scenario (Crétaux et al. 2005).

In Aladin et al. (2005), the impact of the construction of a dam in Berg's Strait in the water balance of the Small Aral as well as on the aquatic ecosystem has been precisely quantified. In Crétaux et al. (2009b), a new water balance of the Big Aral Sea based on altimetry data has been calculated. It was shown that the very narrow strait between the western and the eastern parts of the Big Aral is probably transporting water from East to West, which could explain some of the observations made 30 years ago by a Russian scientist (I.V. Rubanov[3]) regarding deposits of mirabilite on the bed of the Aral Sea in the Middle Ages and which has been found in very specific locations in both the Big and Small Aral seas (Fig. 19.8a and b). Mirabilite is an indicator of the advancement of evaporation of confined waters. Most endorheic lakes, in arid or semiarid areas, precipitate salts such as mirabilite when the water salinity rises above a given value (Rubanov et al. 1987[3]; Crétaux et al. 2009b). A review on Mirabilite can also be found in Auqué 1995.

Recently, a study of the Aral Sea level variation has been performed from a multi-satellite point of view, and an assessment of the quality of the retracking mode made with Envisat has highlighted the effects of ice cover over the Aral Sea on the mode of calculation of the water level from satellite altimetry (Kouraev et al. 2009; Chapter 13 by Kouraev et al., this volume).

Inland sea and lake level variations can presently be measured with great precision. This information, coupled with historical *in-situ* data, should help to further understanding of the responses of central Asian lakes to global climate changes and human economical activities, and perhaps serve to build coherent future scenarios for the sustainable development and protection of these large inland lakes.

19.6
Conclusions

Radar altimetry can contribute to the measurement of lake height variability over the lifetime of the satellite. The measurements can be acquired along the position of the satellite ground track with accuracy ranging from a few centimeters to tens of centimeters RMS. Altimetry measurements are not dependent on cloud cover, and may provide surface heights with respect to a common reference frame. They also are capable of providing water level variation due to continuous satellite operation since 1992 (Launch of T/P) in remote areas where no other information exists. New missions are already programmed by

---

[3] I.V. Rubanov is a Russian scientist who did core drilling of the Aral Sea bottom in the 1980s in order to study the past regressions and transgressions of the Aral Sea. He provided a map of salt deposits in his different core, some of them including Mirabilite.

space agencies to enable this system to continue lake monitoring over the whole Earth (Jason-3, AltiKa, and Sentinel-3), thereby allowing a pluri-decadal survey of lake levels. The altimetry community has made considerable efforts to ensure that the calibration of the products and the corrections used are delivered in a standardized form to the user community. However, the altimetry technique suffers from some limitations with respect to lake studies: it cannot be used to monitor all lakes on the Earth because it is a profiling system. This means that the measurements may be altered by the surrounding terrain of lakes, rendering geophysical correction less precise than over oceans. Moreover, effects such as wind, seiches, high precipitation, or the presence of ice may also degrade the accuracy of the data. Despite these limitations, the combination data sets from different altimetry missions provide a high degree of temporal and spatial resolution. Compilation of data from the Jason-1 (10-day repeat), GFO (17-day), and Envisat RA-2 (35-day) missions can provide a repeat measurement at a frequency better than a single mission alone. While Jason-1 may only view approximately 300 lakes, Envisat can augment this with potentially an additional target of 700 lakes. Such measurements can also be combined with bathymetry and image-derived areal extent to further deduce changes in storage volume. For remote lakes, altimetry can indeed be the *unique* system to determine accurate water balance.

Acknowledgments Authors acknowledge the CTOH in Legos for providing altimetry data in a standard and useful form. Authors thank the ECOS-SUR program, which allocated funds for cooperation between France and Chile for the study of Andean lakes. Authors thank INTAS and NATO science program for supporting research on lakes in Central Asia, which has been used as case studies presented in this manuscript.

References

Aladin NV, Crétaux JF, Plotnikov IS et al (2005) Modern hydro-biological state of the Small Aral Sea. Environmetric 16:1–18. doi:10.1002/env.709
Auqué LF (1995) Greoquimica de las lagunas. Estud Geol 51:243–257, in Spanish
Beltrando G, Cadet P (1990) Variabilité interannuelle de la petite saison des pluies en Afrique Orientale: relations avec la circulation atmosphérique générale. Veille Climatique Satellitaire 33:19–36
Bates BC, Kundzewicz ZW, Wu S et al (2008) Climate change and water. Technical paper of the Intergovernmental Panel on Climate Change. IPCC Secretariat, Geneva, 210 pp
Becker M, Llowel W, Cazenave A, Güntner A, Crétaux JF (2010) Recent hydrological behaviour of the East African Great Lakes region inferred from GRACE, satellite altimetry and rainfall observations. C R Geosci. doi:10.1016/j.crte.2009.12.010
Bergonzini L, Richard L, Petit P (2004) Camberlin, Zonal circulations over the Indian and Pacific Oceans and the level of Lakes Victoria and Tanganyika. Int J Climatol 24(13):1613–1624
Birkett S (1995) Contribution of TOPEX/Poseidon to the global monitoring of climatically sensitive lakes. J Geophys Res 100(C12):25, 179–125, 204
Birkett CM (2000) Synergistic remote sensing of lake Chad: variability of basin inundation. Remote Sens Environ 72:218–236
Calmant S, Seyler F (2006) Continental waters surface from satellite altimetry. C R Geosci. doi:10.1016/J.crte.2006.05.012

Cazenave A, Bonnefond P, Dominh K (1997) Caspian sea level from T/P altimetry: level now falling. Geophys Res Lett 24:881–884

Coe MT, Birkett CM (2005) Water Resources in the Lake Chad Basin: Prediction of river discharge and lake height from satellite radar altimetry. Water Resour Res V40(10). doi:10.1029/2003WR002543

Cohen AS, Lezzar KE, Cole J et al (2006) Late Holocene linkages between decade–century scale climate variability and productivity at Lake Tanganyika. Africa J Paleolimnol 36:189–209

Coudrain A, Francou B, Kundzewicz ZW (2005) Glacier shrinkage in the Andes and consequences for water resources. Hydrol Sci J 50(6):925–932

Crétaux JF, Birkett C (2006) Lake studies from satellite altimetry. C R Geosci. doi:10.1016/J.crte.2006.08.002

Crétaux JF, Kouraev AV, Papa F et al (2005) Water balance of the Big Aral sea from satellite remote sensing and in situ observations. J Great Lakes Res 31(4):520–534

Crétaux JF, Calmant S, Romanovski VV et al (2009a) Implementation of a new absolute calibration site for radar altimeter in the continental area: Lake Issykkul in Central Asia. J Geodesy. doi:10.1007/s00190-008-0289-7

Crétaux JF, Letolle R, Calmant S (2009b) Investigations on Aral Sea regressions from Mirabilite deposits and remote sensing. Aquatic Geochem. doi:10.1007/s10498-008-9051-2

Depetris PJ, Pasquini AI (2000) The hydrological signal of the Perito Moreno Glacier damming of Lake Argentino (Southern Andean Patagonia): the connection to climate anomalies. Global Planet Change 26(4):367–374

De Wit M, Stankiewicz J (2006) Changes in surface water supply across Africa with predicted climate change. Science 311(5769):1917–1921

Fagel N, Boes X, Loutre MF (2008) Climate oscillations evidenced by spectral analysis of Southern Chilean lacustrine sediments: the assessment of ENSO over the last 600 years. J Paleolimnol 39:253–266

Fogt RL, Bromwich DH (2006) Decadal variability of the ENSO Teleconnection to the high-latitude South Pacific governed by coupling with the southern annular mode. J Clim 19: 979–997

Frappart F, Calmant S, Cauhopé M et al (2006) Preliminary results of Envisat RA-2-derived water levels validation over the Amazon basin. Remote Sens Environ 100:252–264

Garcin Y, Williamson D, Bergonzini L et al (2007) Solar and anthropogenic imprints on Lake Masoko (southern Tanzania) during the last 500 years. J Paleolimnol 37:475–490

Garreaud R, Aceituno P (2001) Interannual rainfall variability over the South American Altiplano. J Clim 14:2779–2789

Garreaud RD, Battisti DS (1999) Interannual (ENSO) and interdecadal (ENSO-like) variability in the Southern Hemisphere tropospheric circulation. J Clim 12:2113–2123

Gillett NP, Kell TD, Jones PD (2006) Regional climate impacts of the southern annular mode. Geophys Res Lett 33:L23704. doi:10.1029/2006GL027721

Guzkowska M, Rapley CG, Rideley JK et al (1990) Developments in inland water and land altimetry. ESA contract report 78391881FIFL

Laxon S (1994) Sea ice altimeter processing scheme at the EODC. Int J Remote Sens 15(4): 915–924

Hastenrath S, Nicklis A, Greischar L (1993) Atmospheric–hydrospheric mechanisms of climate anomalies in the western equatorial Indian Ocean. J Geophys Res 98:20219–20235

Hostetler SW (1995) hydrological and thermal response of lakes to climate: description and modelling. In: Lerman A, Imboden D, Gat J (eds) Physics and chemistry of lakes. Springer-Verlag, Berlin, pp 63–82

Kouraev AV, Kostianoy AG, Lebedev SA (2009) Recent changes of sea level and ice cover in the Aral Sea derived from satellite data (1992–2006). J Mar Syst 76(3):272–286. doi:10.1016/j.jmarsys.2008.03.016

Legresy B, Remy F (1997) Surface characteristics of the Antartic ice Sheet and altimetric observations. J Glaciology 43(144):197–206

Marchant R, Mumbi C, Behera S et al (2006)The Indian Ocean dipole – the unsung driver of climatic variability in East Africa. Afr J Ecol 45(1):4–16

Medina, C, Gomez-Enri J, Alonso J et al (2008) Water level fluctuations derived from Envisat Radar altimetry (RA-2) and in situ measurements in a subtropical water body: Lake Izabal (Guatemala). Remote Sens Environ. doi:10.1016/J.rse.2008.05.001

Mercier F, Cazenave A, Maheu C (2002) Interannual lake level fluctuations in Africa from TOPEX-Poseidon: connections with ocean-atmosphere interactions over the Indian Ocean. Global Planet Change 32:141–163

Mistry V, Conway D (2003) Remote forcing of East African rainfall and relationships with fluctuations in levels of Lake Victoria. Int J Climatol 23(1):67–89

New M, Todd M, Hulme M et al (2001) Precipitation measurements and trends in the twentieth century. Int J Climatol 21:1899–1922

Nicholson SE (1996) A review of climate dynamics and climate variability in eastern Africa. In: Johnson T, Odada E (eds) The limnology, climatology and paleoclimatology of the East African Lakes. Gordon & Breach, Amsterdam, the Netherlands, pp 25–56

Nicholson S (2001) A semi-quantitative, regional precipitation data set for studying African climates of the nineteenth century, Part I. Overview of the data set. Climatic Change 50:317–353

Nicholson SE, Yin X (2002) Mesoscale patterns of rainfall, cloudiness and evaporation over the Great lakes of East Africa. In: Odada E, Olago DO (eds) The East African great lakes: limnology, paleolimnology and biodiversity, vol 12. Kluwer, Dordrecht, pp 92–120

Ogallo LJ (1988) Relationships between seasonal rainfall in East Africa and the southern oscillation. J Climatol 8:31–43

Plisnier PD, Coenen EJ (2001) Pulsed and dampened annual limnological fluctuations in Lake Tanganyika. In: Munawar M, Hecky RE (eds) The great lakes of the world: food-web, health and integrity. Backhuys, Leiden, the Netherlands, pp 83–96

Randall DA, Wood RA, Bony S et al (2007) Cilmate models and their evaluation. In: Solomon S, Qin D, Manning M, Chen Z, Marquis M, Averyt KB, Tignor M, Miller HL (eds) Climate change 2007: the physical science basis. Contribution of working group I to the fourth assessment report of the Intergovernmental Panel on Climate Change. Cambridge University Press, Cambridge/New York

Richard Y (1994) Variabilite pluviometrique en Afrique du Sud-Est. La Meteorologie 8:11–22

Rodriguez E, Chapman B (1989) Extracting ocean surface information from altimeter returns: the deconvolution method. J Geophys Res 94(C7):9761–9778

Romanovski VV (2002) Water level variations and water balance of Lake Issykkul. In: Klerkx J, Imanackunov B (eds) Lake Issykkul: its natural environment. Kluwer, Norwell, MA, pp 45–57

Ropelewski CF, Halpert MS (1996) Quantifying southern oscillation–precipitation relationships. J Clim 9:1043–1059

Rubanov IV, Ichniazov DP, Vaskakova MA et al (1987) Geologia Aralskogo More. FAN, Tachkent, 248 p

Saji NH, Goswami BN, Vinayachandran PN et al (1999) A dipole mode in the tropical Indian Ocean. Nature 401:360–363

Small EE, Giorgi F, Sloan LC et al (2001) The effects of desiccation and climate change on the hydrology of the Aral Sea. J Clim 14:300–322

Smith LC (1997) Satellite remote sensing of river inundation area, stage, and discharge: a review. Hydrol Process 11:1427–1439

Stager JC, Ryves D, Cumming BF et al (2005) Solar variability and the levels of Lake Victoria, East Africa, during the last millenium. J Paleolimnol 33:243–251

Stager JC, Ruzmaikin A, Conway D et al (2007) Sunspots, El Niño, and the levels of Lake Victoria, East Africa. J Geophys Res 112:D15106

Swenson S, Wahr J (2009) Monitoring the water balance of Lake Victoria, East Africa, from Space. J Hydrol 370:163–176. doi:10.1016/j.jhydrol.2009.03.008

Tierney JE, Russell JM (2007) Abrupt climate change in southeast tropical Africa influenced by Indian monsoon variability and ITCZ migration. Geophys Res Lett 34:L15709

Verschuren D (2003) Lake-based climate reconstruction in Africa: progress and challenges. Hydrobiologia 500:315–330

Wingham DJ, Rapley CG, Griffiths H (1986) New techniques in satellite altimeter tracking systems. In: Proceedings of IGARSS'86 symposium, Zurich, 8–11 September 1986, ESA SP-254: 1339–1344

Xu CY, Widen E, Halldin S (2005) Modelling hydrological consequences of climate change – progress and challenges. Adv Atmos Sci 22(6):789–797

Zola RP, Bengtsson L (2006) Long-term and extreme water level variations of the shallow Lake Poopo, Bolivia. Hydrol Sci J 51(1):98–114

The Future of Coastal Altimetry

20

R.K. Raney and L. Phalippou

Contents

20.1	Introduction	536
	20.1.1 Initial Conditions	537
	20.1.2 The Challenge	538
	20.1.3 New Technologies	539
20.2	AltiKa	539
	20.2.1 Introduction	539
	20.2.2 Key Innovations	540
	20.2.3 Capabilities in Coastal Areas	542
	20.2.4 Limitations	543
20.3	SAR Mode (Delay-Doppler)	543
	20.3.1 Introduction	544
	20.3.2 Key Innovations	544
	20.3.3 Capabilities in Coastal Areas	546
	20.3.4 Limitations	546
20.4	Wide Swath	547
	20.4.1 Introduction	548
	20.4.2 Key Innovations	548
	20.4.3 Capabilities in Coastal Areas	552
	20.4.4 Limitations	554
20.5	Closing Comments	555
	20.5.1 Commentary	555
	20.5.2 Future Outlook	557
	20.5.3 Conclusions	558
References		559

Keywords Altimeter • Delay-doppler • Ka-band • SAR mode • Wide swath

R.K. Raney (✉)
Applied Physics Laboratory, Johns Hopkins University, Laurel, 20723 Maryland, USA
e-mail: keith.raney@jhuapl.edu, k.raney@ieee

L. Phalippou
Observation System and Radar Unit, Thales Alenia Space, 31037 Toulouse, Cedex 1, France

Abbreviations

CTE	Coefficient of Thermal Expansion
DDA	Delay-Doppler Altimeter
DEM	Digital Elevation Model
DORIS	Doppler Orbitography and Radiopositioning Integrated by Satellite
EM	ElectroMagnetic
FFT	Fast Fourier Transform
GAMBLE	Global Altimeter Measurements By Leading Europeans
JPL	Jet Propulsion Laboratory
KaRIN	Ka-band Radar INterferometer
KaSOARI	Ka-band Swath Ocean Altimetry with Radar Interferometry
POD	Precision Orbit Determination
PRF	Pulse-Repetition Frequency
RA	Conventional non-coherent Radar Altimeter
RAR	Real Aperture Radar
SAR	Synthetic Aperture Radar
SARAL	Satellite with ARgos and ALtika
SIRAL-2	SAR/Interferometric Radar ALtimeter
SRTM	Shuttle Radar Topography Mission
SRAL	SAR Radar ALtimeter
SSH	Sea Surface Height
SWH	Significant Wave Height
SWOT	Surface Water and Ocean Topography
WITTEX	Water Inclination Topography and Technology EXperiment
WS	Wind Speed
WSOA	Wide Swath Ocean Altimeter
WVR	Water-Vapor Radiometer

20.1
Introduction

Applications in coastal environments would be well served by a radar altimetric system that could provide measurements of critical parameters – notably sea surface height (SSH), significant wave height (SWH), and wind speed (WS) – with the same precision and accuracy as has become the norm for basin-scale oceanography (Fu and Cazenave 2001), but having much more dense spatial coverage and rapid revisit characteristics. Alas, that ideal is not in prospect at this time. However, significant progress has been made on certain aspects, the highlights of which this chapter attempts to capture.

20.1.1
Initial Conditions

Conventional ocean-observing radar altimetry[1] provides a reasonable and non-controversial baseline for the discussions. All ocean-viewing altimeters to date have been radar-to-surface ranging instruments whose measurements are confined to nadir, the surface track immediately below the spacecraft as it moves along its orbit. Conventional radar altimeters (RA) are pulse-limited (Brown 1977), whose footprints are determined by the radar's pulse length (typically about 0.5 m) as it intersects the surface. The resulting spatial resolution is on the order of 2 km, which increases for larger SWH. Such pulse-limited resolution is much better than the width of the altimeter's antenna pattern, which usually is more than 15 km in diameter at the surface. Pulse limited instruments have several advantages; chief among them is that they are relatively insensitive to (small) angular errors in pointing the antenna pattern towards nadir.

Ocean-viewing altimeters derive their information from waveforms each one of which corresponds to the averaged time-history of the power backscattered from a restricted area on the surface. Whereas with conventional altimeters that area is pulse-limited, for alternative technologies the area may be determined by other constraints, such as Doppler bandwidth. In all cases, the resulting backscattering area is the altimeter's spatial resolution, equivalently, its footprint, or resolution cell.

Altimeters, as is true for most radars, extract ranging measurements from the propagation of an electromagnetic (EM) wave that is essentially monochromatic. The inherent result is that reflections from many different facets of the illuminated sea surface interfere with each other, causing coherent random additions and/or subtractions within the net signal received by the radar. This "fading," or "speckle," is an unavoidable and natural part of each individual return. The only way to suppress such coherent self-noise is to average the power responses of statistically independent versions of the same measurement. The result is an (averaged) waveform, which is the basis for any altimeter's data products. Typical basin-scale waveforms are produced on 1-s postings, each representing the average of about 2,000 individual contributions.

The uncertainty of an altimeter's measurement – *precision* – is stipulated by its standard deviation, which is a description of the width of the distribution of the constituent data points with respect to their mean. Precision improves (smaller standard deviation) in inverse proportion to the square root of the number of statistically independent samples averaged. Thus, precision depends primarily upon the instrument, processing, and constraints imposed by the observed surface. The mean of the absolute value of an altimeter's measurement – *accuracy* – is the centroid of the data point distribution, relative to the "true" value. For any ocean-viewing radar altimeter, accuracy depends to first order on the effectiveness of path length delay corrections, and on "accurate" orbit determination, especially the radial component, all of which are external to the instrument. An altimeter's accuracy is covered extensively for the open ocean in standard references (Fu and Cazenave 2001), and for coastal altimetry is a recurring theme in other chapters of this book. Precision emerges as the more significant radar characteristic of the two that is pertinent to the topic of this chapter.

---

[1]See Chapter by Benveniste, this volume.

In the open deep ocean, the largest SSH signals are in response to the earth's geoid. The magnitude of these signals have a global variation greater than ±100 m, but at any given location they remain effectively constant (to within a cm or so) on decadal scales. Thus, for altimeters aimed at dynamic phenomena, the background sea surface topography set by these geodetic signals can be established by repeat-pass altimetry, eventually to be removed from the measurements. Repeat-pass orbits (typically having periods between 10 and 30 days) imply that the surface tracks must be widely separated (from about 110 km to more than 300 km at the equator). The implied spatial sampling scale is sufficient for most open-ocean phenomena of interest. In contrast, if the geodetic signal is the objective, then non-repeat orbits (or very long period repeat orbits) are essential (Smith and Sandwell 2004).

20.1.2
The Challenge

The requirements for coastal altimetry[2] differ considerably from those typical of open-ocean applications. Thus, most of the conditions summarized in Sect. 20.1.1 for conventional basin-scale radar altimetry do not carry over into the coastal environment. In particular, typical spatial and temporal correlation scales in the littoral region are much smaller than those that prevail in the open ocean. This implies that the potential for averaging to improve measurement precision is compromised. The immediate consequence is that the radar altimeter itself if it is to be suited for coastal applications should have an intrinsic measurement precision that is better than that of a conventional radar altimeter.

Spatial and temporal coverage are closely related issues. The tidal signal[3] is continually varying, with dominant constituents having periods of 12 h, and may vary spatially on scales of a few kilometers. Observation of near-shore tidal phenomena, or unrelated signals in the presence of tides, would be best served by altimetric data taken on short time frames, and over large areas. Indeed, certain littoral applications require some form of altimetric data over an area rather than sparse nadir traces.

SWH measurements in the coastal region have practical applications, but there are complicating factors. Wave characteristics may change on small spatial (and temporal) scales, reducing the potential for averaging to improve SWH measurement precision. More subtle, waves in shallow conditions have profiles that do not conform to those typical of the deep open ocean. SWH measurements (and related corrections of altimetric data to extract SSH) rely on algorithms developed from open-ocean data. Extrapolation to shallow water wave conditions offers challenges, and perhaps also opportunities.

By definition, coastal altimetry implies the proximity of land, which corrupts the water vapor radiometer's (WVR) measurements (thence compromising correction for wet path delay), and which complicates (or may totally confound) the radar's onboard tracker. A radar altimeter to be suited for coastal applications must have a tracker that is "land-proof," and a radiometer that is designed to minimize the effect of nearby land, or one that does not require WVR data at all.

---

[2]See Chapter by Dufau et al., this volume.
[3]See Chapter by Ray et al., this volume.

20.1.3
New Technologies

For many years considerable thought has been invested towards enhancing the capabilities of radar altimeters, especially their spatial coverage, revisit frequency, and precision. Three advanced concepts are reviewed in this chapter, two of which are being implemented for operational missions,[4] the Ka-band altimeter (on India's SARAL satellite), and the Ku-band SAR mode (on CryoSat-2 and Sentinel-3). The third concept is wide swath, which is of great practical interest, and is being promoted actively, thus deserving of inclusion in this discussion. These three technologies are presented in the order of their increasing technical distance from conventional radar altimeters.

20.2
AltiKa

AltiKa – an Altimeter at Ka-band – is fundamentally a conventional pulse-limited radar altimeter, although it represents an advance in the state-of-the art thanks to several key innovations (Vincent et al. 2006; Richard et al. 2007). The first orbital instrument will be aboard India's SARAL mission (LeRoy et al. 2007), planned for launch late in 2010. The AltiKa concept was motivated by a perceived need for improved height precision and smaller footprint size. In addition AltiKa includes a built-in dual frequency radiometer that shares the altimeter antenna. This architecture helps to reduce spacecraft size and cost. The altimeter design is based on the classical deramp technique (Chelton et al. 1989), and its implementation takes advantage of extensive Poseidon heritage. Several of AltiKa's attributes promise improved performance in coastal waters.

20.2.1
Introduction

The design philosophy behind AltiKa was to simplify the flight instrument while enhancing its performance. This led to the choice of Ka-band (35.75 GHz) as the altimeter's center frequency, rather than the customary Ku-band at about one third the Ka-band frequency. The higher frequency offers several advantages, including a smaller beamwidth (for a given 1-m antenna diameter), eliminates the need for a second frequency for ionospheric propagation corrections, and simplifies the antenna and feed issues to accommodate a microwave water vapor radiometer. The higher frequency also brings with it a significant disadvantage, greater susceptibility to propagation attenuation (or outright blockage) from atmospheric moisture. The radar bandwidth was also increased, leading to a finer range resolution, as well as associated added benefits. In other regards the baseline

---

[4]See Chapter by Benveniste, this volume.

SARAL concept is conventional in the modern sense, in that it includes precision navigation, comprised of an active system (DORIS (Jayles et al. 2004)) and a passive device (a laser corner reflector), and the AltiKa altimeter has a tracker (Vincent et al. 2006) that takes advantage of an internally stored digital elevation model (DEM).

20.2.2
Key Innovations

Four key innovations lie behind the improvements represented by AltiKa: bandwidth, antenna beamwidth, pulse repetition frequency (PRF), and echo tracking.

20.2.2.1
Bandwidth

Conventional Ku-band radar altimeters since Seasat have had range bandwidths W on the order of 320 MHz (Raney 2005) (although larger bandwidths at Ku-band would be feasible from both technical and spectral allocation considerations). Resolution dr depends on bandwidth according to $dr = c/2W$ where c is the speed of light. Thus, AltiKa's larger bandwidth of 480 MHz[5] supports a range resolution of 0.31 m, smaller than the usual 0.47 m. Guided by a standard rule of thumb, that change leads to an improvement in the radar's intrinsic SSH measurement precision by the ratio 0.31/0.47, thus a factor of about 1.5 smaller standard deviation than for a conventional altimeter. Of course, SSH precision is improved by the number of statistically independent waveforms combined for a given measurement, so the improvement of 1.5 accounts only for the range resolution contribution.

One secondary consequence of the smaller range resolution is a smaller footprint. The diameter D_{PL} of the pulse-limited footprint from an altimeter at altitude h over a quasi-flat sea is given by

$$D_{PL} = 2(2\rho h/a)1/2 \qquad (20.1)$$

thus proportional to the square root of the radar's range resolution r (MacArthur 1976). The parameter a captures the effects of earth curvature, whose value is given by

$$a = \frac{R_E + h}{R_E} \qquad (20.2)$$

As a result, AltiKa's pulse-limited footprint is about 0.8 times smaller than a Ku-band altimeter for a given pulse length and altitude.

---

[5]By international spectral allocation agreements, this bandwidth is the largest available for space-based earth observation at Ka-band.

20.2.2.2
Beamwidth

For a given antenna diameter D, the effective one-way 3-dB beamwidth β is proportional to the ratio of radar wavelength λ to D. Thus a decrease in radar wavelength from 2.2 cm (Ku-band) to 0.84 cm (Ka-band) corresponds to a smaller antenna beamwidth by a ratio 2.6. From an 800 km altitude and AltiKa's 1-m diameter antenna, the footprint is about 8.4 km, in contrast to the 21 km diameter in Ku-band. Narrower beamwidth implies higher antenna gain, which is helpful for the radar's link (power) budget. In contrast, narrower beamwidth decreases the altimeter's tolerance to mis-pointing due to variations in spacecraft pitch or roll, or antenna mis-alignment.

20.2.2.3
Pulse Repetition Frequency

Choice of the radar's pulse repetition frequency (PRF) is constrained by several factors. One set of arguments urges for a high PRF, primarily to maximize the number of potential individual waveforms ready for averaging. However, effective averaging presumes statistical independence, which determines the upper limit on PRF. When evaluated, the resulting decorrelation-driven upper bound is on the order of 2 kHz for conventional Ku-band altimeters.

The first estimate of that limit was published by Walsh in a classic paper (Walsh 1982). However, when derived from first principles, the upper bound may be shown to be

$$\text{PRF}_{MAX} = \frac{2V}{\lambda} \frac{D_{PL}}{h} \qquad (20.3)$$

which differs in detail from the Walsh equation.[6] In Eq. 20.3 V is spacecraft velocity, λ is radar wavelength, h is spacecraft height above the earth's surface, and D_{PL} is the pulse-limited diameter (Eq. 20.1). The first term is the Doppler shift generated by the spacecraft, and the second term is the angular sector of the footprint orthogonal to the ocean surface. The product of these terms determines the Doppler bandwidth corresponding to the pulse-limited footprint. To avoid substantial pulse-to-pulse correlation, the PRF must be smaller than this bandwidth.[7] Further insight is available. It is well known that the pulse-limited diameter increases with significant wave height. Increase in the minimum PRF as a function of SWH is captured in Eq. 20.3, in contrast to the classic formula.

Only two parameters in this expression – λ and ρ (which is implicit in the pulse-limited expression) – are open to choice in an altimeter's design, since the others are essentially constants, given that spacecraft height is stipulated by external constraints. In comparison

---

[6]If Eq. 20.3 is recast in terms of temporal range resolution τ, the resulting expression is equivalent to the classic form, except for Walsh's Earth curvature term $1/a$ (Eq. 20.2). Walsh introduced that factor to scale spacecraft velocity V into the surface footprint velocity, which was not a correct rendition of the underlying physics.
[7]From this point of view, the PRF upper bound may be seen as the obverse of the Nyquist sampling criterion, which states that to assure pulse-to-pulse correlation the sampling rate must be larger than the bandwidth of the (complex) spectrum being sampled.

to conventional altimeters, AltiKa has smaller compressed pulse length by a factor of 1.5, but that is overshadowed by its wavelength which is smaller by a factor of 2.6. The result is that the maximum allowable PRF for AltiKa is larger than for conventional Ku-band instruments by a factor of about 2. AltiKa has been designed with a PRF of 4 kHz to respect this more generous decorrelation constraint.

The higher PRF of AltiKa, although only by a factor of two with respect to Poseidon, improves SSH measurement precision by a factor of $\sqrt{2}$. This, coupled with the 1.5-fold finer range resolution leads to an intrinsic precision improvement of about 2, so that the standard deviation of a SSH measurement from AltiKa will be half that from conventional radar altimeters, when compared over the same integration time.

20.2.2.4
Tracking

AltiKa is the second radar altimeter to incorporate a global digital elevation model (DEM) over ocean and land surfaces as a guide to maintaining echo tracking. The Ku/C-band Poseidon-3 on board Jason-2 launched in 2008 is the first altimeter to use this technique. The first Poseidon-3 results using DEM tracking on coastal and continental surfaces are very encouraging regarding tracking robustness and measurements performances. The AltiKa DEM is keyed to the DORIS navigation system, so that the altimeter is prepared at all locations with an open-loop estimate of radar range, thus facilitating more reliable initial acquisition and tracking performance in areas that may have rapidly changing surface elevations such as coastal zones.

20.2.2.5
High Data Rate Mode

As demonstrated successfully with RA-2 on Envisat (Gommenginger et al. 2007), AltiKa will have a mode in which all (complex) data in response to each pulse transmission will be retained, and subsequently made available for processing in ground facilities.[8] Such data will be collected only over small (7 km) and preselected areas.

20.2.3
Capabilities in Coastal Areas

AltiKa's narrower antenna (by nearly a factor of 3) compared to that of a conventional radar altimeter means that its waveforms are much less corrupted by land masses only a few kilometers off-nadir. Stated another way, the antenna footprint diameter is only about four times larger than the pulse-limited resolution, as opposed to a conventional Ku-band altimeter for which the antenna footprint is nearly ten times larger than the (flat sea) pulse-limited diameter. This attribute allows altimetric performance closer to shore than is

---

[8]http://earth.esa.int/raies

possible with a conventional instrument, all other factors being the same. If antenna beamwidth is the dominant source of waveform corruption from land, then AltiKa will achieve about three times better coastal proximity than a conventional Ku-band instrument, to less than about 3 km. The corollary to this feature is that the pointing accuracy of the Ka-band altimeter must be more tightly controlled, especially in the coastal environment.

"Smart" tracking is especially valuable in the near-shore environment. Although the same strategy may be applied to a conventional Ku-band altimeter (as illustrated by Poseidon-3 on Jason-2, and to a limited extent on CryoSat-2), it is especially effective for an altimeter that has a significantly smaller beamwidth, as does AltiKa.

The waveforms from AltiKa may be analyzed by conventional algorithms to retrieve the usual SSH, SWH, and WS parameter values, subject to differences in the sea surface physics between the open deep ocean and the coastal regime, and subject to the availability of sufficiently accurate path length correction data (for SSH measurements), a restriction which also applies to Ku/C-band altimeters. However, retrieval algorithms for AltiKa, as for all other forms of radar altimetry, are subject to re-verification and most likely revision for the special circumstances to be encountered in coastal areas, as reviewed in Sect. 20.5.1.3. Data from the high rate mode should be of particular interest for targeted near-shore areas.

20.2.4 Limitations

As is well known, the major disadvantage of Ka-band is that it is far more susceptible to attenuation by atmospheric hydrometeors (water droplets) than lower frequencies. On average, AltiKa is predicted to lose less than 4% of data, based on analysis of global cloud and rain climatologies (Tournadre et al. 2009), although upwards of 10% may suffer serious attenuation under tropical conditions. As is commonly observed in tropical and subtropical latitudes, for example, the occurrence of cloud and precipitation near coastlines is more prevalent than over the open ocean, so these numbers may be optimistic when extended to near-shore altimetry.

Although the high rate data from AltiKa will be available at full radar PRF, thus corresponding to along-track postings on the order of only a few meters spacing, each such return is comprised of the backscatter from the radar's pulse-limited footprint, typically more than 1.6 km in diameter. Thus it is fair to characterize these data as "high rate," but they are not necessarily "high resolution." However, analytical tools such as along-track Fourier transforms applied to high rate data from selected coastal regions may offer insights not otherwise available.

20.3 SAR Mode (Delay-Doppler)

The SAR (synthetic aperture radar) mode is based on the delay-Doppler technique (Raney 1998a) which was developed originally in response to the challenge of radar altimetry over (sloping) ice sheets. It turns out that its attributes also are responsive to several of the requirements of coastal altimetry.

20.3.1
Introduction

A delay-Doppler altimeter (DDA) produces a footprint whose along-track size is effectively beam-limited, in contrast to the pulse-limited footprints of conventional ocean-viewing radar altimeters. This beam limit typically is about 250 m, which is much smaller than either the pulse-limited dimension or the physical width of the altimeter's antenna pattern. The beam-limited footprint dimension is determined by signal processing, which may be either on board, or deferred to a ground-based facility. For many users, the most prominent characteristics of this new class of radar altimeter may be appreciated by viewing the instrument as "beam-limited" in the along-track (orbit plane) direction, rather than by studying the details of the SAR-based Doppler processing.

The radar electronics of a DDA are essentially the same as those of a conventional radar altimeter; the few exceptions are significant, but most are of interest only to radar experts. The key difference from the user's point of view is that the radar's pulse repetition frequency (PRF) must be higher for a DDA, indeed sufficiently high that the Doppler frequencies within the main beam of the altimeter's antenna are adequately sampled. That leads to PRFs on the order of 15 kHz. In contrast, most conventional radar altimeters have PRF's on the order of 2 kHz. However, the intrinsic similarity between the two classes of altimeters implies that the same radar may be tasked to operate either as a conventional instrument, or as a delay-Doppler (SAR mode) instrument. Such duality characterizes both SIRAL aboard CryoSat-2 and SRAL on Sentinel-3 (LeRoy et al. 2007).

20.3.2
Key Innovations

The processing is based on range/Doppler techniques that are well established for synthetic aperture radar data (Curlander and McDonough 1991), hence the "SAR mode" nomenclature commonly invoked in European circles. The "delay-Doppler" description derives from radar astronomy (Green 1968). The signal processor generates a dozen or more simultaneous beam-limited footprints within the altimeter's real antenna pattern, thus emulating an ensemble of discrete beam-limited altimeters. These footprints (Fig. 20.1), each of which is much smaller in the along-track direction than the pulse-limited dimension, are distributed side by side. After processing, there is one waveform from each of these footprints. The natural waveform rate typically is less than 30 Hz, corresponding to the rate of traverse by the altimeter over these footprints. Each such waveform is the average of all of the processor's observations at a given footprint location. This is in marked contrast to waveforms from a conventional altimeter, each one of which corresponds at a minimum to the pulse-limited footprint diameter of 2.4 km, and which at a 10-Hz rate corresponds to about 3 km.

The along-track footprint size does not depend on external factors such as significant wave height (SWH) since it is beam-limited, hence a constant of the system. This is in distinct contrast to a conventional altimeter for which the pulse-limited diameter increases with larger SWH (in proportion to approximately 4/7 SWH). However, the across-track footprint of a delay-Doppler is pulse-limited, which leads to an elongated two-dimensional

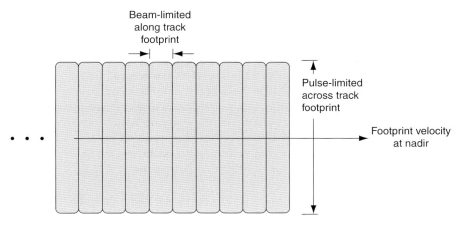

Fig. 20.1 Delay-Doppler (SAR mode) footprints

footprint, having an aspect ratio on the order of 10:1. The across-track pulse-limited characteristic is helpful, in that the resulting sea surface height (SSH) accuracy does not require precise control or knowledge of the altimeter's beam pointing.

The delay-Doppler altimeter enjoys larger signal-to-noise ratio than a comparable conventional radar altimeter (Raney 1998a). This feature is a direct consequence of the DDA's footprint, which is pulse-limited in one dimension and beam-limited in the other, thus a geometric mean between pure beam-limited performance (Moore and Williams 1957) and pure pulse-limited performance (Brown 1977; Hayne 1980).

Averaging is required to reduce the self-noise (speckle) inherent in an individual (single-pulse) reflected return. The residual noise after averaging determines the lower bound on the altimeter's intrinsic measurement precision. A conventional altimeter (e.g., RA-2 on Envisat or Poseidon-3 on Jason-2) transmits pulses ~2,000 times/s. In between each such transmission, the altimeter footprint moves along the surface by about 3.4 m. In effect, the instrument operates as if the footprint were dragged along the surface, with the incoherent sum accumulated piecewise along the way. At this rate, for example, in 1/10 s the footprint would progress about 650 m along the surface, which increases the effective footprint to ~3 km, and more with larger SWH. The effective number of averages ("looks") for this case is equal to the number of transmissions, which would be ~200.

Waveform averaging in a DDA differs fundamentally from that of a conventional radar altimeter. A typical instrument such as SIRAL would operate with a mean transmission rate (PRF) on the order of 18,000 pulses/s or more (Raney 1998a; Wingham et al. 2006). The returns from these are processed in four stages: (1) sub-beam formation through Doppler analysis by (coherent) fast-Fourier-transforms (FFTs), (2) co-registration of the waveforms (in Doppler and range), (3) square-law detection, and (4) summation (incoherent averaging) within each Doppler bin and at each resolved surface location. In effect, a DDA instrument operates as if the altimeter "spotlights" each resolved along-track cell and averages over all data available from that cell as the spacecraft progresses along its orbit. For a typical DDA, from start to finish of the overhead pass, each resolved 250-m surface cell would accumulate on the order of 150 looks, depending on the specifics of the particular design. Many such cells are tracked and processed in parallel, so that the surface sampling keeps up with the along-track satellite motion. Another stage of averaging may be applied, by aggregating adjacent 250-m cells. An averaged waveform at a 10-Hz waveform rate (corresponding to a 650-m resolved along-track cell) would comprise approximately 400 looks, and more than 4,000 looks at a 1-s waveform rate.

These examples illustrate that DDA-processed waveforms enjoy about a four-fold advantage in number of looks over Geosat's performance, and about a two-fold advantage over a modern conventional radar altimeter such as Poseidon-3. Measurement precision goes as the square-root of number of looks, so the precision advantage of a DDA over a modern radar altimeter for 1-s data is on the order of 1.4 (Phalippou and Enjolras 2007).

20.3.3
Capabilities in Coastal Areas

The relatively small 250 m along-track footprint is helpful in several aspects of coastal altimetry. It is better suited to observing sea surface conditions that may have correlation lengths of less than 1 km, a common condition in near-shore environments. For situations in which the spatial scale of variability is larger, the user may opt for finer precision (more averaging), at the cost of a larger effective (along-track) footprint. Thus, the trade-off between resolution and smoother (averaged) waveforms may be made by the user, guided by the specific situation at local areas within the coastal region.

As with conventional pulse-limited altimeters, waveforms from a delay-Doppler instrument support measurement of sea surface height (SSH), significant wave height (SWH), and wind speed (WS). Conventional retrieval algorithms may be used. The precision of each of these measurements depends of course on the amount of averaging employed, respecting the customary inherent trade-off between footprint resolution and measurement precision.

Since the radar returns are transformed into Doppler frequencies, each one of which correspond to beam-limited along-track locations, in principle these data could help to guide the tracker (or re-tracking in subsequent ground-based processing) on sub-kilometer scales. Although not yet demonstrated with real data, the Doppler frequency decomposition potentially offers an enhanced capability for re-tracking and quality control that could be of value in the near-shore environment, especially in combination with DEM-referenced techniques.

If care is taken, and within certain constraints, delay-Doppler data (such as that expected from the SAR-mode aboard CryoSat-2 or Sentinel-3) can be processed to emulate the performance of a conventional altimeter, thus offering the opportunity to look at data products of both types from the same (SAR-mode) pass over an area of interest. This capability may have utility–such as quality control of the waveforms–in certain coastal areas.

20.3.4
Limitations

The delay-Doppler's elongated footprint – beam-limited along track and pulse-limited across track – implies that the altimeter's response in coastal regions depends on the relative angle between the shoreline and the spacecraft orbital plane. The 250-m dimension

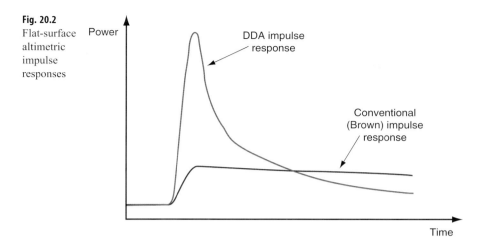

Fig. 20.2 Flat-surface altimetric impulse responses

prevails for coastlines that lie at right angles to the altimeter's path, whereas the longer pulse-limited footprint dimension dominates when the altimeter is passing parallel to the coast line. (The asymmetrical DDA footprint may also be a significant factor when viewing very long swells as in the Southern Ocean, in which case there may be an additional SSH bias as a function of swell direction relative to the footprint's orientation.)

The DDA's flat-surface response function (Raney 1998a) is "peaky" (Fig. 20.2), which is characteristic of beam-limited systems, in contrast to the much flatter plateau of a conventional altimeter's Brown-model response function (Brown 1977). The peaky nature of the DDA response function tends to suppress the off-nadir contributions from the pulse-limited portion of the footprint, but not completely, thus mitigating to some extent the potential effect of land masses to the side of the orbital plane.

20.4
Wide Swath

Wide swath techniques differ markedly from altimeters that seek sea surface information from only the sub-satellite (or nadir) track. Range measurements confined to nadir are relatively robust, since in all cases they correspond to the minimum time delay observable in the radar's reflected signals. Further, these returns often are dominated by a relatively strong component when viewed at nadir, since the mean sea surface is orthogonal to the incoming EM illumination. In contrast, wide swath altimeters gather data that derives from reflections to the side of the nadir track at off-nadir angles up to several degrees. Thus, this class of sea surface height measurement requires triangulation, and may need to adapt to reduced radar backscatter at larger off-nadir angles. These factors introduce new sources of potential error. The benefits anticipated for wide swath altimetry are considerable, especially in coastal applications in which there is a premium on relatively dense spatial coverage, but these promised benefits come with associated technical challenges.

20.4.1
Introduction

High accuracy swath altimetry concepts for ocean and ice topography measurements have been of serious interest for more than 30 years (Brown 1976; Bush et al. 1984; Rapley et al. 1990; Rodriguez and Martin 1992; Pollard et al. 2002a; Phalippou and Guirraro 2004; Enjolras and Rodriguez 2007). The main appeal of the general concept is to improve the temporal and spatial surface sampling using a single satellite. Besides improving the sampling, for a given transmitted bandwidth, swath altimeters are also capable, in principle, to improve significantly the spatial resolution in comparison to the footprint of conventional altimeters. In coastal areas, high resolution and selection of pixels that are not contaminated by reflections from land should contribute to successful near-shore proximity. Finally, ocean-viewing swath altimetry promises to provide a single-pass three-dimensional image of the surface topography over an extended area, in the same manner as single-pass interferometry over terrestrial regions.

Most swath altimeter concepts include an embedded nadir altimeter in the system. Its functions are to provide sea surface height, wind speed and significant wave heights along the nadir track, to collect data otherwise missed in the gap between and right- and left-side swathes, and to support the baseline calibration by providing reference anchor points at orbital crossovers. While the nadir altimeter implies many substantial system impacts and design trade-offs, this section focuses exclusively on key attributes and the intended performance of the swath instrument.

While very challenging technically, critical aspects of swath altimetry have been investigated by airborne and Space Shuttle experiments, and by interferometric radar altimeter studies and development with the SIRAL-1 and -2 instruments on the CryoSat ESA mission (Wingham et al. 1998; Rey et al. 2004; Wingham et al. 2006). The Wide Swath Ocean Altimeter (WSOA) Ku-band radar (Rodriguez and Pollard 2001; Pollard et al. 2002b) was studied and partly breadboarded by JPL to be embarked as a demonstration mission on the operational Jason-2 satellite. However, WSOA was de-scoped from Jason-2 during phase B mainly for budget and development risk concerns. In recent years swath altimetry activity has increased again, in response to interests in surface hydrology and ocean measurements. One manifestation is the proposed SWOT (Surface Water and Ocean Topography) mission, whose main payload instrument is the KaRIN (Ka-band Radar INterferometer) instrument (Enjolras and Rodriguez 2007). The principal technological differences between WSOA and KaRIN are the 35 GHz frequency (Ka-band in place of Ku-band), a larger deployable antenna subsystem, and the addition of a SAR (Synthetic Aperture Radar) mode. The interferometric baseline is increased from 6 to 10 m on KaRIN. The transmitted bandwidth is also increased by a factor 10, from 20 to 200 MHz.

20.4.2
Key Innovations

The key innovation is to bring interferometric area measurement of surface elevations into the realm of ocean altimetry. Swath altimetry is based on well-established (single-pass) radar interferometric techniques (Graham 1974; Zebker and Goldstein 1986;

Massonnet 1996; Hanssen 2001) to generate a three-dimensional image of the surface topography. For example, the Shuttle Radar Topography Mission (SRTM) provided nearly global coverage of the Earth's land masses with a typical vertical accuracy of the order of a 5–10 m (Bamler et al. 2003). It is worth noting that the height precision required for the ocean, either in open or coastal areas, is at least 100 times smaller (typically a few centimeters). From the standpoint of that fine precision, ocean swath altimetry is an innovative concept, presenting challenges for the end-to-end system, including the radar, spacecraft, and data processing. Current conceptual designs are characterized by high spatial resolution over a wide swath, typically 160–200 km.

20.4.2.1
Swath and Resolution

The geometry of swath altimetry is shown in Fig. 20.3. The radar transmits pulses at a given pulse repetition frequency (PRF) alternatively on the right side and on the left side of the ground track. Each backscattered echo is received simultaneously by two spatially separated antennas. Variations in sea surface elevation can be retrieved from the phase difference between these two signals.

A swath altimeter is essentially side-looking imaging radar. Thus the cross-track resolution is determined by the radar's bandwidth, and by the incident angle of the EM illumination on the surface. For instance, assuming a 100 MHz bandwidth and a 3° incident angle, the resulting across-track resolution is 28 m.

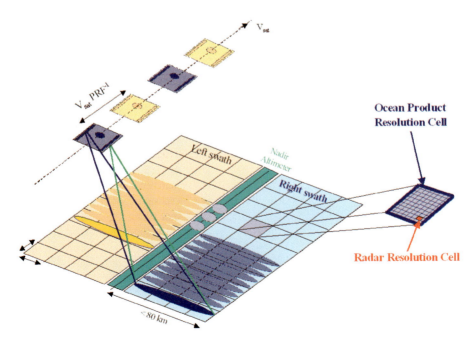

Fig. 20.3 Geometry and resolution cells for a swath altimeter

The along track resolution depends on the way the data are handled in the processor. There are two baseline alternatives: real aperture (RAR) or synthetic aperture (SAR) techniques (Ulaby et al. 1982; Raney 2008), each having performance, resolution, and data rate implications.

In the RAR mode, along-track resolution dx is inversely proportional to antenna length L_A, and, at steep incidence, proportional to spacecraft altitude h, according to $h\lambda/L_A$ where λ is the radar wavelength. At a frequency of 35 GHz (Ka-band, 8.6 mm wavelength) and an 800 km satellite altitude the spatial resolution would be 2.7 km for a 2.5 m antenna length.

In the SAR mode, the ultimate (single-look) along track spatial resolution dx is approximately equal to half the antenna size, thus $L_A/2$. For the 2.5 m antenna, the resulting spatial resolution would be 1.25 m. In practice using SAR mode data, along-track resolution is traded for increased averaging (multi-looking) thus reducing significantly the fading (speckle) noise. The finer resolution of the SAR mode in contrast to the RAR mode is particularly beneficial in coastal applications.

20.4.2.2
Mapping Ocean Surface Elevations

Typical ocean radar altimeters measure the range between the satellites to the surface with a 0.5 m range resolution. Parameter retrieval is based on matching a known model of the sea echo to averaged waveforms, after suitable tracking (or re-tracking[9]) processes. Conventional altimetry achieves typically 2 cm accuracy in SSH retrievals (Jason/Envisat class) for 2-m significant wave height.

In contrast, swath altimetry is an interferometric technique. Elevation interferometry (Fig. 20.4) requires two antennas whose phase centers are separated by a known baseline B.

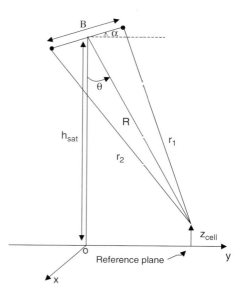

Fig. 20.4 Geometry for interferometric altimetry, where the "flat-Earth" representation (and its relatively simple equations) is adequate for this overview discussion

---

[9]See Chapter by Gommenginger et al., this volume.

In practice the radar transmits a pulse from one antenna, giving rise to an echo that is received simultaneously on both antennas. For terrestrial applications, template or re-tracking schemes are not required to retrieve topographic profiles of the surface. However, there may be added benefits of such techniques when observing the ocean. The effectiveness of modeled responses with interferometric wide-swath data must remain unexplored until actual data are available.

The (interferometric) phase difference φ is given by

$$\phi = k[2r_2 - (r_1 + r_2)] = k(r_2 - r_1) \quad (20.4)$$

where $k = 2\pi/\lambda$ is the wavenumber. Under the reasonable condition that the baseline B is very short compared to the satellite altitude, the phase in Eq. 20.4 can be approximated as

$$\phi = kB\sin(\theta - \alpha) \quad (20.5)$$

showing the important dependence on the incident angle θ, a consequence of the fact that the method is at its core a triangulation. Note that the roll angle α also is an essential and equally-weighted factor in the measurement of interferometric phase. Both baseline length and roll angle emerge as the dominant sources of systematic error that must be mitigated for successful swath altimetry.

From Fig. 20.4, one can derive the elevation z and across-track position y through the pair of equations

$$z = h - R\cos\theta \quad (20.6)$$
$$y = R\sin\theta \quad (20.7)$$

Hence, measurements of slant range R (or more correctly, round-trip time delay) and the angle of incidence θ provide the relative elevation and across-track position of a scatterer.

20.4.2.3
Elevation Accuracy

The elevation error δz is proportional to the uncertainty $\delta\alpha$ of the baseline roll attitude (angle α), and to the distance y of the measurement cell from nadir, according to

$$\delta z \approx R\theta\,\delta\alpha = y\delta\alpha \quad (20.8)$$

This simple equation exposes one key technical limitation of swath altimetry – extreme sensitivity to roll angle uncertainty:

- The required accuracy of roll angle knowledge is very stringent. For example, in order to get less than 1 cm elevation error at the edge of a 100 km swath, the roll angle uncertainty must be less than 0.1 μrad, an extremely low value.
- To minimize systematic elevation errors, it follows that the swath must be positioned as close as possible to nadir. The inner limit of the swath is fixed by the need to avoid radar ambiguities (Curlander and McDonough 1991) between left and right swaths.

- The systematic height error increases linearly with distance y from nadir, which implies an upper limit on tolerable swath width for a given acceptable error.
- Note that any source of unknown differential phase between the two channels contributes an error of comparable magnitude. If such phase differentials are constant, then they should be subject to calibration and hence cancelation. However, if variable, as with temperature or frequency, they could be troublesome.

It has been shown in principle (Rodriguez and Pollard 2003) that the required roll uncertainty can be minimized by correlating along-track and cross-track swath measurements at orbital crossovers, although this presumes supporting statistical stationarity of the ocean's surface over the data collection time, which may span several days. Whereas such stationarity should be supported in the open ocean, the condition may be insufficient for near-shore waters, necessitating extrapolation of open ocean roll calibration to near-shore observations, thus introducing a source of potential error. Finally, even the open ocean crossover data are insufficient to correct for all possible roll errors when the two orbit planes are not orthogonal.

Elevation accuracy depends also on precise knowledge of the interferometric baseline B (Eq. 20.5). In-flight variations of the baseline length are due the mechanical antenna structure's response to the thermal environment along the orbit. Although physically minute, such changes may impart substantial errors on relative phase as the spacecraft passes from daylight to the earth' shadow and back again. Therefore, only materials having an extremely small coefficient of thermal expansion (CTE) can be used for the antenna and for its structure. Studies show that in principle the problem is solvable, yet remains very challenging.

A third source of potentially significant sea surface height uncertainty δz derives from variations $\delta\phi$ in the interferometric phase

$$\delta z = \frac{y\lambda}{2\pi B}\delta\phi \qquad (20.9)$$

Using the previous numbers (800 km altitude, 2 m baseline, 100 km distance from nadir, and Ka-band frequency), from Eq. 20.9 a phase noise of 1.5×10^{-4} radian leads to a 1 cm elevation uncertainty. Such sub-milliradian phase accuracy cannot be met at the radar's finest (single-look) resolution. Many pixels (N) must be averaged together to achieve a useful phase robustness, thus improving the measurement precision (standard deviation) in proportion to the inverse of $(N)^{1/2}$. For the swath altimeter of this discussion, N is on the order of 10^4–10^6 for an ocean mission, corresponding to a 5 by 5 km resolved cell. Typical results are shown in Fig. 20.5.

20.4.3
Capabilities in Coastal Areas

The main advantage of swath altimetry for coastal application is its capacity to improve significantly the spatial coverage. The temporal sampling is mostly driven by the orbit parameters (inclination/repeat cycle) because the swath is limited to a maximum of say 160 km (left to right). However, temporal sampling for a given orbit will be more favorable

Fig. 20.5 Instrument precision limits (SSH) for a 5 by 5 km post-processing resolved cell as a function of off-nadir distance

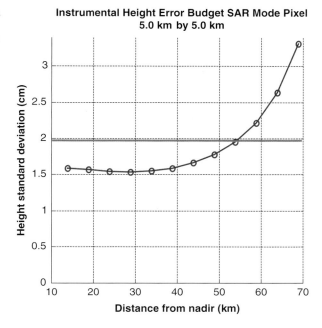

than that obtained with a nadir altimeter as swaths may overlap several times during the orbit repeat cycle, depending on latitude.

Swath altimetry can provide a surface height map of the ocean with a precision similar to nadir altimetry (1–2 cm) for 5 by 5 km resolution cells. The shore line can be traced on a two-dimensional map that may traverse the swath altimetric data at arbitrary angles. Pixels can be selected that have minimal or no corruption from land masses.

A swath altimeter in SAR mode can provide very fine radar resolution cell sizes, on the order of few tens of meters in the across-track direction and few meters in the along-track direction. This fine resolution will offer the possibility to follow the coastline in detail. Of course, fine resolution implies larger speckle noise. During ground processing, the user can adapt the resolution of the ocean product by averaging of the radar cells (multi-looking) to the coast line characteristics. In effect, the post-processing measurements can be matched to the two-dimensional arrangement of the feature that the user wants to observe.

A swath altimeter in SAR mode is subject to range displacement (known as layover in imaging radar jargon), a well-known phenomenon that is most severe for steep incidence viewing (Raney 1998b). This may occur in an ocean-viewing swath altimeter when elevated land features at greater distance from nadir appear at the same range as the ocean's surface. The limits imposed by land layover are described in Fig. 20.6. As expected, the effect is most pronounced for near-nadir shoreline standoff distances. Layover within the nadir pulse limited footprint has also been computed following the same layover model Fig. 20.6. It must be noted that even for high cliffs (say above 100 m) the swath altimeter will provide data very close to the coast. For a given cliff height, it is remarkable that swath altimeter will provide data much closer to the coast than nadir altimeter, this is simply due to the fact that the pulse limited footprint perimeter is at very small incidence angle (typically 0.1°).

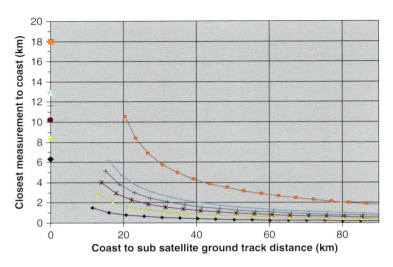

Fig. 20.6 Coast-wise closest distance of measurement as a function of the distance between sub-satellite track and coast line for various cliff heights. Nadir altimeter closest distances to coast is also shown on the y-axis

The layover phenomena will be worse in RAR mode since it will occur not only in the across-track direction, but also in the along track direction. In SAR mode, the layover will be filtered by the SAR processing (or Doppler filtering), and, at first order, only across-track crossover will dominate. This is one argument for requiring SAR mode for swath altimeter in the coastal areas. The portion of the ocean which is affected by layover can be computed using DEM models. A mask can be created to quality control the coastal ocean where land contamination occurs.

20.4.4
Limitations

If accurate SSH information is required in coastal areas, then the usual constraints obtain: one must apply corrections for propagation delays, and rely on precision orbit determination (POD). This is common to all concepts considered in this chapter.

The swath altimetry concept outlined in this section to be successful would demand the maximum abilities of today's technologies. Even so, such a system would be limited to relatively narrow swaths, say of the order of 160–200 km (right to left) to meet centimeter-level elevation error objectives.

Over open oceans, the wide swath interferometric baseline attitudes can be calibrated at points of orbital crossover. This method relies on the fact that the topographic ocean's temporal decorrelation is larger than a few days (Phalippou and Guirraro 2004), the typical time separation between orbital crossover opportunities. More rapid sea surface elevation fluctuations are normal in most coastal areas, thus compromising the method. There are two means of mitigation: extrapolation, and shore-based references. If spacecraft roll rate is slow (and known), then offshore crossover baseline attitude determination could be extrapolated along the radar's orbit to nearby coastal waters. Alternatively, the interferometric image of a known (low-lying and level) shoreline offers a local means of measuring the baseline roll angle, although this method has not yet been demonstrated. Complicating realities in some cases such as land tides or rougher near-shore topography may compromise the results. Therefore data will be corrupted by systematic cross-track elevation errors such as may derive from baseline length or attitude errors. Filtering of the systematic errors remains to be studied.

Accurate determination of the baseline roll angle remains as the dominant limitation of any swath altimetric technique. The technical challenge is far from trivial; the only satisfactory proof of acceptable performance will be through in-flight experience.

There is a corresponding and larger issue: calibration. Wide swath interferometry depends critically on differential phase measurements, whose relationship to the desired sea surface heights could be compromised by many error sources. If these are constant, then they should be subject to calibration, but by what means? A robust (and efficient) calibration methodology will be required, both internal to the radar and external, thus accounting for the antennas, including their respective patterns, their support structure, and their baseline vector. For phase errors that are not constant, they must be bounded if the swath altimeter's performance is to be validated.

In principle swath altimetry can provide a surface height map of the ocean with nominally a 1-cm precision for 5 × 5 km resolution cells. A general proviso applies: precision degrades as the square root of the degrees of freedom in the average. Thus, a swath altimetric height map having finer resolved cell areas of $a \times a$ km would have corresponding height uncertainty (standard deviation) $5/a$ larger, hence 5 cm for 1 km cells, or 10 cm for 500 m cells.

There is no evidence that the significant wave height (SWH) and the wind speed (WS) can be retrieved from swath altimetry. This is a disadvantage in its own right, but corrections to SSH that may depend on SWH or WS will not be available, and so the swath SSH measurements may be compromised as a consequence.

SAR mode swath altimeter operations imply very large data rates and volumes, in the same manner as with an imaging SAR. Although not a trivial factor for the satellite and ground segment design, large data burdens should not detract from the potential applications value of the general concept. By its very nature, any instrument that offers substantially more information than its counterparts must generate substantially more data.

Swath altimetry depends critically on certain spacecraft structural and attitude parameters. Keeping the variation of these parameters within bounds, especially in the context of the relatively large and complicated antenna array, implies programmatic costs and risks.

20.5
Closing Comments

20.5.1
Commentary

Several issues apply to all of the known advanced radar altimetric technologies that are relevant to this chapter. Of these, the altimeter's orbit, constraints on averaging, and retrieval algorithm design and validation are the three prominent themes.

20.5.1.1
Orbit Considerations

Any nadir-viewing altimeter can make its measurements only within limits imposed by the spacecraft's orbit. If that orbit is an exact-repeat pattern in the style of most predecessors, then coastal altimetry would be supported only in a few areas separated by hundreds of kilometers. In contrast, if the orbit were a non-repeat (or long repeat-period) in the manner of the first 18 months of Geosat (MacArthur et al. 1987) or CryoSat-2, then many more coastal areas could be covered, although less frequently at each location. Depending on the intended purpose, one or the other of these two paradigms may be preferred.

20.5.1.2
Averaging Considerations

As noted in Sect. 20.1.2, measurement precision is improved after more extensive averaging, where to be effective, averaging presumes statistical equivalence of the contributing samples to first and second order. In general, an assumption of wide-sense statistical stationarity may be applicable only over smaller areas in the coastal environment than for the open ocean, thus placing an upper bound on the allowable degrees of freedom of the resulting average. For any altimeter concept, users are advised to use caution when applying to their own application measurement precision and averaging numbers that have been derived based on wide-sense stationarity conditions that obtain for the open deep ocean.

20.5.1.3
Retrieval Algorithms

The retrieval algorithms that have been developed through many years of experience with waveforms from conventional altimeters do not necessarily apply to newer technologies when observing coastal waters. There are two issues, differences in the instruments' outputs, and differences due to the sea surface itself. A conventional (Brown) waveform may not be applicable to the output peculiar to a new altimeter technology. Likewise, the small-scale roughness and larger-scale wave profiles of the sea surface in shallow waters, especially in the presence of current flow, cannot be described by deep-water open ocean conditions.

Whether due to a multi-mode instrument, or when combining retrievals from two different instruments, problems may arise if one technology is used in one case, while a different technology is used in the other. What assurance, if any, would a user have that the retrieved parameters from the two modes would provide seamless continuity? If two different styles of re-tracker are applied to conventional altimeter data, it is known from experience that a systematic offset of SSH frequently is observed between the two resulting retrievals. Extending that lesson to SRAL data (Sentinel-3) for example, what will be the relationship between offshore conventional retrievals, and near-shore SAR-mode retrievals? Likewise, for a swath instrument, what is the relationship between the nadir-viewing altimeter's SSH measurements, and the off-nadir SSH retrievals?

The off-nadir coverage of a swath altimeter presents a special case. The surface returns for nadir altimetry are characterized for moderate to calm sea states by relatively strong radar backscatter. When looking a few degrees away from nadir, what is the relative strength of the returns? Are these a function of sea state? Angle of incidence? Significant wave height? Will sea-state bias corrections be independent of incidence? If there are variations that depend on sea state or incidence, can they be modeled or otherwise quantified? To meet cm-scale accuracy objectives across the swath, does the SSH retrieval algorithm have to take account of these factors?

It may be necessary to develop new retrieval algorithms for near-shore observations by any new altimeter technology. Even if standard retrieval algorithms appear to provide acceptable results, their applicability to near-shore conditions will have to be validated.

20.5.1.4
Crossover Considerations

It is standard practice for open ocean altimetric studies to rely on comparative analysis of altimetric data from intersecting orbits to improve precision orbit determination (POD). The crossover method, although robust in most regards, requires that the ocean's surface be at the same elevation (with respect to the geoid) to within a cm or so for observations that may be separated by several days. This requirement will be violated by a large margin for most coastal environments. What might be the means of mitigation to achieve satisfactory POD for near-shore applications?

20.5.2
Future Outlook

Radar ocean-viewing altimeters have provided spectacular and invaluable results, especially in the past two decades. However, their brilliance is most evident on large scale phenomena, especially those having long persistence. As users become more interested in shorter spatial and temporal scales, altimeters as now known become less effective. The situation climaxes in coastal applications. Moving beyond the capabilities of current systems, this chapter reviewed three approaches that show promise, two of which should see operational status within the near future. Given that basis, what extended or new radar capabilities lurk just over the horizon? A recent special issue of Sensors (Chen and Quartly 2006) reviews the possibilities now foreseen. Highlights of two themes are summarized here.

20.5.2.1
A Variation on the Wide Swath Theme

Wide swath altimetry techniques continue to be explored. A good example is KaSOARI[10] (Ka-band Swath Ocean Altimetry with Radar Interferometry), an instrument based on the original concept (Rodriguez and Pollard 2001), but which makes several significant departures. One objective was to eliminate the need for a large antenna structure that would require on-orbit

---
[10]http://www.coastalt.eu/pisaworkshop08/pres/07-Phalippou_Pise_Coastal_Workshop_Nov_08.pdf

deployment. This led to Ka-band (rather than Ku-band), inclusion of either RAR or SAR mode operations and more favorable radar operating parameters. KaSOARI can be programmed to operate either in RAR mode (over open ocean) or in SAR mode, thus adapting its resolution (and data volume) to local requirements. KaSOARI promises to be a significant simplification upon KaRIN, while meeting the ocean mission requirements including coastal areas.

20.5.2.2
Variations on the Constellation Theme

All ocean-viewing altimeter missions to date have been based on solo spacecraft, although as opportunity affords two or more of these may provide data which when take together add value. That said, for decades there has been an ongoing interest in the user community for a coordinated multi-satellite constellation comprised of three to five altimeters. Several candidates exist for smaller yet more capable radar altimeters, each of which could be hosted on a small spacecraft. If the cost of access to space decreases, and if the economy of scale can be applied to the fabrication of a set of radar altimeter satellites, then a constellation would be viable. The driving motivations for such a mission would be coastal applications, together with mesoscale oceanography and (given suitable orbits) oceanic geodesy. An important study (Rapley et al. 1990) concluded among other recommendations that a minimum of three spacecraft would be required, with five preferred. Concepts that have been seriously proposed over the past decade include Gander (daSilva-Curiel et al. 1999), WITTEX (Raney and Porter 2001), GAMBLE (Allan and Ash 2003), and AltiKa (Richard et al. 2007).

20.5.3
Conclusions

Within a few years, the nadir-viewing radar altimeter AltiKa, and the delay-Doppler (SAR) modes of SIRAL-2 and SRAL will demonstrate improved spatial resolution and sea surface height precision. Further, their improved near-shore performance will be exercised. As a result, coastal applications should be supported substantially better by these advanced technologies than by conventional altimeters.

Most coastal requirements would be better served following further major advances, including in particular improved spatial and/or temporal sampling. A constellation of nadir-viewing altimeters would afford a significant improvement in these two regards, but the inherent limitations of one-dimensional (nadir track) sampling still remain.

A swath altimeter is the only approach foreseen that can respond to the spatial coverage requirement, since that technology provides a three-dimensional topographic image. Although the 3-D aspect of the swath data product is promising and immensely appealing, it is associated with non-trivial technical challenges for the host spacecraft, the radar instrument, data processing, and robust SSH retrievals.

There is room for new technologies that would be more fully responsive to the requirements of coastal altimetry. To date, no technology offers in one conceptual mission design rapid revisit, wide-field spatial coverage, and retrievals of SSH, SWH, and WS with accuracy and precision. However, given the progress of radar altimetry over the past decade, one can only be encouraged that in decades to come many if not most of these objectives may be met.

References

Allan TD, Ash ER (2003) The GAMBLE project: a fresh approach to satellite remote sensing of the sea. Phil Trans Roy Soc 361:23–26

Bamler R, Eineder M, Kampes B, Runge H, Adam N (2003) SRTM and beyond: current situation and new developments in spaceborne InSAR. In: Proceedings, ISPRS workshop on high resolution mapping from space, Hanover, Germany

Brown GS (1976) Multiple beam radar altimetry for oceanographic and terrian remote sensing. Applied Science Associates, Apex, NC

Brown GS (1977) The average impulse response of a rough surface and its applications. IEEE Antennas Propagat 25:67–74

Bush GB, Dobson EB, Matyskiela R, Kilgus CC, Walsh EJ (1984) An analysis of a satellite multibeam altimeter. Mar Geodesy 8(1–4):345–384

Chelton DB, Walsh EJ, MacArthur JL (1989) Pulse compression and sea-level tracking in satellite altimetry. J Atmos Ocean Technol 6:407–438

Chen G, Quartly GD (2006) Editorial: special issue on satellite altimetry: new sensors and new applications. Sensors 6(6):542–696

Curlander JC, McDonough RN (1991) Synthetic aperture radar: systems and signal processing. Wiley, New York

daSilva-Curiel RA, Jolly G, Zheng Y (1999) The GANDER constellation for maritime disaster mitigation. Acta Astron 44:685–692

Enjolras V, Rodriguez E (2007) An assessment of the Ka band interferometric radar altimeter for monitoring rivers and lakes with the WatER mission. In: Proceedings, IEEE IGARSS, Barcelona, Spain, pp 3525–3528

Fu LL, Cazenave A (eds) (2001) Satellite altimetry and the earth sciences. Academic, San Diego, CA

Gommenginger C, Challenor P, Quartly GD, Srokosz MA, Berry PAM, Rogers C, Benveniste J (2007) Envisat altimeter individual echoes: new scientific applications for ocean, land, and ice remote sensing. Geophys Res Abstracts 9:08979

Graham L (1974) Synthetic interferometers for topographic mapping. Proc IEEE 62(6):763–768

Green PE Jr (1968) Radar measurements of target scattering properties. In: Evans JV, Hagfors T (eds) Radar astronomy. McGraw-Hill, New York

Hanssen R (2001) Radar interferometry: data interpretation and error analysis. Kluwer, Dordrecht, the Netherlands

Hayne GS (1980) Radar altimeter mean return waveforms from near-normal incidence ocean surface scattering. IEEE Antennas Propagat AP-28(5):687–692

Jayles C, Vincent P, Rozo F, Balandreaud F (2004) Doris-Diode: Jason-1 has a navigator on board. Mar Geodesy 27:753–777

LeRoy Y, Deschaux-Beaume M, Mavrocordatos C, Aquirre M, Heliere F (2007) SRAL SAR radar altimeter for Sentinel-3 mission. In: Proceedings, IEEE IGARSS, Barcelona, Spain, pp 219–222

MacArthur JL (1976) Design of the Seasat-A radar altimeter. Oceans 8:222–229

MacArthur JL, Marth PC, Wall JG (1987) The Geosat radar altimeter. Johns Hopkins APL Techn Digest 8(2):176–181

Massonnet D (1996) Tracking the Earth's surface at the centimeter level – an introduction to radar interferometry. Nat Resour 32(4):20–29

Moore RK, Williams CS Jr (1957) Radar return at near-vertical incidence. Proc IRE 45(2):228–238

Phalippou L, Enjolras V (2007) Re-tracking of SAR altimeter ocean power waveforms and related accuracies of sea surface height, significant eave height, and wind speed. In: Proceedings, IEEE IGARSS, Barcelona, Spain

Phalippou L, Guirraro J (2004) End to end performances of a short baseline interferometric radar altimeter for ocean mesoscale mapping. In: Proceedings, IEEE IGARSS, Anchorage, Alaska

Pollard BD, Rodriguez E, Veilleux L (2002a) The wide swath ocean altimeter: radar inerferometry for global ocean mapping with centimetric accuracy. In: Proceedings, IEEE aerospace conference, vol 2, pp 1007–1020

Pollard BD, Rodriguez E, Veilleux L (2002b) The wide swath ocean altimeter: radar interferometry for global ocean mapping with centimetric accuracy. In: Proceedings, AEEE aerospace conference, vol 2, pp 1007–1020

Raney RK (1998a) The delay Doppler radar altimeter. IEEE Trans Geosci Remote Sens 36(5):1578–1588

Raney RK (1998b) Radar fundamentals: technical perspective. In: Henderson F, Lewis A (eds) Principles and applications of imaging radar, 3rd edn. Wiley, New York, pp 9–130

Raney RK (2005) Radar altimetry. In: Chang K (ed) Encyclopedia of RF and microwave engineering, vol 5. Wiley, Hoboken, NJ, pp 3989–4005

Raney RK (2008) Space-based radars. In: Skolnik M (ed) The radar handbook, 3rd edn, chap 18. McGraw-Hill, New York

Raney RK, Porter DL (2001) WITTEX: an innovative three-satellite radar altimeter concept. IEEE Trans Geosci Remote Sens 39(11):2387–2391

Rapley CG, Griffiths HD, Berry PAM (1990) Proceedings of the consultative meeting on altimeter requirements and techniques, Final Report (ESA Reference UCL/MSSL/RSG/90.01). University College, London, England

Rey L, DeChateau-Thierry P, Phalippou L, Calvary P (2004) SIRAL the radar altimeter for the CryoSat mission, pre-launch performances. In: Proceedings, IEEE IGARSS, Anchorage, Alaska, pp 669–671

Richard J, Phalippou L, Robert F, Steunou N, Thouvenot E, Senenges P (2007) An advanced concept of radar altimetry over oceans with improved performances and ocean sampling: AltiKa. In: Proceedings, IEEE IGARSS, Barcelona, Spain, pp 3537–3540

Rodriguez E, Martin J (1992) Theory and design of interferometric synthetic aperture radars. IEEE Proc Radar Signal Process 139:147–159

Rodriguez E, Pollard BD (2001) The measurement capabilities of wide-swath ocean altimeters. In: Chelton D (ed.) Report of the high-resolution ocean topography science working group meeting, oregon State University, pp 190–215

Rodriguez E, Pollard BD (2003) Centimetric sea surface height accuracy using the wide-swath ocean altimeter. In: Procedings. IEEE IGARSS, Toulouse, France

Smith WHF, Sandwell DT (2004) Conventional bathymetry, bathymetry from space, and geodetic altimetry. Oceanography 17(1):8–23

Tournadre J, Lambin J, Steunou N (2009) Cloud and rain effects on AltiKa/SARAL Ka-band radar altimeter. Part I: modelling and mean annual data availability. IEEE Trans Geosci Remote Sens 47:6. doi:10.1109/TGRS.2008.2010130

Ulaby FT, Moore RK, Fung AK (1982) Microwave remote sensing: active and passive, vol II. Addison-Wesley, Reading, MA

Vincent P, Steunou N, Caubet E, Phalippou L, Rey L, Thouvenot E, Verron J (2006) AltiKa: a Ka-band altimetery payload and system for operational altimetry during the GMES period. Sensors 6:208–234

Walsh EJ (1982) Pulse-to-pulse correlation in satellite radar altimetry. Radio Sci 17(4):786–800

Wingham DJ et al (1998) CryoSat: a mission to determine fluctuations in the mass of the earth's land and marine ice fields (proposal to the European Space Agency). University College, London

Wingham DJ, Francis R, Baker S, Bouzinac C, Cullen R, DeChateau-Thierry P, Laxon S, Mallow U, Mavrocordatos C, Phalippou L, Ratier G, Rey L, Rostan F, Viau P, Wallis D (2006) CryoSat: a mission to determine the fluctuations in Earth's land and marine ice fields. Adv Space Res 37:841–871

Zebker H, Goldstein R (1986) Topographic mapping from interferometric synthetic aperture radar observations. J Geophys Res 91:4993–4999

Index

A

Absolute Dynamic Topography (ADT), 56, 59, 326
Acoustic Doppler Current Profiler (ADCP), 54, 206, 419, 431, 442
ADvanced CIRCulation and Storm Surge Model (ADCIRC), 204
ALtimeter-Based Investigations in COrsica, Capraia and Contiguous Areas (ALBICOCCA), 15, 220, 221, 241, 242, 300, 301, 305
ALTImetry for COastal REgions (ALTICORE), 15, 255, 343, 346, 354, 375, 378, 395, 396, 400
Archiving, Validation and Interpretation of Satellite Oceanographic data (AVISO), 29, 108–134, 155, 158, 222, 230, 231, 302, 304, 312, 317, 319, 322, 326, 375, 396, 425, 435, 437, 457, 484, 513
Australian coastal regions, 93, 94, 495, 497, 498

B

Barents Sea, 389–413
Binary Universal Form of Representation of meteorological data (BUFR), 58
Black Sea, 336, 355, 367–386
Brown Ocean Retracker (BOR), 67, 73, 90, 92

C

Calibration/Validation (Cal/Val), 52, 261, 270, 273, 274, 276, 281, 290, 292, 342–345
Caspian Sea, 26, 41, 225, 331–362, 380, 384, 400
Centre Of Gravity (COG), 75, 76

China coastal seas, 453–470
Climate change, 7, 24, 41, 47, 179, 261, 334, 336, 368, 371, 475, 488, 489, 513–516, 519–521, 527, 529
Coastal anticyclonic eddies, 369, 370, 373
Coastal current, 227, 229, 240, 302, 324, 326, 409, 419, 425, 430, 435, 445, 455–457, 475–484, 488, 503
Coastal upwelling, 56, 356, 369
Coastal zone altimetry, 1–15
Collecte Localisation Satellite (CLS), 88, 122, 123, 136, 137, 139, 226, 227, 307, 309, 317, 462
Continental Shelf model (at 3 times former resolution) (CS3X), 181, 186
CryoSat-2, 7, 9, 10, 292, 504, 539, 543, 544, 546, 556

D

Data correction retrieval, 241
Data editing, 220, 221, 312–314
Data services, 248, 254, 256, 513
Data Unification and Altimeter Combination System (DUACS), 6, 304, 437, 484
Delay-Doppler Altimeter (DDA), 544–547
Détermination Immédiate d'Orbite par Doris Embarqué (DIODE), 67, 71, 72
Digital Elevation Model (DEM), 67, 71, 72, 540, 542, 546, 554
Doppler Orbitography and Radiopositioning Integrated by Satellite (DORIS), 3, 6, 7, 9, 25, 26, 67, 119, 263, 264, 274, 462, 540, 542
Dynamically Linked Model (DLM), 116, 141, 149–152, 172, 173
Dynamic Height (DH), 326

E

Earth Gravitational Model (EGM), 501
Eddy Kinetic Energy (EKE), 434, 439, 441, 443–446
Empirical Ocean Tide model (EOT), 125
Empirical Orthogonal Function (EOF), 437
Empirical retracker, 75–81, 94
Envisat, 4, 37, 52, 64, 108, 149, 185, 192, 222, 252, 264, 301, 338, 378, 395, 461, 475, 510, 542
ESA development of COASTal ALTimetry (COASTALT), 15, 52–54, 57–59, 90, 153, 178, 180, 186–187, 200, 220, 231, 252
ESA development of SAR altimetry mode studies and applications over ocean, coastal zones and inland water (SAMOSA), 15
European Centre for Medium-range Weather Forecasts (ECMWF), 113, 116, 129, 139, 141, 149–155, 159–161, 171–174, 185, 285, 287, 288, 300, 304, 307, 339, 344, 354, 361, 437, 462, 512
European Remote Sensing Satellite (ERS)-1, 5, 6, 14, 43–46, 97, 135, 136, 139, 141, 142, 194, 200, 222, 251, 262, 263, 338, 345, 395, 396, 400, 402, 404, 426, 427, 429, 432–434, 439, 475, 476, 510
European Remote Sensing Satellite (ERS)-2, 5, 6, 43, 45, 46, 64, 68, 72, 73, 91–94, 135–137, 139–142, 194, 200, 222, 304, 338, 345, 350, 373, 376, 395, 396, 400, 402, 403, 411, 412, 433, 439, 454, 475, 476, 494–498, 510, 513

F

Finite Element Solution (FES), 125, 196, 200–202

G

Geoid, 5, 26, 93–95, 97, 106–109, 132–134, 186, 187, 225–227, 263, 264, 267, 272, 280–283, 301, 304, 306, 310, 316, 317, 337–339, 341, 343, 353, 361, 455, 493, 494, 497, 499, 501, 502, 504, 538, 557
Geophysical Data Record (GDR), 25–28, 30, 32–38, 42, 44, 48, 56, 59, 89, 90, 93, 94, 150, 151, 154–158, 160, 161, 172, 222, 223, 226, 250, 251, 285–286, 290, 301, 305, 307, 309, 311, 312, 314, 318, 326, 345, 371, 396, 469, 495

GEOS-3, 4, 263
Geosat, 4–7, 77, 135, 136, 155, 194, 250, 301, 304, 357, 380, 383, 395, 396, 430, 431, 433, 434, 454, 455, 475, 512, 556
Geosat Follow-On (GFO), 6, 7, 23, 33–39, 41, 43, 44, 46, 114, 135, 136, 155, 158, 194, 200, 222, 228, 229, 237, 238, 280, 300, 304, 316–320, 326, 338, 339, 341–343, 345, 354–357, 361, 378–380, 383, 395–400, 404, 408, 475, 476, 485, 510–513, 517, 518, 521, 522, 527, 528, 530
Geosat Follow-On (GFO)-2, 7
Geostrophic Velocity Anomaly (GVA), 323
Global Ionosphere Map (GIM), 34, 119–121
Global Ocean Tide (GOT), 196, 200, 201, 204
Global Positioning System (GPS), 3, 25, 34, 106, 116, 117, 119, 141, 152, 153, 170, 172, 263–265, 267, 269–275, 278–286, 288, 289, 292, 343, 345, 361, 464, 491, 492
GNSS-derived Path Delay (GPD), 149, 153, 158, 161, 173
Gravity and Ocean Circulation Explorer (GOCE), 5, 493, 504
Gravity field, 5, 13, 34, 93, 475, 493–494, 497, 503
Gravity Recovery and Climate Experiment (GRACE), 5, 34, 41, 133, 192, 268, 493, 494, 501, 503, 504

H

HY-2, 7, 13–15, 338
Hamburg Shelf Ocean Model (HANSOM), 186
Hydrology, 4, 7, 338, 342, 510, 521, 525, 548

I

Ice, Cloud and Land Elevation Satellite (ICESat), 48, 192
Ice cover, 335, 338, 339, 341–342, 346, 349–351, 357–362, 392–395, 409, 411, 413, 529
Innovative techniques, 64
In situ instrumentation, 292
Inverse Barometer (IB), 107, 129–132, 136, 140, 141, 177–187, 224, 225, 304, 371, 400, 452, 484, 491, 492
INVerse solution 2 from the original and interleaved T/P ground track (INVTP2), 207–210

Index

INVerse solution from the original T/P ground track (INVTP), 207
Inverted Barometer (IB), 129–132, 140, 141, 224, 225, 304

J

Jason-1, 6, 24, 64, 108, 149, 178, 200, 222, 230, 236–238, 240, 251, 261–265, 268–271, 273, 275, 277, 278, 280–282, 285–290, 292, 301, 304, 326, 338, 341, 342, 345, 346, 354–357, 361, 371, 379, 438, 457, 461–464, 466–469, 475, 485, 491, 510, 513, 518, 521, 522, 528, 530
Jason-2, 6, 43–46, 52, 53, 64, 67, 68, 71–75, 85, 87–90, 98, 113, 117, 118, 120, 141, 142, 173, 178, 199, 222, 251, 252, 261, 263, 264, 268, 270, 273, 275, 277, 281, 282, 288–291, 326, 338, 475, 484, 485, 504, 510, 516, 521, 542, 543, 545, 548
Jason-3, 15, 516, 530
Jason-1 Microwave Radiometer (JMR), 114, 262, 273, 285–289, 354

K

Ka-band, 7, 9, 48, 241, 539, 541, 543, 548, 550, 552, 557, 558
Ka-band Radar INterferometer (KaRIN), 7, 48, 548, 558
Ka-band Swath Ocean Altimetry with Radar Interferometry (KaSOARI), 558
Kara-Bogaz-Gol, 334–336, 342, 346, 352–354, 361, 362

L

Lakes, 3, 8, 9, 14, 22–48, 72, 222, 248, 292, 338, 339, 342, 343, 345, 354, 358, 360, 361, 455, 509–530
Local Inverse Barometer (LIB), 179, 180

M

Maximum Likelihood Estimator (MLE), 86, 94, 95, 97
Mean Dynamic Topography (MDT), 58, 133, 134, 226, 323
Median Tracker (MT), 71, 72
Mediterranean Sea, 69–71, 88, 91, 94, 97–99, 115, 124, 163, 225, 297–326, 368
Merged Geophysical Data Record (MGDR), 290, 371, 396

MicroWave Radiometer (MWR), 5–7, 34, 113, 114, 148–153, 155, 157–163, 172, 173, 283, 300, 307, 419, 448, 462, 512
Minimum Mean Square Estimator (MMSE), 87
Mixed Brown Specular (MBS), 90
MOdelo HIDrodinâmico (Hydrodynamic Model) (MOHID), 186

N

network Common Data Form (netCDF), 58, 59, 250, 253

O

Ocean currents, 107, 179, 240, 248, 430, 475, 484–488, 503, 504
Ocean Surface Topography Mission (OSTM), 43, 46, 277, 290, 291
Ocean Surface Topography Science Team (OSTST), 108, 110, 251, 254, 285, 289
Offset Centre Of Gravity (OCOG), 75–76, 79, 94, 305, 310, 311, 454, 455, 461–469
Ohio State University (OSU), 206, 322
On-board tracker, 64–67, 69, 71, 72, 98
Open-source Project for a Network Data Access Protocol (OPeNDAP), 58, 59, 255
Operational Geophysical Data Record (OGDR), 291
Orbit Minus Range (OMR), 289

P

4-Parameter SSB model (BM4), 121–123, 139
PISTACH. *See* Prototype Innovant de Système de Traitement pour les Applications Côtières et l'Hydrologie
Point Target Response (PTR), 81, 82, 85, 262
Post-processing, 45, 47, 63, 217–242, 300, 301, 311, 312, 317, 326, 553
Prototype Innovant de Système de Traitement pour les Applications Côtières et l'Hydrologie (PISTACH), 15, 52–56, 58, 59, 88–90, 180, 187, 220, 231, 251, 484, 503
Pulse Repetition Frequency (PRF), 67, 540–542, 544, 549

R

Radar Altimeter (RA), 1–15, 22, 24, 25, 42, 44, 46, 48, 52, 79, 117, 219, 241, 254, 262, 272, 277, 305, 318, 338, 341, 346, 351, 357, 362, 373, 396, 454, 493, 494, 503, 512, 537–540, 542, 544–546, 548, 550, 558

Radar Altimeter Database System (RADS), 108, 110, 111, 134, 254, 305, 354, 373, 378, 384, 396

Radar altimetry missions, 5, 6, 8, 23, 360, 521

Radar altimetry principle, 3–4

Reference frame, 22, 32, 34, 106, 134, 135, 225–227, 241, 262, 265, 266, 268–270, 274–275, 282, 283, 311, 529

Regional de-aliasing, 220, 314, 315

Regional tidal model, 206, 404–406, 413

Reservoirs, 8, 19–48, 336, 338, 345, 512–515, 521, 525–527

Retracking, 14, 15, 43, 45, 47, 59, 61–99, 122, 124, 220, 252, 268, 273, 274, 291–293, 300, 301, 306, 309–310, 326, 341, 343, 448, 453–470, 475, 494–498, 501, 503, 512, 513, 529

Rim Current (RC), 369, 370

Root Mean Squared Error (RMSE), 90, 91

S

SARAL. *See* Satellite with Argos and AltiKa

SAR/Interferometric Radar ALtimeter (SIRAL-2), 558

SAR mode. *See* Synthetic Aperture Radar mode

SAR Radar ALtimeter (SRAL), 544, 546, 558

Satellite altimetry, 485–516, 521–542

Satellite radar altimetry, 22–24, 336, 338, 361, 520, 521

Satellite with ARgos and ALtika (SARAL), 7, 9–13, 15, 48, 539, 540

Scanning Multichannel Microwave Radiometer (SMMR), 357, 362

Scanning Point Target Response (SPTR), 262

Sea level, 5, 52, 63, 105, 149, 178, 195, 222, 248, 261, 299, 333, 368, 392, 422, 456, 475, 527

Sea Level Anomaly (SLA), 56, 222, 301, 368, 408, 441, 461, 476

Sea level rise, 57, 63, 199, 234, 334, 348, 371, 475, 488–493, 503, 504

Seasat, 4, 14, 79, 199, 263, 540

Sea-State Bias (SSB), 121–124, 139, 141, 267, 291–293

Sea Surface Height (SSH), 9, 56, 77, 124, 219, 261, 301, 341, 400, 419, 454, 493, 536

Sea Surface Height Anomaly (SSHA), 400, 403–405, 407, 420–422, 442

Sea Surface Temperature (SST), 241, 324–326, 425, 427, 429, 431, 432, 437–439, 485, 486, 520

Segment Sol multimissions d'ALTimétrie, d'Orbitographie et de localisation précise (SSALTO), 230, 304, 437

Sensor Geophysical Data Record (SGDR), 305, 307, 309, 457, 462

Sentinel-3, 7, 9, 11, 15, 48, 338, 345, 516, 530, 539, 544, 546, 556

Shuttle Radar Topography Mission (SRTM), 7, 549

Significant Wave Height (SWH), 9, 56, 57, 83, 85, 94, 97, 121, 122, 124, 271, 291, 458, 466–470, 536–538, 541, 543–546, 555, 558

Skylab, 4

Special Sensor Microwave/Imager (SSM/I), 163, 167, 170, 357, 358, 362

Split Gate Tracker (SGT), 24, 71

Storm surges, 55, 132, 133, 178–185, 187, 195, 197, 211, 263, 334, 346, 348–351, 361, 369, 393, 395, 405, 493

Surface water level, 22, 24, 41

Surface Water Ocean Topography (SWOT), 7, 48, 212, 292, 548

Synergy with coastal models, 241

Synthetic Aperture Radar (SAR) mode, 15, 48, 539, 543–546, 550, 553–556, 558

T

Tide(s), 18, 138–143, 152–156

Tide Gauge (TG), 301, 302, 306, 308–311, 313–315, 317–323, 326, 396, 400, 403–407, 434, 437

Tide-surge modelling. *See* Storm surges

Time of Closest Approach (TCA), 284

TOPEX Microwave Radiometer (TMR), 114, 163, 167, 169, 170, 173, 287–290

TOPEX/Poseidon, 5, 6, 14, 23, 26, 113, 125, 136, 138, 141, 142, 155, 170, 178, 192, 194, 222, 227, 231, 236, 238, 261–264, 268, 275, 278, 280, 301, 303, 304, 338, 357, 370–372, 374, 376, 377, 395, 420, 421, 455, 457, 475

Total Electron Content (TEC), 117–119

Total Electron Content Unit (TECU), 117

Index

Toulouse Unstructured Grid Ocean model 2D (T-UGO 2D), 339, 340, 361
Two-dimensional Gravity Waves model (MOG 2D), 131, 136

U

Ultra Stable Oscillator (USO), 105, 262
User requirements, 51–59

V

Validation, 14, 28–37, 41, 43, 44, 47, 48, 55, 170, 182, 185–187, 193, 201, 204, 206–210, 212, 220, 252, 259–293, 302, 305, 326, 338, 339, 342–345, 351, 358, 359, 361, 375, 395–406, 425, 448, 457, 470, 484, 520, 525, 555
Vienna Mapping Functions 1 (VMF1), 154, 155
Volga River, 336, 348, 361

W

Water Inclination Topography and Technology EXperiment (WITTEX), 558
Water management, 7, 513, 525, 526
Water vapor, 6, 105, 113, 114, 116, 148, 262, 267, 272, 273, 278, 419, 437, 448, 512, 538, 539
Waveform retracking, 63–67, 75, 88, 94, 96–97, 99, 301, 309, 341, 454, 455, 461, 469, 475, 494, 496, 498, 503

Waveforms, 4, 14, 15, 61–99, 121, 124, 141, 219, 220, 222, 230, 272, 274, 300, 301, 305, 306, 309, 310, 312, 326, 341, 419, 448, 454, 455, 457–464, 466, 467, 469, 470, 475, 494–498, 500, 503, 511, 512, 537, 540–546, 550, 556
Wave height, 3, 7, 9, 29, 52, 57, 59, 63–65, 83, 86, 88, 89, 95, 96, 98, 121, 141, 271, 277, 300, 352, 354, 356, 357, 362, 369, 384, 395, 458, 536, 541, 544, 546, 548, 550, 555, 557
Weighted Least Square (WLS), 86–87, 93
Wet tropospheric correction, 5, 14, 15, 26, 105, 113, 149–153, 172, 173, 220, 222, 272–273, 287–290, 292, 300, 301, 307, 338, 339, 361, 437, 462
White Sea, 192, 380, 389–413
Wide swath, 7, 47, 48, 193, 539, 547–555, 557–558
Wind forcing, 303, 369, 392
Wind speed, 3, 6, 52, 59, 63, 97, 121, 163, 273, 279, 280, 300, 349, 352, 354–356, 362, 368, 369, 378–385, 391, 392, 395–400, 413, 536, 546, 548, 555

Z

Zenith Hydrostatic Delays (ZHD), 153–159, 170
Zenith Wet Delays (ZWD), 153–155, 157–162